We Were The Last [illegible]
And the Birth of Americana Music
But Was America Ready For Us?

FAT CHANCE

GILBERT KLEIN

Fat Chance: We Were the Last Gasp of the 60s and the Birth of Americana Music, but Was America Ready for Us?

Published by: Main Frame Press

Paperback ISBN: 978-0-9856790-0-2
eBook ISBN: 978-0-9856790-1-9

Library of Congress Control Number :

Yeah, KFAT had the most colorful cast of characters outside of Hogwarts.

Yeah, the FBI and the FCC kept stopping by to exert some control on us: fat chance.

Yeah, the CIA tried to stop me from doing an interview: fat chance.

Yeah, a couple of cowboys dropped by late one night to complain that we "played too much of the niggers," and the jock threw 'em out. They wanted us to know who was listening, so they left their card. It was the Ku Klux Klan.

Yeah, when the Hells Angels went on a run and the bars closed, they used to come to KFAT, and you can't tell me the Hells Angels ever came to *your* place and nothing interesting happened.

But it was the DJs, discovering daily a new world of music and creating a new format, despite poverty and ridiculous working conditions, who are the heart of the story, and the music is its soul. A lot of heart and a lot of soul.

Yeah, it all happened. I was there.

It was KFAT. It's in the book.
Read about it here.

This Book is Dedicated To

Laura Ellen Hopper, 1950 - 2007

Without Whom There Would Be No KFAT, No Book, And Maybe No Americana Music.

FAT CHANCE

(A Delightfully Disquisitive Adoxology Regarding The Excess Of Freedom And A Concomitant Paucity Of Managerial Control Over Socially Disenfranchised Persons Given Access To The Manipulation Of Frequency Modulation)

IS MADE POSSIBLE BY THE

STATUTE OF LIMITATIONS

Contents (Not for the Faint of Heart)

Instructions, Thanks and the Other Thing

In this book you'll see underlined words or phrases, and those are links to photos, sound bites, news articles, etc. If you have a wi-fi enabled ebook, press on the link. If you're reading this in a paperback, go to www.readfatchance.com to find the links. For the obsessive reader, please go to The Section For Obsessives in the website for even more stuff.

You don't have to go to the links, that's up to you. You'll like some links, others maybe not.

To hear the jocks, you can go to http://audio.lns.com/more_airchecks/kfat/1982_12/ or www.kfat.com and thanks to whoever's running those. Now for the thanks and no thanks:

Sister Tiny said, "Tell it, Gilbert, tell it all," and I have. Well, I told *a lot.* A BIG thanks to those who supported me and the book during all the years it took to see it in print:

Thanks to Johnny Starbuck who reads everything first. Doug and Kitsy helped with a gift, as did Marc & Chuck and Eddy Jennings. Bob Gordon was a friend and legal advisor, whose life and work is the standard of excellence that I wanted to emulate.

Thanks to those who supported the Kickstarter campaign to put the paperback out.

Thanks to all of the KFAT buds and friends who interviewed and shared and encouraged me, and put it on the record. Thanks to Laura Ellen Hopper, Terrell Lynn Thomas, Sherman Caughman, Michelle Busk, Kathy Sully Roddy, Gordy Broshear, Christa Taylor, Lori Nelson, Amy Bianco, Bob Cassidy, Marty Manning, Bobby Eakin, Cuzin' Al Knoth, Steimie, Russ Martineau, Larry Yurdin, Felton Pruitt, Harold Day, Rick Nagle, Annie Mitchell, Mary Tilson, Bill Goldsmith, Jeremy Lansman, Leo Kesselman, Stephen Seaweed, Belinda Phillips, Lorenzo Milam, Lynda Kelly, Long John, Bob Simmons, David Chaney, Jeanette Camelio, Bill Denton, Tom Diamant, Don Mussell, Chris Feder, Rob Bleetstein, Greg Arrufat, Mary McCaslin, Gary Faller, Awest and Patrycja Awest, Hewlett Crist, Michael Turner, Travus T. Hipp, Chuck "Wagon" Maultsby. And a special thanks to Weird Harold for support above and beyond. To Gary McDole for pulling the trigger and putting dinner on the table. Fil and Pete Maresca. Thanks: David Palmer, John Patterson, Barry Porter. To Tom Donahue, for the spark. And big thanks to Fat Friend Dave "DaveBob" Nielsen.

Thanks to the city of Gilroy, who have been gracious and supportive, especially Gary Walton. And it's a much nicer place now. And to Chris Charucki for outstanding friend-ism.

Thanks to Cory Dworken, who found the right spot to fish in.

The book is long because I took the time I wanted to take and wrote the book I wanted to write and this is it. Yeah, I know I'm gonna get slammed for the length, but here it is.

Special Thanks to Terrell Lynn's sweetheart, Cathy Hogan, Fathead Extraordinaire. Cathy came late to the show and got right into it. Extreme thanks. She knows why. Thanks to Dr. Geri De Stefano-Webre for all the psychic input. Carolyn Prior, ace book designer, was a joy to work with. And Dave Bricker. Oh, God- I know there are more…

To everyone else who contributed to this book over the years and I've forgotten them now, and to all those Fatheads on and off Facebook for the emails: Thanks!

It's accepted wisdom that we should graciously dismiss the bad things done to us and give thanks to those we appreciate. Maybe, but Chloe Rounsley and Mollie Glick are the ass end of literary agents, whose dishonesty in the former and callous indifference in the latter turned me off to literary agents. No thanks for you!

Everything in this book is as true as I remember it or as whoever told it to me remembered it. I don't think there are any inaccuracies, but try to corral 35- and 40-year-old memories from a bunch of aged, dissolute hippies. Like me. So if there are errors, it's someone else's fault. This is my book and I don't want to hear any complaints about the contents. You want a different version? Go write your own fucking book.

Introduction

Disquisitive adoxology, indeed. You could look up disquisition, and you'll like what you find. It's so simple you won't really learn anything new, but you'll have fun looking it up. It's a written statement, a thesis. You'll find it in any dictionary. Adoxoloy, not so much.

You could look it up in your dictionary at home and not find it, or you could go to the library's big- ass dictionary and look it up and it won't be there, either. Go to an online dictionary, if you must, and search for it there. You're probably not gonna find it. I know. I've done it. I'll tell you about it in a minute.

"Chewin' the Fat" was the talk show I did on KFAT for several years. The first version of the show lasted four years and was a pre-recorded fifteen minute-long interview show, and it was on during the important period known as "Morning Drive." It paid me almost no money, and if I were to factor in the time and cost of travel to and from the station, plus the hours I drove to do my interviews, and then the hours I spent writing, recording and editing the show and my promos, I'm sure I came out behind. But I never considered it a loss. Most of us with no prior experience in radio, we were a staff of dedicated lunatics, united together to bring you KFAT, and it was loose and crazy and sad and fun and I think I loved every minute of it.

The first staff, the pros, lasted a few months and left. I was there within a week after they left, and I was there at the end, seven years later, when KFAT went off the air.

I remain on good terms with many of the surviving staff, and I'm extremely close with a few of them. I love those people, and we went through something together that none of us will ever forget. We were pioneers, and I think we might have been ennobled because we toiled so hard for so little pay. None of them would admit to that. To their credit.

KFAT wasn't the biggest or the most popular radio station of all time. It wasn't even all that well heard in its own area in its time. It wasn't the most influential in its time, nor the longest– lasting. It certainly wasn't the most famous.

What it was the most of, was a lot of things, and a varied lot they were. KFAT was the most unique station, maybe ever. It was certainly the most fun.

It had to be the loosest, most permissive professional radio station that there ever was. And that wasn't because there was no attempt at control, nor any success at that attempt. No, there was control at KFAT, but it was of a very delicate and ill-defined nature.

People showed up and did their jobs, commercials got written and recorded and put on the air. Much music was played and the dreaded "dead air" was only an accident, occasional at best, and committed not out of malice or due to the lack of available personnel to helm the controls. No, it was only the happenstance, if unfortunate, lack of either sleep or sobriety that deprived the airwaves of the hallowed signal that is the lifeblood of any station.

More of the list of things KFAT was the most of: the strangest, most unprepared group of miscreants ever assembled as a radio staff. Gotta be the most. Also, more of its staff was either arrested or taken in for questioning than most.

Most loyal and dedicated fans ever. Has to be. The station has been off the air for over thirty years and there are still people talking and reminiscing about it. Look for the websites about it. Hell, there are still tributes and celebrations and reunions about it. KFAT memorabilia is traded right now on eBay. Old tapes are played 24/7 on the web. Who the hell is still talking about a radio station that went off the air thirty years ago? Who still misses it and loves to talk about it? Fatheads, that's who.

People who never thought they'd listen to a country music station listened to KFAT, and of course those who always had. Even though it wasn't really a country station. Its owner lied to the FCC and snuck a tower onto the tallest peak in the Bay Area, and people listened in droves wherever they could get it. Stories would come in of heroic sacrifices and monumental efforts to listen to KFAT when conventional antennae just didn't capture the signal. KFAT owned Soledad Prison and brand new Silicon Valley. They listened in San Jose, and in pockets in San Francisco. You could hear it in parts of Marin, Sonoma and Mendocino counties. Inland, it went up into the Gold Country, Yosemite and beyond. It went down the coast into Monterey, Carmel and inlets and pockets as far as San Luis Obispo.

Broadcasting from Gilroy, just south of San Jose, it showed up in erratic pockets, and we'd always hear from jubilant listeners in distant areas. We'd hear that these people had friends up on weekends so they could listen, too. I remember that most of the time I worked at KFAT, I lived in San Rafael, a hundred miles north of Gilroy, in an L-shaped apartment building. I had better stereo gear than the people on the other side of the building, but I couldn't get KFAT, and all my neighbors could. Pissed me off, too.

It didn't always show up in the ratings, but its demographic numbers were impressive. KFAT was listened to by people across all economic and educational levels. An absolutely blue collar station, KFAT consistently rated in the top brackets of education and income in the Bay Area. Silicon Valley grew up listening to KFAT.

Steve Wozniak, who started Apple Computers loved KFAT, and I heard he glued the radio's dial to KFAT in that garage in Palo Alto. The Hells Angels said it was their official station and came by to party- a bunch of times. Neil Young called in with requests and thanked us on an album cover. We know the Ku Klux Klan listened because they came to talk to us about the station, to make a... programming suggestion. They left their card. It's in the book.

Ah, but it was the music. My God, the music! The music on KFAT was unlike any other station. In fact, KFAT's last Music Director, Dallas Dobro, thinks that 85% of the music played at KFAT wasn't heard on any other station in America.

Think about that. Eighty-five percent of what we played wasn't played anywhere else. No, really, that's a remarkable statement and I'd like you to stop reading for a moment and think about that. Eighty-five percent. Think about it. I'll wait right here.

Done? Thanks. It was the music and the jocks. The music was an unpredictable mix of so many styles of music that was made to blend and mix and rock you and move you and amuse you in ways you heard only at the "Wide Spot on the Dial." You never knew what was coming next on KFAT. Think about that. Twenty-four hours of music every day, and you never knew what was coming. And when it came, you liked it. It was raucous and rocking and swinging and crying and laughing and you just never knew what was coming next. They played artists you'd never heard of and that weren't being played anywhere else. Not in America. Not in the world. Think about that- and they made up fake ads and played those with the rest. They weren't nice ads, but they were funny.

There was no room at KFAT for second-rate music. If it was just strange, it wasn't played on KFAT. The jocks and management agreed on one thing without ever discussing it: quality was everything. Mediocre blues? No way. Bad country music? Never. Bad anything? Unthinkable. Unacceptable. The music was everything. Go listen to www.kfat.com and you'll see. Here are some songs you might hear at any moment on KFAT:

A Good Woman Likes To Drink With The Boys
Ain't Living Long Like This
Amazing Grace (Used To Be Her Favorite Song)
Bump Bounce Boogie

Caffeine, Nicotine, Benzedrine
Come Back To Us Barbara Lewis Hare Krishna Beauregard
Disco Sucks
Drop Kick Me, Jesus (Through The Goal Posts Of Life)
Hank Williams, You Wrote My Life
Hauled Off And Loved Her
Hawaiian Cowboy
Honky Tonk Heaven
How Can I Love You (If You Won't Lay Down)?
I Ain't Really A Cowboy, I Just Found The Hat
I Heard You Been Layin' My Old Lady
Lost In The Ozone
Mamas, Don't Let Your Babies Grow Up To Be Cowboys
Mamas, Don't Let Your Cowboys Grow Up To Be Babies
Man Smart, Woman Smarter
Moose Turd Pie
My Baby Thinks He's A Train
My Head Hurts, My Feet Stink, And I Don't Love Jesus
My House Is Your Honky Tonk
Ninety Miles An Hour (Down A Dead-End Street)
No Beer In Heaven
Nothing Sure Looked Good On You
One Less Jogger On The Road
Pissin' In The Wind
Pussy Pussy Pussy
Red Hot Women And Ice Cold Beer
Rode Hard And Put Up Wet
She's Acting Single (I'm Drinking Double)
Somebody Shoot Out The Jukebox
Tell Old I Ain't Here, He Better Get On Home
The Last Country Song
They Ain't Making Jews Like Jesus Anymore
Them Dance Hall Girls
Why Don't We Get Drunk And Screw?

Twenty-four hours a day, every day. And those were just some of the country and oddball stuff- the stuff you only heard on KFAT. We also played the new, the classic, the traditional, the weird, the great, the genuine stuff you'd never heard before and the stuff you'd always liked. This was real music, and when we played country music, we played real country, not that Faith Hill shit that passes for country now. There were so many cuts that were Fat by many and diverse other artists like Randy Newman, Laura Nyro, Billie Holiday,

Big Mama Thornton, the Rolling Stones, Louis Jordan and Fats Waller, the Grateful Dead, Tex Ritter, Django Rheinhart and Stephane Grappelli, Dave Edmunds and Bonnie Raitt and of course Buddy Holly, Louie Prima, Clifton Chenier, Fats Domino, and Carl Perkins, and Robert Johnson and Conway Twitty, and Chuck Berry and Merle Haggard and Guy Clark and Emmylou Harris and Wanda Jackson and Waylon, Willie, and Johnny Cash, JJ Cale, Hank and Hank Jr. and...and... how much time do you have?

And Monty Python and National Lampoon and country comics like Jerry Clower and Justifyin' Justin Wilson and more funny people you never heard of. Whatever was funny.

There was so much more that it makes it impossible to describe it. Many have tried to define KFAT, but none that I know of have succeeded. Go to www.readfatchance.com and take a shot, or read how others have tried to define KFAT.

My God, there were rockers and weepers and jokers and YOU NEVER KNEW WHAT WAS NEXT. And then there were the jocks. They weren't pros, and as Sister Tiny said, "Anyone who ever listened to me knew that I had no training." But they were real and endearing and above all, they made you happy- you hung on their words because they were your friends.

I've always been attracted to unusual, colorful people, and the staff of KFAT were the most colorful people I've ever known. Their lives ran parallel to the music they played, and their stories run the gamut of human experience, from triumph to tragedy. That's not a cliché, that's what happened. And the listeners paid close attention- to the music and to the jocks.

Who didn't chuckle or worry when Buffalo Bob had trouble being sober enough to get through a commercial that he had to read all... the... way... through? Who didn't want to hear what Tiny had to say, or what Sully thought or Laura Ellen told you about the music? Gordy woke up with you and did a damn good job of being cheerful, but sometimes had as much trouble living as you did, and sometimes more. Unkle Sherman was a raspy rascal and you probably loved his show but you wouldn't let your kids listen. They put him on late at night.

Terrell Lynn was the coolest jukebox you ever heard, and he kept you moving smoothly. Cuzin' Al? Who didn't laugh or at least smile (I swear at least a smile!) just listening to Cuzin' Al? I still do, when I hear him on the KFAT online stream.

I used to love to listen to Al's show on Sunday nights, not because I like bluegrass, which I don't, but because his good time is so fucking infectious that I always have a good time listening. I actually laugh with him. How many entertainers can you say that about?

I like to think that my talk show and promo's contributed something unique and flavorful to KFAT as well, but I'm too modest to mention it.

God, it was a good time, and it will forever stay with me as a highlight of my life. I hope you like reading about it.

For my last year at KFAT, I did a two-hour live show on Sundays with guests and listener call-ins. With the station about to go silent, on my penultimate show on KFAT in December of 1982, I got several of the staff from the early, crazy days, back together for the first time since they left the station late in 1979, and into the largest room at the new KFAT, down the street from the original studio. I brought in good coffee, which any staffer will tell you might have been a first; I also brought Danish pastry, because that's what New York Jews do. Maybe some flowers, too. We set up a bunch of mics, sat around in a circle and talked about the station, about its earliest days. Indeed, with Laura Ellen there and Jeremy on the phone, and with Larry Yurdin, who created the format, calling in to reminisce and summarize, we also talked about the times and the forces that brought KFAT and its staff together before it was on the air; not many people knew the pre-KFAT story. And during that show, Larry made an interesting prediction.

He said that while it was a shame that the station was going off the air, that not only had it, at seven and a half years, been a respectably long run for a format as unusual as this had been, and that that long a run was something to be proud of. Then he predicted that the format would not die. No one knew it yet, but he and KFAT had created Americana Music, currently America's fastest-growing music category. Larry remembers hearing the KFAT format referred to as "Americana" back in 1981, but it wouldn't be until 1994 when Fathead Rob Bleetstein went to the "Gavin Report," a trade journal running the charts of all American music. When he complained that there was no chart for KFAT- type music, the owner agreed and offered him the job of creating and running the new chart. He took the job and named the chart "Americana."

But back on that show, when KFAT was about to go silent, Larry predicted that where you'll hear it, where it'll pop up, was in small markets and non-commercial stations. But you'll hear it. And he was right. I know of several stations that are either all Fat music or have block programs of Fat music. And then there are satellite stations that feature Alt Country channels. That Alternative Country, also known as Americana, is a recognized category of music is KFAT's legacy, and for a while, one of them had a former KFAT jock at the helm of their Alt Country channel.

It's out there. It's a chart, and it's a section in your CD store and online. There are Americana sites and concerts and an Americana Association with conferences and a convention. But this was the beginning, the Golden Age of Americana music on radio, right there in Gilroy, California. And it was one wild-ass ride.

Oh, and "Adoxology?" I forget where I saw it, but I know it was in a legitimate medium: a magazine, I think. But when I saw it, it came with a definition, and it stuck with me:

"Adoxology: serious writing about a trivial subject."

You decide if it fits or if it's Fat. Thanks for coming with me this far. Stay on board and enjoy the ride. We sure did.

Part One

Management Gathers

"If we'd have known what we were doing, it wouldn't have turned out the way it turned out."

—Unkle Sherman

Chapter One

Oh, it could have gone any which way. It could have been a catastrophe, or it might have been an unholy disaster. But in the end, it was just crazy enough to work.

It certainly didn't go the way the owners thought it would. The lunatic they hired to put together a staff and develop a format he had invented was Larry Yurdin–brilliant, inspired by the medium of radio, filled with great ideas, and completely without social skills. And the staff that came in from Texas and from the pool of local radio wannabes that he hired to play this weird mix of music saw the dump for what it was: a small-town station with antiquated equipment, inadequate plumbing, moldy walls and stained, worn-out carpeting. It was under-funded, it had crazy people who were *supposed* to be in charge, and the jocks that they hired quickly knew it wasn't going to work.

It almost worked, but in less than four months the checks were bouncing and they'd all had enough, the Texans going home to Texas and the locals back to their dreams. It was a big disappointment for everyone, but especially for the Texans. They were the Pros From Dover, but they hated the studio, the equipment was a joke and they felt stranded in a distant, inconsequential cultural backwater 80 miles south of San Francisco that stank of garlic. Most of all, the Texans loved being in the heart of Texas, and soon they were back there.

The two owners were Jeremy Lansman and Lorenzo Milam, who, by the time they bought the low-wattage KSND in Gilroy in the summer of 1975, had both achieved reputations in the underground radio world as brilliant, rebellious anarchists, free-thinkers and trouble-makers, the both of them. The FCC certainly knew about them, and so did the FBI. When Jeremy and his girlfriend, former hippie runaway Laura Ellen Hopper, joined Lorenzo in California, they had what they thought was big-time money from the sale of their radio station in St. Louis, and a taste to open up the airwaves to their left-leaning brethren and sistren* on the West Coast.

Before Yurdin left, he promised to find his replacement, but he was in a hurry to move on, so he went against his better judgment and recommended an arrogant, coke-snorting New York rock 'n' roller, who knew that no radio professionals would move to Gilroy to work for $3.25 an hour, so he went

---

* I don't know—did I just make up a word? Can I do that?

out looking for the nearest people willing to learn radio and work at the new station. What they wound up with was a ragtag aggregate of drug dealers, scofflaws, a guitar-playing beer truck driver, a working prostitute, a college student, some wannabes, a couple of radio gypsies, a drunk or two or four, and in general, an assortment of miscreants most likely to appear in a lineup. And it worked. It was KFAT.

Do you remember the old, "Hey, kids! Let's put on a show!"? Well, this *would* be like that, if you changed the kids to hipsters, if real money was involved, if there were laws that had to be complied with, and if people needed their salaries to eat and pay rent, buy gas, booze and drugs, and a few other things that don't fit the analogy. But other than those, it was just like that. It was KFAT.

Of course, trying to control this crew was like Jerry Jeff sang, "Pissin' In The Wind," and getting them to show up for work on time was its reward. But what went out on the air was the most joyful, raucous, and unusual mix of music on the planet, and the people who heard it loved it instantly and never listened to any other station. For Fatheads, there *was* no other station.

Jeremy Lansman was born in St. Louis in 1942, and spent his youth there and in San Francisco. That he was a strange and solitary boy has not been contradicted by anyone who knew him. At seven, a family friend gave Jeremy a circuit board onto which he had wired some switches, a buzzer, a light bulb, some transformers and other elementary electronic gadgets to entertain a young boy, and Jeremy was fascinated. Unlike everything else, his interest in this never waned and he taught himself how to wire in more circuits for more gadgets, and he learned more and more as he went. He did it on weekends and after school. Then he did it on weekends and instead of school.

Soon, he was given a simple radio broadcast kit that allowed him to send a signal from upstairs in his bedroom, downstairs to the living room radio. He remembers being amazed that he could send sounds from one room to another, and it just grew from there. He extended the range of these broadcasts and soon he was drifting through school, ignoring the curriculum, but going home to his broadcasting.

His grades were terrible, and his parents split up, not that I am implying a connection between the two. His dad was a professor of mathematics who had those Einsteinian good looks, and (perhaps one will see a pattern emerging) was without discernible social skills. His mother was one of the original Free Spirits: joyous and exuberant of nature, a lover of recreational chemicals and communes, and she needed out of the constrictions of marriage.

At some point, Dr. Lansman applied for work at the Los Alamos Labs where the Atomic Bomb was designed, built and tested, but he was rejected. Jeremy speculates one or both of two reasons was that there might have been a quota of Jews allowed at Los Alamos, and/or that his mother, Elizabeth, had attended some Communist Party meetings in the 1930s, and the review board might have looked without favor at admitting her to their community.

Which was fine with Elizabeth, who had no desire to live in a community that required her to sign in and out every time she left the base. And they probably wouldn't approve of where she went, either. So she went out of the marriage and back to San Francisco.

Now living with his mother in San Francisco, Jeremy recalls that after the fourth grade, he saw no reason to go to school and his grades and his unhappiness reflected that, but he kept going. While in high school, Jeremy took his small home-built transmitter and parked outside a radio station that went off the air at midnight, which was when he would turn on the transmitter and start his own broadcasts on the station's frequency. One night the station's owner was still in the parking lot after shutting down his station for the night, and he heard Jeremy's mix of comedy and weirdness coming out of his car radio on his own frequency. He called the FCC, and Jeremy's mother was *not happy* when FBI agents came to her home. That they were looking for her son, and not her, was small comfort.

Perhaps this was a factor when, shortly before he graduated from high school, his mother gave him an ultimatum: "Drop out of school and get a job before I throw you out of the house." Pursuing his instincts, he looked for and found a job in radio. He got his First Class Operators license, which meant that he knew how and why the equipment worked and how to fix it, and he considered it his meal ticket. Back then, every radio station needed at least one of those license holders on the premises at all times. Jeremy says that with that license, he didn't have to know anything else about anything else. But he knew everything about building and running a radio station.

His first radio job was at an "Easy Listening" station, located in a bathroom at the famous Sutro Mansion in San Francisco. That's right, a bathroom. He still shudders when he remembers the long hours locked in the bathroom with Mantovani, the 101 Strings and the never-inspiring Percy Faith Orchestra. He spent uncounted hours, bored and with adolescent hormones raging, staring at the pretty women on the album covers, and some have said that this might account for the weirdness that followed, but others knew it had already started.

He took another job, much like the first, began to get bored with the dreariness of the job, and soon he was looking outside of San Francisco. A friend of his had a friend who had to leave the state in something of a hurry, and Jeremy had the only car among them. The guy needed to get across the state line, and the three headed for Reno. While passing through Truckee, on the California-Nevada border, they heard a funky AM station they liked, and stopped in to see what was what. Inside, they found a bunch of people in sleeping bags on the floor and they liked what they saw. It was the early Sixties on the West Coast, hippies were starting to show up and beatniks were still around, revolution was in the air, and they had found some people they could relate to. The friend's pal had to keep going, but Jeremy wanted to hang out.

Jeremy's friend drove his pal to Reno and came back for him a few hours later, but by then Jeremy was working at the station. The owner needed a First Class license holder and – Tag! Jeremy was it. The owner was a scam artist who made his money by buying cheap radio stations, fixing them up on the cheap, getting them on the air and selling them to rich widows.

But that came out later, and was none of Jeremy's business, and Jeremy stayed with him long enough for the owner to send him to Hawaii before he was twenty to build a station from the ground up. More experience, and more frustration with radio, and Jeremy was casting about for something a bit more… meaningful. And that's how he met Lorenzo.

Lorenzo Milam used to be rich. Today he is a severely crippled, impoverished man who has no regrets. He has fought the good fight and won some considerable battles. He still holds the record for mail generated to the FCC for his "Petition Against God," and banks in America can no longer take money from your savings account to pay a check without your consent because Lorenzo sued Wells Fargo when they did that to him. Lorenzo fought the lawyers Wells Fargo sent out from New York and Washington, won his case and changed banking laws in America. He's a contentious protector of the Bill of Rights, and as such he's been investigated repeatedly by the FBI and the FCC.

He got into radio because he was sickly as a child, and being in radio allowed him to indulge his neurotic behavior patterns. He liked being in a room by himself, talking or playing music. It was the only way he could communicate with people without them communicating back to him. It was deliciously antisocial and it was a perfect match for Lorenzo.

On the air since 1955 in liberal, free-thinking Berkeley, California, KPFA was America's first listener-supported radio station. That meant that whoever owned it or programmed it could have on the air whatever the hell they liked without being tied to the constraints of commercial radio stations. No sponsors meant no ads; it also meant no one pulling the plug on its operating costs because they didn't like the content of a show. The station had "Block Programming," which means that there needn't be any consistency in what they had on the air. One hour could be a show of South Korean folk ballads, the next an earnest discussion of Apartheid in South Africa. One could hear a book review followed by a reading of French poetry, then a block of the music of the Andes, followed by a monologue of political discontent. Whatever.

The first radio station of its kind, Lorenzo heard KPFA and was intrigued and began listening regularly, fascinated by the potential of putting stuff on the air that matched his anarchic tastes. After a brief flirtation with conformity, Lorenzo was already done with college and marriage, and he started hanging out at KPFA. Because it had so little money, KPFA was staffed by volunteers, and he engineered a show on Friday and Saturday nights. He waited for them to hire him, but they didn't, so he decided to start his own station. Back then, radio stations were free to people, groups, churches, whoever, if the frequency was available, if no one else wanted it, and if you asked the government for it and filled out the forms correctly.

Separately, both Lorenzo and Jeremy did some time at KPFA as engineers, but they wouldn't meet until years later. When KPFA didn't hire him, Lorenzo tried to get a station in Washington, D.C., to be near the battle he was sure was coming and he would be a part of, but the fight to get the station took years, and in the end he was rejected. Frustrated, he wanted to get as far from D.C. as physically possible, so he headed for Seattle, Washington, where he bought KRAB. It was to be the centerpiece of a chain of free-form stations that he would spawn, and would be known as the KRAB Nebula. What he needed was an engineer with a First Class Operators license. He advertised in a trade journal for an engineer who was willing "to suffer nobly for a cause."

Jeremy Lansman read the ad while visiting friends in Yakima, and thought, "what the hell, I'm suffering now for *no* cause..." and he went to Seattle to meet Lorenzo. He was the only guy to show up, and he was hired.

Lorenzo Milam is a major league queer duck, and would that time permitted, I would tell you about his other interests, or the 25 community radio stations he founded in ten years, many of which are still going. Seriously crippled with polio, he is a brilliant, dignified, insightful, fiercely unique individual who cares more for his and your right to express ourselves freely than anyone I've ever known. I spent an afternoon with him a few years ago and enjoyed it immensely. Although his body is badly broken, his dignity and integrity are robust.

Jeremy built KRAB and stayed in Seattle with Lorenzo for almost four years before he set out for one of his two home towns, St. Louis, to buy, with Lorenzo, the non-commercial KDNA and bring them the community radio that they loved and believed in. Lorenzo would be half owner of KDNA, but rarely involved in the programming. KDNA went on the air in St. Louis in 1968 and was sold in 1974, the money providing the capital with which they bought KSND in Gilroy, California, which they were going to rename.

By 1970, Lorenzo was living in Los Gatos, California, near San Jose, running KTAO, another KRAB Nebula free-form station in the Lorenzo Milam mold which he sold in 1974. Which was when Jeremy and Laura Ellen came out from St. Louis to visit him.

This is where the story begins.

Chapter Two

It had taken them six years to get the license for KDNA, and now it wasn't anything like Lorenzo had planned. Lorenzo was half owner, but it was really a Jeremy Lansman Operation.

The station was on the second and third floors, above a Polynesian restaurant in a seedy neighborhood. Jeremy was the programmer and he gave pretty much free rein to his DJs, but they always knew when they had pissed Jeremy off: they could hear Jeremy coming down the stairs from the third floor, and if he was pissed enough, without saying a word, he would take the offending record right off the turntable as it was playing and break it on the floor. And Jeremy was *always* listening.

One of Jeremy's policies was that anyone was allowed to voice their opinion on the air. That would be how the Nazis came to say their piece on KDNA. Then, when the people from the John Birch Society protested the Nazi's, Jeremy had them come in to have *their* say. Of course, the Socialist Workers Party was also happy to have the opportunity to speak their minds. If Jeremy had said that anyone could have their time on KDNA, then he was prepared to let anyone have their time. When people complained about who was on the air there, he let *them* have their say on his station. That way, he thought, no one would be left to complain who hadn't already had their say, and the complaints would disappear. And so would any potential trouble. He thought.

In 1973, the St. Louis press started writing articles about the station, saying that it was backed by communists, and that led to the raid by the FBI. Years later, Lorenzo sued for his file under the Freedom of Information Act. He learned that the FBI had been worried about what was going on and being said at KDNA, and almost certainly began a course of COINTELPRO: agitating, spreading rumors of a communist conspiracy and planting disinformation for the articles.

COINTELPRO stood for "Counter Intelligence Program" and is one of America's more shameful historical footnotes. Beginning in 1956, the FBI began surveilling, and in some cases, planting evidence on citizens. It was largely a "black operation," meaning it did not officially exist in the budget rolls, but a paper trail clearly demonstrates that it did exist. They operated with brazen impunity, knowing that their very secrecy would protect them, but when the paper trail began to surface in 1971, all such operations reputedly ended. Documents suggest that 10-15% of the COINTELPRO subjects were actively interfered with, detained and/or prosecuted. The same documents estimated that 67% of the subjects were interfered with, in the form of planting false evidence, falsified documents, and articles placed in the media based on false evidence that was created by the FBI.

COINTELPRO's stated goal was to "expose, disrupt, misdirect, discredit, or otherwise neutralize the enemies of the State." When Jeremy encouraged the Communists to come in and speak, he primed FBI interest in KDNA. Anarchy, communism, rampant hippie-ism and a general disregard for what those in the FBI saw as acceptable cultural standards surely must have alarmed those in the St. Louis office of the FBI, who were now listening regularly to KDNA.

Perhaps due to COINTELPRO interest in the station, and those articles, KDNA got busted, with the Feds coming in looking for drugs. The charges were later dropped because the drugs allegedly seized, which Jeremy swears were planted, were later taken out of the police storage lockers and used for some *other* drug bust.

Lorenzo never bothered Jeremy about it. So much else was assailing Jeremy, what with the bust, the articles, the rapes that occurred at and nearby his station because the area was so seedy, the building that was falling apart, trying to keep a staff together and keeping something listenable coming out of the transmitter, that Jeremy had enough on his mind. One of the things that Jeremy had on his mind was getting the hell out of there. And then there was someone else to consider.

Laura Ellen Hopper was a local high school dropout and runaway teen. In hippie fashion, she took to the road, traveling all over the country before settling back in St. Louis. She was pretty in an unadorned sort of way: no makeup, no jewelry or accessories, just pleasant, attractive features, light brown hair, worn long and straight, and a soft voice with a pleasing Missouri twang that let you know she was listening to you. She was pretty without the glamour. My mother would have called her wholesome. So would yours. She was a very likable woman who was, unknowable to her, soon to be forced into the tiring, thankless job of being Mama KFAT and riding herd over incorrigible charges. But for now, that was still in the future.

Growing up in St. Louis and much attuned to the hippie community, Laura Ellen had heard about the radio station that offered its listeners free laundry and she went to check it out. It was KDNA. She used the washer and dryer and got to know some of the freaks at the station. Then she traveled some more and got involved in what she described as "a drug dealing episode" where everyone else but her went to prison, but she escaped and went back on the road.

After six months, things had cooled down and Laura Ellen found herself back in St. Louis, and back at KDNA where she got "got swallowed up pretty quickly." She met Jeremy there, and he gave her her first assignment in radio. The job was to take a recorded speech by Black Panther leader Eldridge Cleaver that Jeremy wanted to broadcast, and edit the "u" out of every "fuck" and leave in the "f" and "ck."

What went out over the air left little doubt of what had been said, and yet stayed within the limits of the law. As would any broadcaster, Jeremy knew the law was The Law, and heaven forfend that Jeremy should break the law. Why, that would be... illegal. And doing something illegal meant that you could lose your license.

Jeremy was the boss, there were a lot of "fucks" in the speech, and Laura Ellen got to be pretty good with a razor and an editing block. Welcome to radio, Ms. Hopper.

Jeremy had a mop of brown, unruly hair, he stood at about five ten, and was skinny and shy. His eyes might wander as he looked at you. It's curious that some remember conversations with Jeremy looking at your left shoulder and mumbling. Others say that conversations with him were strange, some remember a dance, or a twitch, but don't remember the eye thing. Everyone remembers that he always seemed uncomfortable when talking to people. He was into electronics, not people, and hadn't developed the social skills that an employer might want to have. Fortunately, Laura Ellen was there for the People Things.

At first she thought he didn't like her, but it turned out that he did. His social skills were seriously deficient, but somehow they found a groove, and they stayed together. Laura Ellen did well to remember that Jeremy was very "here and now—if you're not here, you just don't exist."

One of the cliques at the station was the classical crowd, including the conductor of the local symphony. Another guy, a classical music reviewer for the newspaper, was a bit of a drunk, and Jeremy's mother, Elizabeth, wound up writing most of his columns for him. Laura Ellen wanted to learn more, and be more involved at KDNA, and she tried to learn the classical music canon, but found it hard going. Then she heard her first Bob Wills song, and thought, "Fuck this classical stuff."

So Lorenzo was uninvolved and uninterested in KDNA, Jeremy was getting restless and Laura Ellen was open to whatever came along with those two. Lorenzo had bought KTAO in Los Gatos in 1970, a suburb of San Jose, and went further into debt because of it. He gave KRAB in Seattle to the community, but he needed to get his cash out of KDNA, and Jeremy was ready to move on. In late 1974, they sold the station in St. Louis and each walked away with more than half a million dollars, which was a lot of money back then. It wasn't a successful station in the ratings, but it was a good signal, and worth the money.

Jeremy wanted to move to San Jose and buy a television station, and while KDNA was selling, he spent a lot of time at Lorenzo's house in Los Gatos, planning his next move, and much of their talk was about radio. One night, after consuming dinner and a surplus of their "green goop" jug wine, they got to lamenting a new, disturbing development in broadcasting.

It was 1974, and religious stations started coming onto the lower portion of the radio dial. This was where the government provided free frequencies to educational organizations. That was the plan, but then the FCC started giving stations to religious broadcasters, and that pissed them off. What about the separation of Church and State? What about the Constitution?

The Moody Bible Institute in Chicago had set up some hokey bible school of the air, and they used it mostly for proselytizing. That they were pushing their product was bad enough, but Lorenzo thought it was also bad radio. Bible radio was bad, but bad bible radio was just unforgivable. The frequency could have been used for educational purposes, but instead the church groups put this crap on the air. And for free. Paid for with *his* tax money.

Bad church radio? For free? That was insupportable. And worse, after reading the FCC releases, he saw that more and more of these guys were applying for the free frequencies.

Lorenzo and Jeremy talked about it one night in 1974 and, after the wine and their idealism about the sanctity of free speech had worked them into a frenzy of indignation, they agreed that this bullshit could not be permitted to stand. They made notes, and sketched out a petition which would demand that the FCC put a freeze on all applications made by religious groups or bible schools on the non-commercial frequencies. They wanted the FCC to investigate whether the applicants were going to run educational programming as they maintained the government had intended, or a narrow type of religious program.

Lorenzo asked his FCC lawyers to send him a previous petition, so he could see the form it should take, and he and Jeremy typed it up over the weekend. Lorenzo asked the lawyers to touch it up, but they wouldn't go near it. They offered to deliver it for them, but then Lorenzo and Jeremy would be on their own, and they paid the $35 to file all fourteen copies.

Almost instantly, the FCC started getting letters protesting against what became known as "The Petition Against God." At first they got dozens of letters, then hundreds, then thousands, and then millions. Apparently, all the sermons that are read on Sundays in churches all over America aren't written by local pastors. It appears that some of these sermons are sent around and shared from church to church. Sermons all across America began urging those listening to write to the FCC to stop the petition that they said would ban all religious programming from the radio.

The FCC acted faster on this petition than on any other document in its history. Was it approved? Hell, no.

Submitted in December 1974 and perhaps envisioning the ensuing firestorm of public disapproval they might ignite by supporting this document, in August 1975 they refused to rule on the petition.

In their statement, they said that they couldn't get involved with programming in any way, whatsoever. But still the sermons denouncing the Unholy Petition continued, and still the letters came pouring in.

Years later, Lorenzo had cause to be in the Dockets Room at the FCC, where the mail comes in. A clerk pointed Lorenzo out to another clerk, who came up to him and asked if he was Lorenzo Milam. As he was, the clerk asked him "Are you the one that filed this petition, Mr. Milam?" When he said that he was, the clerk suggested to Lorenzo that "the next time you get an idea to do something like this, don't do it. Okay?"

They were sick of all the mail, at one point getting over a million letters a year. And they still come, piling up in their mailroom. They stopped counting at 10 million letters, and it has to be at least 15 million by now, although I've heard estimates as high as 25 million. Lorenzo finds it amusing that all this effort has been put into trying to stop something that had already failed.

Like you, Lorenzo takes his amusement where he finds it.

As a further aside, in the website you will find the entry from www.snopes.com, the site that investigates rumors to corroborate them or debunk them. The "Petition Against God" entry re-surfaces periodically and so it is periodically refuted on the site.

But back to our story. Jeremy and Lorenzo had sold KDNA and walked away with a half million each. Laura Ellen had a small percentage.

In today's terms, that half a million would now be about six million, so that was real money. Laura Ellen remembered sitting in a conference room in Washington with a bunch of FCC lawyers during the sale of KDNA. She remembered the incongruity of being a hippie girl in torn jeans (before that was cool), a peasant blouse and sandals, dealing with the lawyers with their suits, their attitudes and all that money. People talking in different rooms, back-and-forth offers, increases in bids, back-and-forth negotiations, both wheeling *and* dealing. She also remembers being amazed at walking away with a million dollars for the station. (Kids: a million dollars used to be a lot of money.)

Lorenzo used to read the FCC releases, and in early 1975, he read about KSND in Gilroy, about a half-hour south of San Jose. KSND was a Class A station, which means that it was rated at a low wattage and could go no higher. The release noted that the owner of KSND had applied for a change of frequency and power, and this was how Lorenzo learned of it.

He called his FCC attorney, wanting to challenge the power increase in the hopes that he could batter them into submission—and a possible sale to him. His lawyer told him that he also represented the owner of KSND, and suggested that they come in and work out a deal. The guy who owned KSND wasn't much of a fighter, the lawyer knew he was fading in his desire to be in the radio business, and he thought Lorenzo could buy it for cheap.

"Is it for sale?" asked Lorenzo, and his lawyer said, "It is now." Lorenzo knew that the key to the value of the station was that it might get its power increased and its transmitter moved, which would seriously increase its coverage area, which would seriously increase its value.

Now really interested, Lorenzo had Jeremy look over KSND, who saw it and knew that it was so badly engineered that the signal wasn't getting outside of Gilroy, and with little more than a screwdriver the power could be boosted to its legal limit, which it wasn't nearly hitting now. With that slight adjustment, KSND might reach all the way into San Jose and Santa Cruz, both lucrative commercial markets. It seemed that the signal was interfering *with itself,* which meant that whatever signal it broadcast actually canceled itself out, preventing the signal from going more than a mile. That could be easily fixed, and if the transmitter could be moved, it was conceivable it could reach San Francisco, Marin and Sonoma Counties, and who knew how far to the north, and Monterey and Salinas and who knew how far to the south?

Yes, the key was getting permission to boost the power and change the location of the transmitter. If the permit was approved. But he'd have to act fast.

Because he constantly read the FCC releases and knew the FCC laws, Lorenzo knew that the rules of radio station ownership were about to change. With the new rules in place, a case could be made that the station no longer served its listenership area. It was a gamble, but he and Jeremy thought it would probably go through, although one never knew. Even if it didn't go through, correcting the error to boost the signal to its legal limit would make the station worth way more than whatever the guy was going to ask for it; it would certainly be a good deal either way. Lorenzo liked his chances, and he liked the idea of being a player in the Bay Area. Maybe he still felt stung by KPFA, who refused to hire him all those years ago. If he could get the transmitter moved to the top of Loma Prieta, the highest hill in the Bay Area, a couple of hilltops away, he could cover the entire Bay Area and then he'd have a serious station.

Lorenzo wanted to roll the dice, and when the owner asked for only $150,000, he was both pleasantly surprised and ready to pounce.

By this time Jeremy, Laura Ellen and their brand-new daughter, Elsbeth, had packed things up in St. Louis and were preparing to storm Bay Area television. They moved to Los Gatos, near Lorenzo, to buy Channel 48, a public-access TV channel in San Jose.

He'd been visiting the San Jose area when looking at Channel 48, and had stayed with Lorenzo during these visits. About this time, Lorenzo went into what he recalls as his "first nervous breakdown" and asked Jeremy if he was interested in KSND instead of Channel 48. Jeremy wasn't all that excited over abandoning his dream, but by this time, he realized that the television station was going to be beyond his budget, anyway. And radio *was* his first love...

Both men had owned radio stations, but those were non-commercial, and therefore non-profit radio stations. This move they were considering would put them into the real world of commercial, profit-making radio. You know, like the Big Boys.

Lorenzo's original ideas for KSND—or whatever it would be named—were similar to those that he had envisioned for KRAB, KDNA and KTAO: free-form, listener supported, with one and two-hour segments, or "blocks" with whatever that jock specialized in. Music from Japan, classical, chamber music, interviews, a lot of volunteers. But whatever Jeremy wanted to do with it was fine with Lorenzo, as he was pretty busy with his nervous breakdown. He says he "couldn't stand the shit you had to do to run a commercial station," and if Jeremy thought he could put up with it... cool. He was too neurotic just then to deal with it, and he recalled to me that at the time he was mostly occupied by staring at a wall.

Back at home, Lorenzo liked to cook, and with Jeremy and Laura Ellen newly moved into Los Gatos, he armed himself with *the New York Times Cookbook* and many a dinner ensued. One night at dinner (Jeremy favored spaghetti), Lorenzo asked Jeremy if he'd like to be a partner in KSND, and Laura Ellen, who wanted to get back into radio said, "Yes, he does!"

Lorenzo: "How much, Jeremy?"

Laura Ellen: "He wants half."

Lorenzo: "Is that true, Jeremy?"

Jeremy: "Mumble, mumble. Mumble, sure, Okay."

This was in 1975, and KSND was a small-town radio station in every sense of the phrase. The equipment was from the 1950s, the station, designed originally to be a dentist's office, was small, the rooms were small, the town was small, and before San Jose's urban sprawl spread those thirty miles south, Gilroy was known for only one thing: garlic.

Garlic wasn't just grown around there, it was also packed there. In fact, garlic from all over the vast agricultural region that is the Santa Clara Valley, and from the inland Imperial Valley, came to Gilroy to be processed. In time, and in large measure because of KFAT, Gilroy has come to be widely known as "The Garlic Capital of the World."

We'll get to that, but depending on the season, other fruits and vegetables were also processed around the clock in several processing plants in and around Gilroy, and some aroma will envelop you whatever time of day or time of year you travel through Gilroy. And travel through or around Gilroy you would have: there was no reason to even slow down. But when KFAT was radiating,* there was a reason to stop there.

Trying to get over his nervous breakdown, and leaving Jeremy and Laura Ellen in charge, Lorenzo took off for Dallas to work on KCHU, yet another of his listener-supported community stations, this one most famous locally for getting celebrities to sneeze for station ID's. (KCHU. Say it out loud.)

By this time the Great Radio Experiment of the 60s was over, and rock was increasingly formatted. Radio was boring, but this was *their* station. But what would they put on the air? After talking with Lorenzo, Jeremy and Laura Ellen were open to whatever developed, but what Jeremy wanted was to tinker with the electronics. His managerial style was to start a station, get a staff, have a basic idea of what to broadcast, then see what develops. But Laura Ellen still remembered that song by Bob Wills, and what that had done for her.

What was going to be their format? What music would they play? How would they get a staff in an area where they knew nobody? Couldn't they hire someone to make all these decisions and do all this for them? Where would they find someone to run it, and what was going to be their format? How would they get a Station Manager who thought *like them*?

Enter Larry Yurdin, stage left. Very left.

* Well, I couldn't say "radio-ing" could I?

Chapter Three

Ah, this is the chapter I've been dreading. I have to say some things I'm going to be sorry I said. I'll regret them, even though they'll be true, because they might be hurtful to a man I love and respect. But there's dirt to be dished here, so fuck him.

In doing the dozens of interviews I did for this book to appear as if real journalism had been practiced, two of the people I spoke with had the same memories, on separate occasions, of meeting Larry Yurdin for the first time. Both Buffalo Bob Cassidy and Christa Taylor remember meeting him in the earliest days of KFAT, standing in a room with records covering every inch of floor space save for that beneath his feet. He was six feet tall, pale-skinned, pear-shaped, with a dazed expression on his face, unkempt thinning hair, a big belly, bulging eyes that went in separate directions, stained, mismatched, wrinkled clothing, his un-tucked shirt bearing the remains of recent meals, and a high, bulging forehead with a two-inch square scar at its peak that said to even the casual observer: "I have a metal plate in my head."

This would be because he was a strange, unkempt, pear-shaped slob, and he had a metal plate in his head. But he was also a genius.

But, having said those admittedly awful things about him, let me add that anyone who knows Larry is as aware of his brilliance and sweetness as much as of his dishevelment. He is Linus. I don't believe he has a malicious bone in his body; he so clearly wants to be liked, and his genuine-ness is so endearing that there's no reason not to like him. Unless you've been fired by him, and we'll get to that. So: a slob, yes, but also brilliant and sweet.

Born in 1944 in Newark, New Jersey, Larry Yurdin was always interested in radio. He was devoted to the medium before he ever saw a television. As a tyke, he would explore around the dial at night to find exotic programming, but found only one station that interested him: WBAI in New York City. WBAI featured what Larry soon found out was called "Block Programming," which meant that there was no single sound or format that represented the station. While the local rock and classical stations played the same thing all the time, WBAI played whatever was scheduled for that block, be it one hour or two hours for each program. It could be anything from classical music to political discourse, poetry or the music of Botswana.

WBAI was started in 1956 by a wealthy paper tycoon, and one day Larry read that he was donating the station to the community. The idea that a community could own and run a radio station amazed him.

Intrigued by the programming at WBAI, Larry wanted to know who called the shots, who scheduled the shows, and why. He was so interested in programming, in fact, that at age nine, Larry wrote to *TV Guide,* telling them that his family subscribed to the *Guide* in New Jersey, and asked how could he obtain issues from Chicago, Los Angeles and other cities. He explained that he was interested in programming, and he gave them his age and they sent all the magazines that he asked for.

WBAI had extraordinary programming for its time, featuring blues, folk, jazz, classical and experimental music, and talk shows that Larry listened to avidly. When he was eleven, he took a train into New York City, found his way down to the exotically bohemian enclave that was Greenwich Village, went to WBAI and volunteered to lick envelopes so he could see what was going on at the place where he tuned his radio to at night. He volunteered on weekends during the day, but he had to be home by dinner. Hey, I told you! He was eleven!

While his peers were rockin' and rollin' to Buddy Holly and the Coasters, Larry's interests lay with Woody Guthrie, Leadbelly, and others of the folk ilk. Already considered peculiar at school, he remembers being thought of as "beyond strange."

At twelve, he took a bus into the City to the Folklore Center, which at the time was one of the centers of the growing folk music scene the Village. Izzy Young ran a store where you bought your instruments, strings, picks, magazines and other folkie stuff, but it was also a center where folk musicians and those who appreciated folk music gathered for readings, concerts, gossip and general chit chat. This night, there was going to be a party for folksinger Jean Ritchie. Ms. Ritchie wasn't some Kingston Trio, Peter, Paul & Mary, white bread sort of folksinger, this was a *bona fide* Appalachian folk-singing genuine article, who had captured Larry's heart when he heard her on WBAI, and he went there to get her autograph. That a twelve-year-old boy would be interested, much less take a bus from New Jersey to attend this party, totally blew away the people there. They surrounded him and talked to him, and Larry found a place where his eccentricities were not only accepted, but praised and marveled over. He liked the attention as well as having met people like him, and now, here and at WBAI, he had two places to go to talk about what he'd thought only he was interested in.

He told me, "Other people had imaginary friends, I had an imaginary TV station. WBBB. Anything I came across that interested me in radio or books or music, I created a program for it on my schedule."

At twelve, he was making lists, creating programs and fitting them into a schedule. He was programming an imaginary station. Of course his parents were in a quandary about this. On one hand, their son was doing something harmless and legal, but on the other hand they wondered about his ability to earn a living at it. And they'd like to see him outside a little bit more. They also worried at the solitary nature of it. Maybe little Larry should have a friend. But it was legal, he was passionate, and they let him go his strange way.

Have I mentioned that Larry is brilliant? Everyone who knows him knows it, and now you do, too. After graduating high school in 1962, he chose Bard College in upstate New York. It was among the most liberal of the Liberal Arts colleges, it was small, and they had a radio station that hit all the dorms. No one else was interested in operating it, and Larry took it over, spending all his time there.

He approached some of the more interesting people on campus to come in and play their own records, and he recalls getting an invaluable education in music. He also got good feedback on his show from people on the campus. He wrote to the labels that put out the music he liked, and asked for service from them, which means that the record companies send you their new releases, or their back catalogue, if you asked for it. He may have used the college letterhead to approach the labels, but it was his money, and the collection was his. He had never said they were for the station library. Larry says, "There was no pretense," and when he left Bard, he took his collection with him.

Instead of doing his schoolwork, he spent so much time doing radio that he almost flunked out. But before they bounced him, he went to a school counselor and signed on for Bard's two month work program, which was designed to facilitate students' exposure to whatever field they might pursue after college. Larry went straight to WBAI and volunteered.

He started his first day there by editing tapes, but although he said he knew how to do that, he really didn't, and he got so frustrated that he almost didn't come back the second day. Not to worry: come back he did the next day, and a guy said to him, "Ya know, they're giving me the overnights the first week of January. How'd you like to be my producer and help me do the show?"

That conversation took place in December of 1962, and the guy who asked the question was Bob Fass, who went on the air in January of 1963 with Larry as his producer. Fass soon became a legend in underground radio by playing and saying what others didn't play or say. Fass clearly had the proverbial gift of gab, but he had no records, and Larry had about a thousand. So for their first month on the air, Fass talked and played Larry's records exclusively. He called it free form, it was somewhat revolutionary, and it was good.

Fass was the first person to use the phrase, "Free-Form Radio."

Free-form Radio? Well, maybe it's not a big deal for you, but for me, yeah, it's a big deal. It was the start of an exciting new approach to radio programming, and without it, you'd be reading something else right now.

Have I mentioned that Larry was a bit awkward? In the physical world, that manifests as clumsy, and Larry was responsible for Fass getting fired. It seems there was this flamenco guitarist who was playing live in the studio, and Larry was so involved with the proceedings that he knocked over a lamp, which fell on a turntable and broke it. The station was so strapped for money that they couldn't afford to replace it, which caused a crisis, and Fass was let go.

Larry and Fass stayed in touch, and eventually Fass was back on the air on WBAI and he asked Larry to be his producer again. Larry was getting his degree during the day and staying up all night doing radio.

By this time Larry saw that Bard College was a haven for spoiled rich kids, and not very exciting, so he chose instead the small, experimental Goddard College in Montpelier, Vermont, leaving Fass and his first radio gig behind.

He got a radio station going on the campus, and kept up with what was going on in the culture, the better to know what to program, should anyone ask.

What was going on out on the west coast on KMPX and later at KSAN at this time was an experiment in radio, but this experiment wasn't in some obscure New England college town, this was in San Francisco, where cultural change had been fomenting since before Jack Kerouac got there with his "beat" friends and added their influence to a post-war rebellious fervor. Tom Donahue turned KMPX, a Filipino station with almost no listeners into the first alternate, underground rock station, and started a revolution in American radio that continued after he moved his crew to KSAN, where they dominated northern California rock radio for almost a decade.

Tom Donahue had been a successful AM jock in Philadelphia and then San Francisco. Tired of the incessant drivel that withered all hope of hearing any good music—music that mattered—on the AM band, Donahue quit the Top Forty thing in 1965. He kept his hand in the music business with a tipsheet, in which he alerted radio stations who subscribed to his sheet which "hot" songs those stations should look for, to add to their playlists. He also enjoyed modest success with his record label, and owned what would turn out to be the country's first "psychedelic nightclub". So he'd quit. At night, Donahue would get high with his friends and listen to the new music that was happening in the mid-Sixties, the stuff that wasn't played on any radio station, and one night in the spring of 1967 he had a brainstorm. He thought, '*This* music should be played on the radio!' Then he thought, '*I* should play this music on the radio!' Then a friend mentioned that FM had a stereo signal, which AM did not have. Brilliant!

KMPX was an instant success, playing full-length album cuts by Dylan, Hendrix, the Doors, and a long list of exciting rock pioneers whose music could be heard nowhere else on either the AM or FM bands. Radio people all over America heard about the experiment. My God, they played whole album sides by the Beatles and Iron Butterfly and Pink Floyd and Frank Zappa and all the freaky new music. Plus, there was this developing scene in San Francisco of exciting local bands like Jefferson Airplane, the Grateful Dead, The Quicksilver Messenger Service, Big Brother and the Holding Company, Santana and all the others that were blowing the doors off the conventions that had stagnated the music industry in America, and you could hear all the hip new bands only on KMPX. Those bands and their fans were listening to KMPX, and followed the staff when they moved to KSAN. In concert, those bands were playing for its listeners, and it was making sonic waves all over the country, becoming known as "The San Francisco Sound." Radio stations and record labels were paying attention, and willing to take chances to increase record and ad sales.

Soon, stations all over America were changing their formats to play the new music. They were hiring a new breed of DJs and playing a new kind of music. The revolution was on, and it was exploding all over America. And Larry was immediately aware of it.

Needing to see what was happening first-hand, Larry went out to San Francisco in the famous "Summer of Love" in 1967. Everyone he met there was listening to KMPX, which he felt was doing some amazing stuff. He went to KMPX and showed his letter of introduction from Bob Fass, who they knew and respected as a fellow radio pioneer, and he got in to see Tom Donahue.

Larry and Donahue spoke the same language, and that night Donahue offered Larry the Sunday overnights, but Larry was afraid of losing his draft deferment if he left school, and he went back east and enrolled in the New School, another small, experimental Liberal Arts college in New York City. After he graduated, he continued working with Fass, who was now calling his show "Radio Unnameable."

Back in New York, Fass and Larry were at the meeting that organized the Yippies, the loud, smart, media-seeking vanguard of the counterculture, and Yippie ringleader Abbie Hoffman was a regular guest on the show, along with radical humorist/social commentator Paul Krassner. Krassner's magazine, *The Realist,* was the literary equivalent of the new FM sensibility. It featured counter-culture art, commentary and humor, and was at the forefront of the revolution. Everyone who was interesting in, or coming through, the City, came on Fass' show. Bob Dylan made his first public appearance on their show, with Larry producing.

Organizing and revolution were the bywords of the day. It was a time of daring optimism, and change was almost palpably in the air. On the air, Fass and Larry called for a "Sweep-In" and 10,000 listeners showed up with brooms to sweep up the Lower East Side. It was a revelation to all and to none. It was around this time that Larry heard WFMU, a radio station from Upsala College across the river in New Jersey, which made it all the way to his radio in the Village. Vin Scelsa was there, playing folk music records starting at midnight.

Larry went to see Scelsa and told him about Donahue and what was going down out west, and said, "I have an idea for you. What's keeping you from bringing in pro radio people and going 24 hours a day with free-form? You can reach Manhattan. You're good, let's get people who are equally good, and go all day."

Larry offered to find the staff, Scelsa said yes, and soon they were on the air full time, mixing folk, folk-rock and other new music. They were good, and the station and the format were an instant hit, but a small crisis erupted at the college over this outsider coming in and influencing the station. It was supposed to be a college station, and Larry was accused by a few disgruntled people of stealing the station, but he also got lots of mail supporting him. It was chaotic and exciting, and for Larry, what was important was that he now had his first taste of programming a real station, and he felt good as he moved on to his next project with a sense of accomplishment. By now, he had flunked his physical for the army and the draft was no longer an issue, so he bounced around, "spreading the seeds of anarchy," and worked at or programmed several stations before he found himself back at Goddard College in February of 1970, where he began to teach a course in media.

The counter-culture revolution was in full swing by then, and the hip youth of America were in the front lines. Larry's enthusiasm for "free-form radio" was infectious, and when he gave his class a choice of four options for their class project, they chose to put on a conference for alternative media. For "alternative," read: "hip." Larry wrote out a press release announcing the "First Gathering of the Alternative Media Project." He got Fass and Donahue to endorse it, and he sent out mailers to all kinds of underground media. Whether it was music, radio, stage, film, comics, art or newspapers, if they were alternative, Larry wanted them to participate. Whatever could be considered an alternative medium was welcome.

As this was in the dark ages before the internet, the various people doing alternative media had no way of knowing what the other alternative media guys were doing. What Larry wanted from this conference was to connect them and put an end to that separation.

The mailers went out in March of 1970, with the conference scheduled for June 19th and 20th. It succeeded beyond his wildest dreams. He told the DJs he contacted not to talk about it on the air, because space was limited and he was expecting almost a hundred people. But he'd be happy if he got forty or fifty. To avoid getting crushed by too many people and hangers-on, he wanted word of the conference kept out of the newspapers and off the air. He told the jocks and editors and artists and label people that if they thought someone should be there, let him know and he'd invite them, but this had to be kept under control. Fat chance. Finish the chapter now or later, but there's more on the conference.

It's interesting to note that at the time of the conference, Goddard College was between presidents, and supervision of what was assumed would be a small, inconsequential event, was lax.

Then Crosby, Stills, Nash & Young heard about the conference and volunteered to play, and Larry said of course, but word of that leaked out in a hurry. Larry had to thank their label and cancel the performance, but by then, the press was on to it.

Invitations came back: Yes, we'll be there. Then, more and more people said they'd be there, and then people who hadn't been invited heard about it and asked if they could come. Larry had a budget for up to a hundred people, but in the end, two thousand people showed up, and Larry had to find places for all of them. On a budget for a hundred. Luckily, a friend of a friend, a millionaire, believed in the project and gave them some money, but there was never enough.

Some had hotel rooms, some camped in tents, some were placed in the dorms. Larry was frantic, but madness and chaos suited him; he was in his element and he found a place for everyone.

One of Ken Kesey's original Merry Pranksters was Hugh Romney, a.k.a. Wavy Gravy, and he and his group of activists, the Hog Farm, smelled something significant in the air and showed up. They were used to organizing and mobilizing events, and their help was invaluable. Larry gave them a meadow and they blasted the Grateful Dead from a huge sound system, fed the people, and gave some order to the proceedings.

A DJ from Detroit who didn't attend but recognized that this was a great shot at some free publicity, sent a band that he was representing, the Mighty Quick, to play at the Conference. Perhaps it just wasn't a hometown audience, or perhaps it was because they sucked, but their performance was met with a stony coldness, and in a desperate act to appear "alternative" and to gain audience acceptance, the lead singer pulled out his cock and displayed it for the already hostile crowd. Immediately apparent to the crowd was the obviously diminutive size of said penis, and the crowd started to laugh at what the unfortunately endowed singer had been so proud of. One fellow in full hippie regalia was sufficiently inspired by peace and love—and possibly some chemicals—to get on the stage, take the microphone, swearing that "anyone's got a cock that small oughta get his ass kicked," and proceeded to do just that right on the stage.

A fight broke out, but order was restored and the stage was cleared for a performance by Dr. John the Night Tripper, who was much better received and who apparently remained properly, if bizarrely, attired. Also properly attired, the J. Geils Band came in from Boston to play.

Also in attendance at the conference that weekend were three people who were clearly in the new wave of alternative radio: Jeremy Lansman and Laura Ellen Hopper of KDNA in St. Louis, and Lorenzo Milam of KDNA, KRAB in Seattle and KTAO in Los Gatos, California.

Desperate to find rooms for the unexpected masses and unbeknownst to the college administration, which wasn't paying much attention, Larry had charged a bunch of airplane fares to his convention from all over the country, to the school. As I said, the administration was in something of a between-the-presidents turmoil and hadn't been closely monitoring Larry's event. But when they saw the bills, whoever wound up in charge at Goddard College saw neither the wisdom nor the humor in it, and Larry Yurdin was summarily disconnected from his college teaching career.

While he had never made any money so far in his radio career, Larry was still young and footloose, although arguably his fancy wasn't quite as free as he might like it. But he knew the burgeoning underground music scene was spreading all over America, and he knew most of the players. By now, even the corporate giants knew a change was in the air, and were scrambling to increase their hipster cred, and there had to be a place where he could take that knowledge.

Tom Donahue got hired by Metromedia to create a hip station in LA, and had been successful down there, so they planted the format in other cities, where it kept succeeding. Someone at ABC knew about the conference in Vermont, and soon Larry found a home at ABC, who hired him to consult all of their FM stations so they could catch up with what was, in hipster parlance, "going down" in the cities and campuses across America.

ABC had been doing an automated thing in Pittsburgh, and they wanted Larry to put together a staff for them. He did, and when it was up and running, he went on to hire staffs for their stations in New York, Detroit, Houston, San Francisco, Los Angeles and Chicago.

ABC had Larry flying and driving around the country, listening to radio stations and scouting interesting jocks. Several times he came through St. Louis and took Jeremy Lansman and his staff at KDNA out to dinner on the ABC tab.

By the spring of 1971, ABC was going in a different direction and Larry was out of a job. They wanted to maintain their, uh... hipness, but they had seen the chaos and lack of discipline that hiring freaks had caused, and they wanted their, uh... hipness... tightly formatted. This was the beginning of what we now call the "Classic Rock" format. The Age of the Corporate Consultant had begun, and it smelled to Larry like rigor mortis.

Station owners rushed to hire these consultants and soon a new blandness, a new homogeneity, took over the airwaves in America. Once again corporate rock ruled the charts. Consultants chose the songs and carefully and strategically placed them into a "clock." The clock separated the songs into categories, and told the DJs which songs went in what order.

It was the antithesis of free-form.

Oh, some of it was good, there is no doubt, and there was always room for an innovator or two to sneak onto the charts. But the revolution and the Great Experiment In Radio was over. The direction ABC was going held no interest for Larry. He found a job at KMET in Los Angeles, and settled in as the News Director.

Then two things happened in rapid succession: First, Larry was visited by two guys who owned and operated the Armadillo World Headquarters in Austin, Texas. The AWH was a "big hippie barn" that would accommodate 1,500-1,600 people, had live music and was run like a collective. Austin was where Willie Nelson was amazed to find a whole town full of long-haired, pot-smoking freaks who were totally into his music, and a music scene grew around Willie. The two guys told Larry that as good as the live music scene was back in Austin, what they needed was a good radio station, and as the two guys with the most pull in their town, they asked Larry if he'd be interested in coming out to take a look.

The second thing happened to Larry on the night after his last day at KMET. He took a woman to the Hollywood Palladium, where Chuck Berry was opening for Black Oak Arkansas. Larry was backstage with his date, watching as Mick Jagger and Keith Richards, unplanned and uninvited, got on the stage with Chuck Berry and tried to jam with him. This show has become legendary for Berry physically throwing Jagger and Richards off the stage, and it would be one of the last things Larry remembered for several weeks.

On the way home from the show, a teenager with no driver's license and no insurance, who had been drinking and driving his brother's car, ran a red light and drove right into Larry's car. Larry wound up in the hospital with a fractured skull. Thus, the metal plate in his forehead.

Finally out of the hospital and in recovery mode, Larry and a friend got in the friend's car and took off around the country, visiting some of the more interesting radio stations. He wandered into a station at war with itself in Denver, he met a man who owned a station in Phoenix, and met a production guy there named Marty Manning. He spent a day with Joe Walsh of the James Gang and later of the Eagles, then headed for Tucson, where his friend headed back for LA and Larry went east to Texas. Larry spent a week hanging out in Austin with the two guys who owned the Armadillo World Headquarters in Austin, where a serious music scene was indeed gathering.

In the fall of 1972, Larry was living in a cabin in the mountains outside of Santa Cruz, California, when he got a call from the guys from Austin. Another friend of Larry's had been there and together these three hatched a plot to get

Larry to come to Austin. They told Larry they had a job for him at KRMH, which was named after its owner, R. Miller Hicks, and called "karma" by the locals. But karma had nothing to do with it.

R. Miller Hicks was a farmer and an upright, Bible-reading, God-fearing man and a deacon in the Methodist church, and unbeknownst to all, that was going to be a problem.

The owner gave Larry a free hand to hire and play what he liked. Larry wanted the hip new country music that was outside the staid conventions of the cookie-cutter shit that was coming out of Nashville. There were hip artists recording great music that no one was hearing, because no radio stations were playing it. Larry wanted to change all that.

One of Larry's first hires was a talented production man named Marty Manning, who Larry had met when he had been in Phoenix a few months before. Manning heard the call of free-form radio and headed east to Austin. He'll later go to Houston with Larry, then Lake Tahoe, and still later, show up at KFAT.

The free hand Larry was given proved to be too much for Deacon Miller, what with Kinky Freidman singing "They Don't Make Jews Like Jesus Anymore," Bobby Bare imploring, "Drop Kick Me Jesus Through The Goal Posts Of Life," and Jimmy Buffett singing "My Head Hurts, My Feet Stink, And I Don't Love Jesus," plus a lot of other sacrilegious shit. Deacon Miller felt it was his duty to protect his legions of God-fearing, right-thinking church-going folk. You can almost hear the Deacon asking: Who *are* these people? *What's wrong with them?*

Miller became increasingly concerned over the perversions passing for music being broadcast on his station (a series of sins, I believe, that as the owner of said station, might land him in a lonely position when the Rapture got there). Miller's concern turned to alarm when Bob Simmons, one of Larry's pals from San Francisco who would later work at KFAT, came to Austin. He brought a documentary that played on KSAN, and played it on KRMH. The documentary was called "The Texas Special" and it featured, as one might expect, the new music from Texas. In it, there was an uncensored statement by someone that included the word "shit," and while that might have gotten by in that West Coast cesspool of sin, in Austin it would not stand.

I think the phrase I'm looking for here is "mass firing." Yeah, that's the phrase, and Larry was once again looking for a place to ply his unholy trade. It was the spring of 1973.

The loonies had been in charge of the station in Austin, and it was a lot of fun. The fun lasted two months.

The only guy who didn't get fired was Marty Manning, the production guy. Production is a separate skill from being a DJ. Being a DJ means playing records on the air in a manner that flows pleasingly. You don't want a guy who plays six slow songs in a row any more than you want him or her to play six headbangers in a row. It's gotta flow. It's gotta entertain without making the listeners turn it off. But while the playing of the records is a skill, as any DJ will swear, DJs are an interchangeable commodity, and it's the guy who makes the commercials, does the promos and thus keeps the station in revenue, that keeps his job. And at KRMH, that guy was Marty Manning

Marty told me: "The basic rule of radio is: you don't fire the guy who does the commercials. I've been in the biz for over thirty years and I've never been fired 'cause I'm the guy that does the commercials. If you can't do commercials, then you're just another guy saying, 'here's another Celine Dion record' and they can get a machine to do that."

Marty Manning stayed at KRMH after everyone else was gone, enjoying a paycheck for a couple of months and then he got another call from Larry, who'd been sitting in the Armadillo World Headquarters after the firings wondering about his next gig when he ran into Mitch Green, who he'd met at his big Alternative Media Conference back in Vermont.

Mitch had just left KPFT in Houston. KPFT was another member of the Pacifica chain, whose New York City station, WBAI, had intrigued Larry all those years before. At this point, Pacifica had stations in New York, Los Angeles, Berkeley, and this struggling station in Houston. Green said that they were about to hire a guy to be the Station Manager, but that they hadn't hired him yet. Green thought Larry knew a lot more about programming than this other guy, and could probably walk in and steal the job. KPFT had the usual block programming, was listener-supported, and it wasn't working. The ladies of the Lesbian Poetry Hour were fighting with the Experimental Jazz guy, and everyone was fighting with everyone else. The station had been relevant and exciting in the past—exciting enough to have been bombed twice, one of those times by the Ku Klux Klan.

It seemed that when the station had been under attack, the various elements representing the varied blocks of programming banded together to form a united front. But, once the element of fear had been removed, the station quickly sank into chaos. The public support vanished, and the various factions within the station made for each other's throats.

When Larry heard about the job, the Pacifica Foundation was sick of it all. The owners were thinking about letting their license expire, and the station was about two or three weeks from going silent. It was either hire this other guy, or go silent.

Larry came in, pitched a bold experiment, and got the job. He wanted to do the progressive, alternative take on country music that he had started at KRMH. He wanted to play alternative country music without commercials. He'd done it briefly at KRMH, it was new, it was exciting, and he said it couldn't miss.

His challenge was to create a staff and a sound that was so good that people would phone in and pledge money to listen to it. He suggested that it would sound like, like... well, here's Larry suggesting a fifteen minute musical excursion: "You play something from the Peer Gynt Suite, some Vivaldi violin concerto, then go into an art-rock piece, like Emerson, Lake & Palmer, then from that to something by the Grateful Dead, then to the New Riders Of The Purple Sage, to Hank Williams."

The idea was to "take people on a musical trip without the listener knowing they'd been taken anywhere."

Larry got Marty Manning to come over from Austin, and he put the rest of a staff together, but there was no money to pay anyone, or pay the electric bills or keep the equipment running. Larry took advantage of the rich musical environment from Houston and Austin and called in favors from the artists that he had supported at his stations. Willie Nelson, Jerry Jeff Walker, and Michael Martin Murphy were among the "Cosmic Cowboy" folk who put on a series of benefits to keep the station running and get a new home. Willie and the others happily played until they had the money they needed, then said goodnight and thanks.

The station featured free-form radio done as a non-commercial station, with a strong Texas orientation; roots music and blues and alternative country and neo folk-rock and humor with good production values. Also, please keep in mind that no other radio station in the world was playing anything by Willie Nelson or the others who later became the "Outlaws" at this time. They didn't fit into the Nashville mold, and they were too strange to be played on country stations. Imagine: a revolution in country music was under way and the folks in Nashville didn't know it yet.

The station was so successful, recalls Larry, that Houston's major rock station simply lost their male 18-34 year-old listeners almost overnight and Larry had a hit on his hands.

A success in the ratings, perhaps, but it was still a non-commercial station, and there was no money coming in from advertisements, which put a strain on everyone.

The stress reached a critical point, and somehow the management thought it best if Larry departed. Ever the gentleman, he offered to find them a new Program Director, which he did, and then he was off to Lake Tahoe.

A rival station, KOKE in Austin, was trying something similar. It was playing straight country artists, but using the more progressive cuts from their albums, and it was limping along. Larry's version was a lot more daring and a lot more fun. Larry's approach to country music was a hit, but a strange one, and perhaps suited best for where it was. In fact, it was damn strange, and Larry knew that the only broadcasters in America who might have heard what he did, who might appreciate his experiment, and who might give him another shot were Jeremy Lansman and Lorenzo Milam.

Chapter Four

KPFT was a success, but it had no advertising and money was always a crisis. It was exhausting, and by the end of 1974, Larry moved on.

During his tenure in Houston, he had been going to Dallas to visit Lorenzo Milam at his new non-commercial station, KCHU. KCHU? God bless you, Lorenzo. Milam was a fellow radio crazy who commiserated with Larry, and together they would indulge in flights of intellectual fancy to an ideal world where they would have complete control of their stations. They were fun, those flights, and Larry was much enamored of Lorenzo's anarchistic rants, which made the most sense when Lorenzo's level of frustration reached uncharted heights.

Larry was just out as PD, Program Director of KPFT when he got a call from a friend who was starting a station in Lake Tahoe, and Larry lasted almost six months before Lorenzo called again.

Lake Tahoe was a tiny village with hardly any listeners and no revenue. It was fun, but Larry was used to larger audiences.

In June of 1975, Lorenzo called from Dallas and asked Larry to come out to see him, and he went. Lorenzo asked if he was interested in going down into California to "work this new station he was buying." He didn't know what he was going to do with it yet. He was interested in Larry's new take on country music. Lorenzo was intrigued, and he asked Larry if what he was doing was applicable to a commercial station in California, and Larry said, "Absolutely!"

Lorenzo knew that the call letters for the new station had to pass FCC scrutiny, but he wanted them to be unlike anything else on the dial. Larry recalls that both Jeremy and Lorenzo were looking to thumb their noses at convention. Larry liked that.

Jeremy was in Madison, Wisconsin, for a meeting of the Association of Community Stations. In a few hours, Larry got a call from Jeremy, who had just spoken with Lorenzo. He asked Larry if he was able to get to Madison. Jeremy wanted to see him about their new, as-yet un-named station. Larry took a train to Chicago, then a bus to Madison, found Jeremy at the convention, and the two men went looking for a private place to talk. They found a quiet stairwell, and they sat on the steps and discussed Jeremy's new station.

The two men knew each other from Larry's conference in Vermont and his visits to KDNA in St. Louis. They were of a similar breed in many ways, and the two men quickly got into an animated conversation. When both men talked about radio, animated was the only speed they knew. Jeremy's idea was to do a straight, obnoxious country station, and he thought that the listeners would know it was a parody of a country station. Larry thought that would turn listeners away, especially those with the musical sophistication of the people of the Bay Area, and he told Jeremy his idea of doing a "Living In The Country Out Of The Box Gonzoid Radio Experience" format, as opposed to a country station.

Gonzo was a term invented by Hunter S. Thompson, whose dispatches for *Rolling Stone* magazine struck an indelible mark on Seventies journalism. Characterized by all manner of excess, Gonzo came to mean wild, crazy, hip and rabidly anti-establishment. It may have been a calculated style, but it had an immediate impact on journalists and journalism. The Gonzo approach could apply to any medium, and Larry wanted this new station to be under some kind of control, but to sound like it wasn't. The *apparent* lack-of-control thing appealed to Jeremy, who wanted to thumb his nose at the establishment and make a buck in the process. And he wanted to have fun. Both men were borderline insane, of course, and at the very least, Jeremy liked Larry's irreverence, and Larry was fun.

Larry was worried about Jeremy's need to thumb his nose at convention and how that might sabotage what he saw as a great opportunity, and he told Jeremy that the listeners in California were too hip to support Jeremy's idea of a parody; Larry knew he had a good format, a unique format, and he wasn't letting go of it.

He said the difference was that he wanted a country station for a citified listenership, as opposed to one a bit more… rural. Jeremy got it. Jeremy liked it, and asked Larry to come out and run the station in Gilroy. The station, still KSND, was a mess. The owner hadn't cared much about discipline or profit. Every once in a while, when he needed money, the owner would go out and hit on his friends to buy ads for their businesses, and thus the staff got paid. When KSND changed ownership, only two people were kept. One was Gordy Broshear, an overweight, over-eager but not overly ambitious and non-confrontational small-town disc jockey, whose principal qualifications for radio were his affability, his willingness to come to Gilroy, and that he knew how to turn the station on. The other staff member who could stay was Enid Wingate, the alcoholic wife of a friend of the former owner who had recently killed himself, and Jeremy hired her to run the office and do the traffic, by which the commercials were logged and scheduled. She also answered phones and took messages. She may have stumbled a bit in an alcoholic mist, but she knew the station, and she completed the few tasks required of her. Enid was the office staff.

Jeremy understood that his money would only go so far, and he had to replenish it sometime soon. He had to make a profit to keep this station on the air, and you don't fire the person who runs the office and keeps the books and routes the money, even if she was a bit too dazed to know what was going on at the time, much of the time.

Larry came down to Gilroy, and was appalled and challenged by what he saw. He knew the equipment was antiquated and the studios were shamefully inadequate for what he had hoped, but listeners never saw the studio. Also, this was California, and Jeremy had plans to change the site of the transmitter, and thus the coverage area of the new station. He had a chance to make an impact in a major market and get paid for it.

Although Jeremy doesn't recall, others remember that he said he planned to own the station for the three years the FCC required, boost its signal, and sell it when it became more valuable; he says his plans always morphed into something interesting, and that that had always been fine with him.

Jeremy's plans for KSND included moving the transmitter from its present location on Mt. Madonna, just outside Gilroy, to Loma Prieta, thirteen miles away. Loma Prieta was the highest point in the Bay Area, and if he could slip this move past the FCC...

Jeremy sort of forgot to mention to the FCC in the application to relocate the transmitter that on the other side of the hill from Loma Prieta lay San Francisco and the East Bay. Jeremy bought the station hoping he could move the transmitter, as moving it was crucial to reaching into far-off corners of California and making KFAT a valuable signal.

Why submit the real data, which would only make the FCC nervous *and* alert the San Francisco stations who had the right to protest the move?

In early discussions about the format-that-was-to-be, Jeremy brought in his attorney and friend, Leo Kesselman, who will appear in a much more prominent role later in the story. Leo hated the idea of a humorous, alternative take on country music, and tried to talk Jeremy out of it. Larry remembers Leo as being a real sleazebag, which nobody, including Leo, would dispute. If you like self-aware and self-deprecating sleazebags, you'll like Leo Kesselman when you meet him. Kesselman likes himself a great deal, and so did I.

Jeremy was going back and forth between Gilroy and St. Louis, getting rid of KDNA's last bits of business, and he couldn't be reached for days at a time. When in Gilroy, much of his time was taken up with doing the engineering to give the new station a clear, strong signal. When in St. Louis, he was sending checks to cover whatever costs were most pressing. As this was when the checks were still clearing, let's call it The Calm Before The Storm.

The Experiment Begins,

Hipsters Move Into Town, And

Americana Music Finds A Home

In The Garlic Capital of the World

Chapter Five

Where were we? Oh, yeah, Lorenzo and Jeremy, with Laura Ellen as a minority owner, bought 250 watt KSND in Gilroy, about a half-hour south of San Jose. Gilroy was a sleepy agricultural community of about eight thousand souls. Wait! I forgot about the junkies, so let's say there were 8,000 people, and about 7,900 souls. Cruel? D'ya know any junkies?

The principal industry in town was the processing and packing of garlic, tomatoes, peppers, both bell and hot, Brussels sprouts, artichokes, tomatoes, cauliflower, and in the fruit department, cherries, apricots, prunes, pears, plums and whatever produce I may have forgotten. What made all that processing so obvious was the smell, the constant smell. Ah, the Gilroy smell... In fact, with all the people I've spoken to about KFAT, when we reminisce about it, the one thing about Gilroy that everybody brings up, is the smell. That constant, and constantly changing smell. I know some people who have fond memories of the smells; others, not so much.

But as long as there was processing going on—and there was always processing going on—Gilroy would smell. I never thought it was so bad, but then I like garlic and I was only there for a few hours at a time. I was there to do my work, and I could take anything for a while. I think it was the tomatoes that most people had a problem with. Processed tomatoes have a certain... sour pungency. And that cauliflower smell! No, the cauliflower was the worst.

What was being packed or processed, of course, depended on what was being harvested, which in turn depended on what time of the year it was, and at all times of the year, you would smell something. And nothing had no smell.*

Other than that, the town was a typical quiet, rural small farming town, much like you'd find anywhere in rural America in the 1950s, even though it was the 1970s. Change hadn't come to Gilroy, as it had to most of the rest of the state. Who had time for the revolutions in music, politics and fashion that were going on up the road in San Francisco when you had to be up at dawn to run a tractor or operate an industrial dehydrator or a canning machine? These were mind-numbing jobs, and these were by-gum working people. These were church folk, you know.

Except for the drunks, the bookies, the dealers and the junkies who hung out downtown at a bar called the Green Hut.

---

*I know it sounds awkward, but I've read it a dozen times and it makes sense. Try it again.

What passed for a downtown was a four-block strip along its main street, Monterey Street. There were a couple of bars, a few antiques shops; there was a newspaper and a Western Wear store. There were a few cheap motels, none of them of the chain variety; there was a Jack-In-The-Box, a couple of Mexican restaurants, a vacuum cleaner store, a gun and sporting goods store and a handful of other, nondescript stores. In the middle of the downtown hub was an optometrist's office. And above that was an office that had been built to be dentist's office, but was now a radio station, and was barely scraping by.

Neither Jeremy nor Larry remembers the deal that Jeremy gave Larry to hire a staff and run the station. According to Bobby Eakin, who will appear in the next chapter, Jeremy figured that the station cost about $150,000, and would cost another $100,000 to set up and put on the air.

Jeremy estimated that the station would cost him $25,000 per month, and he thought they could bill $50,000 in advertising revenues. Of the money the station earned in advertising sales, half would go back into the station, paying costs, and the other half would go to Larry, who was going to give some to Laura Ellen and Jeremy, and from the rest, take what he wanted and give the jocks a piece of what was left. It would be free-form capitalism. Call it hippie radio.

Lorenzo is a brilliant man, but he was severely crippled, and I'm not sure if he ever even got up the stairs to see the new station more than once, but Lorenzo didn't care how it looked. He didn't care about the carpets or how old the equipment was. As long as it put out a signal, it was worth the price he paid for it. It was the idea of a new radio station that intrigued him, and this would be their first commercial station; a chance to make money.

But it was never money that drove these two. It was the chance to talk to a place, to imprint on an area the message that they wanted to send. Both men were of the belief that better music was available than was being played on commercial radio because it was not economically viable to put that music on the air. They would put that music on the air. Also, maybe a talk show of worthwhile, perhaps enlightening discourse. This was their idea of public service.

They were kids with toys, and as this was their first commercial station; if all went as they hoped, they'd move the transmitter and have an amazingly valuable signal for less money than anyone should have paid for it.

They'd looked over this station that could hardly get its signal heard across the street, and Jeremy knew what the main flaw was, and any engineer who knew what he was doing and had a screwdriver would know how to fix it. Jeremy had a whole case of screwdrivers.

Please remember these screwdrivers, as they will become a big part of KFAT lore.

Jeremy knew that with the signal-canceling problem resolved, the station would get into Santa Cruz and San Jose, both lucrative markets for advertising revenue. And more important, both men knew that if the FCC allowed them to move the transmitter to Loma Prieta, the highest point in the Bay Area, then who knew how far the signal would go?

Both men also knew that the owner had already applied for the move and an increase in signal strength, and that the application had gone in a long time ago. Jeremy and Lorenzo figured that the longer the application had been in process, the sooner it would come up for approval. Both knew that when dealing with the FCC, especially with what the FCC already knew about them, nothing could be taken for granted. But if the transmitter could be moved, why, they'd really have something, wouldn't they? And all for just a bit of the money that they had in their accounts from the sale of KDNA in St. Louis.

Now Lorenzo, Jeremy and Laura Ellen had a radio station. KSND played a normal, routine mix of country music, the sort that lets people listen and not think, which was probably a good thing for the agricultural workers in Gilroy. Its DJs were of the typical small town variety. They had few enough records there, and that was all there was to play. There were no budding stars at that station, just good, rules-following journeymen jocks who spun their records and ran the few ads on the schedule. No one there was looking to make waves, and in that they were succeeding.

Lorenzo didn't much care what Jeremy did with the format, and Jeremy's thought was to do a pretty routine commercial thing, boost the signal and coverage area, and sell it after the three-year holding period that the FCC demanded. Buying and selling radio stations was called "Flipping," and that was precisely what Jeremy intended to do. But before he could flip it, what would he put on the air? This was commercial radio, and to sell ads, he needed listeners. To get listeners, he'd need a format and a staff. And records. Jeremy knew that things had a way of taking another course when you don't exert any discipline. Discipline, you will have surmised, was not Jeremy's strong suit. He remembers that he'd had a plan for his last station, but he knew if he didn't interfere too much, it would always turn into something interesting. KDNA had done that and he was open to whatever KSND turned into.

Jeremy liked what Larry Yurdin had in mind for the station, and Larry promised that he could staff the place with the only people who knew this new format. He knew the right people from Texas, and while some were already employed in other gigs, he was confident he could persuade them to come out to California. And he'd find others. Soon the deal was struck. Larry Yurdin would move down to California from Lake Tahoe, hire a staff and run the station. It was the fateful collaboration of three crazed but like-minded men.

Lorenzo and Jeremy wanted to make a statement with the call letters for the new station. They toyed with a few names, including KPIG, and came up with KFAT. Meanwhile, Larry was making arrangements to move down from Lake Tahoe to put it all together.

There was, of course, already a complete staff at KSND. They'd probably all have to go, but Jeremy and Larry wanted to see who was there, who was doing what, and who was capable of doing what Larry had in mind. Foremost of the old staff was Enid Wingate, the alcoholic who did the books, the traffic, and in general acted as the station manager, mainly because no one else was doing it. Enid might be drinking, but she wasn't stumbling, and she got her job done in a relatively timely manner. Of course, it wasn't like there was a surplus of advertising going out over KSND. There was no sales force to speak of; they were running only two or three ads per hour, and apparently Enid could handle that. Who said alcohol and work don't mix?

Larry said that Enid could stay. This meant that someone was on board who could generate the first revenue stream into the new station. Enid was safe for the time being.

Gordy Broshear remembered KSND as "the worst adult country station ever. It had no record service, and only had 100 singles, which we played over and over." He'd wanted to be a DJ since junior high school, and he had stopped by KSND so many times that they finally gave him a job there. He had been there almost two years when the morning DJ came in and said, "Well, that's it- start looking for a new job. The place got sold."

Gordy was upset because this had been the only radio job he'd been able to get. He was a fat, not overly dynamic guy who was afraid that his radio career was over. He was sharing a house in San Jose and didn't socialize a lot, so he didn't need much money. He didn't own a car, so he took the bus down to Gilroy every morning. He felt comfortable there, and he hated to think of this easy job as going away. No one at the station knew anything about the new owners, until one afternoon while he was on the air, up the back stairs came a procession of four barefooted hippies. Two of them were Laura Ellen and Jeremy.

Enid said, "The new owners."

Jeremy was a lot more interested in the equipment than the staff, and after watching Gordy work for a few minutes, he assured Gordy he could stay.

In a couple of weeks, Larry Yurdin showed up, and he was a sight. Not quite obese in the way that Gordy was, Larry was… gelatinous. He was a bulge-y sort of man, wearing cheap, food-stained clothes that knew neither north, south, east or west, and his hair went in unexpected directions. His clothes had no sense of style or maintenance, but style wasn't what Jeremy was looking for. It was Jeremy's station, but it was Larry's gig. And Larry knew the right guys to hire for this gig.

Be afraid, Gilroy. Be very afraid.

Chapter Six

There was one place where Jeremy's tinkering resulted in thunderously good results.

For those of you with a technical mindset, this is how it plays: KSND was permitted 3KW of power at 300 feet of elevation. The change the owners had applied for would give it 50KW of power at 500 feet of elevation. The change would move it from a Class A station to a Class B. Class C was the highest category, but was not permitted in California. The change would also move the frequency from 94.3 to 94.5 on the dial. All this made it valuable if it was granted.

Jeremy said that he and Lorenzo knew that the owners of KSND had applied for the change "a long, long time ago." And they knew that he had given up waiting on the application, wanting to get out of the radio business, and he needed money, which was why he sold it. The signal sounded good but it was weak. That had to be an engineering problem, and that was Jeremy's meat and potatoes. He couldn't wait to dig in.

The station could barely be heard even at the transmitter site. The station was heard strongly only within a few blocks of downtown Gilroy, and in odd spots elsewhere. Jeremy knew that it was an interference problem before he bought the station, and he knew where to look. Now that it was his, he spent as much time as he could at the transmitter, and eventually he hit on an idea: it might be the relationship between the FM studio-to-transmitter link (STL) frequency and the over-the-air frequency. Then Jeremy explained to me some electronic crap that I didn't understand and cannot translate into English, but the end result was that he made a few adjustments with one of his screwdrivers and fixed the problem; and just like that, KFAT was going into Santa Cruz and San Jose.

Jeremy had just tripled the value of his new radio station. Now, if it was true that Larry knew the right guys for this gig…

Chapter Seven

Larry knew the right guys, all right. It wasn't gonna be just about the records at this new station, it was a sensibility thing. He wanted to take what he had done in Austin and Houston and expand on it. Sure, he wanted a wide range of music: blues and bluegrass and western swing and rockabilly and traditional country and new country and Hawaiian and old rock and new rock, and country rock and folk and… and… and whatever worked. He also wanted to insert a bunch of novelty and comedy tracks to keep it fresh and keep the listener guessing.

That was part of it- the listener would never know what was coming up next. Keep 'em happy, but keep 'em off balance. Like that "musical trip" that Larry talked about in Chapter Three.

Marty Manning will play a big role in this story. Twice. Trust me. Larry had met him while passing through Phoenix and later hired him to work at KRMH, the station that lasted two months until one of Larry's San Francisco friends showed up and played a radio documentary about Texas music with the word "shit" in it. There were other problems at KRMH, like the songs they were playing, and the Methodist deacon who owned it and just couldn't understand why these kids were so hateful, so disrespectful of God. That attitude got everyone fired except for Marty, because he was the Production Guy who made the commercials.

Marty stayed in Austin for a few months until Larry convinced him to move to Houston to work at KPFT, which Larry promised would be a "bold experiment."

In September of 1973, Larry's idea for the non-commercial station was to put together a staff of talented people and make it sound so good that people would call up and send money to keep listening to it. They'd play progressive country, and he'd call it "alternative country" with a heavy Texas influence.

But coming by money was always a struggle. Getting by on his weekly pittance wasn't always possible for Marty, and he learned not to ask any questions when a man he assumed was a drug dealer would show up and hand him $1,500 in twenties and fifties.

When the money wasn't there, Larry put on benefits, calling on local talent for their support. One of the first of these was "A Tribute To The Cosmic

Cowboy," with Michael Martin Murphy. The Cosmic Cowboy idea was poking fun at the hippies and dope-smokers who gathered in Houston and attended the shows at the Armadillo World Headquarters. This was the same group that so shocked Willie Nelson when he found these country hipsters digging his music, when the people in Nashville had no interest in him.

Other artists chipped in their time and did benefits for the station. There was Willie, of course, and Jerry Jeff Walker, and Jimmy Buffett, David Allan Coe, and Goose Creek Symphony. Willie would show up and ask, "How long do you want me to play?" Sometimes they had to do ten hour shows, but KPFT managed to stay on the air.

KPFT played music with a strong Texas orientation. KPFT was a Pacifica station, like WBAI, the station in New York City that Larry had so loved and been so influenced by as a youth, but this was a heretical deviation from the Pacifica template. But desperate times called for desperate measures, and Larry was the go-to guy when you're desperate. They had given him the job and told him to do what he wanted.

None of the jocks at KPFT was in it for the money; they wanted to play good music and reach as big an audience as they could. Money wasn't a factor for them, thank God.

But KPFT wasn't a commercial station and it was hard as hell to survive. The first year, the station made twenty thousand dollars, and was barely able to stay on the air. The next year it made thirty thousand, and things were looking better. Then the shit hit the fan, the staff scattered, and they lost twelve thousand that last year. But by then Marty was gone.

Larry never made any money at KPFT, and he lived on the insurance money from when the kid ran into his car in Los Angeles. Marty was broke and burned out. A friend had a new house in Lake Tahoe and wanted Marty to go there to chill out for a bit. The friend had gone to Tahoe, fallen in love with the place, and bought a house by the lake, then bought a radio station in Lake Tahoe. Larry staffed it with talented burn-outs and refugees from other stations and other formats.

Also at this time, Larry had just spoken to Jeremy in Wisconsin, then had another meeting with him and Lorenzo in Los Gatos, and gotten the job. He went out to Texas to recruit for the new station in California. He wanted Marty, but he knew Marty was chilling in Tahoe, and he'd get to him soon. For now, he was in Houston and he wanted Bobby Eakin. Bobby was one of the guys who worked for Larry at KPFT, and was now working at a newspaper in Austin. Bobby was an easy-going Texan with a cool, unruffled demeanor and a good handle on the music and how to make it flow.

Bobby had been attracted to the looseness and craziness of the Houston station. They were playing music that he liked and no one else was playing.

It was a special station and Eakin appreciated that, and it was apparent from the response they got from their listeners and the local musicians that they knew it, too. Eakin loved Larry and his madness.

In Texas to recruit jocks for the new station, Larry took Bobby out to dinner. Bobby recalls Larry as "such an intense, crazed, comic genius guy that you'll follow him anywhere. I think we had dinner on Friday night, and I went in and turned in my resignation, and Monday morning I had my car packed up and was on my way," he told me.

"I was in Houston putting some stuff in my storage and was at KPFT, and Marty Manning called and heard I was in the building, and they put me on the phone. He said, 'I hear Larry's got this station out in California and you're going.' I said 'Yeah.' And he says, 'Well, I'm in Tahoe, come pick me up on the way there.'

"I asked if he'd talked with Larry and he said, 'No… it's fine.' And sure enough, by Monday morning I had loaded up my pickup truck with about a thousand records. It was June or July, and it was so hot I had to drive at night and rest during the day, and I picked Marty up in Tahoe, and we went to Gilroy."

While waiting for Bobby to get to Tahoe, Marty got the number for the station in Gilroy and called Larry, who said, "Oh, great. You're gonna *love* this place!"

As Marty and Bobby spent ten hours together driving down to Gilroy, I asked Marty if he asked Bobby what Larry had promised him about the station, and Marty said that they didn't have that conversation. He was promised he'd get paid, but that was about all the details he'd been given.

Marty emphasizes that it wasn't about the money. "It was a job. It paid some money. I was young and didn't give a shit. I was on my own, completely loose. I was looking for adventure. If it didn't work, I'd just leave."

He knew that Larry attracted the maverick, both in stations and in staff. "If Larry was involved, it was a maverick deal, and I was a maverick guy. It had to be interesting."

Marty also knew that it was going to be a challenge to work with Larry, because Larry was "a difficult person to work with over a period of time." Larry was "brilliant, but so manically driven that it's hard to keep up with him, and he doesn't have a calm side. He would look at you over the top of his glasses, in the craziest moments, and you would know that he knew that this was one of those moments, and it's okay, we know it's crazy and we'll move on from here."

Bobby Eakin said, "I would have done it as a volunteer. To work with Larry, and to do what he was dreaming of doing… At that time of my life—I was 22—I would have jumped in my car and gone for anything." Then he said, "The real incentive was to work with Marty and Larry again. I had worked with Larry in Houston, and knew what he had done in Austin, and I knew that the guy was a brilliant genius, and given a station to do with whatever he wanted was some place I really wanted to be. I felt lucky. I'd have gone up there for nothing."

Whatever was going to happen in Gilroy, and after whatever *did* happen in Gilroy, Bobby Eakin would always have the utmost respect and affection for Larry. Of course, after he'd left Gilroy and was back in Texas, that didn't mean that there wasn't a certain relief not to be working for that crazy son-of-a-bitch.

Bobby and Marty drove all night and got to Gilroy in the morning, before it got too hot. First, they smelled Gilroy. Then they saw Gilroy. Then they walked up the stairs and saw the station, and then they weren't happy. They had just walked into a Jeremy Lansman Operation.

Marty remembers, "They had funky gear at Tahoe. But KFAT was the worst gear I'd ever worked on."

KFAT had the oldest control board they'd ever seen, and there'd be hell to pay if the Board of Health ever saw those carpets… the only way to get any air moving in the control room was by cranking open the windows, which also let in the smell and the mournful cry of the passing trains. Marty thought it was almost charming at first. But the production room, where he'd be spending much of his time, was impossibly amateur. There was a turntable and a mixer that was one step ahead of what you'd use if you were a home hobbyist on a budget. It was going to be a burden to work with this gear, but it was what came out of the equipment, not the equipment itself, that made the station. It was old stuff, and harder to work with than he'd hoped for, but it was a poor craftsman who blamed his tools, so he sucked it up and went to work.

Marty didn't care. He found the place charming enough. He loved the bowling pin lamp and thought the studio was cool. One thing he didn't like were the cassette machines Jeremy had rigged to play the commercials. They were standard cassette machines, rewired by Jeremy to play at double speed, which increased the fidelity of what they played. Jeremy put them in because they were cheap. No other radio station in America used this system, but if it was cheap and it worked, it was going to be used at KFAT. This was a Jeremy Lansman Operation.

It was frustrating to work with shitty gear, but it was an adventure in good radio and that was why Marty and Bobby had come to Gilroy. There was still some work to do to get this puppy on the air, and Jeremy was always around and ready to fix something.

Chapter Eight

Enid Wingate was the bookkeeper, accountant and Traffic Manager. This involved the process of writing the information down and giving it to someone who would make sure the ad got written and recorded, and then she would put it on the schedule. The schedule of when ads were to run was also known as the Log. Making sure the ads got played at the right time was the job of the Program Director (PD).

Even whilst imbibing, Enid could do her work, and those who remember her brief tenure at KFAT all enthusiastically recall that her wandering around the station in a slight alcoholic haze hardly ever precluded her from her appointed rounds. So, Enid could stay until Jeremy and Laura Ellen figured out what they wanted to do with her and her position.

When Larry got to Gilroy, he walked up the stairs and got his first look at the station, and we are left to imagine his dismay. The town smelled, the carpets were filthy, the equipment was old and shaky, the woman running the office was old and shaky, the air staff was useless, and there wasn't a record in the house that he would allow to be played on his new station. But this was another opportunity and Larry saw through all of that. He was in California and he stood a real chance of having a signal that carried his format into a significant portion of a major market. It was time to roll up his sleeves and do some firing.

It was simple. Almost everyone would have to go. Enid and Gordy Broshear would stay for the time being, but everyone else was out.

He had hired Marty Manning and Bobby Eakin from Houston, and Fermin "Speedy Gonzalez" Perez was coming from Texas, too. Perez had been a jock at KOKE, and an important hire for the new station. Larry also knew that if the FCC didn't approve the transmitter move, they'd need a bridge to the local community, and a Latino jock would be a help. He'd be the Program Director (PD).

That Speedy was a dishonest, contentious, obstreperous, confrontational, shifty, gin-swilling sticky-fingered drunk who would soon cause problems for almost everyone at the station was as yet unknown to them. But not for long.

Larry rented an apartment in Santa Cruz and began scouting for local talent. One Saturday, listening to Santa Cruz's listener-supported station, KUSP, he heard "Road Apple Radio," which was similar to what Larry had in mind for KFAT, and he knew he had a recruit. Johnny Simmons would fit right in. You know a road apple is a horse turd, right? Larry found him, talked to him, and Johnny was happily on the air the next day. He'd be Genial Johnny.

Lorenzo wanted Larry to hire Cece McGowan, who Lorenzo remembered as a success at KSAN up in San Francisco. KSAN was already a legend, and easily the most prominent rock station in northern California. It would be good to have a pro with a built-in following at the new station. So Cece was hired.

Another local hire was Christa Taylor. Christa was an Air Force brat who'd lived all over the country, spending much of her youth in Louisiana, where she paid attention to the local music scene. It was rootsy, blues-based music, with a sound and a swing all its own.

Christa had traveled a bit and landed in San Francisco at just the right time. It was July, 1966, and within a week she was one of the lucky souls to be at Ken Kesey's second Acid Test. The Acid Tests were the litmus of the new freedom, with music, slide and light shows, and a seemingly endless supply of free LSD. The Acid Tests were leading the Revolution, and Christa arrived in town just in time for all of it. She looked all over the Bay Area and settled in the mountains outside of Santa Cruz, ninety minutes south of San Francisco, and almost a world away.

Running a record store in Santa Cruz, in 1974 her friend, impressed with the breadth of Christa's knowledge of music, suggested that she try out for a radio station in Carmel, just down the beautiful coastal road.

She got the job, moved to Carmel, and settled in until the engineer, Don Mussell, told her about KFAT. Mussell was a friend of Jeremy's and was helping him put the newly re-jiggered KFAT on the air. He told Christa that KFAT was a "kind of take-off on a kind of country music." KFAT wasn't on the air yet, and deciding not to wait, Don and Christa went to Gilroy to talk to Larry, who had just gotten there. She says her first visual of Larry still stays indelibly in her mind: "I went up the back stairs and walked into a room covered every inch with 45's all over the floor, and there was this fellow in the middle, brown earth shoes and nasty socks falling around his ankles, and his pants that were shiny gabardine- shiny from being old and worn- and a large belly, with his T-shirt with all his food from the last three meals on it."

"One eye going one way and one eye going straight, he was pulling on the side of his hair and tweaking on something on the side of his chin, and his huge baldness and the scar that said, 'I've got a metal plate in my head and I'm crazy.' And that's exactly what he was."

Don and Christa went with Larry to Santa Cruz and "had the pleasure of watching Larry eat two roast beef dinners. Larry told us his plan for KFAT." Larry liked her energy and she was hired.

She got there a week before KSND changed to KFAT, and because she had managed a record store and knew how it worked, they put her in charge of calling record labels and getting records sent to the station. KFAT needed records. Most of the labels and most of the artists she contacted were so obscure they were astonished that a radio station in California knew of them and wanted their artists. They loved the idea of KFAT, and sent them everything they had. It was exciting.

Christa remembers the sounds of the trains in the background, the smell of garlic, and that when the wind would blow just strong enough, it would blow the needle off the record. But it was summer and you had to leave the window open. The train whistles were nice, though…

David Chaney was an eighteen-year-old from Watsonville, California, which is darn close to Gilroy. He had done some radio in Oregon, and when he heard the new station, he came by and talked with Perez, who said he could do a show that weekend to try it out. Larry liked what he heard and hired him for weekends, overnights, or whenever Larry or Speedy needed him to be there.

David Chaney will play a significant role at KFAT a lot later, but for now he wasn't given much air time or much direction for his shifts. No one got involved with his shows, and he was only told to "concentrate on regional artists, and play station favorites and listener requests. Try to keep it going." Chaney listened to other jocks and picked up on new songs all the time, and each week he stretched a little more in his show.

While Larry was the Station Manager, Chaney knew that Speedy was the PD and the musical pace-setter at the station, and he took his cues from him. He didn't see Larry much, as Larry wasn't there during his overnight shifts. Speedy had been a major player at KOKE, and Chaney had heard KOKE when he'd been through Austin the summer before. He liked the music a lot and was excited about learning as much as he could of the new format, which was kind of similar to KOKE, but also way different. It had a more sophisticated sensibility and they played a much wider range of artists at KFAT, and the jocks had a much larger palette of music and oddities to choose from.

Once Jeremy fixed the malfunction in the transmitter, the people of Santa Cruz tuned in. Then they started telling their friends about KFAT. Chaney remembers that at first it was "pretty much a Santa Cruz station." He also remembers the equipment, and having to use "paper clips and scotch tape, and we'd wedge toothpicks in the buttons in the transmitter remote control panel. We were going off the air all the time, but it was still a kind of a modern miracle that it seemed to work and we still had listeners."

Santa Cruz was KFAT's main support network, and it's a good thing to know that Santa Cruz was a lot like San Francisco was in the mid-to-late 1960s. It was a very laid-back, long-haired, take-it-like-it-comes kind of a town.

It was a cool town, right on the beautiful Pacific coast, where the surfing was good, the weed was righteous and times were slow and easy. Today, Silicon Valley money has come into Santa Cruz, so real estate prices are scary-high, and electronically, the town is as wi-fi'd and up-to-date as your nearest Starbucks. But mellow still abides.

But back in the Fall of 1975, everyone at the station knew about Jeremy's plans to move the transmitter, and that was also exciting. Everyone knew that if they kept their shit together and made the station work, by the time the transmitter moved and the signal improved, they would be heard beyond just Santa Cruz, maybe heard all over the Bay Area. San Francisco was in everyone's sights. There was a palpable excitement in the air at the new KFAT, and new records were coming in the mail every day.

The ad schedule was light, but Jeremy or Larry or someone would find someone to sell ads and the money would start coming in. But when they boosted the signal, well, that was worth working for in these dire conditions. That would be worth it, then.

Everyone who worked there in the early days remembers a strong sense of camaraderie among the air staff. They were working for peanuts under horrible, hot and smelly conditions, but they were playing an exciting new format and they loved the music that affected everyone who heard it, and if they stuck it out, they'd all make money when the station made money. It was going to happen and it was going to be fun.

No one had any money and they all worked long hours. Have I mentioned that it was August and September, the two hottest months of the year and hot as hell? People were always sleeping at the station, either in a back room or wherever they fell. Jocks were playing music for each other on the air or in the production room.

Genial Johnny was earning less money than when he was driving a cab, Speedy was pissed off all the time about having to wait to get the money he was used to. Cece was doubtful, Christa was still hopeful, and David Chaney was willing to wait and see. Marty and Bobby were starting to get a vague sense of despair, but the station started to sound really good. People in Santa Cruz were abuzz about the new station and listeners were calling in with comments and requests. It was getting better all the time, and the optimism at KFAT was off the scales. But optimism is a thin gruel for a steady diet.

Once Lorenzo learned what sort of format Jeremy had in mind for their new station, he told Jeremy that he wanted no part in making the decisions. After its purchase, Lorenzo would be a no-show in the shaping of KFAT.

When Larry got there, he listened to the talent he had at the station and thought Gordy was one of the worst people he'd ever heard on the air, but he needed that one last fill-in guy. Larry told him that when the new staff got there, he would not be hired with them, and he gave Gordy a variety of shifts all over the clock.

Gordy told me, "It was terrible working for Larry. He never knew exactly what he wanted, and when you didn't do it, then he'd yell and scream." He said that Larry was a bad communicator. They had long talks, and when he did what he thought Larry wanted, Larry would say, "No, no, no! You've missed the whole idea!" Larry terrified Gordy.

Gordy remembered Larry coming into the control room one Sunday afternoon after a month at the new station, and putting his arms around his shoulder, saying, "Gordy, why don't you go home and we'll call you, okay?"

He waited, but no one called. After that, he knew he should have gone out looking for new work in radio, but he was too depressed, and because he couldn't hear the station all those thirty miles away in San Jose, he had no idea of what was happening at the new KFAT.

Soon after he got there, Larry had hired a guy who walked in off the street who intrigued him. Sherman Caughman had done a little radio, and he didn't know what Sherman would sound like, but "he was so off the wall he intrigued me." He'd been living in San Jose, but he was from Arkansas, he knew the music Larry wanted to play, and he was one of the few people who understood what Larry was trying to do. And that voice. It was wizened and rough like old, weathered sandpaper. It was craggly, and sounded like he was just barely winning the battle between chaos and control.

Larry liked that Sherman "looked seedy and had a raspy voice and was a larger than life character. He knew why the disparate elements would make a whole picture." Larry also liked that the sound of his voice made listeners wonder what was going on and who the hell was in charge at the station. Larry had him on weekends at first, but by the end of the story, Sherman would be one of the longest-lived and most-loved and most-fired of the DJs at KFAT.

From the first time he heard about KSND, Jeremy's plan to change the location of his transmitter and boost his power put him on a collision course with the San Francisco rock giant. KSAN broadcast at 94.9, and with the boost, KFAT would move from 94.3 to 94.5. For the plan to succeed, he had to do three things: Submit the request to the FCC, omit pointing out that the signal would be positioned so that it was aimed at San Francisco, and get the application past the owners and lawyers at KSAN. Jeremy filed part one, hoped for part two, and wished for part three.

To this day, Jeremy is amazed that he got his approval, and KSAN let the switch go through with no protest. They had a right to protest the switch*, and they could have easily done so.

Even Jeremy knew that their protest would have merit because KFAT's new signal would over-ride KSAN's in some areas, and many longtime, faithful KSAN listeners would turn on their radios one morning and hear whatever madness was being on played on KFAT.

---

* I could indulge my love of puns here and say that KSAN must have been asleep at the switch, but we have become friends now and I wouldn't do that to you. But I could have. Asleep...at the...switch. Thank you

Chapter Nine

Michelle McGovern lived in one of those pockets of Santa Cruz where KSND made it over the hill. It was her favorite station.

She knew a lot about country music, and this station wasn't all that great for the music, but it did have two qualities that she liked: it came in clearly, and she didn't have to think while listening to it.

She knew the songs and she knew the jocks by their voices, and she was comfortable with the station. Then one day it was gone. She checked her radio to see if it was broken, but other stations came in and Michelle figured they were having temporary technical difficulties and they'd be back soon. That had certainly happened before. She tuned back to where KSND should be, forgot about her radio, and went on with her day.

The next morning she was in her kitchen and she heard someone talking in her living room. There shouldn't be anyone there, so she went to see who it was. The room was empty, but a new voice with a laid-back Texas twang was on the radio, saying that they were a new station and he hoped "y'all like it."

Then he played some Texas Swing that knocked Michelle on her ass, and she listened to the new station the rest of the day.

It was August 6th, 1975, and KFAT was on the air.

Chapter Ten

KFAT went on the air with Marty Manning at the controls of the morning show.

Larry had an apartment in Santa Cruz and a staff to run all the shifts, Jeremy had his technical problems to tinker with, Laura Ellen had a baby to deal with, and the station was up and running.

Christa knew the way the record companies worked, and she was on the phones, asking labels for record service. Enid Wingate was getting about and assigning what ads there were to Marty Manning to write and record the spots that would run on the air. Speedy was the Program Director (PD) and set the pace for how the station should sound.

They cast about in Santa Cruz for sales people to sell the station, and they got a series of guys who thought they'd give it a try. Experienced radio sales people weren't flocking to this unlikely gig, and any that they approached knew it was going to be a hard sell. In fact, this problem would plague KFAT for its entire existence.

The problem was, in a phrase: what was it? Any experienced sales person would know that to sell a station, they would go to a client or an agency, and said client or agency would invariably ask: What was it? Meaning: What was the format? Was KFAT a country station? Was it a rock station? Was it Sports? Oldies? News? Jazz? Talk? Christian? Farm reports? What was it?

At that time, radio stations played formats, and there were just so many formats around. They were AOR (Album Oriented Rock), MOR (Middle of the Road), Urban Contemporary (UC), Classical, Jazz, Talk, Country, Oldies, and maybe a couple that I forget. The point is that radio ads were sold to clients who knew what demographic, what group, they wanted to appeal to. Major companies wanted to devote X% of their ad campaign to this format, Y% to that format, and Z% to another.

If you had a sports talk show, which appealed to a young and middle-aged male demographic, you sold spots to beer makers, sports drinks, athletic shoes, cars, whatever would appeal to that group. If you had an Oldies format, you sold products that appealed to older, maybe middle-aged people, like the Baby Boomers. You get the point.

KFAT was always successful with the local merchants who supported it.

KFAT's listeners were supportive of the local sponsors like restaurants and a hardware store, and these local sponsors were always happy to support KFAT. But then, as now, the money was with the national advertisers: the beer makers, the car companies, the airlines, etc. When a sales person got an appointment to see the agency representing these major accounts, the first question that was always asked about KFAT was: What format was it? Was it AOR? MOR? UC? What? KFAT was none of those. So how would a salesperson describe KFAT? It not only fit into none of the categories anyone had ever heard or heard of, it was almost impossible to describe. In fact, KFAT's format wouldn't have a name until the mid-90's, when a Fathead named Rob Bleetstein suggested to a magazine that published the charts that a new category was needed, got the job and named the music "Americana."

But back in 1975, how would they categorize a station that played rock and folk and Hawaiian and country and blues and western swing and comedy and bluegrass and the Grateful Dead and more music that no one had ever heard of than any other station in the country? In the most literal sense possible, KFAT was off the charts. This made for interesting listening, to be sure, but it made for baffling sales presentations. I have always imagined a sales person getting an appointment with an ad buyer, going in to his or her office and being asked the inevitable question: What's the format? I think the answer would have to be: How much time do I have?

KFAT was impossible to explain, and if you needed to explain your format, your problem was magnified. And if you needed to explain your format and you couldn't…

For the sales people, defining, explaining KFAT, was always a problem. The money was with the major accounts, and with KFAT's success, some of them would take a chance on this strange, unclassifiable hybrid, but for the most part, the major accounts would avoid KFAT for the more quantifiable entities elsewhere on the FM dial. KFAT was weird and they played too many songs with profanity in them. This was the reason KFAT couldn't attract any experienced sales people in the early days, and so in the first weeks and months that KFAT was on the air, they resorted to desperate measures and went with whoever wanted to take a whack at the job.

For the most part, the hipsters they hired to sell ads had no experience; they thought they'd give it a try, and they went out to find people willing to spend money on advertising on KFAT. One of the first ads sold for the new station was for an adult book store.

The ad was paid for, but what would it say? Do you say, 'Our books are dirtier than the dirty books at other dirty book stores, so come on in?' Or, 'If you're not getting any, and you're horny, why don't you come in and buy something to take home to jerk off with?'

Marty was looking at the sheet that told him who the client was, and he was perplexed. They had a chuckle over it at the station because it was a funny situation. No one remembered ever hearing an ad for an adult book store on the radio before. Christa asked him what he was going to do with it, and he said he didn't know. He said he'd come up with something, and went in to the Production Room and turned off the lights. He just sat there, thinking. Then Marty made some notes, went out to the record stacks and got a record. He came back in, turned on the lights, stuck a tape on the reel-to-reel deck, put the record on the turntable, and opened the microphone.

Twenty minutes later Marty came out with a cassette and played the spot for Christa.

"Uhh...what are you doing with yourself right now? If you've got time on your hands..." and a song came on. It was Mark Spoelstra's "Dirty Movie Show," which went, "The best ol' time I ever did see, went down town to the dirty mov-ie," and Marty talked about the store, its location and its large inventory. Satisfaction was assured. It was definitely Fat, but it wouldn't encourage those national agencies to be associated with a station that did *that*.

Getting ads was a problem. A series of salespeople went through KFAT in the first weeks and months, but none turned out to have a talent for the job. They came, they went. One of these serial sales people at the new station was Lee Kelsen. Kelsen was a Santa Cruz guy, and he bragged to Jeremy about how good a salesman he was and got the job. But his only real experience in sales was in selling drugs. He lasted about two weeks at KFAT, not making much money, supporting himself with his other trade, which varied from pot to hash to LSD.

In the summer of 1975, Bob Cassidy was a fixture in Santa Cruz bars, a self-described "unemployed sort of a guy" who drank a lot and hung out a lot, and supported himself in various illegal ways. He knew everyone that he thought mattered, and he was happy getting by in the small world that was Santa Cruz.

Bob Cassidy was a presence. He was about 5'10", with broad shoulders, a bushy mustache and long sandy colored hair that he wore tied back in a pony-tail. His face had been put back together from when a satchel-bomb had exploded it in Viet Nam. All the pieces fit... almost... perfectly, and you'd look at him and you might not get what was wrong with his face, only that something was... slightly... off. Other than that, he was a fairly handsome fellow. But it was his voice that you noticed.

He had *that voice*: those deep, round, rumbling tones that took a word and made it resonate. It was an impressive voice, and it compelled you to listen. Was he a nice guy?

Well, I like him. I like him a lot. But then, I'd also never leave him alone in a room with my children or my pets. He always seemed to be angry at something, and even when he was happy, he seemed to be angry about something. And when he was drunk, which was often, you gave him a respectful distance.

Cassidy knew Lee Kelsen "via recreational chemicals" and was drinking and bullshitting one night at a bar in Santa Cruz, waiting to meet Kelsen to transact an LSD deal. When Kelsen came in, he had with him an attache case and four stoned, bedraggled hippies. The the case held ten thousand hits of the famous purple LSD, the hippies were members of the Charlie Manson family and Kelsen said, "Bob, let's talk." Bob remembers, "We proceeded to close down the bar, locked all the doors, and gave everyone a hit of acid. We talked."

Kelsen wanted to be out of the radio sales business, but he liked the station and he liked the people at the station, and he knew that Bob knew everyone in Santa Cruz and that Bob was… was… well, you could look at it like Bob was persuasive or you could look at it like Bob was intimidating. In any event, Bob Cassidy knew so many people in Santa Cruz that if he went to some of them and told them to advertise on KFAT, they would. That was what Kelsen wanted Bob to do, and Bob was looking for something to do. Kelsen brought Bob to Jeremy, who told him he had the job. Not a job, Jeremy said, "an opportunity."

One night Bob was out drinking and the next day he was selling ads for KFAT. He didn't listen to KFAT. He knew people in Santa Cruz were talking about it, but when he heard the word "country," he tuned it out. He didn't like country music. For those who don't believe in foreshadowing, about two weeks before this encounter, Bob had been sitting on a deck with a friend, thinking about his next career move. He thought he'd been wasting his life in drug dealing, and the friend said, "Well, Bob, you've got a great radio voice. You should be in radio."

He didn't give it much thought then, and didn't give it much thought when Lee Kelsen asked him to sell ads for KFAT. He had never heard KFAT, and was only vaguely aware of it. But he felt he could make some money hitting on his friends. He determined to go only to those people he thought would benefit from the ads, and he set out to make some sales.

He says he "didn't know anything about the station, I just told my friends this was a good deal, and they bought the time." So far, so good, and he liked the idea of a steady, honest income for a change, something he could tell his family about. If it didn't pan out to be either fun or financially rewarding, he'd move on. What the hell.

He sold some ads and he got some money for the station, but there was trouble at KFAT that he didn't know about.

Chapter Eleven

Back at the station, trouble was brewing.

Jeremy had fixed the malfunction and the station was now getting out of Gilroy. All the shifts were covered, and it seemed like a stable staff. Listeners were calling from San Jose and Santa Cruz, and a few of the other outlying towns where the station was heard. It was starting to be played in a few shops and restaurants in Santa Cruz. The Texans had been meshing with the local talent; they were all talking about the music all the time and letting each other know about some cut they'd found. Except for the money, this was fun…

But this was Jeremy's first commercial station and he was used to getting high school volunteers, community people and other free talent, and that just wasn't going to be the case here. The staff wanted to be paid and they wanted to be heard. Everyone knew about the plans to move the transmitter, which would give them a larger coverage area and also give them more money from the increased ad rates the increased coverage area would generate.

The station was costing more money than Jeremy had anticipated, and the ad money he was counting on wasn't coming in. Bobby Eakin remembers, "there's always a lag there. You sell it this month, it runs next month, you bill it next month, and it comes in the next month." The formula for dividing the incoming money was useless without the incoming money, and running this thing without ad revenues was going to bankrupt him. Jeremy just had no cushion. He had no float, and deeply rooted in his makeup was his distaste for paying bills and for dealing with disgruntled employees, or humans in general. Also, he had to spend more time back in St. Louis wrapping things up, which kept him from his beloved work on the gadgets at the new station. He was getting frustrated and the staff was getting agitated.

Marty, Speedy and Bobby were particularly disgruntled, although that's not the word they'd have used. As anyone who has heard any songs about Texas knows, Texans love Texas, and they missed it mightily. And no one sold Lonestar Beer out here, not even that small town booze trough down the street, the Green Hut.

Now they were in a dusty backwater with nowhere to go, no one wanted to know them, drink with them or fuck them. They were bored and lonely, the equipment was a joke and the money was pitiful. The Texans were forlorn.

And when they weren't forlorn, they were pissed off. And so was pretty much everyone else except for Gordy, who was happy to be one of the cool guys at last.

Jeremy wanted to have good relations with the town, and he and Larry knew that if the FCC didn't approve their application to move the transmitter, they'd better have built a relationship with the local people of Gilroy. A cocktail party or barbeque would get the locals out, but it would cost too much. They decided to have a breakfast for the merchants of Gilroy who were members of the Chamber of Commerce. Larry had Christa spend a week addressing 300 invitations to the big breakfast. On the day of the Big Breakfast, Laura Ellen made sure Jeremy had clean clothes, and they and Larry went there and… one person showed up.

That was Gary Steinmetz, who ran the Green Hut, the bar five doors down from KFAT. Steinmetz, known as Steimie, was the bartender there, but he was also the town bookie, and his bar hosted the local drug dealers as well as the local constabulary. The Chief of Police drank there, and so did the Fire Chief. Steimie knew everyone in town, and where all the bodies were buried. He loved the new station and the new jocks who rolled into town, and he made them feel welcome immediately. The jocks were happy to have a place to hang out, and were lucky to have a place so close to the station. The Green Hut always had its radio on, and the radio was always tuned to KFAT.

Over the years, many of the jocks would put on long cuts and race to the Green Hut for a couple of drinks. Sometimes when a jock called, or just sounded dry on the air, Steimie would send his girlfriend, a nurse, up to the station with drinks for the jock. Steimie was a pal to the staff, and it was a mutual love affair at first sip. And who didn't like having a uniformed nurse show up with a free pitcher of Screwdrivers?

Oddly enough, the only other friends the new staff had in Gilroy turned out to be a surprise. But the friendship didn't start out well.

The station had been on the air for three weeks, and Christa was doing her Saturday afternoon shift and living and commuting from Monterey, about an hour away. She drove her VW bug to the back of the station, but the small parking space, which held two or three cars at most, was being repaired (never to be repaired again) and she couldn't park there. She drove next door and parked in the lot of the Continental Phone Company, thinking it was the weekend and no one would mind. She did her shift and when she got off, Marty, who had just come in to replace her on the air, told her that there were a whole bunch of cops down in the parking lot, and they wanted to see her.

She had occasionally used a bit of cocaine to lose weight, and that day she had bought (she swears) her first-ever bindle of it, and she put it in her back pocket with her keys. She didn't know what the cops wanted, and went downstairs more curious than concerned. When she got to her car she found four police cars, two fire trucks, plus eight cops and six firemen standing around her car. Assuming they were going to ask her to move her car from the phone company's lot, she pulled out her keys, and the bindle, which had become entangled with the keys, fell onto the ground.

She tried to be as casual as possible when she stooped down and picked up the bindle and put it in her back pocket, and walked calmly over to the cops and asked what was up?

They told her she had a gas leak and said she needed to do something about it, and she agreed. A cop asked her what she had picked up and she said, "Nothing." The cop asked to see it, and Christa suggested that nah, he didn't want to see it.

There ensued more talk about the leak, and then the cop asked again to see what she had picked up. "Well, church is out," Christa said to herself, and she showed it to him. There were now eight cops huddled over the bindle and Christa stood there with shaking knees, visualizing her next several years in prison, and one of the firemen asked her who her insurance agent was.

"I was thinking two-to-ten, and he wants to know my insurance agent?" She went to her glove compartment, muttering, and a cop asked her what she was saying. She spoke really fast, saying "it's not going to do any good, but it's not mine. I was at the Carmel, and there was this guy, and he didn't have any pockets, and I said I'd... and he said... so I... and do you think I'd be so stupid as to come down here to talk to cops with this in my pocket if I knew it was there?"

The cop agreed that she had a good point. They took the bindle, but not Christa. She thought, "Hallelujah! I don't want it!" But they told her she had to fix her car before she could leave, and they went back to their cop cars. She went back upstairs and called the local 76 station and then called Cece's daughter to come get her and take her to Monterey. She was still shaking from the encounter and waiting out back at the top of the back stairs for the tow truck to arrive, when one of the cops drove up and beckoned for her to come down. Christa said "Nope."

He said, "Christa- *come here!*" and she shook her head, "uh-uh." She remembered that she had never told him her name, and his knowing it was not a good sign. When he said to get in the car, again she said "No!" Now he was getting pissed, and he raised his voice and said, "*GET IN THE CAR!*" and Christa thought it was all over. She had to go or risk being dragged down the stairs. She went.

He said, "I just came by to see how you were." She asked why, and he said, "It just looked like I was going to have a crying girl on my hands." She asked if he wouldn't be upset, too, and he said that that might be true, but that he wouldn't have dropped it in the first place.

She asked why he didn't turn her in. "You just looked like you didn't need the hassle."

Christa thanked him and said that if there was anything they at the station could do for the police, to just come on by—or calling would be better—and they'd do it. She thanked him again.

Then he told her that he had been training the other seven cops. He wanted them to see the stuff and observe the testing process. He said it was what they'd thought it was, and it was pretty good stuff. It occurred to Christa that it would have been a pretty quick test…

After that, the police became good friends with the station, coming by occasionally to hang out or request a song. Later, one of them told her that they saw all kinds of hard drugs around Gilroy, but there was never one time when they thought that the people at the station were involved with any of that. Their position was to leave KFAT alone as long as there was no trouble there.

But the trouble would come from a different source. The Texans were in a bad mood and calling home a lot, and the phone bills were freaking Jeremy out.

The money trickling in from the new sales guy, Bob Cassidy, was clearly not enough. Another salesman, Jim Shugart, was having some success. But again, it was just a trickle, and Jeremy was having a bad time with it. Jeremy, Laura Ellen and Lorenzo just hadn't calculated the income-versus-outflow thing well. Lorenzo wanted nothing to do with a commercial station and left Jeremy and Laura Ellen to deal with it. He certainly wasn't about to put any more of his money into it, and that whatever would happen, would happen.

It had already tripled in value, so whatever…

Laura Ellen had Elsbeth, who was about one, and tried to run the office as much as possible, but she was living in Los Gatos, almost an hour north of Gilroy, and couldn't be running back and forth a few times a day. Enid was getting her job done, but when that job got more demanding, as they hoped it would, she was going to have to sober up or go. As it was, Enid had no idea who these people were who had shown up and taken over the station that she had known so well for so long. These were certifiable lunatics, and, and… well- *what was happening*? They were strange people, and some of them were strange, angry people, and the anger was building.

Chapter Twelve

Clouds were gathering like a darkening bruise over the town of Gilroy. The locals, Christa, Genial Johnny, David Chaney, Sherman Caughman, and Gordy Broshear were getting frustrated. But the Texans, Marty, Speedy and Bobby, were especially upset, as they had come the farthest and had the greatest expectations. All were disgruntled with the circumstances, but Speedy was the most vocal in his disapproval of the way things were turning out.

The Texans were used to going out and meeting friends, seeing all the live music that Austin, Houston and Dallas had to offer. They loved Texas and they hated Gilroy. Of course, in fairness, Gilroy *was* a dusty, over-heated, lifeless backwater. They missed their friends and they hated the old equipment and they hated the isolation and they didn't find the smell endearing. The lonesome sound of the trains in the background only made them miss Texas more. Have I mentioned the constant smell of garlic and whatever else was being processed? Marty Manning still remembers the sounds of the trains as endearing, but it wasn't enough.

The jocks regularly discussed the proposed transmitter move, and they speculated on when that would be, and what that would be like. They knew that when the transmitter moved—if it was approved—that would translate into better pay and more perks, like concert tickets, T-shirts, free drinks and whatever else was around for the taking. Like sex. These people had never been into radio for the money, but if they were going to do commercial radio, and money was going to come in because of their work, then they wanted to be paid for that work. It's called capitalism, and it's pretty cool when it works for you.

But it wasn't working.

Bob Cassidy was, depending on your relationship with him, either convincing or scary, but he *was* selling the ads. Shugart, alone among sales people in those early days of KFAT, actually had some experience in radio sales, and he was becoming more successful in selling the station.

More and more people in Santa Cruz were listening and talking about the station. The jocks were becoming known in Santa Cruz. But Santa Cruz was too far from Gilroy, and was a pretty small place to be a star in, anyway. Also, feeling like a star was damned difficult when you couldn't pay your rent.

And they couldn't pay their rents because their paychecks started bouncing.

Timing is everything, and Bobby Eakin was affected by the bad timing more than the others, and it pissed him off. Enid would bring the checks around during his afternoon shift, and while the others raced out to the bank to cash them before the money ran out, Eakin was stuck at the controls until after the banks had closed. By the time he got there the next morning, the money was often gone. David Chaney remembers everyone racing to the bank and whoever's check cleared bought the beers for those whose check bounced. That was funny once, *maybe* twice.

Sherman Caughman remembers driving to Los Gatos to get paid, and was told as he was handed the check, "cash it *now* while there's money in the account."

It was an untenable situation, and the jocks were getting more pissed off by the day.

Larry saw the problem. Jeremy and Lorenzo had never owned a commercial station before, and they were used to high school kids in the office and volunteers on the air. The volunteers always thought themselves lucky to have a show, and typically got there for their shifts prepared, on time, and happy to work for free.

But these people, Jeremy found, wanted to get paid, damn it. Jeremy didn't want to put any more money into it. He'd be happy just to fix the broken signal so that it got into Santa Cruz, hope the FCC approved his transmitter move, hold on to the station for the required three years, and then sell it.

There are those who maintain that Jeremy didn't care what was on his air; as long as he had a signal that reached into San Jose and maybe San Francisco. But he still had to pay those people!

Another problem was that Jeremy was often in St. Louis and couldn't be reached when Larry was calling him in a panic about the paychecks, and he wasn't calling in. Back in Gilroy, everyone was pissed off. It was all coming apart in front of Larry's eyes, and he couldn't reach Jeremy! Larry couldn't tell the staff that the bouncing checks were a mistake, that the money was there. He couldn't tell them that the money would be there any time soon. He didn't want to lie to these people, and his position was unbearable. The format was a success, but the promise and the dream were evaporating right in front of him and he was powerless to do anything about it.

Jeremy could see that money was going to come in, but he needed to put up more money now, and he couldn't do that.

He was used to running a station on the cheap, and the bills kept mounting. Then the phone bill came in showing that the Texans were calling back home and Jeremy freaked out, pulling the phones out of the station as the overnight jock looked on in horror.

Now the listeners couldn't call in with requests, and this also put a huge strain on the sales and office staff. No more could they call in with sales information or station business. Now, they had to have a runner who drove to the station from Los Gatos at least once a day, which was an hour-long commute each way when the traffic was light, and commensurately longer during peak hours. It was a big pain in the ass for everybody, and now everyone was even more pissed off. Jeremy had unilaterally fucked up the station. It seemed to happen often when Jeremy had to deal with humans.

Jeremy loved his gadgets and his tinkering, and I have yet to find anyone who doesn't acknowledge his genius in that area, but he wasn't all that good dealing with people. Laura Ellen was better at that, but she was stuck in Los Gatos with an infant and the office work that Enid couldn't do.

Jeremy felt besieged, Laura Ellen was bored in Los Gatos, and Larry and the staff were increasingly demoralized. It wasn't a happy time for anyone. And discontent bred more discontent.

It was now just before Christmas, 1975. KFAT had been on the air for four months and Larry was as frustrated as the staff. He'd spent his life in radio and had dealt with incompetence before, but his frustration was now at a new level.

After two of Larry's paychecks in a row bounced and he couldn't pay his rent and he couldn't reach Jeremy, his frustration rose to a newer, unprecedented level. Then he got a call from the people at Earth News, who wanted to know if he was interested in going to San Francisco to start a news service called "The Daily Planet." It sounded like they would pay him and it wouldn't be insane.

But Larry was then, and is now, an honorable man. He would not leave this station and this staff without providing a replacement for himself. He told Jeremy what he had in mind and Jeremy was just glad that Larry wasn't going to abandon him. Larry knew the landscape of available people all over the country, and neither Jeremy nor Laura Ellen knew anyone, anywhere. Lorenzo was still having his nervous breakdown in Dallas, but even in good mental health he would have wanted nothing to do with the problems at this commercial station.

Larry told Earth News he'd be there in a week, but he wouldn't leave Jeremy and Lorenzo completely stranded. The bridges were burned with anyone in Texas, of course, but he knew a man from Long Island named Kenny Kohl who could do the job, and Larry called him. Kenny wasn't interested in coming to California for what sounded like a very chancy deal, but he knew someone who might fit the bill, and the guy was already living in California.

Kenny had worked with this guy in radio on Long Island, and together their small college radio station beat the big rock stations at their own game. Their station kicked the ass of the biggest rock station in New York City, and now he was floundering in Marin County, just north of San Francisco. His name was Chris Feder, and Kenny had his phone number right here...

Chapter Thirteen

In December, 1975, Chris Feder was living in Corte Madera, a sleepy, sophisticated town about ten miles north of the Golden Gate Bridge. He had worked with Kenny Kohl in the late 1960s at WLIR, the station at C.W. Post College, where Kenny was PD and Chris was the Music Director (MD). Together, they took over the daytime station and turned it into a full-time 24-hour–a-day rocker. They produced over a hundred live concerts and Chris remembers their success came from a mix of music that was free of corporate influence. They both felt that business and music shouldn't mix, and that the influence corporations had in determining what was being played on commercial stations was killing good radio.

They were rewarded by blowing the doors off WNEW, New York's biggest rock station. Feder knew he had a good ear for music, and was flush with his success when he saw a local band called the Good Rats. He knew they had the talent to become stars, so he talked them into signing with him as their manager.

The day he signed the Good Rats to a contract with Warner Brothers, he and his wife moved to California. They were living in Corte Madera when Chris went on the road with the band, and he met a woman and started an affair which broke up his marriage, and his girlfriend moved in. They were just settling in to their new life when Kenny Kohl called him and said he had an opportunity that Chris "was gonna love."

"These guys," Kenny explained, "want me to be PD of KFAT." But he didn't want to move, and he wanted to know if Chris wanted it. Chris recalled that Kenny didn't know good music; he was a good manager, but it was Chris' ear that set him apart, and he wanted this job.

Larry wasn't all that sure that Feder, being a rocker, was the right guy for the job. This format, remember, was Larry's baby, and Larry didn't think that Feder knew the music or understood the concept. He already felt guilty about leaving KFAT, and he didn't feel this was the right guy, but he wanted to start his new job, so he swallowed his instincts and recommended Feder to Jeremy and Lorenzo, who wanted Feder to come to Los Gatos. Perhaps this was to hide the station from him until he had signed on, but whatever the reason was, it wouldn't matter. Chris Feder wouldn't come to them.

Feder wanted Jeremy and Lorenzo to see his living room. It was where he was comfortable and it was where he displayed his trophies: his collection of rock memorabilia, his impressive stereo system and his huge record collection. He wanted Jeremy and Lorenzo to see that he was a serious music guy.

Jeremy and Lorenzo came up to Corte Madera, and over wine and joints they talked about music. They knew that Larry was leaving and that the staff was mumbling about leaving, Lorenzo wanted nothing to do with the operation of the station, and Jeremy wanted to be left alone to tinker with the electronics. They were desperate to have someone who could hire a staff and run the station. Feder was a strong, forceful presence who could command respect. Jeremy and Lorenzo were desperate, and Feder had clearly made an impression on them.

Feder was confident at that meeting, as he seemed to be about most things. He told them, "I can do this job cold, but it's gotta be my way." He couldn't have conflicting opinions. He said, "We'll have fun and do great things, but it's gotta be my way." They said they'd think about it.

Did I say that Feder was a forceful guy? When I hung out with him at his home in Corte Madera, he was mellow, engaged, interested and a fun conversationalist, but when he got to Gilroy, and fueled by a constant supply of cocaine, he was an arrogant New York Jew with the aggression and assertiveness that Californians have come to love in such fellows. Feder became pushy, intractable and goal-oriented. He wanted things done his way, and was intolerant of those who did it wrong. He was an argumentative clash waiting to happen with the hippies at KFAT, both management and staff. Later, Feder would admit that perhaps his natural aggressiveness, his implacability, and his use of cocaine impacted negatively on his effectiveness at KFAT. He no longer uses cocaine, and he apologizes. Feder didn't hear back from them for the next few days. No problem. What could happen in a few days?

A week later Lorenzo called and gave him the job. The deal was that Chris would get two hundred dollars a week, cash, no taxes. But first, Lorenzo wanted Chris to come to the area to listen to the station. It had gone "amuck."

The week Larry had asked for from Planet Earth was almost up, and Larry had almost fulfilled his obligation to KFAT, but before he left, he had one more thing to do: he called a staff meeting.

Chapter Fourteen

Larry was almost out of KFAT. He had the Daily Planet lined up, and after he finally reached Jeremy, who had just come back from St. Louis where he had been incommunicado, Larry went to see Jeremy in Los Gatos that Sunday night. The next day he called a staff meeting. He'd heard Jeremy's position, and now he had to tell the staff that the station had to cut expenses, that everyone was going to have to take a pay cut. Please try to imagine how *that* was received.

At the meeting, Marty Manning, whose salary was $550 a month, had a suggestion, saying, "Well, I'll help you out. You can take five hundred and fifty dollars right off the top. I'll see you later," and he walked out. Now KFAT had no morning guy.

The next day Speedy spoke up. He'd never made an effort to control his fondness for Bombay gin, and he'd frequently clashed with the other staffers, who had grown to distrust him. He'd also repeatedly clashed with Larry and Jeremy. He was pissed off at the owners, they were fed up with him, and Fermin "Speedy Gonzalez" Perez was fired. Now KFAT had no PD or afternoon guy.

The night before the meeting, Sherman Caughman was doing his Sunday afternoon shift when Larry, just back from his meeting with Jeremy, came in and put his arm around him. He told Sherman that he sounded tired and ought to go home. Larry would finish his shift and he said he'd call Sherman on Monday. The next day Larry called and told Sherman that he didn't want to use him on the air any more, but would keep him on for production work. The next day Sherman tuned in and KFAT was silent. It was just before Christmas in 1975.

Sherman now believes that when Larry told him to go home right before the Texans and everyone left, he'd been cleaning house. Sherman feels that Larry had argued with Jeremy and Laura Ellen over the way the station was being run, and knew the Texans were about to leave. The staff was pissed because Jeremy hadn't delivered on what they had been promised, and when most of them left, Larry was already gone.

Larry had told Jeremy that he'd be there until a replacement arrived. In the middle of December, Chris Feder, who'd met with Lorenzo and Jeremy the week before, got a call from an increasingly agitated Jeremy, who told him to come to work the next Monday. Feder went to Santa Cruz that Sunday night and checked into a motel and tuned to KFAT. He says the station made him crazy as he sat and listened. It was all over the place, with no central tone or theme. It sounded like every record store in America had been programmed for "Random Select."

The next morning, he drove over the hill into Gilroy, found the station and went upstairs. Christa was there and Chris remembers that she was "acting as PD, kind of a den mother." She asked him if she could help him. He said he wanted to see Jeremy, but Christa said Jeremy was gone for the day.

Gone for the day? It was nine o'clock in the morning. He told her he was the new PD, and she asked what he meant by that.

"Jeremy just hired me. I'm the new Program Director." Christa was incredulous. Feder now sees Jeremy's not being there at this crucial moment as one of several examples of Jeremy's unwillingness to deal with people and their problems. He now thinks that if the problem wasn't electronic, Jeremy would do whatever he could to avoid it.

Feder thinks that Jeremy wanted Chris to do what Jeremy couldn't. What was needed was a strong hand and Jeremy wasn't strong in this way. He was a brilliant engineer, Chris remembers, but…

Chris explained to Christa that he had been hired to replace Larry, and at that, Christa and the rest of the staff present went nuts. There was screaming and yelling and so much hostility that Feder felt that this was not the time or place to resolve it. It was incredible to him that Jeremy would have put him in this position, and he didn't like that he was one of the targets of the staff's rage.

He maintained again that he was the new PD and Christa shot back, "That's not gonna happen here!"

As Feder remembers, no one had been paid and "their attitude was that if they weren't going to get paid, then they could play whatever they wanted." And that clearly left no room for direction from a stranger like Chris Feder, who left them ranting and yelling at him, and told them, "Do what you want, and I'll find Jeremy and come back tomorrow."

At the end of that miserable day, Enid distributed the checks and Bobby Eakin's check bounced. He was furious and he said to Christa, "I'm outta here. You want to come?" She said yes, and David Chaney said he had a friend he wanted to visit in Albuquerque, and asked if they could drop him off there.

The next day they had packed up their belongings and loaded up Bobby's truck. Bobby left a note for Jeremy, asking him to pay him for his records, and he, Christa and Chaney were on their way to New Mexico. They dropped Chaney off and headed for Austin, landing at a record store where Bobby had worked. As there was still no telephone at the station, they called the office in Los Gatos, and found out that everyone else had left the station as well.

Chaney was in Albuquerque, Bobby stayed in Austin and Christa caught a bus for Louisiana, where she had relatives. Marty was in Phoenix, and there is some dispute as to where Speedy went, but you could make easy money betting that he didn't stay in Gilroy. Genial Johnny went back to driving a cab in Santa Cruz, and Cece moved to New Mexico.

Gordy and Sherman had already been fired by Larry, and were back in San Jose.

Back in his motel in Santa Cruz, Feder was listening to KFAT. He got in his car the next morning at 7:30 to get to the station before more trouble could erupt. KFAT was playing its quirky mix of music, and he listened as he drove. At 8:00 a.m., while he was still driving and listening in his car, KFAT went off the air. Feder thought: "Oh, shit."

When Feder got to the station, Jeremy was sitting in front of the control board with a screwdriver in his hand. Laura Ellen was with him, and they were sad. There was no one else at the station, and there were no records on the turntables. "It's all over," Jeremy said.

"What do you mean?"

"Everyone quit. They think I'm an asshole. They said some very unkind things to me, and I'm just gonna take the station apart and give up."

Feder told him they'd better think about that. "First of all, it's illegal. You can't just dismantle a radio station and pull it off the air without telling anyone about it and petitioning the FCC. You're pretty much stuck with being on the air."

"How can we be on the air? There's nobody to work."

"That's all right. We've got plenty of records. We'll manage. I can do this. This is too good an idea not to do it."

KFAT went on the air that day, Tuesday, for twelve hours, Feder making Jeremy play records for four of those hours and Laura Ellen for an hour. Feder was there early again the next morning, doing a shift, when he turned around and there was "a big fat guy" in the doorway to the studio. He looked scared.

"Can I help you?" Chris asked.

"I'm Gordy."

"What does that mean?"

"I'm the morning guy."

"Didn't you quit?"

Gordy Broshear explained that he'd been the morning guy for a couple of years at KSND, and had worked at KFAT until a while ago. "Didn't you quit with all the others?" Chris asked.

"No, that wasn't me."

Jeremy had called him and told him to come back, and he wanted to work there again. Chris said he'd speak to Jeremy about him and be in touch. Jeremy said he was "reliable."

Gordy would also be an important piece of the puzzle because in the old days KSND had been going off the air at 2 a.m., and he knew how to turn it back on when he came in for his morning shift. That had been a marketable skill in those days, and now it was valuable again.

Feder and Jeremy played music again the next day, playing cowboy songs, Willie Nelson, the Grateful Dead and Bob Dylan. Those were about as country as Feder knew, and he stuck with them. His longest shift at KFAT was thirteen hours.

Chris Feder says he taught Laura Ellen how to play records and he put her on the air for four hours every day.

Chapter Fifteen

Chris Feder needed a staff, and fast. He played DJ for nearly eight hours that first day, drafting both Jeremy and Laura Ellen into service, but that was good for today and maybe tomorrow. With whatever free time he had, he had to find a full-time crew, and, as this was part of his job description, he set about to find them. In his spare time, one supposes.

Bob Cassidy, who had been selling ads for KFAT for about two weeks, came to Gilroy to turn in some ads he'd sold, and hang out at the station, where he met Feder. There was no one else at the station. Feder heard Bob's deep, resonant voice, and dug his macho confidence, and when Bob asked him what he could do to help, Feder asked him if wanted to go on the air. "Sure," Bob said, and Buffalo Bob and the "Bison Boogie" were on the air. With a twist: Bob Cassidy hated country music.

He detested country music because he detested the people he knew in the army who listened to it. He thought the people he'd met in the army who listened to country music were "one step ahead of a rock." In the short time Cassidy had been selling ads for KFAT he'd met some of the Texans at the station. When he thinks about the Texans now, Bob remembers them by recalling a Terry Allen song called "Cocktail Desperado" which had the lyric, "How'd you get such a great big mouth on such a tiny little head?" I know: it's sort of sweet.

Bob Cassidy remembers that KFAT was playing "Texas bar-room music: rockers and weepers. It wasn't even close to any other station." But it wasn't his kind of music and, as a DJ, everything would work out if they'd leave him alone to find what he liked. He says that he enjoyed the learning process and "started playing everything I could grab."

Buffalo Bob's sets were always more rock 'n' roll than the other DJs because he always picked out the most raucous cuts. It was a much more rockin' show, and they gave him the overnight shift. He remembers it was a "very casual" way of getting a job; he doesn't think they even had his social security number for his first year and a half at KFAT. To his recollection, he never signed any papers.

He remembers Jeremy as a "cadet" with "wild hair and a wild look in his eyes. He was a wall-walker, he used to do weird calisthenics in front of you during conversations." Cassidy remembers that Jeremy didn't look you in the eyes as he spoke with you. With his previous work experience in the drug trade, Bob Cassidy had lived among the space cadets in Santa Cruz and was used to dealing with them, and he didn't care.

Jim Shugart was the other salesman at KFAT, and he didn't know what was going on at the station. He'd sold some spots the week before, and he came in to record them so they could schedule them for airplay. He had the copy written, and he knew Marty would do the vocal and the production, but he was surprised to find that everyone was gone. There was a new guy, Chris, in charge, and that scary fuckup, Bob Cassidy, was now a jock. Well, it was all the same to him; once he sold the ads and oversaw that the spots got recorded by those lunatics up there, he could go out looking for more money for himself and the station. That was his job and he liked it. Santa Cruz was his territory, he knew it well, and he knew where to go to sell this station that everyone was talking about.

He was a bit discouraged by the changes at the station, but once they sorted it out, he'd do better than before, as more people were listening all the time. All over Santa Cruz, people were talking about KFAT. And if the transmitter move was approved…

For these spots he'd wanted to record, he wanted some live guitar picking under the written copy. He knew one of the jocks would record the text of the ad, but he knew that no one there could play guitar for him, so he brought in an acquaintance, Terrell Lynn Thomas.

Terrell Lynn was a tall, laconic cowboy-type of fellow. He had long dirty-blond hair, jeans and cowboy boots, and a soft voice. He spoke in a quiet, lilting tone that hovered enough above a whisper to draw you in and make you listen. He was handsome in a rugged, outdoorsy way, and when he spoke you got a sense that he was speaking to you exclusively, and he meant what he said. He was, in a word, of a sincere nature.

Terrell Lynn liked the station and had been a big fan and a constant listener from the time it went on the air. His cow-folk parents used to go to the Roller Rink in Salinas on Saturday nights to dance to the country and western swing bands that played there. They'd bundle little Terrell Lynn up in his pajamas and a blanket, and nestle him into the back seat of the car and go inside to dance. Once they were inside, Terrell would slip out of his swaddling, turn the ignition on and listen to the radio broadcast of what his parents were dancing to inside.

Invariably, when his parents got back to the car, Terrell Lynn was asleep on the front seat and the battery was dead. He'd fallen asleep listening to the music, the battery had drained, and his parents had to ask their friends to push the car to get it started. But they never once scolded Terrell Lynn for this, and to this day he has fond memories of the music that he fell asleep to in his youth. Those songs and that music stayed with him all his life, and he knew the music well. He heard a lot of it again for the first time in years on KFAT. Yeah, Terrell Lynn listened to KFAT all the time.

He'd worked for the telephone company in San Francisco, then drove a beer truck in Santa Cruz and gigged as a guitar picker whenever he could. His only experience in radio had been as a teenager in the late 60s when he'd go to KSAN's studio in San Francisco after a night of music and LSD at the Fillmore Auditorium or the Avalon Ballroom, where he'd gone to see the happening bands of the day. Tom Donahue (see: Music Sucked…) would invite him and his friend into the booth and ask them to report on the shows. They were crashing on acid, and Donahue would open the mic and they'd tell the listeners about the shows.

Terrell remembers that Tom Donahue was so mountainous a man, and the studio so small, that to allow Terrell Lynn and his buddy into the studio, Donahue would have to leave the studio, let the two in, and then come back in himself. Terrell Lynn liked going to KSAN to hang out, but remembers thinking at the time, "I never, ever, ever want to do this."

Shugart and Terrell Lynn had worked the famous light shows at the Fillmore, and they had lived on a commune for three years. Terrell Lynn had just lost his job delivering beer and was living off of his unemployment checks when Shugart asked him to come with him to Gilroy to play some music. It wouldn't pay much, but he had nothing else to do and this was a way to earn a little money for himself the way he'd dreamed of: playing guitar.

At KFAT for the first time, Terrell Lynn recorded the music tracks behind two commercials that afternoon, and when Shugart made a mistake on the third and needed a few minutes to make an adjustment, Terrell Lynn wandered out of the Production Room and into the record library. The library was just outside the control room, where Laura Ellen was frantically trying to deal with keeping one turntable spinning, getting the other turntable cued up with a record and ready to spin, and the wandering, chittering proclivities of her infant daughter, Elsbeth.

Terrell Lynn saw that she was frazzled and tried to keep out of her way as he looked through the records. The library consisted of what came with them from St. Louis and what Bobby Eakin had brought with him from Texas. But it was a small space, and fate had it that Terrell Lynn was in Laura Ellen's line of sight.

He was really excited looking through the records, and he knew that they had a bunch of records that he loved but that they weren't playing. Laura Ellen noticed that he was taking out specific records and studying them knowingly, approvingly. She knew by watching him that he knew them; she knew she didn't. She asked him if there was anything he wanted to hear and he said, "'Pick Me Up On Your Way Down' by Charlie Walker," and in a brusque frantic tone, she pointed to the library and snapped at him, "Well, go get it!"

This was not Laura Ellen on her best day. The pressures and the long hours had made her frantic. Terrell Lynn jumped to it, found the record, and brought it into the studio. She took the record from him, cued it up and hit the play button. Laura Ellen had only gone on the air a week before, when everyone else walked out, and she was still new to this type of music. There was so much music in the library that she had never heard of, and had no idea what was on most of the records in the stacks. She played Terrell Lynn's request and liked it, and asked him, "You know any more of these?" He said "sure" and she snapped at him to "go get 'em!" When he came back in with several records, she said, "Sit down here for just a second" so she could use the bathroom. She showed him the volume controls and told him which switch to flip for this turntable and which to hit for the other. She showed him which switch was for to cueing a record, and what switch was for playing it, then she left and she never really came back. She'd come in every once in a while to say she was really busy, and just put another one on. Terrell Lynn remembers, "I played music for four hours, and I kind of had it down."

He also had a job. It paid sixty dollars a week, and he couldn't have done it if he wasn't getting those unemployment checks.

Michelle McGovern was a pretty, slender brunette. Pretty in a trailer park sort of way, but alluring and playful. She had a smile that was spontaneous and infectious. She liked men, but her judgment in this regard rarely ran as fast as her enthusiasm. She had a wariness about her that showed as a hesitation before she'd believe anything you were saying, but she liked that you were trying. She lived in one of those pockets of Santa Cruz that got KSND, and she used to listen to it when the weather or the wind brought it to Santa Cruz. And then one day it was gone. She knew the radio was on because she checked. It was on and it was tuned to the right frequency, but no one was talking and there was no music. She forgot about it and went about her day, and the next morning, suddenly she was startled to hear someone talking in the living room. She ran in to see who it was, and it was the radio. They were back on the air, but it was different. It wasn't the usual crew of DJs. Instead, some guy with an accent was saying that he was from Texas and this here was a new station and he hoped y'all liked it. Then they played music that knocked her fucking out.

She listened all the time, and then a few months later, they were gone again.

Michelle was working in Santa Cruz as a, ummm… uhhh…at a trade of dubious legality, and felt the need to get a more stable, less arrestable income, as she had two small children and the possibility of losing them if she went to jail was on her mind. She was telling one of her tricks, a man named Leo Kesselman, of her desire to straighten out her life, and he told her that the staff had just walked out of KFAT, and she had a good voice, she seemed to know a lot about music, and she ought to go over there and ask about a job.

"Yeah, right," she said, and forgot about it.

The next day she got a call from some guy named Chris. He wanted to know if this was Michelle, and if she wanted to be on the radio. She asked him if this was a joke, and he assured her it was not. He said he was "a friend of Leo, and there was this station and do you want to work there?" She told him she knew the station and liked it, but she didn't know how to be a disk jockey, and the caller said, "There's nothing to it."

Michelle's trick, Leo, was a friend of Jeremy's and he knew about the chaos at KFAT, and knew they were looking for DJs. Leo had asked Jeremy what he was looking for, and Jeremy said that he wanted a brassy, honky-tonk gal who was "ballsy and had a good voice." Leo told him about Michelle, and Jeremy told Chris Feder, who called. Leo will reappear significantly, but much later.

The next day Michelle loaned her kids to a friend and hitch-hiked over the hill to Gilroy. The man who stopped to pick her up was also on his way to Gilroy, and in fact was on his way to KFAT for his second day of work as a DJ at the station. It was none other than Bob Cassidy, who told her how much they needed a staff, and to talk straight to whoever she met there. Cassidy had no idea who'd be there, of course, and he took her to the station and introduced her to Chris Feder, who liked her voice, liked that she had kids, liked that she was a honky-tonk sort of woman, and he wanted to hire her on the spot. Feder told her to put it together somehow so that it sounded connected, and he showed her how to listen, then led her to the studio, put her in the DJ chair, and had her face the board, showing her a bunch of on-off, up-down switches and how to cue up a record. Then he told her, "Find another song that sounds like the last one. Not the same, but similar. Similar beat or subject, whatever. Just make a connection for yourself." He told her to look for some similarity between songs, be it the tempo or the subject. "And no dead air!"

Then he put the headphones on her and left to go to the Green Hut. Feder needed a drink almost as much as he needed a DJ, and, as right then he *had* a DJ...

Michelle stayed at the controls and played music for a couple of hours until Feder came back. Feder knew she had the instincts, and he told her to come back Saturday night at eight o'clock for a shift. One day she was in one dubious trade, and the next day she was in another. Sister Tiny was on the air.

Now we get to one of the most important people in the story, and perhaps one of the most important people in the world. And they don't let you say that in a book if it isn't true. May I introduce Gilbert Klein…

Chris Feder had many responsibilities at his new job. One of them was to meet the FCC regulations, so he had to put on a show of community interest. This was called Public Affairs back then, and every station had to have such a program to address the needs of the listening public. Most Public Affairs programs were an hour in length, and buried in the wee hours of the weekend mornings, when no one was listening and advertising was scarce.

Feder had to come up with a person who would conduct a talk show for KFAT. He knew just the guy.

Gilbert Klein was a free form sort of fellow who lived in San Rafael, about fifteen minutes north of the Golden Gate Bridge. I had a friend in New York City who told me that *he* had a friend who lived nearby in Corte Madera, and as we were both from Long Island and into music, I should look him up. The friend was Chris Feder, and we became friendly, if not close. My business card at the time said cryptically, "Sometimes Something Original" with the explanation below it: "Custom Designs and Occasional Inspiration." It meant that "What I did kept changing, but not how I did it." Uh huh… I have promised to explain that to me sometime. Below that was my name and phone number. No area code because, "If you didn't know the area code, you had no business calling me."

I had been making wooden plaques for sale to tourists at Fisherman's Wharf, had done some carpentry, and been a trafficker in various substances. I was currently engaged in taking thin slabs of beautiful stones, like agates, and shaping them into belt buckles. I gave the stone buckles to a friend who made the leather belts that attached to the stones, and sold them to tourists at Fisherman's Wharf.

I had lived in the Bay Area in 1966 and seen the real Summer of Love (see Summer of Love), and moved back and forth between Long Island and Marin County for seven years, getting my bachelor's degree, driving a cab, doing carpentry and assorted other trades which entailed no reportable income. I moved back to Marin permanently in early 1973, enrolled at Sonoma State College, got my Master's degree, and started teaching as a substitute. I was hoping to be hired as a full-time teacher, but there had been a recent round of firings of teachers in the Marin County schools, and when the school district did get around to hiring teachers again, they would hire back those whom they had let go, and so I continued cobbling a living together from substitute teaching and whatever sources presented themselves.

To make the belt buckles, I needed to operate my little all-purpose stone cutter and polisher. To operate this, I needed a steady source of water. To do that, I asked Chris if I could pay him a few bucks a month to work out of a sink in his basement. Feder knew that I wasn't burdened with an excess of ambition, and wouldn't be around that much. Feder figured I'd be there maybe one day a week, a few hours at most, and wouldn't bother him, so we agreed on twenty-five dollars a month.

Then Feder got a new job and began spending his weekdays down in Gilroy. I liked having my weekends free so I needed access to Feder's basement during the week when Feder wasn't there, and I went to Corte Madera in early January to talk to him about it. If I didn't reach an agreement with Feder about a key, I'd be out of work for another week until Feder got home again.

But I was also a music fan and knew a lot of the nuances and the trivia of music, and this was what Feder and I usually talked about over joints and records at Feder's house. Feder knew that I was a bright, articulate guy with an active curiosity. I read a lot and always wanted to know more. I asked a lot of questions, good questions. I loved the radio, and while I had never once entertained a thought of doing radio for a living, I had been a good teacher, and was comfortable as a communicator and public speaker.

When I told Feder I was excited for him about his new job, I also asked if I could "come down to Gilroy one day and read a commercial on the air?" Feder said he had a different idea. "How would you like to be Public Affairs Director and have your own show?"

It wouldn't be live. It would be a pre-recorded fifteen minute talk show played during Morning Drive. Feder would show me how to edit tapes and help me get a show ready for broadcast. There would be no restrictions on who I could or could not interview. The show should be as eclectic as KFAT, and could reflect my own interests. As long as no one said "shit, fuck or cunt," there'd be no censorship.

Feder recalls: "There was a need for a Public Affairs program, and you were the perfect guy for it. You liked to talk to people, you weren't afraid to go out and find people to interview, and you only wanted to be on the radio one day a week. It was perfect. You would bitch and moan about the commute, but you'd do it."

I would have a two-hour commute to the station each way. I said I'd like to think about it, and spent the next twelve hours asking myself: "Can I do this?

Should I do this?"The next morning I called Feder and told him yes. Feder's girlfriend, suggested a name for the show, and "Chewin' the Fat" would start as soon as I could get a tape recorder and find someone to interview.

Feder had spoken to Jeremy about Gordy, then called and gotten him back for the Morning Drive shift. Defining "fat and happy," he was glad to be working again. He loved the new music, and he even got along with Chris Feder, one of the few people who did. Gordy knew the station needed more bodies, and he put up a notice about KFAT on a bulletin board at KFJC, the radio station at Foothill College, near his apartment in San Jose. It said that KFAT needed a staff and to call and ask for Gordy.

Kathy Roddy was a pretty, black-haired girl who looked as Irish as she actually was, and had a beautifully sing-songy voice that always sounded like she was laughing with you about something. It was a wonderful voice for radio, but she hadn't started out to be a DJ.

At college, she was something of a political activist, and the anti-war group she hung around with gravitated towards the college radio station so they could broadcast their views. Then she started doing news at KFJC, in between doing her schoolwork and going to her job as a desk clerk at a local motel. At her job, she had a lot of time to do her homework and listen to the radio, where she hated the crap they played on rock and Top Forty stations, so she found the two country stations on the dial. She was pretty dissatisfied with everything on the air, but at least with country music, unlike rock, you could hear the words, the songs made sense, and it didn't hurt to listen to it. But there was something missing. Of course, as a bright and observant student of the culture, she was very aware of the musical revolution that had begun in the mid-60s, and which was just then petering out into crap again. While sitting at her desk in the motel office, Kathy thought that if someone were to take country music and approach it as the rockers had after the initial British Invasion, the period where music got stretched into serious and unexpected directions… if someone could be as serious about country music as the innovators of progressive rock had been, why, then you'd have something. Then she saw the notice on the bulletin board and tuned to 94.5. It was KFAT, and she wanted to work there. It was love at first listen.

She called, asking for Gordy, who told her what had been happening at the station and he encouraged her to come down and talk with Chris Feder. She drove down 280, onto 101, stopping every few miles, asking, "Is this Gilroy?" No, it was San Martin. "Is this Gilroy?" No, it was Morgan Hill. Eventually, of course, it was Gilroy.

She talked with Gordy, who told her that the previous staff had just bolted a matter of days before, and he had just gotten back. That Gordy had hoped for a good connection with the woman who was driving down to see him and the station was probable, but he immediately knew he'd never make it with this woman. She was damn near pretty. No, she *was* pretty! And very nice, indeed. She was 5'6" and slender, with long, rich, full black hair, blue eyes, and you won't believe me but I'm telling you the truth, because I knew her then and I know her now, and those eyes actually sparkled. Maybe it was the ready grin or the ever-present sense of humor, but her eyes actually sparkled. That's what I remember, and it's my book. There was, and still is, laughter in her voice.

After a few minutes of listening, and despite the looking, she told him she wanted to work at KFAT. Gordy sent her to the office, where Jeremy was struggling with some paperwork, and they talked for a while.

She told Jeremy that she wanted to work there, and Jeremy told her he was looking at applicants and he'd be in touch. In the next few days, she'd call Jeremy at the station and, she now admits, whine about wanting the job, and Jeremy, uncomfortable in conversation with humans and anxious to avoid such interactions, eventually gave in and told her to call Chris Feder, and to tell him he said to call.

It was just before New Years Eve, 1975.

The next day, Kathy Roddy showed up and asked for Chris. She said that she had been sent by Jeremy, and Chris sat *her* down and put her on the air, and said he'd be back. Kathy was startled to be on the air this suddenly, and she was equally as surprised when Feder left for parts unknown. Feder wasn't worried, because *he* knew where he was going and so do you. He went five doors away to the Green Hut, where he could drink and listen to her. Kathy wanted to do well in this unexpected audition, but she didn't know the library, didn't know the music, and soon another problem arose.

After an hour or so of DJing with no idea of where Chris had gone or when he'd be back, Kathy started sweating mightily in the DJ chair because she had hypoglycemia, and she'd begun to get light-headed. It had all happened so fast, this audition thing, that she hadn't thought to tell Chris about her condition, but now she needed to eat, and she still had no idea about where Chris had gone, or when he'd be back.

She was starting to get dizzy, but she wanted the job and she stuck it out, focusing on the mechanics, the physical stuff, until Chris finally came back and assigned her to the graveyard shift from eight until one a.m. She listened to what she was being told, woozily, trying to understand what he was saying, wanting to impress Chris by appearing alert. She pretended to understand what he was saying, she agreed to it all, and then she bolted out of there in search of food, hoping that she'd remember what Chris had said about a job…

The world being as small as it is, it turned out that, unknown to her, one of her classmates was already there. He got to stay when the others left because he had just gotten there, he didn't know anyone; he came in, did the news and left, having little interaction with either staff or management.

Karl Hess, known at KFJC as "Karl J, the Rock of the Bay", was doing the morning news at KFAT. He had a voice that made him sound like he was having fun, even that first day when he was shitting in his pants before he went on. He remembers he came for an interview with Larry Yurdin, who he found standing in the hall smoking a cigarette nervously. Larry seemed mightily distracted as, unknown to Karl, Larry was about to leave, and the staff was furious about something and on the verge of walking out.

There was a dog wandering around and a little girl with no clothes on, and people eating lunch in the hallway, and he told Larry who he was, and why he was there. "You're hired," Larry said, "Go and talk to Jeremy, you've got a newscast in fifteen minutes." He thought: "This is it! Professional radio! The career has been launched!"

Larry directed him to a back room to see Jeremy, who he found in a filthy office in the back of the station. Karl remembered that Jeremy was laying on a mattress "so filthy that no reasonable person would be laying on it." He was told he'd be on in fifteen minutes, and he asked how they "were going to formulate this."

Jeremy said, "Well, what I'd like is for you to get a couple of communist papers in here, and a couple of things from the John Birch Society, and just kind of wing it."

Karl said, "Sure. I'm already getting those."

That was bullshit, of course, but he *was* going on the air in his professional debut in a few minutes, so he sat down and thought of a few timely things to say. He walked into the studio and some guy pointed to the mic, flipped a switch, and his first broadcast was on the air.

This certainly wasn't anything like he thought real radio would be. He was way out of his element, his confidence was shot, he was nervous, and he recalls, "I can't speak real clearly, and I can't speak real fast, so "I'm Karl J with the Morning News" came out as "I'm Karl J with the Morning Nudes," and Jeremy liked it. "It sounded unprofessional", Jeremy said, "Let's leave it in."

They moved Buffalo Bob somewhere else on the clock and gave Kathy Roddy the graveyard shift, from midnight 'til 6 am. But they gave that shift to everyone who was starting out there. They moved Kathy around to several shifts until they gave her the shift following Gordy's Morning Drive, which was from 6 a.m. until 10 a.m. Kathy started calling herself Sully and she worked 10 a.m. 'til 3 p.m. Chris Feder handled Afternoon Drive from 3 p.m. until 7 p.m. They were shuffling around shifts a lot in the earliest days of the new staff, and Terrell Lynn, Sherman and Buffalo Bob filled in where they were needed. Shifts lasted anywhere from four to twelve hours in those days, but it was new and exciting to everyone.

Then there was Frisco. Ah, Frisco. Mystery Man. Genius. Tragic. Out of control. Drugs. Alcohol. Whatever. Excess. There is much dispute and little agreement as to when Richard "Frisco" Bo arrived at KFAT. He had been a brilliant production man at several top stations in major markets and fired from all of them. The last one was in Denver, where he had been booted out again for his out-of-control alcoholism. Broke, depressed and strung out, he came back to live with his mother in her trailer in Gilroy, and with nowhere else to go in town, he was drawn to KFAT like a kamikaze moth to a blowtorch.

There was nothing else going on in Gilroy, and living with his mother was making him even more depressed. He heard the new station and was drawn to its insanity and lack of discipline. No one remembers who hired him, not Larry, not Chris, not Laura Ellen or Jeremy, but he appeared at just the crucial moment when the original staff walked off, and there was no one to do production. Frisco fell into an environment that was a lot more tolerant of his habits than at any other station, and he immediately added KFAT's production work to his daily regime of drinking and drugging. No one remembers hiring him, but he just sort of moved in and started tinkering in the production room.

He lived at home in his mother's trailer, but spent all night at the station, often passing out in an alcoholic stupor after his work was done. At first, he got his work done and then passed out, but eventually his drinking and his work co-existed to the extent that no one at the station was seeing him sober. Somehow, when the jocks came in in the morning, the work was done. Over the years, Jeremy would send him to dry out and he'd return to Gilroy and to his old ways. But he always got his work done and it was always of the first order.

Everyone at KFAT loved Frisco, but no one knew what to do about him.

Let us also consider Awest: artist, free spirit, dedicated contributor to the Fat sensibility.

Awest was the youngest-ever animator at the world-famous Disney Animation Studios. While stuck inside working with the older, established animators, he saw the revolution that was going on all over the world, especially in California, and he was getting restless to be a part of the New World Order that was happening outside the sheltered, old-school Disney environment. He wanted to be an underground cartoonist. He wanted to contribute to the Movement. He chose Santa Cruz as his location, quit Disney, packed up his girlfriend Wanda and his parrot Jedediah, hopped into their VW bus and headed north.

They found a small attic studio to live in that was above a, uh… massage parlor, and set up housekeeping. By day he'd commit his art to paper, and by night he and Wanda would cruise the hippie redoubt store by store, always looking for the odd thing to collect.

One evening, while sitting in the bathtub, he heard something strange. It was duck calls. Non-stop, one-after-the-other duck calls. He knew there were no ducks in the eaves above his head, knew there were no ducks in what passed for a front yard, and none of the neighbors had any ducks. Naturally of a curious nature, he soon discovered that the source of the duck sounds was the water in which he sat.

Apparently, the water acted as some kind of transducer, and the radio that the uh… ladies… downstairs were playing was somehow weirdly amplified by the water, and what he was hearing was the radio. Being of a curious nature, he wanted to know two things: how the water amplified the sound, and who the hell was playing duck calls on the radio.

After about ten minutes, and with his bath water cooling, the duck calls gave way to some of the swingin'-est music he'd ever heard. He was diggin' it! He couldn't believe he was sitting in a tub of cold water, swingin'.

Now rabidly curious, he went downstairs and asked the ladies what they'd been listening to. It was KFAT. The duck calls had been Jeremy doing his Fat Dork Show, about which I will soon stumble badly while trying to explain, but the music had gotten Awest hooked.

Determined to be a part of whatever the hell weirdness was going on there, Awest learned that the station was in Gilroy, and the next morning he set out for Gilroy, wherever that was, to get himself involved.

He drove over the hill, found the station, and went up the stairs. He found Jeremy there, holding a screwdriver, and asked who he should talk to about contributing to the station. Jeremy said that he was the owner, and asked what Awest thought he could do for them. Awest told him he was an artist and asked if they needed any artwork. Awest showed Jeremy some of his drawings, and, smiling, Jeremy said that yes, they needed everything: bumperstickers, stationery, letterhead, a logo, posters, T-shirts.

He suggested that Awest design all that stuff and print up a bunch of T-shirts. Jeremy suggested that Awest give the station some of the T-shirts, sell a bunch, and keep the money for himself.

Delighted, he drove home and told his girlfriend the good news, that he was now working for the great new station. When she asked him how he was going to make any money there, he thought about it, and then admitted that he didn't know. So the next day he drove back over the hill, found Jeremy, and asked him again. Jeremy explained it again, and now Awest thought he got it. He'd work on spec, and make some money from the sale of T-shirts. Cool. He was working at KFAT. So to speak.

Out in the hinterlands of California's Central Valley, Mark Taylor had been fired from yet another stupid, boring, inconsequential radio station, and he was wondering what he was doing in a business that was frustrating the bejeezus out of him, when he heard about a weird new radio station. He heard they were loose, unorganized, and playing an unheard-of mix of music, and they needed a staff.

The Ozone Ranger was on his way to Gilroy.

Intermission

Chapter 16: A Brief Lesson In Cosmology

By this point in the book you will have met many people and heard about many radio stations. Thank you for your attention and your patience. Now it is time—briefly, and for just this once—to defer to my mother by invoking a piece of her wisdom.

My mother once told me that I should learn something new every day. And I have. Except for the days I took off. Now it's your turn to learn something new, and what you will learn is that many an accomplished cosmological theoretician has opined that the universe, unless acted upon by an outside force, tends to entropy.

For those who do not wish to learn any more today, please skip to the next chapter. Others may read on.

In my dictionary "entropy" has four definitions, which are (and I quote):

1. A function of thermodynamic variables, as temperature or pressure, that is a measure of the energy that is not available for work in a thermodynamic process.
2. A measure of the loss of information in a transmitted signal.
3. A hypothetical tendency for the universe to attain a state of maximum homogeneity in which all matter is at a uniform temperature.
4. A state of disorder, as in a social system, or a hypothetical tendency towards such a state.

Learned theoreticians have postulated that the universe leans toward entropy. What can we learn from this? That all systems, unless acted upon from an outside source, tend toward disorder. Chaos, as it were.

But this book is about KFAT, not cosmology, so now you should be asking yourself how this applies to our story. I'm glad you asked. Observe:

Definition One: **A function of thermodynamic variables, as temperature or pressure, that is a measure of the energy that is not available for work in a thermodynamic process** tells us that, given all possible situations, someone was not going to show up for their shift. Need a warm body at the board? Have a substitute nearby. Or money for bail.

Definition Two: **A measure of the loss of information in a transmitted signal** tells us that somehow, despite the best available engineering, and the best of intentions of a jock to make it to the station, something is not going to get played on the radio. In the case of KFAT, it could be that a jock was too fucked up to get to work, or a car broke down. Or that Jeremy was working on the transmitter, or that a wayward wind blew the needle off the record, or a heavier gust moved the antenna, blowing the station off the air. Or someone hit a wall near a turntable, or jumped in just the wrong spot and the station went off the air.

Definition Three: **A hypothetical tendency for the universe to attain a state of maximum homogeneity in which all matter is at a uniform temperature** clearly demonstrates the hypothesis presented in the "Forbidden" online chapter Music Sucked...- that, if historical American social and economic trends were to prevail, then someone or some group is going to exert an influence that dilutes the quality and originality of the music exposed to the public, making it all sound alike. When artistic and economic forces met in opposition at KFAT, the result was chaos, as predicted. See below:

Definition Four: **A state of disorder, as in a social system, or a hypothetical tendency towards such a state** postulates that with no outside regulating force, no external enforcement of discipline, everything will devolve into chaos and disorder.

Learning is fun, isn't it? For an illustration of this Theory of Entropy, please see Chapters 17 through 151. You may begin now.

End of Intermission

Part Two

Americanus Eruptus

...and it was probably one of the most outrageous radio stations you ever heard."

—Lori Nelson

Chapter 17: A Legend In Its Lunchtime

Without knowing it, Chris Feder was already in too deep. When he got the call from Jeremy and Lorenzo asking for a meeting, he was leading a stress-free life in a nice house on a shady, tree-lined street in Corte Madera, a mellow town in notoriously mellow Marin County. The 60s had lasted a little longer than usual in Marin, but the people who lived there were cool with that, man.

From this aerie of tranquility and reflection, Feder chose- no, *allowed himself*- to be thrust into the roiling cauldron of anxiety and disorganization that I think of as a vortex.

You'll remember that Chris had checked into a motel in Santa Cruz the night before he went to the station for the first time. Chris was a Rock 'n' Roll guy, a hip guy. Of course he stayed in Santa Cruz the night before: it was a hip place. I hope he got a good night's rest, because as always, came the dawn, and after several hours spent listening in dismay to the station the night before, he got into his car and drove over the Mt. Madonna Pass from Santa Cruz by the ocean, to Gilroy in the valley. This trip soon became commonplace for those who worked at KFAT, and was known as "going over the hill."

Chris went over the hill and into Gilroy, and here is what he found: Gilroy was not hip. It was a lonely place for a hippie or a rocker in December, 1975.

The town was a dusty backwater, bereft of modern influences, either culturally or architecturally. There weren't any interesting shops along Monterey Street, Gilroy's four-block main downtown thoroughfare. Did I say "downtown?" Did I say "thoroughfare?" That exaggerates the reality of those few uninteresting, uninspiring colorless storefronts whose insides housed inventories specializing in the mundane. And damned little of that.

A lot of Gilroyals* worked at Goldsmith Seeds, one of America's largest suppliers of seeds, both commercial and retail. But the majority of people who lived in and around Gilroy worked in the fields or at the canning factories and processing plants; they toiled, picking, transporting, packing, canning and again transporting the never-ending flow of produce that went through Gilroy daily. Produce-bearing trucks rolled into Gilroy night and day from all

*The people of Gilroy called themselves Gilroyans. They should call themselves "Gilroyals." It's so obvious… And how's this for a new slogan for a garlic-rich town: Gilroy: No Vampires, Ever!

over central California's vast growing regions, discharging their cargoes onto the loading docks of the giant Gilroy Foods and the other, smaller, processing plants in town. All manner of fruits and vegetables were then processed and packaged, canned, put into stacks or packed into cardboard and wooden cartons and crates which were then excreted back onto the loading docks for distribution to warehouses and supermarkets around the state, the country, and around the world.

The packing plant smokestacks belched out a continuous stream of acrid, malodorous smoke, which settled on Gilroy like an invisible but inescapable soul-dampening fog. Will Rogers once observed that Gilroy was 'the only town in America where you could marinate a steak by hanging it outside on a clothesline.'

The odors would change with the season and with the variety of produce being processed at the various plants in and around Gilroy. Tomatoes and garlic were processed almost the year 'round and were the most common elements of the olfactory stew; one jock remembers a "marinara period," while another fondly recalls that her favorite was the season that smelled like minestrone. No one liked cauliflower season, which had a sour aroma and the jocks had to suck it up a bit to keep the music flowing.

The stores in Gilroy were all of a practical nature, like the western wear store. No python or ostrich-skin boots there, hombre. The liquor store sold the staples: cheap wine, beer and malt liquor. No fine wines, no liqueurs, no single malts, no foreign beers.

You could buy a used or a new vacuum cleaner or get yours repaired at the vacuum cleaner repair and sales store.

The Leedo Gallery sold what I guess you'd call "modern collectibles," the new stuff that was tacky but might pass for an antique, or something someone might call an heirloom in about fifty years. You know, cut glass vases and bowls made of colored glass, replica paintings of pastoral scenes in imitation good frames. Cheap is cheap. "Lido" might have been correct, but would have confused the locals. High-falutin' didn't fly in Gilroy.

There was an antiques store in the middle of town that specialized in old oak furniture and lace. There was a barn-shaped building at the north end of downtown that was a pretty fun place, the only store in town where anyone hip enough to shop for collectibles would find anything worth collecting.

The motels were the King's Rest and The Oaks, neither of them part of a chain, the restaurant was the Longhouse, which was about as good as any Denny's, with pretty much the same menu. The only chain place in town was a Jack-In-The-Box.

There was a pretty good restaurant in town, the Harvest Moon, and the wealthier folks in Gilroy ate there.

Augie's Mexican restaurant was two blocks away, and the food there was "muy auténtico." Good salsa and chips, cheap beer, and as much atmosphere as cheap chairs and vinyl tablecloths, recorded mariachi music and dim lighting would allow. With an endless, renewable supply of free chips and salsa with your order, most of the staff ate there as often as they could afford. It was close to the station, the food was good, they spoke English there, which not everyone in Gilroy did, and the KFAT staff was welcome, which was not true everywhere in town.

There were a couple of banks on Monterey Street and the biggest building in town was the office of the Continental Phone Company. The next door south of the phone company was the vacuum cleaner store, and next to that was an optometrist's office.

The optometrist's office was right about in the middle of the town, and five doors further southward, was the Green Hut, the bar that quickly became the hangout for the radio staff when not at work or at the station. Or during a long song.

There wasn't really anywhere else to go in Gilroy, and as the townsfolk were a little uncomfortable with KFAT's hippies, the places where they were welcome were few. And where else would they find free drinks and KFAT blasting through the radio? They could hang out and relax at the Green Hut. Other than Augie's, where else could they go?

But back to that optometrist's office just up the block from the Green Hut.

In previous incarnations, the place had been a hair salon and a vacuum cleaner store, but the vacuum store had moved next door and now it was an optometrist's office. As you faced the building and the optometrist's office was in front of you, the door to his office was to your right, and to your left was a glass door, through which you could see a stairway going up. Above the door were brass numbers, 7459. There was a softball-sized sticker of some kind in the middle of the glass door. You'd go through that door and you'd know: up that long stairway was KFAT.

Remember the song, "Stairway To Heaven?" Well, forget it. It was nothing like that.

If you weren't there to get your eyes examined, and went to the other door, the one with the sticker, and opened it, you heard the music immediately. There was always music, except for commercials, jocks talking, or the talk show. Music, music, music. Sometimes weird, always great, anything-goes music. The carpeting on the stairs was worn, but I doubt whether anyone ever thought about that. No, when you walked through that door you'd think about what was upstairs, listening as you climbed. The magic that was happening up there was all you thought about when you climbed those 23 steps. It was twelve steps up to a four-foot landing, where you could catch your breath, then take a step or two before taking on the last eleven steps.

That might not sound like much of an effort to you in-shape folks, but over the years, many an attempt to scale that stairway failed in those last eleven steps. And many remember someone bouncing down most of them.

All of the jocks always listened to KFAT at the Green Hut or at home or in their cars as they drove to or from the station. Once there, they'd sit in their cars until the song ended and a new one started so they could hear the segue, get out—and for God's sake remember to lock up their cars!—and get upstairs in time for the next segue. They always wanted to hear the next segue. No one wanted to miss it.

That's what it was always about: this song and the next song. There was always music. There was the music and the segue.

Music + Segue = Flow. It was always all about the Flow.

At the top, you found yourself on another landing, this one with a doorway. I think there was a door in that doorway, but no one remembers it, and certainly no one remembers that, if it was there, it was ever closed.

You got to the top, and the wall that had followed you up the stairs on your left continued into a hallway straight to the bathroom. But if you turned right when you got there, there was another, shorter, hallway, and if you turned that way, on your right there were two doors. The nearest door was the entrance to the two rooms that comprised the offices. Across from that first door on your right, on your left, was another doorway, where the record library was on your left in a narrow passway, and on your right was a doorway into the studio, where you could see the jock plying his or her trade. I like that: plying by playing… You can use that.

The room was 11 feet by 12 feet, and about a third of the space in that room was taken up by the equipment. There was about a 6' X 8' space to move in. The turntables and work surfaces were at right angles to each other along the left and the far walls as you walked into the room, and the control board was right in the middle at right angles to both walls. (Look at the photos, dammit.) The control board was from the fifties, and the knobs and meters were already old when the first of the KFAT jocks wandered in and began their careers in broadcasting. Any experienced jock or engineer at that time would see this equipment and scoff. If a pro had to work on it, he would despair. But to these jocks, it was all new and exciting, and it was the music they cared about, not the gear.

On top of the control board was a flat surface that held three cassette players next to each other, and a slanted, lectern-like thing with a short piece of wood attached at the bottom which held the three-ring binder where the ad copy could be read from.

In the middle, between the cassette decks and the copy book was an adjustable boom that held the microphone. (again: the photos, please.) The boom arm was just like those lamps you buy at an art supply place that clamps on to a flat surface and the arm moves up and down and right and left, and had springs and it swung this way and that to adjust to the height and position of anyone speaking into the mic that it held. The mic was unremarkable, of no particular design, year or origin. It just was, and no one there gave it a second thought.

As you faced the board, there was a turntable to your left, and one to your right. On your right was a three-foot by four-foot window built into the wall that looked into another room of the same size and dimension as the studio. This was the Production Room, and we'll get into it in moment. Back in the studio, on the wall, above the window into the Production Room, above the left-hand corner of the window, was a clock. At the lower end of the left side of the window was a lamp made from a bowling pin with a tiny, cheap red-and-white-checked gingham shade. You could see the pole sticking up out of the top of the bowling pin that held the shade. It gave off precious little light, and that was the point. At night, if you turned off those glaring overheard fluorescent lights, which would harsh your buzz, the bowling pin lamp gave off just enough light to function. Darker was better than brighter for those late nights.

The lamp looked like it had been made from a bowling pin by some hobbyist whose wife had thrown it out. But it was Fat, and it had a home. Christa from the first staff bought it at a second-hand store, and left it behind when she left for Louisiana.

About 20% of that window into the Production Room was covered with bumper stickers. There was the yellow and red first KFAT bumper sticker that said "KFAT" and underneath it, spread out, "94 ½ FM." There was a bumper sticker that read "Paul English for President." Mr. English was Willie Nelson's drummer, and he had been with Willie since the beginning. Others said, "This Here's Cowboy Country," "Fat's Where It's At," "Grand Ole Opry" and a couple more.

Two 7-Up cases held the cassettes on which the commercials were recorded, and they were on the wall to your right, and there was a reel-to-reel deck on a case just below the cassette racks.

The window into the Production Room had a curtain on the Production Room side.

The walls and ceiling were the standard cheap off-white acoustic tiles with the pock-mock holes which led to the temptation of sticking something into them with a pin. Or a knife. These are also the soft ceiling tiles that people of a certain age will remember as having pencils mysteriously stuck in them, perhaps by bored people, perhaps from some period when the gravity bill had gone unpaid.

The cassette decks were a Jeremy innovation, and not used at any other radio station in the country. Why did KFAT have them? Because Jeremy got them cheap. He was cheap, all right. What? You want to call it innovative? All right, I guess he was. Jeremy was… innovative, yeah, that's it, but he was also a purist who wanted the very best sound he could have, as long as he could have it cheaply. The decks cost him almost nothing, but the sound would be improved if he increased the rate of speed that the tape would play, so Jeremy doubled the speed, and this increased the fidelity of the sound to an acceptable broadcast level.

Most stations employed "carts", or cartridges, which were in fact just like the eight-track cartridges you remember if you are old enough, but were designed for radio use. They had enough tape in them to play at double speed for a minute or a half-minute- the standard lengths of ads, and they would play out their loop and then reset automatically, so the next time a jock would put them in the deck they were already cued-up and ready for the "play" button. But Jeremy had gotten a huge carton full of blank cassettes cheap.

Why three cassette decks? Because you needed at least two operational decks all the time, and one was usually broken.

For the most part, the jocks didn't know or wouldn't care whether they inserted cassettes or the more common cartridges that the other radio stations had, until Jeremy's cheap-o cassette decks suddenly acquired the nasty habit of going into Rewind in the middle of their play cycle. Bad for the jock, bad for the advertiser, and bad for the station that wanted to collect the money for a commercial that hadn't run all the way through. But it only happened sometimes, Jeremy would work on it soon, and that was life at KFAT.

To the right of the rack of cassettes, behind the jock, was a window that looked out onto the roof, and the wood-plank path someone would take to get inside the station from the back parking lot. This window had to be open to admit any wandering breeze that might wend its way through that poor, dried-out dustbin of a town during the summer months. The open window was also the source of one of the few delights afforded both jock and listener at KFAT: the mournful moan of the passing trains as they rumbled through Gilroy on their way to more important places.

The two small speakers that played the music into the studio were overhead, to your left and your right, and it sounded pretty good.

In its overall effect, the word I'm looking for is "cheesy." It was all so cheesy. But the jocks didn't notice. It was radio, and they were on the side of the mic that the listener never saw, and it was what came out of the listener's speakers was what counted.

When the jock was facing the board and the mic, behind him on his left was a wall with another, bigger window. This looked into the hallway you'd have walked in if you had turned sharply right when you got to the top of the stairs. Down that hallway there were two doors on your right and a door at the end of the short hallway, about 15 feet away. The first door on your right went into the front office, and there was another, larger, room behind that. If you went past the first door on your right and went into the second door, you entered a large-ish room with another doorway into another room at the back wall, and a doorway on your left that went into a mid-sized room with a doorway into a last, smaller room behind it. (See Diagram: 0ffice 1, 0ffice 2, Meeting Room, Rooms A, B, and C)

If you came up the stairs and turned right and walked past the offices and went straight to the end of the hall, the door there would lead you out the back, onto the roof of the building and along the plank path to the back stairs, which took you to a small parking space that would fit maybe three cars. If it was daytime, that was where you parked if you didn't want to park on Monterey Street, taking the chance of forgetting the meter again and getting another ticket. You parked out back at night if you were willing to take the chance of having your tires, battery or radio stolen.

Some of the local hermano's needed the batteries to make their cars bounce up and down, as that was cool there then. Other brothers needed the batteries to sell for a fix. Front or back, day or night, parking was always a crapshoot.

The parking lot was small, and in poor repair. The macadam was cracked, pocked and uneven, and you daren't walk on it without shoes. Which was bad news only for Jeremy, who generally resisted all manner of footwear.

We can deal with the offices and the assorted back rooms and who worked in them and who lived in them at a later date, but for now let's start again at the top of the stairs and walk straight forward.

The wall that had been on your left as you climbed the stairs was still on your left, and lasted about 25 feet and ended at the bathroom. There was only a toilet and a sink there, but that was enough to intimidate the timid souls among us. For many, one look and one whiff of the inside of that closet-sized repository of bacteriological mutation and suspicious fungal growth was enough to suggest that if one needn't empty one's bladder or bowels right then, a saner mind would elect to wait until less riskily situated.

For the staff, one of the necessities of the job was a belief in a certain immunity, acquired over time, to the dangers of however brief a sojourn was required in that bio-hazardous environment. Whatever lurked in there, the regulars believed themselves to be immune. Flushing was both required and expected, but no one ever thought that such a simple act as flushing would eliminate the problem. The problem was never what was flushed; the problem was what remained behind. On walls, carpeting, porcelain, in the air.

The one attraction to the toilet was the graffiti. It was usually clever, rarely sophomoric, and always entertaining. As no one ever painted over the old graffiti, the way to stay entertained and ignore whichever bacteriological entity was currently looking for a suitably habitable environment, was to look for any recently un-graffiti'd spot that had now been filled. This was amusing as long as you weren't even slightly germophobic. Of course, with the bathroom at KFAT, it would help if you had made an accommodation with both your visual and your olfactory sensitivities.

If you wisely stopped in front of the bathroom and did not enter it, on your right was another doorway, and this was where I spent most of my time at KFAT. This was the Production Room.

Like the studio, it was 12 feet by eleven feet. On your left was a short wall that had the toilet on its other side, but as you got into the room, you were surrounded by funky green, yellow and brown-striped carpeting covered the floor and all the walls. (See Diagram: PR)

The far wall had two reel-to-reel tape decks and a small desk-like area with a cassette deck, a turntable and a built-in mixing console. The mixing console would have been old in the '60's, and would have been used by a home hobbyist, if he didn't have any sophisticated use for it, and only needed a cheap mixer.

The table on the right was your standard institutional affair, with metal legs and top. It was about five feet long and three feet wide, and was below the window that looked into the studio. The microphone on the console was on an adjustable boom, like the one in the studio. The ceiling was the same soundproofing tiles, and it worked, for the most part.

All of this would have been a depressing first sight for a man who had seen and could compare this shabby equipment to the major stations in New York City, but Chris—remember Chris?—walked into all this and was met by Christa and a group of whichever two or three remaining staff were left, and she screamed at him and told him that Jeremy was gone for the day and there was *no way in hell that he was going to be the new Program Director!* His new pal Jeremy was gone, knowing what Chris was about to walk into. Alone. With no support.

So Chris left, telling them to "do what you want, and I'll find Jeremy and come back tomorrow morning."

And by the next morning, when he returned, he had no staff at all.

Fuck!

Remember?

Chapter 18: Taking Control

They were a rag-tag bunch at best, every one of them. Cast-offs, outlaws, reckless ruins and the detritus of unrequited radio love. Some were bedraggled by life, but none were yet bereft of hope, energy or enthusiasm. They were warm bodies, they wanted to be there, and KFAT needed them. Over the first weeks of January in 1976, they all came stumbling, unsure, anxious and excited, into their new jobs.

I could have said "careers" instead of "jobs," but I don't think any of them thought in terms of careers. The new-ness of... *everything*, was so powerful, so overwhelming, that all they thought about was the next segue, or the next shift, or the next rent, or how to get through the week with enough money left to eat with, or buy enough gas to get to where they could eat or sleep. It was a bit ragged, living like that, but they were young, it was exciting and it was fun, and it was a constant challenge to keep their heads above water, and to keep the fun happening. It was a mission and a calling, and they loved the cause, and doing it well was what drove them. All of these eccentric, individualistic people were suddenly part of a team, a crew, a happening collective. Individually and collectively, they were the sound and the voice of KFAT, and people in Santa Cruz and parts of San Jose were talking about it. To do it at all was a privilege for all of them, and doing it well was what drove them. They were always aware of the listener requests, but they played their shifts for each other.

As their leader, Chris Feder had imagined such power back on Long Island, and now it was his. His staff was fresh, unschooled and in need of direction. He moved them about the schedule as they came into this new world, first into the overnights or the graveyard shift, and eventually to a place of relative permanence. Each according to his skills, and each to where Feder thought they would best fit his vision.

Gordy would do the mornings, as he had always done the mornings. Of average intelligence and wit, but affable, Gordy Broshear had a chuckle to his style and an easy manner, and he would do well in Morning Drive. He was easy to wake up to. Never a rocker by nature, the overweight San Josean knew more about the softer songs that would be best played while the folks were waking up and getting used to the idea of another day of.... whatever they did. Yeah, Gordy for Morning Drive.

Gordy had to be proud of being the only member of the first staff and also had been there before the Great Texas Purge (which some called the Great Texas Retreat), but he never mentioned it. He had been fired by Larry Yurdin, had been despondent about it, and now he was happy to be back. He would tow the line, Feder had no doubt. If he could get past Feder's bullying.

Feder picked on Gordy the most because he was the easiest, most obvious target. He was fat, insecure, and had been a journeyman jock at a nowhere station. If he wanted to make it in the new world, he would have to prove himself. He had to *get* the new format, and that would be a challenge. Larry used to yell at him for not getting it, and then came Feder's constant yelling, which Gordy seemed oddly able to take complacently. Gordy burned with wanting to be there, and he put up with a lot. Feder took note.

Michelle McGovern wanted to be known on the air as Tiny, her childhood nickname, and Terrell Lynn called her "Sister Tiny" from the start, and it stuck. Feder recognized her as a good-time, hard-partying woman who'd known her share of hardship. She was somewhat suspicious by nature and had those hardened edges, and Feder saw that she had a strongly independent nature, which meant that she'd put up with a lot and wouldn't be a Prima Donna. He liked that. She had two toddlers, Finnian and Amber, six and eight, and he liked that she was as protective of her kids as a mama lion, and he would use that. He wanted a honky-tonk gal for his staff and Tiny was the real deal. And Tiny was *so glad* to be doing something legal. And yet, she always seemed to be looking over her shoulder, looking to see where the next problem was coming from. Feder took note of that, too. Bob Cassidy recalls that she had the most soulful shifts of the group. And if Buffalo Bob says something nice about someone, you must take notice.

Kathy Roddy had the sweetest voice, with a playful chuckle to it, but she was also very bright, and she was always looking ahead to see where this was going to lead. In college, she had been into politics, and she was willing to be led, but not abused. She was of a feminist turn, and Feder took note.

Kathy wanted to use her KFJC college radio name, "Sister Kate," like in the song from the 30s, but there was already a "sister" there, so she chose her mother's nickname, "Sully." Her mother had been a Sullivan.

Frankly, I'd be afraid to pick on Bob Cassidy, and so would you. Chris took note. Ergo, Bob got picked on less than the others, but he was a whole other sort of challenge for Feder. Chris recognized in him a man much like himself: Bob was a bully and he liked to party. But there was another similarity, and this was very big: Feder knew little of the new KFAT music, as his passion had always been rock, and Cassidy hated "country" music, and his shows always featured the rockin'-est cuts available to him.

Bob remembers one time during one of his first shifts at KFAT, a night when his shift followed Chris, who had just put away his records and was putting on his coat, about to leave for the day. Bob was about to open the mic for the first time to announce what songs he had just played, when Chris came up behind him and whispered, "Always remember: Led Zeppelin is God." A few cuts by Led Zeppelin were Fat, but Bob was a rocker, and there were plenty of rockers in the library.

Cassidy was a natural DJ and Feder liked his musical tastes and put him on at nights, warning him only about being too drunk to do his work. Cassidy also had the best connections in Santa Cruz for weed and cocaine, as much of it came through him, and that made Chris more forgiving, and perhaps even a bit conciliatory towards Cassidy.

Lacking in both sobriety and kindness, Bob Cassidy got along just fine with Feder, which was that much less for Feder to worry about. Feder took note: except for excesses in alcohol, leave Bob alone.

Terrell Lynn Thomas was just as easy-going in life as he sounded on the air. His voice was soft, and you almost had to lean into your speaker to hear what he said. Okay, maybe you didn't lean, but you felt like you did. His voice drew you in; it was soft, personal, but loud enough never to be a problem. Always soft, he sounded… intimate and sincere. I understand the ladies liked it. He knew more of the traditional country music than anyone else at the station, and he knew the rare stuff, the old stuff, the stuff that he knew from his childhood, and which gave the new station so much of its flavor. The others listened to his shows and made notes.

Well, they all listened to each other's shows. That was almost all they did, wherever they lived and whenever they could. They listened for new songs and for good segues. They listened for songs that would work when played with a different song that they had in mind for their shift. In their heads, they played imaginary segues, and they thought up things to talk about during the breaks. Feder had told them all: talk only if you have something to say, otherwise shut up. And everyone at the station listened to everyone else at the station, so they were performing for themselves, for the other jocks, and for Chris, who was always listening. And they wanted to keep these jobs.

Karl Hess kept his job as the morning news guy, and after Jeremy's approval of his nerve-wracked on-air audition blunder, he called his news broadcasts "Karl J and the Morning Nudes." His voice had an upper-register lilt that made him sound happy, as if he was always telling a joke. In fact, Karl sounded vaguely like Snidely Whiplash, the villain in the "Dudley Do-Right" cartoon. He told me that he tried to read the news as if reading a children's story. People heard the news as if they were waiting for a punchline, and it worked. Gordy and Karl would banter, until Feder told them to keep the banter, but shorten it. Tighten it up. They did it, of course, and it made the morning show better.

Feder and the new jocks were always reaching for some new music, trying a new segue. The range of selection at KFAT was unprecedented, but there was so much untried, unknown music in the stacks and they always wanted to play the new song, try the new segue. And new stuff was coming in the mail daily. They'd get to everything eventually. No one else was playing most of what KFAT played. It was all new to Feder, it was all new to the jocks, it was all new to the listeners, and more people heard about it and more people listened.

Feder was taking control. KFAT was working, and he liked it.

It had been an unholy mess, and Chris Feder had rolled up his sleeves to tackle the great chore ahead of him, which included finding a staff, and who he found was… *these guys!* He needed a new staff, but he had almost no money to pay them with. It was intuitively obvious that professional radio people would laugh at an offer to work at KFAT, thus rendering any easy transition to a functioning staff that much harder, but that was his job.

Laura Ellen was doing weekend shifts, and Chris thought she was terrible, but Jeremy insisted on her being there. Laura Ellen would learn. As we will see, she learned a lot.

He now had Michelle, Gordy, Kathy, Terrell Lynn, Bob Cassidy, myself and a few fill-ins, but he needed more bodies, more airshifts. And that's how Sherman Caughman came back into the picture. Sherman had been doing fill-in shifts at KFAT when it first went on the air with the Texans, until the day Larry came in to see him during his shift and told him to go home, he'd call. When he did call, he said he wanted Sherman only for production, and he'd call back. The next day the station was silent.

Larry had fired some of the local guys right before the big staff meeting that led to the meltdown with the entire first staff. Now Sherman was calling Jeremy and asking to come back, and to do justice to Sherman, I am going to need a whole new chapter.

Chapter 19: Sherman Caughman 101

Sherman Caughman is the most colorful person I know, and the only man I know ever to have begun a conversation with, "As I was hitching a ride to Arkansas, on a break from my job as a Snake Woman…"

He is a former: Snake Woman, pimp, ladies' shoe salesman, second-story man, carnie, roustabout and thief. Currently, he is a DJ, serial imbiber of alcohol and smokeable vegetation, player of dirty songs, and he is both a rapscallion *and* a scalawag. He is a Civil War re-enactor, a make-believe 49er gold miner, an actor, an educator, a pretend train-robber, and the possessor of the best cackle heard since Old Nick pranced, cloven-hoofed, on what were then pastures.

Sherman's face is a wonder to behold, both terrible and fascinating. To say that his face is ravaged does scant justice to the concept of ravage. Imagine the roughest grade of sandpaper that you can buy: the kind with the biggest sand chips, and so coarse and rough that you want to wear gloves to hold it. This describes that which is visible of Sherman's eyes, forehead, nose and lips. The rest is beard, which covers most of his face and all of his chest. But what you can see of his face, except for his lips, which are normal, is excessively endowed with those tiny bumps, pebble-y growths and wrinkles.

His voice is very much like that sandpaper: low, raw, raspy and ragged. His voice conveys a crazed urgency that defies description; you have to hear it. It's a voice that makes you look for the emergency exits. Larry Yurdin heard the sound of anarchy in Sherman's voice; Larry loved the *sound* of anarchy, but loathed anarchy itself. Sherman sounded to Larry, simply, larger than life. But that face! Ow!

The bumps on his skin are both tiny and large: tiny in that there are millions of them, and large in that they are prominent. He dresses in the manner of a gold-miner from the Gold Rush days, wears the floppy hat, pants tucked into his boots, and suspenders of that era, and when you see him, you look for the mule that is certain to be nearby. He is a character. No, he *is* the character that you see in front of you. I've been with Sherman at bars when he approached a woman and asked her if she was a "rich widder-woman" who might like to take him home with her. But there's always a twinkle, a cheery note in Sherman's voice; there's always a joke nearby, and offense is rarely taken.

Look here to see more Sherman.

He was born in Arkansas, and never fit in with a family whose parents ran a Christian radio show on Sunday mornings, and when the show was over, they'd go to the jails to hand out Christian pamphlets. Sherman was not long for that world: he liked the radio part, but the rest was strictly for dipshits.

Beside the radio show, his parents sang gospel songs in the area, and afterward his mother preached. They had a gospel group and they made records as the Caughman Family back in the 50s. If you find one of those records, I'll buy it from you.

By happy circumstance, Sherman's friend's father did the afternoon show at the same station, and it wasn't a religious show. It was a show of popular music, and Perry Como was starting to give way to Jerry Lee Lewis. The father couldn't hang with the new stuff and gave the show to his son, but the son wasn't passionate about it, and when he got tired of doing it, he gave the shift to Sherman. Now it was a rock 'n' roll show, and Sherman played Chuck Berry and Fats Domino and Jerry Lee and Elvis. He was 15 when he got the show, and all the kids at school listened. When his shift was over, he'd shut the station down for the night. You know we're talking about rural Arkansas, right?

He did that for a while, until he discovered pussy and beer and didn't want to do the show any more. Then there was some juvenile crime involving something about theft and destruction of property, both public and private, and his parents were getting sick of him. His father had remarried, and his new stepmother cared not a whit for this young troublemaker. Then came the night of the unfortunate confluence of a borrowed '35 Chevy Coupe, four gallons of moonshine and a police cruiser, and Sherman was given a choice between jail and the military. Having burned his bridges at home, he chose the Navy, where he parlayed his experience into doing some time with Armed Forces Radio.

Out of the Navy, Sherman tried selling ladies' shoes in Los Angeles, and then moved to San Francisco. Always drawn to the right crowd, one night Sherman and Chocolate George* of the Hells Angels were chipping away at the roof of a Safeway supermarket in San Francisco's Mission District. Their intention, of course, was to break through the roof, drop down into the store and rob it. They'd been chipping away at the roof for three nights and Sherman got tired of it. He reasoned that he'd rather just run in there with a gun and take the money. He said, "Look, man, if I wanted to work this hard, I'd get a job. I don't wanna do this, George. I'm out of here."

George stayed and kept working, hours later making a hole big enough to climb down into the store, where the police were waiting for him. It was San Quentin for Chocolate George, and back to ladies' shoes for Sherman.

Selling ladies' shoes had paid the bills in Los Angeles while he went to

* Hells Angels on their bikes in massed formation on San Francisco's Market Street. It has to be one of the most, if not *the* most famous picture in the Hells Angels archive. Someone even made a poster of it. This assemblage of many chapters of the Hells Angels was gathered for the funeral of… wait for it… Chocolate George.

acting school, eventually becoming management, in charge of the exclusive I. Magnin stores in Palm Springs, San Mateo and Beverly Hills. The Beverly Hills gig gave him hope of making a living as an actor, and he got a few parts in TV, but not enough to live on. When that didn't look as promising as it might, he moved back to San Jose to go to college for broadcast classes, hoping to get a radio gig that would afford him the time to pursue the acting thing.

For spare cash, Sherman was also picking up 16 year-old girls at the Greyhound depot in San Francisco, and selling them to a whorehouse in Reno. But there was no long-term future in that, and he kept up on his classes at broadcast school, so he could get back into radio. He had just "got his radio chops back up, and then KFAT happened."

Larry Yurdin loved Sherman's "seedy, larger-than-life character," and used him on weekend mornings. Larry liked that listeners would hear Sherman's raspy voice, consider what weird things he was likely to utter next, and wonder who was in charge at the station. Then one day, Larry came in during Sherman's weekend shift, told him "he looked tired," and said that Sherman should go home and rest, and that he, Larry, would finish his shift. He said he'd call.

That was a Sunday, and on Monday Larry called and told Sherman that he liked him, but only wanted to use him for production for the time being. The next day Sherman tuned in to KFAT, and it was off the air.

He was offered a gig in Grand Coulee, Washington, way the fuck up north, doing Afternoon Drive. He put everything he owned into his car and moved up there. After a few weeks on the air, the owner called for a staff meeting. While the staff gathered for the meeting in the station's lounge, they listened over the speakers to what going out over the air. They heard the owner in the next room stop the music and tell the listeners, as well as the staff in the lounge, that it had been nice being there, and that this was the end of the station. He said good-bye to the listeners, turned some switches to the off position, and in the sudden, stunned silence, he came into the lounge, paid the staff in cash, and went out the door after them, locking the door behind him.

Outside, the owner's station wagon was already loaded full of stuff he had traded out for ads that would never run, then he got in the car and drove off, leaving a confused group of strangers in the parking lot, speechless.

There were guys there from Portland, Seattle and the Bay Area. Still in the parking lot when the shock wore off, they talked about it, and then they all chipped in and rented a giant U-Haul trailer. They drove to each person's house, loaded up the trailer, the farthest first, then drove to the nearest guys' destination, dropped off him and his stuff, and proceeded to the next nearest guys' location and dropped *him* off. The last guy would return the trailer. That was Sherman, and he returned the trailer in San Jose.

Back in the Bay Area, Sherman called Jeremy and started bugging him about work. Jeremy was resistant at first, and, freaking out about all the bills and needing money more than DJ's, he suggested that with all Sherman's knowledge and experience in sales, that he would better serve the station by selling ads for the station.

Living back in San Jose and taking courses in broadcasting, and needing an income, Sherman got out the one suit he had left over from his sales days, and went to a Wherehouse Records store and asked for the manager. The Wherehouse Records chain was a large one, and might be a real score for Sherman. Luckily, the manager had not only heard of KFAT, he'd been informed by his staff that people had started to come in asking for records that they'd heard on the new station. They had almost none of those unusual records, but he was a smart guy and took note, of both the requests and the new station.

By his own admission, Sherman was no "closer" of ad sales, and he told Jeremy of the potential new client, maybe a big one, and Jeremy said that he'd help him out.

Sherman had his suit on when Jeremy picked him up in his normal torn, faded jeans, torn T-shirt and no shoes. Jeremy almost never wore shoes. Jeremy also had his daughter, Elsbeth, who was then about a year old, and had on only a diaper. Jeremy spent half the time at the store playing "peek-a-boo" with Elsie, who was hiding behind a Coke machine, and half the time talking with the manager, who looked at Jeremy like he was an alien.

No sale was made that day, but Sherman bugged Jeremy until he relented and gave him an airshift. Jeremy told Chris about the new overnight guy, and now Chris had most of a staff. By the way, Sherman's hair was still dark at this time, and he wore it in an "afro" and there is one photo of it in existence. It's what Sherman used as "Head Shot" an audition photo you use when applying for a job in media. As a photo to recommend you to a potential employer, it's, uhh… unusual. See his Head Shot. Note the open beer and the roach clip.

Sherman came back to KFAT and found Gordy Broshear, Kathy Roddy, Buffalo Bob Cassidy, Terrell Lynn Thomas and Sister Tiny. He started by relieving Tiny, who was on until One. He was on until 6 a.m., when Gordy came on. But no one could do that seven days in a row, and that was when Mark Taylor called.

Chapter 20: Wisp

Wisp is a wonderful, under-used word. You don't hear it much anymore, maybe because everyone's gotten so fat. I don't know when it entered the language, but if it hadn't been invented yet, it would have to be invented to describe Mark Taylor.

He was a wisp of a man. At 5'10", thin of arms, legs, face and torso, even his thin, brown, receding hair was wisp-y. Need I add that he had no butt? He might appear delicate, but there was an impish sense of fun about him, just a little less than a chuckle, and no one wanted to feel sad near him. He had a cheerful disposition and he spread it wherever he went. As a child it might be endearing, but with Mark Taylor it was infectious.

From the time he was ten and found himself with a transistor radio under his pillow, Mark Taylor knew that radio was his dream, his future.

Born and raised on the outskirts of Los Angeles, he'd been some trouble to himself and others, particularly his mother, and at sixteen he'd "flipped out pretty bad" on LSD. It wasn't his first go-'round with the chemical, and his mother couldn't take it any more and sent him to live with his sister, who was in college in Santa Cruz. She took him in and got him into high school, where a helpful teacher learned of his passion for radio and got him a gig at local rocker KRML in Carmel. They saw some natural talent there, and wanted him to do weekday mornings, but he was in high school, and couldn't. They gave him a Sunday morning slot, and he was new at the high school and his new classmates listened to the station. Pretty soon after moving to Santa Cruz, he was known, and he was cool. That was cool.

Clint Eastwood had just finished shooting "Play Misty For Me" at his station, and Mark got to go to the premier with his new girlfriend. Now he was *very* cool. And, of course, the drinking and drugging continued in Santa Cruz as it had in Los Angeles. After a while, what with all the pot and the booze, it was a burden to be at the station at 5 a.m. on Sundays, and so he started bringing his friends to the station to smoke pot and listen to music with him. And they say we had no "support networks" back then…

Before his music shift started at six, he had to play a one-hour pre-recorded religious show called "Revival Time." The smoking and the giggling commenced just before he hit the "Play" button, and they heard: "Revival Time!

Across The Nation! It's… Revival Time!"

He had the gig for over a year, but eventually he missed too many mornings when there was no one to play "Revival Time," so he was fired. He hung around Santa Cruz until he graduated from high school, did a semester at the college there, sold some pot, and thought about his future.

He knew people who had gotten degrees in communications, but couldn't get jobs in radio. Station managers didn't want to hire these graduates because they thought they knew it all, and because they always wanted to jump to larger markets as soon as they could. So he said screw it, he left college and started working for a non-comm in Santa Cruz while looking for a paying job. It had to start somewhere…

It started with KOAD, at what he recalls as the "lowliest station in the San Joaquin Valley." The station was in a bullet-riddled trailer in the middle of a beet field. He was the only white guy at an all-black station, and racists drove by periodically and took shots at the trailer. He and his girlfriend lived in a nearby trailer with no bullet holes in it… yet.

In rapid order, his girlfriend tried to commit suicide, and he got fired. Then it was KBOS, then KGEN in nearby Tulare, somewhere else in that damn valley. He liked this station because it had carpeting and an employee lounge, which made him feel like he was moving up in his career. AT KGEN, the Station Manager, Smilin' Jack, lined up the singles for him to play, and he played them in that order. He was doing a country shift in the afternoon, and a rock shift at night. Drugs and alcohol were the norm until Smilin' Jack went into the production room while Mark was on the air, called him on the request line and fired him. Over the phone. From the next room.

Then it was KONG in Visalia, "The Most Livable City In The Valley!" Visalia? Tulare? Are you recognizing any of these names? Exactly.

While at KONG, his girlfriend left him and there were several incidents of drugging and partying at the station with underage high school girls, which some might have seen as irresponsible, immoral and illegal, but Mark saw as one of the few perks of the gig. Then, a girl he met over the phone came over to the station and gave him some pills. He remembered nothing after that until a friend called him the next morning and asked him what drugs he'd taken the night before. He didn't know.

For some reason Mark Taylor was fired from KONG in Visalia, "The Most Liveable City in the San Joaquin Valley!"

Maybe it was the drinking and the drugging, or the underage girls, or the perpetual lateness. Management can be real assholes about that stuff.

He remembers: "I was 19 years old and living cheaply, doing what I love… I lasted seven or eight months. Parties with high school girls. I was young and wild. I picked up a girl over the phone and she gave me these pills and my friend called me up the next day and asked what the hell I'd been doing the night before. [He said] I'd been playing songs over the baseball postgame show. I had no recollection. It all added up, and I got fired. Drinking beer and smoking pot with underage girls…"

He had no money, no job, and was sleeping on a friend's couch. He heard KFAT one night and called the next day "because they were doing some interesting shit."

Jeremy answered the phone and told Mark, "Our afternoon DJ (Sully, that day) had just had something happen to her eye, and if I came right down I could go on the air." With all of his experience in radio, he'd never heard of *this* before, but he drove right over to Gilroy and went on the air.

Jeremy thought he "sounded way too slick for what they were doing. I had just come from Top Forty, and was tight, which was the opposite of what they wanted. Plus, I really didn't know country and was leaning toward the rock side of the library, 'cause that was what I knew. I played Hot Tuna and Grateful Dead and Southern Rock."

Jeremy said: "Way too slick and way too rock 'n' roll."

"Jeremy said I had to loosen up, and somehow I had another airshift, plugging a hole for someone, and I drank a quart of Colt .45. I got loose and played a bunch of country shit that I didn't know, and purposely ran a loose thing, and Jeremy said it was closer, so I listened to the station all the time and got to know the music. The music was a revelation to me."

"I always winged it, never wrote any notes down." Like everyone else, he started doing overnights, and that was where he stayed. He liked the night and he liked the night-crawlers who were out during his shift. He was now the Ozone Ranger.

Because the Ozoner had more radio experience than anyone else on the staff, Laura Ellen made him Production Manager, and he would do this work after he got off at 6 a.m., when Gordy started his shift. He kept this job until Frisco showed up and showed him up in production skills, but as no one really knows when Frisco showed up, we'll say the Ozoner did it for a few weeks.

No one called him "the Ozoner," mostly he was called OZ, pronounced Oh Zee.

Chapter 21: Big Fat Mess

Chris Feder was a Noo Yawkuh. He was loud, brash and confident. He spoke fast and he was demanding, his cocaine use exacerbating his already abrasive personality. He was pushy, imperious and intractable. He had been in the World of Rock back east, where he'd beaten the biggest rock station in New York City at their own game, he was the manager of a hot band that was signed to a major label, and he was the boss here now. But he was an alien to the hippies at KFAT.

No one got along with Chris at KFAT. No one liked him, but Jeremy had needed a strong guy to pull the station together, an alpha male, and Chris Feder was The Man.

Feder had found a mess on his hands in Gilroy, but it was his mess and he could handle it. He would shape this amorphous clusterfuck of personalities into a working, functioning staff. He had to: it was all he had. He loved the concept of the station, he trusted his ear, and as an avid listener of music, he knew that he had his hands on something special. It would work, by God. He would make it work.

Terrell Lynn remembers: "Horrible times... tons of hours on the air... seven hour shifts."

Sherman says, "Ten hour shifts."

But let's go back to that part about it being a mess. He had some seriously unique people, all of whom had only two things in common: they were almost all inexperienced in radio, and they were all undisciplined, independent types, and not any good at taking direction. Some couldn't, some wouldn't. But it was *his* mess, and he'd make it work.

That first day he had been on the air for almost 8 hours. He got Jeremy to play records, then Laura Ellen. He thought she was terrible, so he used her on week-ends and as a fill-in. He needed a staff fast, and so they filtered in.

Gordy had quickly learned to get there as early as possible for his shift, because he "never knew if the guy doing the overnights would be there" when he got there, and he could turn the station on, which was valuable back then, and Chris took note.

Now, Gordy was doing the morning shift, and everyone else was moving around. Kathy, now Sully, moved around from evenings to overnights and fill-ins, sometimes to the slot after Gordy, doing 10-3, and that was working. Chris did the Afternoon Drive shift until 7, partly because he knew almost no country songs or artists, and the Afternoon Drive shift was best for rockers: uptempo stuff to get you home in a good mood. More than the other shifts, it was a party. And Feder liked to party.

Buffalo Bob came on after Feder, 7-10, then Tiny until One, then the overnight with either Sherman or Ozoner. Terrell Lynn moved around where he was needed.

Everyone moved around. Buffalo Bob was a great DJ, and as long as Chris was there to control his drinking, he was cool. Of course he got drunk on his shifts. Chris told him that he could pretty much do as he wanted, as long as he came in sober and lasted through his shift. Bob agreed, and for a while he came in sober.

Chapter 22: Heroine On Board

Laura Ellen Hopper is the heroine of this story, but no one could have guessed that yet. She was a pleasant, quiet, unassuming young woman, dedicated to raising her daughter and riding out whatever storms Jeremy created and needed guidance out of.

She was the quiet voice of reason that had propelled Jeremy out of St. Louis and out to the West Coast, to try some bigger tricks. She had been a runaway and a drug courier, and had skipped town when a drug deal went bad. When the heat was off, she came back to St. Louis, met Jeremy and started working at KDNA. She was surprised to learn that Jeremy had harbored similar feelings of attraction for her, and they had awkwardly fallen into a relationship that had now brought them the crown jewel of that union, their beautiful daughter, Elsbeth.

It was Laura Ellen who answered Lorenzo's question to Jeremy about whether he wanted to be part owner of the new station he wanted to buy in Gilroy. She had answered that yes, he did, and Jeremy now owned almost half, with Laura Ellen owning a small percentage.

She was a quiet sort, but she was smart, focused and determined. But she was stuck being a mom in Los Gatos, where she cooked, cleaned, mopped, dusted and kept the house in order. There was a constant stream of visitors to the house; people were forever coming by to talk about the station, and she had food at the ready. Feeding them at home was more homey, and cheaper than restaurants.

She was also doing about half of the office work for the station on her dining room table, a big, dark thing, one of the antiques she'd brought with her from St. Louis. When I first met them in Los Gatos, I remember a lace doily in the center of the table, but that was rarely seen anymore, as the station's paperwork spread out and often covered the entire surface.

The sales office for the station was also located on that table, and on it she did the books, marking what spots were sold, which ran, and who owed what, when. She was doing that when the Texans and the locals walked out and left them with no one to play the music. Almost immediately, Chris Feder showed up and put Jeremy on for a few hours, and put Laura Ellen on for a few hours. Anything to keep the signal flowing.

Laura Ellen did her weekend shifts, took care of Elsbeth and much of the office work, and Jeremy did what Jeremy does. He tinkered with the station.

Chapter 23: Mid January, 1976

Chris Feder was in charge, a new staff was assembled, and Laura Ellen was a bit player.

So what sort of a crew did Feder have to work with? Tiny was still a single mom living in Santa Cruz; she was a willing student and a good prospect, but her car was always broken and she had to hitch-hike over the hill every day with her two kids. But she got there every day, she was starting to get the idea of the format, and she was going to be good.

Chris had hired Tiny knowing that she had no experience in radio. He needed a staff, she was there, and Chris knew that she had the instincts.

Bob Cassidy had a car and he made it to work every day. Drunk, or on the way to drunk, he always did a good show. And that voice: deep, resonant and rumbling, with round, mellow undertones. That voice. And it came from *that guy?*

Gordy took the bus to Gilroy from San Jose, and usually got there on time and cheerful, ready for his early morning shift, and Kathy was typically late for hers, but just a little. Kathy was getting better and needed less and less attention, but Gordy wasn't on the program yet, and Chris was riding him hard every day.

Karl J showed up on time, did his news, and stuck around for a little on-air banter with Gordy before he left for the day. Karl had married an ambitious woman who didn't like the people at KFAT; she had plans for Karl, and she wanted him to have a real job.

It was Karl Hess who many credit with the national identification of Gilroy as "The Garlic Capital of the World." There had been a sign that declared this very thing on Highway 101, but it was a small, faded sign, it had been up for decades, and no one, even in Gilroy, paid much attention to this grandiose claim.

Karl had been generously endowed with a sense of humor, and some who knew him from those days still think of him as the funniest guy they'd ever met. Sometimes he'd hang around when his work was done, and if the Production Room was empty, he go in and make a station ID.

Having time and looking for material one day, he did a promo and ended it by saying that "This is K, F, A, T, Gilroy, the Garlic Capital of the World," followed by a massive chorale singing, but I'll get to that later, and for now, being in the Garlic Capital of the world was *something* to be proud of about their otherwise desolate location, and everyone at KFAT started playing Karl's new ID and calling Gilroy the Garlic Capital of the World. Soon even the city fathers of Gilroy took notice. But I'm getting ahead of myself again.

Back at the station, Terrell Lynn was almost always on time, and needed little direction, which was a help, because everyone else needed watching. He knew the music better than anyone else at KFAT, and he went from overnights to graveyard, 8 or 9 p.m. until 1 a.m. when Tiny couldn't make it. Was Chris Feder much of an influence on Terrell Lynn? "Never saw him," he says. "I got hired, and saw him once after that. He said to me, 'just keep playin' records, kid, and shut up. If you have nothing to say, keep your mouth shut.'"

Sherman was doing whichever shifts he was needed for, and so Feder put him on the overnights where he would be free to indulge his taste for explicit and sexually-oriented songs without offending too many people. Chris' influence on Sherman? "Never saw him," Sherman says.

Chris thought Laura Ellen was terrible, but she was the boss' old lady, and Jeremy wanted what Laura Ellen wanted. Chris put her on weekends, and she spent her weekdays in their house/office in Los Gatos, doing whatever office work was needed, making phone calls and caring for Elsie.

Chris had the Afternoon Drive slot, while the rest of the day he worked at being the Station Manager, Program Director and Music Director.

So there was a schedule, and there was a staff.

But Chris was a yeller. He yelled at Gordy incessantly. He would come into the booth while a jock was working and tell him or her what they were doing wrong. Almost all the jocks now tell me that the way to do this was to wait until after a jock's shift, then talk about it. Chris would come in screaming whenever he heard something that offended him, which was often. But this format was a new, never-done-before thing for everyone, and it rattled them all at first, and then less, but still often with Gordy, who put up with it gamely.

For most of the staff, having a job was, in itself, something that needed getting used to, and they were all feeling their way through a massive learning curve: the music, the job, the mic, the controls, the meters, the meter readings, the logs, the commercials to produce, the commercials to record, the commercials and Public Service Announcements to read, live. There were records to go through and preview, notes to take, production work, and more stuff to do than the listener hears on their radio. Jeremy felt that if they were being paid, they were being paid for an eight-hour day, and they should work eight hours.

The pay was shit, the hours were long, and the station kept going off the air. And when were they going to move the transmitter?

Working a full shift is a pretty tiring thing, but these things would get better. Gordy, who knew the equipment, had to learn a whole new canon of music. *Everyone* had to learn a whole new canon of music- jocks and listeners, both. It was a busy, tension-filled, exciting time, everything was new and Chris was yelling to blow off steam. In retrospect, he now feels that perhaps all that yelling was the wrong management style for this group.

But at least there was now almost a complete crew there. More were coming.

I had gotten a tape recorder and a microphone and was casting about, looking for people to interview. I had come down in early January to check out the station for the first time, and I was in love. I liked the equipment and I really liked the music and the people were all freaks and that was right up my alley. I'd always loved radio, dug easy intellectual pursuits and weird people, and I was excited about doing a show. Chris told me to contact Public Relations firms in San Francisco and introduce myself, which I did.

I'd been nervous about these calls at first, but was surprised at how happy these people were to hear from me. I told them about KFAT and its format, and that I'd like to interview some of their clients for my show.

Any PR person lives for the moment when a new media contact calls them up and asks to put their clients on the air. "KFAG, you say?" they'd ask regularly. "No, KFAT- K. F. A. T. With a T." And then I'd try to explain KFAT. I'm still trying, and when you're ready, why don't you go to the website and take a shot at it?

I quickly learned how to get interviews. I had to: I had five shows a week.

I came down to KFAT for the second time in the middle of January, and Chris showed me how to use the equipment. I would record my interviews on my new portable cassette recorder, and there was a cassette machine on the console in the Production Room; I transferred the interview from the cassette onto the reel-to-reel deck in the Production Room, and edited it from there. The tape on the open reel was larger and easier to work with. I'd record my shows and promos on reel-to-reel, leave my shows on five-inch reels and transfer my promos onto a cassette. The finished tapes would be labeled "CHEW- show" and "CHEW- promo" with appropriate running dates, and left in a slot in one of the two slots in the 7-Up cases in the studio marked "CHEW." Woo hoo! I mean it!

Chapter 24: Getting Acquainted

So it was a ragtag group that slipped, tripped, and stumbled into their new jobs at KFAT. Careers? No one was thinking like that.

The town was boring and it smelled funny, the studio was a mess, the equipment was crappy, old and cheap, the staff were all amateurs, the boss was a screaming jerk, and the owners were possibly insane. But God, it was fun! Smokin' pot, drinkin' beer and Jack, and playin' music.

Chris Feder believed strongly in this new format. He didn't know much of the music yet, but what he was hearing was knocking him out. Willie Nelson? Willie was incredible, and no else was playing him. The two brief experiments in Texas that Larry Yurdin attempted had played Willie, and KOKE in Austin was playing him, until they went silent a few months later. But even though he was more "country" than anything else, Nashville was ignoring Willie, but not KFAT.

Willie's pal Waylon Jennings had had some brief, interesting exposure when the Grateful Dead asked him to open for them on a tour in 1973. But these two guys weren't "country" in the way that "Country Stations" played "Country Music." And my God, there were more!

Jimmy Buffett was great, and who else was playing him? No one. What was he? This was before all that Margaritaville and "Parrothead" shit. Merle Haggard was country, but his songs on the country stations were his hits, the commercial stuff, ignoring the enormous body of his more interesting music over the years.

Feder already knew and loved the Flying Burrito Brothers and Poco, and of course Dylan and Leonard Cohen and Emmylou Harris and Bonnie Raitt and pretty much the rest of the east coast folk scene, so he was getting into learning a lot of new music, stuff he was really liking. The job was crazy, sure, but Chris liked this job.

Jeremy's deal with Chris included a small ownership position if he could pull a staff together and give it a coherent sound, based on what Larry had started. He knew he could do that. He'd taken over WLIR on Long Island and with Kenny Kohl's help had made it an astonishing success. He'd do it alone this time. And he liked the music; it was hip beyond all belief, and he saw that people were hungry for it. He had a cool pad in Santa Cruz for a hundred bucks a month, and he liked the music, the people and the power.

He was getting in at seven a.m. and working the phones. He got record service from some labels, and he got KFAT listed in the WALRUS Report, which was then the bible of what was left of underground radio.

Jeremy was around a lot at first. He spent a few hours a day at the station, mostly at night, but he was in and out during the day, sometimes sleeping in a back room. Jeremy loved to tinker with his electronics. He seemed to have an understanding with those circuits that must have sounded to them like: "I know how you work, so let's work together and we'll do cool stuff."

I said earlier that Jeremy was a genius, and I did not say that lightly. I didn't mean it in the way that you thought the smartest kid in your class was a genius, but in the calibrated, quantifiable, certifiable way. He would test in the genius level, and anyone who spent any time with him would know it. Any small snippet of conversation with Jeremy would demonstrate that his mind did not work in the normal way, and what it usually came up with was either brilliant or crazy, and it was always a little of one and a lot of the other. Any snippet.

And he lived to tinker. He wanted his station to sound good. No, really, to *sound good*. I was surprised to talk with radio engineers and learn that they know that all radio stations *sound different* from each other. Not just in what they played, but how they *sounded*. The tones and stuff. Jeremy was an engineer and he wanted KFAT to sound so good that other engineers would be impressed. That was his passion, his trademark.

So Jeremy tinkered continuously, sometimes at the station and sometimes at the tower, always hopeful that the application for moving the tower would be approved. If he was going to move the tower, he'd keep it working for now, but he'd tinker with it seriously after he moved it, or when he found out he wasn't going to move it. He had his reasons.

Chris was around a lot. He wanted to point his DJs in a direction and let them find their way, but he was always listening. He listened at the Green Hut or in his office. Office? More like his… area. He had a desk in Room A.

Jeremy didn't intrude much on the DJs, and seemed to prefer not to talk with them at all, if possible. He was always friendly and helpful if spoken to, but his preference was for electronics, and that is a lonely endeavor, which he liked, and sought.

The staff was new and unsure about their relationship with Jeremy, what with him being the boss and all, but they quickly learned his strengths and his weaknesses. Circuits, good; people, not so good. When Joady Guthrie, Woody's son, came to visit, and Terrell Lynn introduced him, the first thing Jeremy said was to ask him was if he was going to die from the same disease that killed his father. No 'hello?' No 'how ya doin'?' No 'Nice to meetcha?' Terrell Lynn was appalled, as was everyone else at the station, and he remembers being dismayed that he was working for someone like that; one supposes Laura Ellen vowed to keep Jeremy away from regular people after that. Again: fat chance.

Everyone knew that Jeremy meant no harm, he was just... Jeremy. No one at the station exactly avoided Jeremy, and everyone knew that he sort of liked people, but just didn't understand them. He was always willing to help, but he didn't seem to have the gene for understanding people, so he hung out with his circuits instead whenever possible, and that was okay with everyone.

The one thing Jeremy was obsessive about with the new staff was the station ID. The law called for the announcer to identify the station by its call letters and its location at the top and bottom of every hour, and Jeremy wanted it to be K-F-A-T, Gilroy. Not K-fat, and not "in Gilroy." You could play other IDs that said it differently at other times, but it had to be this way at the top and bottom of the hour.

That wasn't all that hard, and everyone was new and willing to please, and that issue was settled shortly. Now he could go back to his circuits, transistors and resistors: things he understood.

Also, this was Jeremy's first commercial station and he was unused to the constant drain of funds for unexpected needs and breakdowns. All these people wanted to get paid, and suddenly he was responsible for so many people's lives, what with their rents, their cars and their food, and one of them had *kids,* for God's sake, and there were always new and insistent needs, and money was always going out and not nearly enough was coming in, and it was his deal with Lorenzo that the money for running the station from now on was *Jeremy's* problem, and now there were only *his* shallow pockets, which were almost always empty, and his checks were bouncing and everything cost more than he thought it would, which was more than it *should have*, and all these *goddam people* had the *goddam nerve* to be *unfamiliar* to him, which always added to his discomfort *every time he had to talk to these people*, which was becoming more frequent and more necessary all the time, and back in St. Louis everyone took care of themselves and showed up for work prepared and ready to do their shows, *and that was for free*, but now there were *all these people* that he didn't know and that he was *suddenly responsible for* with all their *problems* and their *lives* and their *constant need to be paid* ...and, and...*oh, goddam it!*

Other than that, he was cool.

Laura Ellen was spending half her time running all the office work that she could from their "office" in their dining room in Los Gatos, and being a mom to a toddler and a guardian to a teenage ghetto kid named Raymond (Chapter 32).

The other half of the time she was at the station, almost an hour's drive away. Blonde, curly-headed year-old Elsie was with her constantly, as was Bernice the dog.

Chapter 25: Gilbert

Growing up on Long Island, Gilbert Klein was a bright, genial fuck-up, getting by by getting by. I'd lived in California a couple of times, did the hippie thing and a little bit of the Army thing. I'd traveled and taken my time before graduating from college, eight years after high school. I'd gotten a B.A. from Adelphi University on Long Island, but in doing so, I'd caused something of a controversy. It seems that Adelphi University, like most reputable universities, had a limit of nine credits for an Independent Study project that a student could apply toward the 120 credits necessary for a bachelor's degree. University policy stated that Independent Study meant a school-approved course, with a reviewed and approved course structure, and taken with the guidance of a school-approved advisor, but outside of the classroom.

Being slightly older, and having traveled and lived a bit more than my fellow commuter students, I needed 30 more credits for my degree. With the average class earning 3 credits, I went to see the dean, and argued that the school, who gave classes to students who did not know what they wanted from an education, had no business denying me access to the education I wanted when I *did* know what I wanted from an education. I was a year and a half away from my degree, and I wanted the dean to waive the 9-credit limit and give me permission to devise my own curriculum, all of it in Independent Study, and allow me to select an instructor, then leave me alone to complete the course. I had 30 credits left, and I didn't want to be in any more classes with the other students.

After asserting some clever platitudes about "educational values," I asked for a waiver of restrictions and the special program, giving the school no details whatsoever of my proposed course of study, informing the school only that I called my proposed course "The Culturally Oppressed Tuna." This was in 1970, and with the spirit of the Sixties still upon him and, twisted up in my tautological reasoning, the dean inexplicably acquiesced, and so I spent the last year and a half of my undergraduate years completing those last 30 credits with a professor I liked, and who was happy to be paid for what would have to be easy money. In the end, no discernible work of an academic nature was ever produced, but both of us had gotten high a lot, talked a lot, and had a good time. For my Final Exam, we two scholars took LSD together. Gilbert likes to say that he "passed with flying colors," but I wouldn't try that lame joke on you.

Looking back, we may all, myself included, be amazed that such shit would fly in a respected institution of higher education like Adelphi University. I asked an Associate Professor that I'd liked to work with me, and the guy agreed. He was an English teacher, but he had an outside practice as a Jungian therapist, and I had been drawn to that aspect of the man. Mysticism. I was going to be studying mysticism. And the illusion of time. I was offended at the illusion of time and wanted to talk about it. No further information on this subject was available at press time.

The above-mentioned problem came to light when it was time to graduate. What degree would be presented to the young scholar? The dean supposed that whatever the troublesome bastard had done, it was closest to… philosophy. Yeah, it was philosophy. Okay, we'll give him a degree in Philosophy. He made a call, but the Philosophy Department had used none of its staff to instruct me, they had no idea who I was or what I had done, and they wanted to award me a degree in Philosophy like they wanted leprosy. Excuse me, now it's Hansen's Disease. But still, *no way*, dean!

The English department knew that what I said I'd done with their teacher had nothing to do with the English department, so they said "No" to the Dean's request for a degree for this un-definable course.

Art department? No, thanks! This wasn't art.

Sociology? No way, man. This wasn't sociology; this wasn't even *social.* They said the guy was a loner and a weirdo. Again, no!

Psychology? The Psychology Department at Adelphi was one of the best anywhere on the east coast, and they were justifiably proud of their reputation. Now, psychology may be an imprecise science, but it *was* a science, and there were specific course requirements that they demanded, and I hadn't taken *any* of them. *Absolutely not!*

The dean was frantic. He had allowed me to do… something. He'd signed off on it, he'd allowed me to pay for the course, and they'd cashed the checks. He didn't know what I had done, but I'd paid my tuition, somebody from the university had given me passing marks, and now the dean had to give me a degree in *something.*

Ultimately, the Dean went back to the English Department and laid down the law: I had taken more courses from that department than from any other, and the guy I'd "studied" with, if any studying had actually occurred, had been an Associate Professor in *their* department, and so the English Department was very reluctantly persuaded to grant the degree.

Ladies and gentlemen: Gilbert Klein, Bachelor of Arts.

I'd been a pot dealer and a taxicab driver on Long Island for years while getting my B.A., then I moved upstate to the Catskill Mountains to learn carpentry. I was terrible at that, and after a few months I moved back to California's Marin County, where I dealt pot and was a carpenter and a craftsman, selling clever wooden crafts to tourists at San Francisco's Fisherman's Wharf, doing whatever it took to get by while I got my Master's in education from Sonoma State College. I was going to be a teacher.

After graduating from Sonoma State with a Master's in education, I was unable to get a full-time teaching job, and became a substitute teacher. Then I starting working on beautiful slabs of crystallized rock, turning them into belt buckles to be sold at Fisherman's Wharf, and then I met Chris Feder and needed a place to work, and you know the rest.

While teaching high school, there was a beautiful blonde girl named Susan in my class. She was smart and gorgeous and mature and unbelievably sweet and a lot of fun. I was in love. Lest anyone from the Moral Majority read this disapprovingly, we hung out a few times as friends for two years after she graduated from high school, and she was thus long out of my classroom before we started dating. If we had dated while she was still in my class, that might be cool in a lascivious sort of way, but it would be wrong. And wrong is… wrong.

Not what anyone would mistake for tall at 5'7", slender, with brown hair and glasses, I was pleasant looking, cheerful and extremely likable. At least, *I* think I'm likeable. Why? What have you heard? Whatever…

I saw Chris Feder in the first week of January at his house in Corte Madera, not far from where I lived in San Rafael. I was using Chris' basement, and had gone to see him about a key when he asked me to do the talk show. I said yes, and then went down to Gilroy to check it out. Naïve about such things, I was impressed with the radio station.

On my second trip to Gilroy, Chris showed me how to use the reel-to-reel decks and how to edit tapes, using a grease pencil, a razor and editing tape to do the splicing. My first guest was going to be a guy who was running for some small-time county position, and was driving around Santa Clara County in a station wagon plastered all over with his posters. He'd contacted the station, and Chris had forwarded the request for an interview to me. The interview was going to happen in the studio, and I'd tape it and edit it later.

I was nervous about this first interview. This was *somebody*. He was *doing something*. I hoped I wasn't going to embarrass myself. I got to the station early to prepare for the interview, but when the guy finally drove up in his old station wagon, I was still nervous. In setting up the equipment for my first real interview, I noticed that my hands were shaking.

What a schmuck.

Chapter 26: The Early Shifts

It was ragged but right. Chris had a staff and a plan, but the plan kept changing.

Now Gordy was on at 6 a.m. until 10, Tiny from 10 until 3, Chris was on from 3 until 7, Bob was in from 7 until midnight, Sully moved around and filled in, Terrell Lynn did some fill-ins and overnights, Laura Ellen, Sully, Sherman, Mark Taylor, Reno X. Nevada on weekends.

Reno X. Nevada?

Born Dalton Hursh, Reno had been a part-timer at KSAN in San Francisco during its heyday as the most influential radio station in America in the mid-to-late Sixties. Chronically given to alcoholic excess, he was now a worn-out, drunken shell of what he had been. Jeremy knew him by reputation and from a brief stint at KSML up in Lake Tahoe, and he wanted Chris to give Reno a job. Chris was skeptical about it, but he was drunk when Jeremy asked him to hire Reno. Chris needed bodies on the weekends, and he hired Reno.

Gordy remembered Reno, too. He remembered the time Reno called the station from somewhere up in the Nevada foothills, maybe five hours away if he hustled, but he didn't sound like he was hustling. He was due on the air an hour ago, and he told Gordy he didn't think he'd be there on time. By the time he got there, three things occurred simultaneously. The first thing was that Gordy set a new personal best and a new staff record in the Long Shift category at thirteen hours, the second was that when Reno finally arrived, he mumbled something un-intelligible to Gordy and stumbled into a back room to sleep "it" off, and thirdly, the DJ who had been scheduled to relieve Reno showed up for work on time. Thus, the music never stopped.

Sherman remembered Reno as the laziest jock he'd ever seen: "He'd do sets where he'd play the third cut on each of the albums."

Bob Cassidy remembers Reno, too. He was doing a shift and waiting to get relieved by Reno, who was typically late. Often, the staff never knew who was coming in after them, especially on weekends, when Chris had gone back up to Corte Madera. It could be someone new or someone they knew. Pissed off and wondering where the hell whoever was going to relieve him was, Bob heard a series of increasingly distant-sounding thumps in the direction of the stairs to the street.

Bored, drunk and pissed off, he'd run out of patience with doing a good show, and this wasn't even his shift anymore, so screw whoever got blamed for it. He'd been putting on the longest cuts and waiting for his replacement. Getting up from the board to see what the hell *this* was *now,* he walked to the top of the stairs and looked down.

At the bottom landing, between the door and the first step, was a pile... of Reno X. Nevada. He moaned a couple of times, covering his head with his hands, and Bob had seen enough to recognize the symptoms: he was drunk and passed out. Clearly, he'd been too drunk to make it all the way up the stairs, he'd fallen down to the bottom and lay there. Bob was now even more pissed off. Hell, Reno could even be hurt, but Bob didn't give a shit about him. But inside Bob's crusty exterior, he loved KFAT, and he kept spinning records for a couple of hours until Reno woke up and crawled upstairs. When Bob saw him, he threw him a record and left, not knowing if Reno could handle what remained of his shift.

Bob did not listen to the station as he drove away, he never heard a word about what happened after he left, and he forgot about it until we talked about Reno years later. There just weren't any normal people at KFAT, and there weren't the normal repercussions, either. Bob took note.

Chris Feder remembers Bob running up the stairs once, drunk and stinking of booze, just in time for his shift, with a couple of Gilroy cops right behind him in hot pursuit. His car had been seen wobbling on the road, and he hadn't stopped when the police car put its siren on. Instead, he told Chris later, he was so close to the station when the cops saw him that he thought he could out-run them and get to work with no one on his tail. It had seemed like a good idea at the time, but somehow, it hadn't worked. The cops were there, and they were going to arrest this asshole.

Chris talked the cops out of taking Bob, telling them that he would be safe at work under his supervision for the next few hours, and they had his word that when Bob left, he would leave sober. The station was new in town, Bob was new in town, the cops liked the station a lot and Chris was a silver-tongued devil, so they relented. Buffalo Bob would owe Chris big-time for this. If he remembered it.

Another time, Chris didn't have as much luck with the Santa Cruz cops as he'd had with the Gilroy cops. Bob Cassidy was a regular fixture in the criminal life of Santa Cruz; he knew everyone in the drinking-and-drugging circuit, and the cops knew about Bob Cassidy. He was what the cops would now call "a person of interest" in the trafficking of certain substances, and they were interested in talking with him. In addition, he was also a noted scofflaw, owing untold money from uncounted parking tickets that had gone unpaid for so long, with penalties accumulating, that no one really knew what he owed anymore. The cops just knew that it was a lot of tickets and a lot of money, and they wanted to talk to him about some of those other things, too. But first, they had to find him. Then they had to catch him.

You may not know anyone like Bob, and I daresay you don't, but back in that day, he was of a type, and he had a network of friends who hid him whenever the cops came looking for him at his usual haunts. It went on like this for a year; he seemed to be everywhere and nowhere at all, and the cops were getting frustrated. They just couldn't find him. Then they heard he was working at KFAT, and the cops tuned in to verify that. Now the cops knew where he was every day. But KFAT was just settling in and the shifts were shifted around.

One morning, shortly after Bob got hired as a DJ, the cops showed up at KFAT, found Gordy at the controls and asked if he was Bob Cassidy. No, he wasn't, but if they wanted Bob, why, he'd be back for his shift at 7:00 that evening. Bob had always been mean to Gordy; he picked on him, demeaned his work, and was rude and dismissive to him. To the end, Gordy swore it was not revenge, but rather a regrettable lapse of memory wherein he simply forgot to get a message to Bob that the cops had been there looking for him. Why, anyone could forget something like that. Prudent or forgetful, Gordy was long gone by the time Bob showed up for his gig, and settled in for work.*

Bob got to the station with almost ten minutes to spare before his shift began, he went to the library, pulled a stack of records, and settled in for a long evening of what passed for work in Buffalo Bobville. That night he was sober and on his best behavior because his sister was going to be listening to him for the first time. He'd been the family fuck-up for so long, and he finally had a steady job that he was proud of, so he wanted his show to sound good for his sister. He wanted to sound sober and play the best music that his sister had ever heard. He was in control of the music and he was proud, and a little nervous. He wanted to impress his sister, and everything was going well. For about twenty minutes.

He'd planned his first four or five songs on the way to work, he'd played them and they'd sounded good. He cued up his next song, put on the headphones and listened to the end of the last song of his first set. For once, he'd planned out what he was going to say, and he composed himself for his first vocal break of the show, and leaned into the mic. He flipped the switch, took a breath, and started to talk when he felt a tap on his left shoulder. Annoyed, he turned around. It was the Santa Cruz police.

"Are you Bob Cassidy?"

"Uh, yeah…"

"Would you please come with us, sir?"

At this point, Bob still had the microphone open, and through the headphones he knew the listeners were hearing everything that was said in the studio.

---

*Memory lapse or revenge? Discuss.

"Uhhh... you know I'm working here, right?"

"I'm sorry, sir, but you'll have to stop what you're doing and come with us."

"Hey, I'm working here. Go away."

"I'm sorry sir, you'll have to come with us."

All of this was going out over the air, and I'm sure the listeners were paying attention, not knowing if this was a joke or not. KFAT pulled stunts and all, but Bob was not known for prankish humor, and neither was Bob one to go gently into that good custody, either, and I believe that the listeners soon understood that it was for real as they listened to Bob yelling and scuffling with the cops all the way down the stairs, fighting and screaming at every fought-for step of the way down, and out the door.

That's a lot for a Station Manager to deal with, but Chris liked Bob and he liked his shift and his drugs. Bob was as close to a friend as Feder had at the station, and Bob would have his job waiting for him as soon as someone could bail him out of jail. Chris liked Bob, which was not true of Reno. Chris was passionate about radio, and Reno had long since left passion, and then even caring, behind. Chris remembers Reno as "the laziest, most cynical prick I'd ever met." He was a radio gypsy, too old and too dissolute, taking what was left of his skills and his name as far as they would carry him, wherever they led him, from station to station. And now he was in Gilroy, working the overnights.

Wherever Chris was, he was listening, and he hated Reno's shifts. One night he was listening, and he couldn't believe what he thought he heard Reno doing, so he drove back over the hill at midnight to see it for himself. He snuck up the back stairs and watched through the back window as Reno worked. That lazy bastard was using two records, going from one to the next and then back again. He had a case of beer and a pizza from the Green Hut, and he'd go from record one, track one to record two, track one, to record one, track two to record two, track two, working his way through the two albums. Chris waited for Reno's shift to end and he came in and fired him. Reno said, "You can't do that to me! I work for Jeremy."

"No, you work for me, and get the fuck out of here and don't come back!" Reno left and called Jeremy, who then called Chris. Feder told Jeremy that he had given Reno three or four chances, and he "couldn't work with someone that just doesn't care." Jeremy supported Chris, and Reno was gone.

Was Reno the first guy fired from KFAT? It's a little hard to know that, because as many came as went in those early days. Assigning shifts was a catch-as-catch-can deal, and no one was keeping track of who was there, and when. Some careers at KFAT lasted a few shifts, some only one. Some careers lasted a few weeks, some only one.

Larry had hired a black guy and asked him to call himself LeRoy Rogers.

That guy didn't like to play music; he liked to talk. He lasted a couple of weeks, maybe. Then there was the guy on the graveyard shift with green teeth. He liked to open the mic and have people call up and masturbate to the music. And he'd put them on the air doing that.

Kathy came in one morning after listening to this on her way to work, and she was appalled. The guy told her, "It was so beautiful… we were playing music and this woman called and said 'my husband's in jail and I'd like to send something out to him,' and she went on the air and masturbated, and we put her on the air masturbating, and it was so beautiful, so her husband could hear it…"

Kathy was wishing there was someone who could come in and clean the studio before she went on the air, but that luxury was not available. The guy was a friend of Jeremy's, who'd thought it was great. But Green Teeth didn't get the KFAT format, and soon enough Chris fired him. He stole a pair of shoes and a jacket from Terrell Lynn and disappeared.

So things were smooth at KFAT, but not much.

Chapter 27: Tiny, Shifts and New

Sister Tiny had been doing a daytime shift and it wasn't working for her. She was living in Santa Cruz with two kids and a broken car. Hitch-hiking to work every day with the kids had made everyone cranky, and hitch-hiking home to Santa Cruz at night with the kids was scary. She went to Jeremy and told him that Sully would be better on days, and she would be better at night. They took Sully off the graveyard shift and put her on after Gordy at 10 a.m., Terrell Lynn started after Sully from One until Four, and Sister Tiny now took over after Buffalo Bob's Bison Boogie at Nine p.m., and she was on until One a.m., when either Sherman or the Ozone Ranger began the overnight shift. Tiny soon realized that she would have been better off where she had already been, but by then the shifts were stable, it was finally all working, so to speak, and Chris didn't want to change anything.

Tiny needed to work and live at the station to make things work with her kids. By the time she would go on the air, the kids were ready for bed, she'd tuck them into a mattress in a back room, and she'd do her shift, then she'd go to the back room with her kids, falling onto a nearby mattress and getting to sleep. She'd wake up when the kids woke up to the glorious new day, eager and ready for another day at KFAT.

Sister Tiny lived in that back room with her two kids, Amber and Finnian, twice during the three and a half years she worked at KFAT, once for three months, and once for nearly a year, eating however they could. There were a lot of vegetables cooked out on the back deck, and for a while there was a trade-out for one meal a day at the smorgasbord place.

Terrell Lynn was now doing a regular afternoon and occasional graveyard shift, and Sherman and the Ozone Ranger were sharing the overnights.

As Mark Taylor, he'd had a somewhat nondescript career in radio, but as the Ozone Ranger, he was having a ball. He was a hard-partying guy, and he was always up for whatever good time he could coax out of an evening. He was on from the early hours until Gordy got there just before 6 and took over the studio. These were the spooky hours, and the Ozoner knew that only the spooky people were out then, which was fine with him.

I don't know you, of course, but I think you might be surprised at how many people are out and looking for something to do at three in the morning. You might not want to know those people, and you are probably grateful that they are not out and about when you are. With that strangeness as a given, I don't think you'd be surprised at what these night-crawlers were willing to do for something to do. Who knew about this? The Ozone Ranger knew.

Tiny remembers several nights when she'd wake up from her shallow, tense, restless sleep in her room in the back. That her two kids were on one mattress and she on another nearby, with a crazed, perpetually intoxicated lunatic partying and playing loud, raucous music just a few yards away was a good reason to be tense. Add to this the fact that the walls were thin, the doors did not lock, and said lunatic was inviting everybody listening in those wee hours to come by and visit, and this added up to several good reasons to be tense.

On her way to the bathroom in the middle of the night, she'd pass the studio and usually she'd see several strangers in the studio with the Ozoner, and sometimes others elsewhere around the station, doing lines of coke or crank, drinking beer, whiskey, or whatever, smoking cigarettes and passing joints and stinking up the place. Tiny had two toddlers one thin door away, and returning from the bathroom, she found it hard to get back to sleep. And she needed her sleep.

Some nights, as Tiny padded, groggy, bare-footed on the filthy carpet toward the bathroom, she looked through the window into the studio to see the Ozone Ranger in his seat, cueing up a record, smoking a joint while enjoying a friendly blowjob from some woman who had stopped by to enjoy the ambiance at KFAT. The Ozoner welcomed a bit of company, and he would say so on the air. And people were listening. All kinds of people.

Tiny hated having to live at the station. Her previous job had been had been a tough gig, and she'd had to struggle all her life, but she was a dedicated mom, and first and foremost she was the provider and protector of those kids. She had never had an opportunity like this radio thing before, and she was damned if she wasn't going to make it work. She also had another reason to stay at the station, and his name was Ken.

Ken Walker was a severely damaged Vietnam vet who drank way too much and had a serious anger management problem. Tiny had good reason to fear him, and yet she was his old lady. Tiny's judgment of men and her selection thereof probably had a lot to do with her self-esteem, which was hovering ominously at the low end. She'd had a rough childhood, and it showed, but everyone who knew her knew that she was a dedicated mother and always wanted to do her best for her kids.

Her judgment in men got in the way of that a few times, and this time was one of the worst. Ken was a dangerous man when he was drunk, and he was usually drunk when he was awake. Right after Tiny got to KFAT, Ken came to see her. He was drunk again, and started hitting Tiny, but Laura

Ellen, Terrell Lynn and Sully were there, and someone called the cops. The cops came and got a look and a listen to Ken, and soon there was a lot of screaming by Ken, by Tiny, by the staff and by the cops, and before anyone knew it, Ken had some new hardware on his wrists and he was being put into a cruiser. Tiny suddenly pulled away from the cop who was restraining her, jumped on Ken and started screaming and beating on him. So they took Tiny away, too. Laura Ellen went down to the jail and bailed Tiny out, and everyone went back to their jobs. It was another day at KFAT.

Another time, Tiny was afraid that Ken would come, so she locked the downstairs door to Monterey Street, maybe for the first time, but the door was made of glass, and Ken broke it to get at Tiny. He left that time, with no further damage to Tiny or to the station, so Tiny told Laura Ellen that some drunk had done it, which was basically true, and the door was fixed and the incident forgotten. But not by Tiny.

Another time, Ken came to see Tiny and menaced her throughout her shift. She remembers: "[He] tried to kill me. He terrorized me while I was on the air. He was like this (face close) and I had to do breaks, and he wouldn't let me make a phone call, and he was like, crazy and drunk and threatening me and… horrible.

"Frisco hid me in the back room for a while and he finally went away." Frisco had to hide her in a back room more than once. Sully remembers Tiny describing her relationship with Ken: "It's like having three kids, but one of them is retarded."

Of course, every day wasn't exciting in an assault-the-cops kind of way, but it was always interesting, and almost always fun. All day, the music was non-stop, and it always sounded good. The place was abuzz that the antenna and transmitter might move soon and the station would have more, further-away listeners.

Chapter 28: Cuzin Al

Al Knoth is another curious fellow, another oddity in the KFAT mold. Born in San Jose in 1937, Al was the oldest guy at KFAT. He came from several generations of Italian farmers and ranchers who'd lived and worked in the orchards around San Jose.

If you know San Jose today, you might be surprised to learn how recently farming dominated every aspect of life in this then-small city, almost an hour south of San Francisco. Luckily surrounded by the most fertile valley on earth, San Jose provided a comfortable living for those who tilled the soils and tended the trees, vines and orchards. Al went to the same junior high and high school as his mother. His mother was Sicilian and his father German, and as America was at war with both the Italians and the Germans at the time, he made a point of telling the kids at school that he was an "American."

His father gave his mother a radio in 1936, and as of this writing, she still has it, and it still works. Loving all kinds of music, Al played that radio all the time, listening to everything, especially the Western music that his parents favored.

He played drums in a band and tried college, but didn't like it. He dropped out and married an artist, who repeatedly told him that he wasn't worth anything because he had no artistic expression. He was a Hi-Fi enthusiast, but his wife complained that playing records was not a talent. He "drifted" into folk music and "found" bluegrass. He decided to take up the five-string banjo in 1962 after hearing Earl Scruggs and Grandpa Jones on the Grand Ole Opry.

He was married, he had two children, and he was a deacon in the First Baptist Church. He collected records and listened to the banjo greats, and on vacation in Hawaii in late 1971, he heard about KTAO in Los Gatos. Hell, that was right near San Jose, and he went to check it out. He wanted to meet his fellow bluegrass fans. The station was in the garage at Lorenzo Milam's house.

Al remembers: "I always loved radio and wanted to be on the radio. I walked in with my sport jacket and tie, and a military haircut. There was Lorenzo, and I asked him about the bluegrass show. We talked, and Tiny Freeman, who was doing the show, had come down from Tacoma in mid-December and meant to go back, but it was January and he was still there and wanted to leave."

Lorenzo asked Al if he wanted to do a radio show, "and I laughed. He asked if I knew anything about bluegrass music, and asked if I knew anyone to do it, and I said I'd see. I'll look around. Does it pay anything?"

"No, it's all volunteer, but you gotta start somewhere."

"I went home and told my wife I was gonna do this, and went back the next day and told Lorenzo. He said it's Thursday nights. I hadn't ever cued a record. Tiny [Freeman] showed me a few tricks, and it was real simple, with about 4 knobs. He told me to relax. Then he said he was going to have a cigarette break and he left for half an hour, and while he was gone, Lorenzo called and told me to come back next week."

That was in 1972, and one night that first year Lorenzo had two visitors from St. Louis: his old friends, Jeremy Lansman and Laura Ellen Hopper. That night, those two heard Al's show, enjoyed it, and when they had this new station in Gilroy, they got in touch again. He wasn't sure about doing it, but a friend reminded him that it was commercial radio, he'd be paid, there'd be a bigger listening area, and the new people were crazy and it would be fun. It was February, 1976.

He listened to the new station and it blew him away. Al had never done any drugs—had never seen a joint in his life—and these people on the air were pretty far out, him being a deacon of the church and all... but the music was just outstanding and he'd have a two-hour bluegrass show on Sunday nights, so fuck all those weirdo's.

It turned out that Al was more ready for a change than he'd thought, and he took to KFAT and its ways immediately. Buffalo Bob turned Al on to pot, and he dug it. A lot. So Bob took out an album cover and taught Al how to clean the pot and roll a joint, and the transition to "Cuzin Al" was under way.

Unbeknownst to Al Knoth, he had been playing with the vortex, and now it had him.

Al has two features that make him stand out. He's about 6'2", and although you'd never think of him as fat, he has a truly impressive gut that hangs over his cowboy-belted jeans. It's a *protuberant* sort of a gut. The other feature that you'll never forget is his laugh. It's a belly-busting, happily raucous bellow, followed by a self-contented "heh-heh-heh" kind of chuckle that lets you know that he enjoyed the joke or story as much as he assumes you did. And you probably did. His laugh is infectious as hell and I always smile when I hear it.

Both as a salesman working his territory, and as a vacationer, Al always seemed to be spending time in Hawaii, and I can still hear him bellowing at me: "Howdy, brah!" Al plays a lot of Hawaiian music on his show, the "slack key" stuff, along with the bluegrass.

I like bluegrass music for almost a whole song, and then I've had enough, but I could listen to Al's show just to hear him talk and laugh. He's clearly enjoying himself immensely, and that makes you enjoy it, too, even if it's against your will. Having had his 40th year Anniversary in 2012, KPIG's Cuzin Al's Bluegrass Show was easily the longest-running bluegrass show in American broadcast history until he took it off the air a year ago. In addition, and for the record, the author wishes to state categorically that he loves Cuzin Al.

Chapter 29: Laura Ellen Wasn't Happy

Laura Ellen wasn't happy. As we will discover, Laura Ellen is an intelligent, focused person who was stuck at home in Los Gatos while all the action was happening in Gilroy. Being a mom was rewarding, certainly, but she had gotten into radio back in St. Louis and liked it a lot. But that was in St. Louis, and now she was stuck in a small, hip, upscale town in California, which was a huge change in her life. She was a young, un-married woman, with an infant and an inner-city black teenager (Chapter 32) in a rich, sunny enclave where she knew no one, and she had an uncommunicative old man who was obsessed with the electronics of the station and spent all of his time tinkering with gadgets, half the time sleeping at the station or at the tower.

Laura Ellen was as excited by all the new music as the staff and the listeners. She loved the music, but it was all happening down in Gilroy, and nothing exciting was happening in Los Gatos. Office work! Housework! Babysitting! Cooking! Uchh!

Laura Ellen made the trip to Gilroy almost every weekday to do the work that couldn't be done at home, and Chris needed her for the weekend shifts. She kept Elsie with her while she worked the controls, played the records and cued up the next record and went looking for the next record, often with Elsie squirming on her hip, their faithful dog Bernice always asleep somewhere nearby. Another problem was that the station didn't come into Los Gatos, so she never heard the music that she needed to play when she was on the air, and not listening to the other jocks was putting her behind in the learning curve. As the owner's old lady, and as part owner herself, she could not afford this disadvantage, and she wanted more air time and less office work. But Chris Feder was a problem.

Chris was impatient, intractable and assertive, undoubtedly exacerbated by his use of cocaine. It was hard for Chris to do what he wanted to do and still care about what other people thought, or about their feelings. He focused only on the station, so that it ran efficiently. Most of the shifts were filled now, and every morning he'd get there at seven and do all of the management work until 4:00, when he'd settle into his Afternoon Drive shift. If he mentioned the Green Hut on the air, they'd send over a pizza and some beer. He'd drink, play some music and take his mind off the problems he had to deal with the rest of the day.

Feder thought Laura Ellen was the weakest DJ on the air, and she didn't fit into his vision of what a female jock should be. Tiny was a honky-tonk gal, and Kathy was all right and getting better, but Chris and Laura Ellen just didn't get along. Laura Ellen was a soft-spoken, insecure Mid-western girl with an aversion to confrontations, and Chris was a brusque, outspoken, confrontational New Yorker who liked his power. Laura Ellen disliked confrontations; Chris was aggressively confrontational.

Laura Ellen remembers one thing fondly of Chris Feder: she told me that educational radio in St. Louis never taught her how to be a DJ, an entertainer, and Chris was the basis for her learning that skill. Chris sat with Laura Ellen when he first got there and worked with her on her programming, teaching her how to select cuts that flowed well together, no matter how bizarre it might look on paper. It was about the flow. It needed to take the listener somewhere pleasing, perhaps unexpected, and it needed to flow, to sound good. Laura Ellen was getting it, he supposed, but Chris held out little hope for her. And he didn't like her.

And Laura Ellen wasn't used to being spoken to like that. Chris' arrogance prevented him from being aware that Laura Ellen hated him. He had no idea that she had fallen in love with her shifts and wanted to be a personality. She wanted to be "Cousin Laura."

Chris used Laura Ellen on weekends because he needed those spots filled, and he kept her there because Jeremy had asked him to keep her on the air. She wanted a weekday shift, but Chris was unmovable on this, and he wouldn't give her what she wanted. But Laura Ellen was the owner's old lady, and everyone at KFAT could see what was coming.

Except Chris.

Chapter 30: The New Transmitter

KSND, and now KFAT, had been broadcasting from atop Mt. Madonna, about a half-hour away from the station, on the way "over the hill" to Santa Cruz. It was an okay hill, as hills went, but Jeremy wanted to move the transmitter and tower to the top of Loma Prieta, the tallest peak in the Bay Area, and he was still waiting to hear about it from the FCC. Other stations had their transmitters up there, and the tower had several stations broadcasting from it. Then the FCC wrote, asking for readings.

The FCC needs information from you when you apply for a change of power or location. They want readings. They want to know the coverage area that would be affected with the new power, transmitter, and tower location. The change alone from Mt. Madonna would make a significant difference, but with added power, KFAT's coverage area would grow substantially, and the FCC wanted to know how far it would go, and which existing signals would be affected.

Any station affected by the change had a right to complain to the FCC, who would then take its sweet time exploring the problem and making a decision about whether or not to give final approval to the change. Jeremy was most afraid of resistance from KSAN, San Francisco's dominant rocker. KFAT was broadcasting at 94.3, and that would change to 94.5 with the move, and KSAN was the next station over, at 94.9, and they would certainly complain. If they noticed.

When you file for a license, you submit a "Proof of Performance" form, which assumes that you already have your transmitter built and in place, and you've done the engineering tests to show that it meets all the technical requirements. Yes, he would need hard data to submit to the FCC, but this was a Jeremy Lansman Operation and how he ran his station wasn't taught at engineering school.

Late one night, he shut the station down and, with a spare antenna from a friend, put his transmitter in the bed of a borrowed pickup and drove up to Mt. Madonna to take readings.

"Mt. Madonna," you ask? "Don't you mean Loma Prieta?" Stay with me.

He tied the antenna to the tower that was already in use by other radio stations on Mt. Madonna, did the tests, undid it all and put everything back to how it had been. He could now submit the test results and receive the final OK from the FCC. Jeremy knew the FCC would decide on the change without ever knowing which site Jeremy had used for his data. He was supposed to test the gear at Loma Prieta, but the FCC would review the data and never know where it came from.

Why the undercover shit? Why the subterfuge? Mt. Madonna was a good location, but Loma Prieta was the best. It was the highest you could go in the Bay Area, and if he could broadcast from there, a strong signal into San Jose was a piece of cake. The signal would go up the coast, into San Francisco and who knew how far north? It would go down the coast to the farmers in Salinas and Monterey, and who knew how far south? Or east? Or west, out to the Pacific Ocean and the local fishermen.

KSAN would hate it if it happened, but if they didn't know about it, he was golden.

Knowing KSAN would complain if they knew about it, Jeremy snuck his way up the wrong mountain and did his tests in the middle of the night, and no one noticed. He told the FCC he'd put his tower on Loma Prieta and pointed it at San Francisco, and if no one complained, then it might be approved. It would make his station worth ten times what he'd paid for it, and make him a player in a major market for the peanuts he'd spent.

Late that night, at the tower on Mt. Madonna, with most of the work done, Jeremy was tweaking his equipment to get the best sound and the best coverage from his signal, and he wanted some feedback from the station. Tiny remembers the call: "He used to spend the night up at the tower, you know, and he called me and he goes, 'Okay, I just made the signal blah blah blah stronger. Where do you want to point it?'

"I said, 'You're asking *me?*" and he goes, 'Yeah. Where do you want me to point it? San Francisco or Fresno?'

"I said, 'I can't decide. How about both? I want both.' He goes, 'Okay, both it is.'

North and east it was, not west. Fuck the fishermen.

Luckily, no one complained after that middle-of-the-night test, and now the FCC would almost certainly approve the move. Excitement was rampant at KFAT. The jocks knew they were going into the big leagues, ready or not. Many of them had been there for over a month, and they all thought they were ready. Chris was readier than most. This was his moment.

Within six weeks of Chris' arrival at KFAT, they were fully staffed, on the air 24 hours a day and selling $5,000 a month in advertising. That was great and all, but that wasn't what they were all waiting for. The tower was going to move to Loma Prieta. They could feel it. The big time was upon them, and they were all waiting to hear back from the FCC.

Chapter 31: More Early Stuff

Let me ask you a question: If you lived in a studio apartment and had some dogs, a couple of cats, a ferret, some birds, some tropical fish, a monkey, a wolverine and a marmoset, could you ever say that things were under control at your place? Observe:

Most of the shifts were filled, but things were never easy. Bob was drunk a lot, and Chris was drinking more than usual—but keeping his balance by doing cocaine—and he couldn't or wouldn't control that. Tiny had a crazy drunk boyfriend who had a habit of knocking her around, she had perpetual car trouble and often had to hitch-hike to Gilroy with her kids who she brought with her to the station every day and who needed attention. She was steady, but insecure that this golden opportunity would go away as quickly as it had materialized, and she was willing to do what Chris asked of her. She was getting it and sounding better, anyway, and she seldom felt the brunt of Chris' wrath. Unbeknownst to Chris, it was because she was so insecure that she rarely spoke on the air. Chris wanted to take advantage of her honky-tonk nature, but that would come, he felt, when she got more comfortable with the gig. She just needed to get to work on time, but hitch-hiking to Gilroy from Santa Cruz took as long as it took. Someone would have to extend their shift until she got there. They all did. That's what you do for family.

Terrell Lynn tried doing a shift on LSD and couldn't keep it together, and with that experiment in radio under his belt, he was a steady, dependable jock. Pot, beer, Jack Daniel's, whatever came his way during the course of a shift, became part of his routine, but never to excess, and he always held it together. He knew the old stuff, he was spreading out into the new, he needed little guidance, and it was getting so that Chris only needed to talk to him occasionally. He knew the music, he was easy-going on the air and easy to work with, and he was good.

Mark Taylor, the Ozone Ranger, was getting better. Rare among this staff, he had several years of radio experience, and he was an eager student of the new format. Of course, he stayed stoned for most of his time at KFAT, but his good radio instincts coupled with his willingness to learn fit right in at KFAT. Everyone was learning the music back then. The fact that he usually showed up within a reasonable time for work, and the lack of complaints from listeners and sponsors, kept the wrath of Chris at bay.

Chris had to sleep sometime, and that typically occurred while the Ozoner was on, and if Chris didn't *need* to listen to him, he didn't. Thus, Chris let the Ozoner work in peace.

Sherman… well, it was best just to leave Sherman alone.

At first, Kathy, now "Sully," kept playing songs that Chris thought were inappropriate and he was pissed at her, but he needed her to stick around and get it together. Now she was getting better, she charmed you when she talked about the music, and he loved her playful voice. He was getting mad at her only once a shift now. Progress was everywhere.

Six weeks in, and KFAT was reliably on 24 hours a day. Chris had a full staff, a public affairs show, a gospel show on Sunday mornings, a bluegrass show on Sunday nights, and the station was paying its own way in advertising sales. Then Jeremy came up to him and said, "You have too much power. We need to have some rules."

"Well, I don't think so," Chris said. "If you want to change anything, I'm out of here."

"No, no, I don't want you to leave. You're really good at this, but you're making a lot of people unhappy."

"Like who?"

"Laura. And she's making me unhappy."

"Well, Jeremy, you'll have to manage that situation. It's not for me to manage." Chris knew that Laura Ellen was unhappy, but he didn't know how unhappy she was. Jeremy stood there, stumbling, mumbling, doing his "weird calisthenics," and said something again about "too much power." Chris suggested to him that he, Jeremy, should manage the station, and that he, Chris, would just do an air shift.

Jeremy said "No." He didn't want that.

"Well, then," Chris said, "get out of my shit."

Two days later, Feder told Jeremy again that they should try the new arrangement, and he became just a jock, which he really enjoyed. No more getting there at seven a.m. and staying until 10 p.m. No more calls for record service and worrying over contracts and all the paperwork and extraneous bullshit that went with being the Station Manager, the Music Director and the Program Director. KFAT was staffed and making money. They were over the hump. His deal was for an ownership in the station if he could pull it together, and it was all but pulled together. Sure, there were rough spots, but with this crew, rough was the new smooth, and it would only get better, the more experience these people got under their cowboy belts.

The approval to move the transmitter was ever closer, and then the station would really make an impact. This time- now- was the trial run for the new format, and secretly, Chris was grateful for the out-of-town tryout before they hit the big-time. They weren't ready yet, but soon they would be. And it would be his show and he would be a part owner.

It was September, 1976, and while NBC's "Saturday Night Live" had been on the air for almost a year, these guys, these players from Gilroy, they really *weren't* ready for prime time. But with Chris' guidance, they'd be ready.

Then the FCC notified Jeremy that his application to move the tower and transmitter was approved, and he could move it as soon as he was ready.

Everyone had been waiting for this, and they were all ready, ready or not.

Chapter 32: Moving the Transmitter and Raymond*

It took about a week to get all the parts and all the schematics together, but one night (maybe) Chris or (probably) Bob and Jeremy and Raymond* loaded the transmitter into a pickup truck and drove up the long dirt road to the new transmitter site in the pouring rain. It was Jeremy's unbridled determination and his inability to recognize the discomfort of others that kept them from going home.

It was an unrelenting, drenching rain. And cold. They were miserable. But this was a big night for KFAT, and Jeremy was going to move the transmitter tonight. Tonight!

Jeremy knew the site, knew it well. He'd been there before. Remember, this wasn't regular radio with regular radio people: this was a Jeremy Lansman Operation.

At the site, Raymond had been elected to climb the tower, and with the wind ripping at his clothes and shaking the tower, Raymond was turning paler than you would expect from a black guy. He was terrified, but Jeremy kept shouting, driving him and (maybe) Chris or (probably Bob) on.

It would take all night, but the morning sun would shine on a completed job. On Monday morning, KFAT would be broadcasting from atop Loma Prieta, and a new life was going to begin for everyone at the station. And Chris would own a share.

The sun came up, the rain stopped, the work was done, and Jeremy and Chris were having a beer at the engineer's shack. They should have been jubilant, but Jeremy began to cry. "What's wrong?" Chris asked. "Everything's working out well. We moved the tower, we're going to blow it up in San Jose."

Jeremy told him, "You've worked so hard and it's not going to work out with you and Laura Ellen and Laura Ellen's my wife, and…"

He went on about how sorry he was "about all this," and Chris was stunned into a rare silence. He thought for a moment about what he had just heard and said, "What are you talking about? Part of my deal is that I would get a piece of this station if I made it work, and now you're telling me that I'm not?"

"Well, I'll have to go back and talk to her."

But Chris knew it was over, that Laura Ellen "had already decided that I was gone. They thought they didn't need me any more."

Chris was blown away. He "got them a staff and trained them, and the format, and gave them everything you could want at a radio station."

It was over for Chris.

When he came back to the station for the last time to collect his stuff, the staff that were there, most of whom he had hired, asked him what they should do, and Chris told them, "keep working. It's a great gig and you're having fun." Laura Ellen and Jeremy asked him for the records that Chris had acquired while at KFAT, and he told them to go fuck themselves.

The next day, he packed up his apartment in Santa Cruz and went back to Corte Madera.

Who would be KFAT's new Music Director? Who would do Afternoon Drive?

Economics demanded that whoever it was, he or she would have to come from the existing staff. He'd have to know the rockin'-est cuts for Afternoon Drive. The new Music Director would have to be forceful. Forceful? Laura Ellen must have felt a deepening sense of foreboding as she narrowed the list down to one.

---

*Who is Raymond? I'm glad you asked.

Back in St. Louis, KDNA was in a truly seedy part of town, and Raymond was a ghetto kid who hung around the station. Laura Ellen thought he probably stole stuff from the station, but nothing major ever went missing, so they allowed him to hang out and do odd jobs in exchange for food and whatever. His mother had a lot of kids, Raymond wasn't getting enough food or attention, and one day his mother turned him over to the authorities because he had broken into the family refrigerator and eaten more than his share of the food. Laura Ellen and Jeremy had taken a liking to the boy, and they worked with the Social Services worker to keep him from going into the system.

That happened while they were selling the station and packing things up for the move to California, and it didn't look good for Raymond's chances if they left him behind, so they worked out a deal with his mother and the social worker, and they took Raymond with them to California. He was almost as old as Laura Ellen, and there was no formal agreement, and no papers were signed. Raymond just packed up and left with Jeremy, Laura Ellen and Elsbeth.

Jeremy and Laura Ellen were completely unprepared for dealing with a troubled urban teenager not much younger than they were. Straight from a slum, they put him in school in Los Gatos, and as the only black kid in a wealthy, sunny suburban California town, it was a maximum of culture shock for young Raymond, and neither Jeremy nor Laura Ellen had any idea what to do with a kid in high school. The relationship gradually fell apart over the years, but although he got in trouble with the law soon after leaving Laura Ellen and Jeremy, at least he went to the University of Santa Cruz and graduated with a degree in something. No one I spoke to has any idea where Raymond is now, or what he is doing. Raymond: get in touch!

Chapter 33: The Dispute About Moving The Transmitter

Buffalo Bob says he helped Jeremy and Raymond move the transmitter. Chris Feder says it was *him* and Jeremy and Raymond. Jeremy doesn't remember, no one's seen or heard from Raymond in years, so I don't know. One point each for Bob and Chris.

Both Bob and Chris tell the same story about putting the transmitter, which both remember as about the size of two refrigerators, into a pickup truck (Bob says it was his truck, so one point for Bob) in the rain, and driving up the badly maintained muddy fire trails along the spine-like ridge of hills that lie between the Pacific Ocean and the inland valley.

There was a 30-foot tower there, and several stations were already broadcasting from the site. KFAT had been approved to join the antennae already there, so Jeremy and crew were going to attach his antenna but not interfere with the broadcasting of any of the other stations using the tower.

Both Bob and Chris remember the muddy roads, the potholes and the occasional gate that someone had to get out of the car in the constant downpour and fierce wind to unlock, then get back into the truck, so one point each. Both men remember Raymond being elected to climb to the top of the tower in the rain, with the wind so powerful that it howled, and that he turned pale with fear up there. Both men remember all that, so points to both.

What Chris remembers was what happened right after this, but that would place it back in February, and most reports claimed it happened in September. One point off for Chris. Except if Chris was with Jeremy at an earlier date, maybe for a test. That would account for it. Except for the part about Raymond, which both men remember, and it would then be necessary for Raymond to have been frightened enough to drain the blood from his face, twice, each time on that tower. Judges say: half-point off for Chris.

Cuzin Al remembers that the transmitter moved in September, 1976, and everyone agrees that "Cuzin Al is good with dates." In any event, the tower and transmitter were now on Loma Prieta, the power was cranked up, and that was what everyone had been waiting for.

Now KFAT was going into San Francisco. Not everywhere in San Francisco, but suddenly they were getting phone calls from there, so they knew. All of the callers were surprised, most were happy, and a few were angry. It seems that in some of the outlying areas where KSAN was heard, KSAN's signal had been crowded out by the new signal, and those listeners woke up one morning to hear an entirely alien form of insanity, and many of them wanted to know what the hell was going on.

The KFAT staff was delighted, delirious, excited and motivated. They were in the big time now, and there were smiles and laughter in Gilroy. There was a stable staff for the weekday shifts and the weekend shifts were all covered. They had a talk show and a gospel show, so their obligations to the FCC were all met, and now they could hire a sales staff and make some money to pay for it.

It was September, 1976, says Cuzin Al, so Chris was gone. Buffalo Bob wins.

Chapter 34: The New Regime

When Laura Ellen moved to Los Gatos with Jeremy and Elsie and Raymond and Bernice the dog, she thought she "wanted to be a mom and stay at home and cook three squares a day and mop the floors." She "mopped floors until I couldn't stand mopping floors any more, and when Larry Yurdin got thrown out on his ear, I was thrilled to death. I could go back to work."

Thus, Chris Feder had inherited an ambitious woman who was a part owner and living with the principal owner, and who wanted to do the afternoon drive shift. Chris kept her on weekends and fill-ins, but she wanted more. Now he was gone, and she was once again eyeing the afternoon slot. But she was also working the office, doing what Enid couldn't: traffic, billing, answering phones and doing whatever else needed doing, and there was always something else that needed doing. In her heart, she knew that getting it all done by 3 or so, and then doing a four-hour shift, all the while being the primary caregiver of her toddler daughter was going to be too much.

Besides, the occasional label rep was now showing up, and they needed to be dealt with. If she had to be back at the station by 2:30 for her shift, she was going to miss out on a lot of key connections, not to mention the lunches and meetings. There was just so much promise with this new station. She knew that the possibilities out-weighed her desires for the Afternoon Drive shift. Casting an appraising eye over her group of in-house miscreants, she realized that for dealing with the outside world, she was probably the best person to represent the new station, the one best-suited to talk to the suits.

She was tired of being at home, and she was showing up at the station every day and taking care of its business. Still in denial, she wondered who would be the new Music Director? Sully was too unsure of herself, and Tiny, with her two kids and all her other problems was clearly not the answer. She didn't want Gordy to do it, and Terrell Lynn just wasn't forceful enough.

Chris had been good at his afternoon slot because that time of day called for more rockers and less mellow stuff. Chris didn't go real deep into the library at KFAT, and he played the rockin'-est cuts, and he was a forceful presence on the air, so that worked. The obvious replacement for Afternoon Drive would be Bob Cassidy, and reluctantly, Laura Ellen knew it. She needed someone forceful for MD. She needed forceful? It had to be Bob Cassidy, God help her.

Buffalo Bob Cassidy? Scofflaw, bar crawler and brawler? Bully, criminal, serial abuser of intoxicants? Angry, troubled and usual suspect? Sure, but Music Director? No, really: please, God, please help Laura Ellen.

Bob took over Afternoon Drive, calling it "the Bison Boogie" and also became MD, and he doesn't remember anything he actually did with the title. He remembers talking about it one time with Laura Ellen, but he generally paid no attention when he spoke with her, and although he knew he was now the Music Director, he swears he didn't do anything to justify the title. The new job didn't increase his pay any, but neither did it take away any of the time that he normally put into his drinking. He didn't make any phone calls, didn't hold any meetings or issue any directives. He didn't talk to anyone on the staff about their shifts or their work, and he didn't send out any notes. He didn't like most country music, and as long as they left him alone and let him play the rockin' cuts, he didn't want to tell anyone else what to play.*

Without any direction, the jocks started writing comments about songs on the backs of the album covers.

Bob remembers that his employment at KFAT was so loose that he doesn't think they even had his social security number for a year and a half. What's Latin for "Loose By Any Standard?"

*In college, they teach you that this is called "Laissez Faire." I only mention this here because I took the trouble to learn it, and forty years later, only one situation has ever arisen where that knowledge has been of any use, and this was it, so I used it. Personally, I am amzed that I even remembered it, but thanks for reading this anyway.

Chapter 35: The Learning Curve & Family

It was the Time of The Learning Curve.

Laura Ellen was taking Larry's idea of Texas music with a little bit of progressive country, adding a lot more progressive country, plus blues, folk, bluegrass, roots music, western swing, off-the-wall stuff and comedy. Those were her ideas when Chris Feder had been in charge.

Chris had listened, as much as he listened to anybody, and agreed with her. He told the staff to add the progressive stuff, the folk, the blues, all of it.

With Chris gone, Laura Ellen made Bob Cassidy the Music Director, but she knew that he would do little with the job, and she began to exert control. Let us consider, however, that there is a distinction between *exerting* control and *having* control. Laura Ellen was a very intelligent woman, and she quickly deduced that control at KFAT was going to be a very delicate, moveable thing: hard to define and harder to see with the untrained eye.

On an elementary level of professionalism, most radio stations would be shocked to have a DJ show up for their gig less than an hour before airtime. At KFAT, showing up *by* airtime was a blessing, and Laura Ellen could breathe again. It meant that, for a few hours at least, she would have no frantic phone calls to make, trying to locate said DJ, and it meant not having to have with her, for another few hours, her list of emergency fill-in people, so she could now devote her time to other pressing matters, of which there was never a scarcity.

In the first months of the new staff, copy for ads was being driven in at odd hours, and there always had to be someone there to produce the spots when the paperwork got there so the ad could go on so they could get paid for it. She could worry about that.

Or, when she first got there in the morning, she could worry about what the jock had done the night before. This could mean a phone call from an irate sponsor, an angry women's rights group or a representative of some recently maligned organization, an indignant minister, or somebody pissed off about something. Or something else.

Or she could worry that she was low on cleaning supplies again and she'd have to find someone to send out to buy some or that smell would be there all day.

Her daughter, Elsie, needed to be attended to, fed and amused.

That Tiny's two kids were often there playing with Elsie was either a mixed blessing or a mixed curse, and if she chose to, Laura Ellen could worry about that. But mostly she worried about the bills. They never stopped coming in, and KFAT was only now showing the *promise* of paying for itself.

Laura Ellen was only sometimes getting home to Los Gatos for some rest, and sometimes she passed out on a couch in one of KFAT's back rooms. Usually, this happened during the daytime when Elsie was in daycare, so the jock had no trouble keeping the chaos level down in deference to the boss being passed out nearby. She needed her rest, and everyone saw Laura Ellen as a new hybrid of friend, family and boss.

She was management, there was no doubt, but she was a sweet-natured woman of the hippie persuasion; she was bright and passionate about the station and about the music. She cared deeply about the staff, and they knew it, and she successfully mixed the boss with the friend. The KFAT staff was becoming a family, a weird sort of a family. Was it a business or was it a commune? Was it about the music or about the money? Was it about good radio or good profits? Everyone cared passionately about the music, and even the staff worried about the profits, but they had no say in the ads. Everyone was getting along and it was a gas to be at KFAT. Even visitors felt it, and felt that way. KFAT was loosely run, but whatever central governance there was at KFAT, it was centered in Laura Ellen.

Everyone respected Laura Ellen except for Bob Cassidy, but Bob was grudging in his respect for anyone, and as much as everyone liked Bob, everyone had at least some problem with him. He was irascible, confrontational, loud, a drunk and a bully. I remember Bob having the Ozone Ranger in a headlock one day outside the stacks. Which was different from the time Bob had him by the throat in the hallway. And everyone knew the story of when Laura Ellen woke OZ up when he was an hour late for his shift and the record was going ffftttt....ffftttt... and when he got there he found Bob passed out on the console, his left hand on the left turntable, clutching a gun and his right hand holding a bottle of Jack Daniels. OZ turned down the monitors and tip-toed through the next two hours, playing records on one turntable. No one was gonna wake up Buffalo Bob with a gun in his hand.

But what a great guy! I really liked him, and hung out with him quite a bit. We got along great, and I used to call him "Gruffalo Bob." Behind his back, of course.

Sherman says, "I love Bob like a brother, but he was more of a problem for everybody, and left shit in his wake. Not showing up, or showing up shit-faced, or violent. Busted all kinds of holes in the walls. Never hit anybody."

Sherman: "He kicked and bitched and moaned about playing the country stuff. He wanted to hear Jeff Beck. Did a sloppy show... dead air between cuts. Laura Ellen had a tough time controlling Bob."

Once, I was talking to him in the studio just before a song ended and the segue was coming, when all conversation with a jock must stop. He was sitting in the DJ chair, concentrating on the end of the song, waiting to start the next record, and I stood to his left and, just as the song ended, I asked him a question, ending with, "What do you think of that, Buff?"

Buff?

He spun around and looked at me like he'd just heard a gunshot, raised an eyebrow, and scowled "yeah?" Keeping the scowl, he leaned toward me until I stepped back, and then he turned back to the turntable and started the next record, ignoring me. I did that once.

Other than Bob, everyone else liked Laura Ellen, but she was never one of the working stiffs at KFAT, because no one was making any money and she was management. Jeremy drove an old beat-up van and Laura Ellen had a dusty, dented old yellow Volkswagen hatch-back. No one had a fancy car, except for the sports car that I drove. It was a Volvo P-1800, perhaps best remembered as the car driven by Roger Moore in "The Saint" television series. I was from New York and had a Master's degree, and I talked real good. Besides, the car wasn't worth all that much, it was just different. I was different, but I was definitely one of the crew at KFAT. I was struggling, working the angles to keep my show on the air like everyone else. I was making sacrifices, and my shows were Fat. I was just different. I had a gorgeous blonde girlfriend, Susan, who was a lot of fun. We were both part of the family.

Cuzin Al was also more a part of the family every week, hanging out more, and smoking more pot.

Except for Cuzin Al and I, the staff lived in and around Gilroy, and there was nowhere else to go other than Augie's and the Green Hut. Everyone was either at home sleeping or at one of those two places, or at the station, so they saw each other all the time. It became very insular for the staff, and because money was so tight, anywhere they'd go, they'd go because it was a KFAT event or at a restaurant where the station had a trade-out and they could eat one meal a day for free, so they even saw each other whenever they went out. There was serious bonding going on, and Laura Ellen and Jeremy and Elsie were part of the family.

Talk at the station was all about the station, constantly. It could be about this or that song, or about who was fucked up, or who had said what to whom, or what cut had knocked them out, or what trouble this guy was, or was in, or what are you gonna do without the money for this or for that? Or it was about the transmitter or the move or the power or the new ads or the meters or the logs or the new salesperson or the event coming up or the event just passed, or the new bumper stickers or what someone asked for on the request line or some trucker stopping by or the new crepe at The Crepe Place or what someone had said on a Fat Gram, or it was about Frisco, or the new station ID, or it was too bad about Karl J moving on, or about people not putting their records away, or about Gordy's regular on-air bitching about his illnesses. Or something else.

GK: So once you guys were happening, your shifts were up and running, was anyone giving you direction?

Sully: "Once in a while they would try to, like, well, Laura Ellen would pick songs to put in the hot box, remember?"

TLT: "Oh, yeah."

ST: "I'm sure that she told me I was playing too many slow songs…"

TLT: "Well, we'd tell each other that. When we had meetings. And Jeremy would come. Hell, he'd get a bottle of Jack Daniels, and some beer and shit, and marijuana. Fuck, I don't know where he'd get that." Everyone remembers that Jeremy didn't smoke the pot or drink the Jack.

It was Laura Ellen who tried to run the meetings. The typical meeting was loose, informal and difficult to define. Meetings were called for a back room, but they also happened in the hallways, in the Production Room, at the Green Hut, a concert, wherever and whenever the jocks ran into each other. Whatever it was, it was all about the station, all the time.

Naturally, love found its way to the staff, and various members of the staff formed romantic relationships that came and went, sometimes lasting days, weeks or months, sometimes lasting only a night.

Another way that KFAT was different was that no one there was paying attention to the charts, which was almost unprecedented in commercial radio in America. Radio stations have a format, and they constantly monitor how songs are doing in other markets, so they can add songs to their playlists that are proven winners. When a station that reports to a trade journal adds a cut, programmers all over the country read about it and pay attention. When the cut gets a good response and goes into "heavy rotation," the station reports that, and other programmers take note, sometimes adding that song to the playlists of the stations they consult. Stations hire format programmers to make recommendations for which records would get played, then it would be up to the MD or the PD to determine how many times a day that record was played. Sales reps from the record companies would call or come by to push their new product. That was radio in America. But there wasn't nothin' like that at KFAT. A *record rep* telling a KFAT DJ what to play? *Fuck them!*

There was no one telling them what to play. Laura Ellen changed what Larry had set up and Chris had tweaked. She had her traditional country along with her progressive country, and she had the swing, the comedy, the folk, the blues, the everything else. No one in the country had a handle on what was happening at KFAT, but the format-that-could-not-be-named was making a name for itself.

Musicians were telling other musicians about KFAT.

Truckers up and down the Pacific coast were telling other truckers to tune to 94.5 when they got to King City going north, or Sonoma or Marin when going south. Musicians were telling their audiences about KFAT from the stage, and calling in and introducing themselves when their nearby gigs were over. Sometimes they'd stop by if they were anywhere on 101, which went through Gilroy at first, and then right by Gilroy when they finished the new section of 101.

KFAT was so popular with the truckers that Jeremy installed a CB radio in the studio so they could call in with requests. The jocks were always responsive to requests, and many say that they let whole shifts go by being guided by requests. They'd play a request, and then follow that wherever it led, until they'd run the course of that series, then take another request and run with that for a while.

Except for Bob, who refused to let anyone tell him what to play. But even Bob would sometimes listen to a request completely before hanging up on the caller and ignoring the request. Sometimes, Bob told them he'd think about it. That was Bob being polite.

The jocks talked about the requests and speculated on the tastes of the people out there, as that is what the people out there were, "Out There." The jocks were becoming "Here," and the listeners were "Out There." For the KFAT jocks, it became an increasingly insular world of the people at the station and the people not at the station. The Us and Them became clearly defined, but it wasn't so clear about Laura Ellen and Jeremy.

No one had any money, including the owners, and that made it easier to be family with them. The jocks were unhappy with their $3.25 an hour, but who knew if Jeremy and Laura Ellen even made *that.* They were all in the struggle together. And they were cool people: they cared about each and every member of the staff and treated them like strange family. They were hippies and anarchists, those two, and some of the shit that the staff was doing was all right with them, which everyone knew wouldn't be true at any other radio station in America.

Laura Ellen and Jeremy were family, like everyone else at the station, and it certainly wasn't like they were the creepy, unappreciated part of the family that every family has to put up with. No, that would be the sales staff.

So Laura Ellen and Jeremy were family. They were in this with everyone else.

And everyone on the staff felt good about their jobs. They were doing what no one had done before, they were doing it well, and they were doing it for practically no pay. Who else would or could do it? Sully says, "I remember it was like in waves. When I first started there, there wasn't very much in the library, and we were all getting kind of educated at the same time, and somebody would come in and go, 'Look! These guys are really good! They're rockabilly.' And I didn't know anything about rockabilly, and we'd go through waves, like everybody would get really into rockabilly and we'd all run out and buy all these rockabilly records and then run into the Production Room and listen to them, and then two weeks later somebody'd go, 'Wow, have you heard about Zydeco?' and then we'd be out running around buying Zydeco records and running in and listening to 'em."

Terrell Lynn: "And a million times I'd sit there and have it on cue, and go, 'No! Another one would be better!' And I'd take ten seconds and I'd run into the library…"

Cuzin Al: "And we talked about music every time we were together. We were always talking about music. And this guy (pointing to Terrell Lynn) was out finding records and at least once a week he'd call and say 'look what I found!'"

TLT: " 'Jim and Jesse Live In Japan.' Wanda Jackson on a Sears record."

Sister Tiny: "I remember when I brought the first John Prine to the station, it was my album."

GK: "So how many records did you buy, and how many records…?"

ST: "How many records could we afford?"

With the sense of excitement and the permission to experiment, musically and chemically, the staff may not have had money, but they had each other and they had their jobs, and soon they understood that they weren't about to lose these cool new jobs, that this was gonna be it for a while, and they began to settle in, doing good radio.

Chapter 36: Fat Grams

Back in St. Louis, Jeremy's position had been that everyone should have equal opportunity to state their ideas on the radio. KDNA was a non-commercial station given to him by the government, and he felt that anyone who had something to say should have a place to say it. That was why he invited the communists, the Nazis, the John Birchers, and whoever else wanted to say their piece, onto his airwaves.

KFAT was different, of course, but Jeremy still wanted to allow the public to have a voice on his station. He and Chris Feder came up with the Fat Gram, which was a phone line with an answering machine.

Jeremy's policy was that the jocks were being paid for a full day, and if their air shift lasted only four hours, then they should do other work to fill up the day. That work included production, recording, scheduling, cleaning, answering mail or some other chore, like writing up Public Service Announcements from the flood of flyers that came in to the station from schools, clubs, groups, organizations, bureaus and departments. The jocks took the flyers, wrote them into announcement form, and put them in the book under "PSA," and they read one every time "PSA" appeared on the log.

Jeremy hadn't stiffed the Santa Cruz phone company yet, so a Santa Cruz number was arranged, and an answering machine was installed. Promo's were cut inviting the listeners to call and leave a message. Terrell Lynn was assigned the job of editing and transferring the messages onto reel-to-reel tape, and then creating a cassette with about a minute's worth of messages, ready for broadcast.

Fatheads still remember the intro to the Fat Gram. Whenever it was scheduled, listeners would hear an annoyingly strained upper-register voice asking: "Is there anybody there? I've got a Fat Gram for ya! … Fat Gram… Fat Gram…" Then they'd hear a listener with their message, followed by the announcer again giving the number to call to record a Fat Gram, which were scheduled on the logs for 12-15 times a day.

The Ku Klux Klan called and left a message once, and Terrell Lynn put them on the air. In the Fat Gram, they threatened violence against people, and so many Fatheads called in and told them to shut up in other Fat Grams that they never called again.

Few people called with profanity or joke messages- and they never got on. The listeners responded immediately, with surprisingly few prank or bogus calls. People offered items for sale, gave criticism or praise to a jock for a song or an attitude, announced events and fund-raisers like car-washes, book signings, bake sales and the like. Commercials were not allowed, but it was pretty open, and it was up to Terrell Lynn to choose what to put on the air. It was a popular feature and Jeremy got some Public Affairs time on the schedule.

The only problem was Cosmic Lady, who drove Terrell Lynn nuts. Cosmic Lady lived in Santa Cruz, and called the Fat Gram line ALL THE TIME. Terrell Lynn remembered that she "rambled on forever. She left long political, conspiracy theorist crap. It was her world. Cosmic Lady wasted whole tapes." Terrell Lynn would go on the air and plead, "Please, Cosmic Lady! I'm the guy that has to edit these tapes. We're trying to let other people on the air. Please shut up. But she never did."

Chapter 37: Dork To Demento

Jeremy was a bit of a freak. He was constantly barefoot and wild of hair. His eyes could be looking at you or bouncing around. Gentle and sincere, he was often sincerely confused. Dealing with people made just too little sense, as opposed to, say, dealing with transistors, resistors, watts, amps and ohms. The rules governing electronics were satisfyingly consistent, but who the hell knew what would come next with these people?

Jeremy was never as much at ease as he was with a soldering iron and a schematic drawing. Or better yet, a soldering iron and an *idea.* Logic, physics and common sense were the rule of law with electronics, and having to talk to people was a guessing game at best.

It was his station, and he wanted to play a bit. He knew that commonsense rules applied to running a station, and he couldn't just walk in and take over someone's shift whenever he felt like it: bad for the listeners, bad for the advertisers and bad for the jock. He'd have to have a regularly scheduled time for his playtime, and so he gave himself a Sunday night slot when most people wouldn't be listening, anyway. He knew that was best, because he knew from experience that most people wouldn't be interested in whatever he wanted to play, anyway. And for those hours, he wouldn't be paying anyone. It was win/win, except for the listeners.

One of the interesting aspects of my research for this book was the different memories people have of Jeremy. Some don't remember this, but others remember that when talking with Jeremy, he'd look into your right shoulder, and mumble. Some remember him looking them in their eyes, others remember the shoulder thing.

Bob Cassidy remembers an early conversation with Jeremy standing in front of him, looking at Bob's collarbone, mumbling and doing "weird calisthenics." Others remember him twitching. Terrell Lynn and Sherman both remember an incident when they were visiting Jeremy and Laura Ellen in Los Gatos, and Jeremy invited them all out to dinner at a nearby restaurant. Barefoot as usual, Jeremy got to the restaurant by walking on cars, fences and whatever was handy to avoid touching the ground. Terrell Lynn recalls that, except for the crosswalk, Jeremy walked the several blocks to the restaurant without ever once touching the ground. He did it blithely, naturally, without a trace of "look at me." It was just Jeremy being Jeremy. Terrell Lynn remembers seeing this and thinking, "uhh... I'm working for this man?"

So, you get that he was strange, right? So what would you think his radio show would be like? Music? News? Opinions? Sports? Okay, I shouldn't have even suggested sports. I apologize, and I would take it back if I could. But the rest? What would you think he'd play? Perhaps the question should be: why would you even guess at what Jeremy would play on his show?

Would you have guessed anything to do with Tibetan Temple gongs? Or the sounds of various adhesives being peeled away from various surfaces? Duck calls? Did you really guess any of those? Yeah, sure you did, but you already knew about the duck calls!

Jeremy played farm sounds and the sound of trees rustling, frogs, crickets, coyotes, foxes and other wild animals calling to each other, rocket countdowns and launches, obscure, weird music from the 30s. Okay, I understand someone playing the first-heard blues from Alan Lomax's 1920s and 30s field recordings in the Mississippi Delta, but field recordings of...fields? Did the listeners really want to hear an album side of the mating calls of various insects in a Kansas wheat field?

World War Two airplanes? All of them? All different, all droning, all sounding the same after a couple of minutes? Indy race cars on the track?

Dedicated Zen scholars might debate the sound of one hand clapping, but KFAT jocks debated how long listeners could listen to this unendurable shit before they'd turn it off.

Tiny remembers tuning in one Sunday night and hearing some scary, unidentifiable screeching coming out of her radio, and she called up the station and told Jeremy to take that goddam thing off the air. Then she heard a needle scraping across the surface of the record, and it was gone. Whatever replaced it couldn't have been as bad, but she had turned the station off before the next... whatever, anyway. She was deeply embarrassed.

I was spending most of my Sunday nights down in Gilroy, putting my shows together. I was a substitute teacher up in Marin County, a hundred miles away, and my schedule was such that Sundays were the only whole day I could give to my work at KFAT. I had five shows a week and I needed a whole day to do as many of them as I could, and I remember many a Sunday night working in the Production Room next to the studio, while Jeremy was doing his show. This was such a new gig for me, and everything was new and strange, but the sounds coming out of the next room were incomprehensible.

Jeremy was my boss, and I didn't want to say anything, but I couldn't understand why this man, who had to be successful—he owned a radio station!—why was this man putting that unlistenable shit on the air? Who liked it? Who was he playing it for? Who was listening? What was going on?

My schedule was as full as I could handle, so I did my work, cutting, editing, writing, recording and assembling my shows with the studio monitor off, trying not to hear whatever infamies Jeremy was foisting on whichever unfortunate shut-in was still listening. I remember Cuzin Al coming into the Production Room one night to chat with me after his show, which was on before Jeremy's, and asking Al what the hell was going on. Al had no idea, he was embarrassed that this shit followed his show. He "just didn't get it," and was unable to speculate. He had his show to do, and making it good was all he was about. Then he was out of there, back to his family. But first, he believed he'd stop by and see Buffalo Bob in that back room and have one of those... joints.

Those visits to Buffalo Bob in the back room would turn out to be part of Cuzin Al's learning curve.

Tiny told me that she never listened to the Fat Dork Show, and I later found out that neither did any of the other jocks. Ever.

Sunday nights might be a wasteland on KFAT, but they always found a jock for the Sunday afternoon show, usually Laura Ellen, and they had Cuzin Al's bluegrass show before the Fat Dork Show. Al's show was on from 6:30 until 8:30 on Sunday nights, but after two months, there had been so many calls to complain about the senseless caterwauling after the bluegrass show that Laura Ellen asked Al if he would like to start a half hour earlier and stay a half hour later. After scant persuasion, Jeremy was willing to give up his show.

Cuzin Al's Bluegrass Show was now on from 6 until 9, and after it, Jeremy manned up and paid for a syndicated show by a Los Angeles lunatic who called himself Dr. Demento, who specialized in weird, off-the-wall songs. Dr. Demento's show was, and possibly still is, a lot of fun, and a successfully syndicated show in several markets in the U.S. It may have been a syndicated show, which was not Fat, but the stuff that guy played was definitely Fat.

The differences between Dr. Demento and the Fat Dork were immediately obvious: people wanted to listen to one and did not want to listen to the other. Dr. Demento became a regularly-scheduled Sunday night feature. Now, Jeremy just had to pay for it.

D'oh!

Chapter 38: Sunday Nights

Al Knoth was a straight, church-going family man. He was a deacon in his church, a husband, a father and a salesman in 1962 when he found bluegrass music.

Cuzin Al knew his bluegrass, he came in for two hours on Sunday nights to do his thing. No problems there. Chris Feder was no fan of bluegrass, but Jeremy and Laura Ellen wanted him, and that meant two less hours for Chris to worry about.

Cuzin Al didn't do schtick. That country drawl you heard on the air was what you heard at his dinner table. He was a bluegrass fan and he had a big, really big, presence. He was a laughing, happy country guy on the air. No one had as big a belly-laugh as Cuzin Al. Maybe his belly laugh is so big because no one has a bigger belly than Cuzin Al. Check the photos He started in Lorenzo's garage at KTAO in Los Gatos, in 1972, and started at KFAT in the first week in February, 1976. His shift was Sundays from 6:30 til 8:30 p.m. At 8:30, Jeremy had a show that helped set the low-water mark at KFAT for what was strange.

Jeremy was, and is, an electronics freak, and he didn't do nearly as well with people as he did with tools and circuits. But this was a way to play with the radio, while not being expected to engage in a two-way conversation. It was a way to deal with people without having to talk to them or have them talk back. And he could have his kind of fun. It was *his* station, after all, and he had his collection of music and assorted weirdness from KDNA, which was probably all he knew of the KFAT library, anyway. And while he was on the air, he wasn't paying anyone else to be on the air. For Jeremy it was, as the kids say, all good.

How good was it for the listeners? Or anyone? I can't remember one jock at KFAT who ever told me that they listened to Jeremy's show. A professor of Communications at San Jose State College called up the Fat Gram line and said that he was appalled at what happened to the station on Sunday nights after the bluegrass show, and to "please get that fat dork off the air on Sunday nights."

It was Terrell Lynn's job to make the Fat Grams, and in an act of solidarity with the listeners, he put that message on the air as soon as he heard it. Perhaps Jeremy was amused, or perhaps to show that the management of the station was responsive to its listeners, Jeremy elected to keep his idea of fun on the air, but he started calling his Sunday night flights of indulgence "The Fat Dork Show."

Listeners turned it off in droves. One jock recalled walking on Pacific Avenue, the main drag in Santa Cruz, one Sunday night and hearing KFAT coming from every store, bar and restaurant on the street. At 8:29 he was hearing bluegrass, and at 8:31 he was hearing a mixture of different stations up and down the block, as everyone tuned from KFAT to some other station. Any other station. What was the problem?

In fairness, sometimes Jeremy was working on the station during his show, fixing, testing, tweaking, and he only wanted a signal—any signal—going out so that he could read the meters to tweak something. He would be watching meters, not listening. Sometimes he didn't care what he played, he just needed to make an adjustment or take a reading, and he could play anything. But for God's sake, Jeremy, why not put on a side of Willie or Waylon? Or a comedy album? Something people could listen to?

It was the duck calls, if you'll remember, that Awest heard while sitting in his bathtub above the massage parlor in Santa Cruz that impelled him to come over the hill the next day and get involved, and now Awest had plunged headfirst into the vortex. He had come to the station to look around, get a feel for things to help him a design a logo, which he could use for a bumper sticker, stationery, posters, business cards, and the T-shirts that he was going to use to make to make some money.

Jeremy had asked him to design a bumper sticker for the station, and he had. He found, well- he stole, no… he *adapted* a logo from an old Bob Wills album cover. It was perfect. Awest drew a fat, happy, neckerchief-wearing cowboy, with sun's rays of yellow and blue behind him. In front of him was one of those big old-fashioned microphones, and he wore a white hat. It had a black, roped-in area with the station's call letters and "94 ½" and "Gilroy, California" in it.

Awest wanted to use it to make some T-shirts and stationery for the station, and he wanted to show it to Jeremy, and to hang out and get to know the people there. Don't forget he was new in the area and didn't know anyone outside of his girlfriend, Wanda.

Sherman still remembers meeting Awest. He was out on the back deck smoking something when he saw a pair of upside-down legs and some kind of tight-fitting red coat. Sherman saw only two legs and the back of the jacket, and precious little of that, as someone was head-first, feet-up-in-the-air, buried in the dumpster out back. It's called "Dumpster Diving," and we know that homeless people do it, and we've all seen people do it for comic effect on television. But who was this guy?

Sherman, pissed off at another dumpster-diving bum who would go through the garbage and leave a mess behind, called out, "Hey! What're you doing!" and this guy backed out of the dumpster with blond hair past his shoulders, a Sergeant Pepper coat, and an innocent look, like he knows it's OK to be doing this, but he knows that other people would find it strange. He had a sheepish grin, and he wore a garish red band jacket with big, round brass buttons and yellow braid. His voice boomed out a cheerful, "Hi, I'm Awest!"

Chapter 39: The Inimitable Mr. Bo

Of all the incomplete and contradictory facts, stories, suppositions and allegations that I've sorted through in researching this book, none has as little resolution as how or when "Frisco" Bo came to work at KFAT. But everyone knew why.

There were three staffs at KFAT. The middle staff: Terrell Lynn, Laura Ellen and Buffalo Bob and Sully and Sister Tiny, Sherman, Ozone, myself, Cuzin Al, Rocket Man and others, will always tell you wild stories about Frisco. Whatever insanity at the station is being recalled, no one ever has no stories about Frisco. He and KFAT were made for, destined for, each other. But Frisco took it to the limits, and his brakes were going to fail.

Everyone knew that he was driving too hard, but what was anyone to say? By the time it got too obvious to ignore, he was no longer reachable. Then he'd crash and Jeremy would send him to a cheap place to dry out. Everyone pitched in and did a little more, and KFAT got the work done for a couple of weeks until Frisco came back and the cycle started again.

Everyone, without exception from that second staff at KFAT, has a special place in their hearts for Frisco. He was wonderful, he was brilliant, and he was tragic. As DJs are wont to express their sentiments in the songs they choose, I believe the consensus would agree on "Ninety Miles An Hour Down A Dead-End Street." Ladies and gentlemen, meet Richard "Frisco" Bo. Yes, Bo was his real last name.

Frisco had worked in some of the biggest radio stations in the country, both as a Top 40 DJ, and had also distinguished himself by doing production work of an extremely high level, and production work of a high level became his career. But other levels were always high, too, and that drinking thing got him fired from all of those stations. His last station had been in Denver, and they, too, loved his talent, but then his girlfriend went off with another woman, and that had pushed his already excessive drinking into the out of control, forcing yet another station to let him go.

Broke, broke-down, dissolute and depressed, he had nowhere else to go but home, to live in the tiny spare bedroom in his mother's trailer in dismal, depressing Gilroy, California.

Frisco had been born and raised in Gilroy, and had escaped from there as soon as possible, working at radio stations all over the country. His work was always first-rate until his drinking got in the way. He was used to big cities and big city bars and clubs, and he knew there was nothing in Gilroy, and he'd hate it. He had to do it, he knew, but he'd hate it, and hate what his mother would say. He had a small room in a small, depressing trailer, and it was all he had. He had nothing, and nowhere to go. It was going to be a miserable time for a deeply damaged man, but it was his only option, and he slunk into town on a bus, walked home and crawled into bed.

Some may speculate as to whether it was the forces of light or darkness that guided Frisco to KFAT, but it had to be some other-worldly guidance for a sodden, beat-down ex- production man who was already close to the edge, to find the one place in the universe where there was a place for him, where both his talents and his indulgences were permitted. Some might speculate on other points about the unfortunate Mr. Bo, but one of the few indisputable facts about him was that his mother's trailer happened to be in Gilroy, and that happenstance led him to KFAT. It might be as simple as that. But I doubt it, if you consider karma. Which we will not.

Imagine his delight when, crawling out of bed, but still in despair and desperation, he turned on his mother's radio, poked around the dial to get to that boring little station in town, KSND, to see what they were doing. Did they have any ads yet? Did they have anyone doing production? Might they need someone to do production? Anything? Imagine his delight when he heard what was going on over there. Frisco put on his coat and shuffled on over to KFAT, forthwith. He knew just where it was.

Mrs. Bo's trailer was about three blocks away from Monterey Street, and then another three blocks to KFAT. In the event, Frisco Bo found KFAT. As he was already spinning at the same rate as the vortex, Frisco simply slipped in and became a part of the family.

Frisco wasn't what you'd call a day person, so he must have drifted over at night, and we are left to ponder his dismay when he saw the equipment in the Production Room. While we're at it, we might ponder his reaction to the inexperienced freaks who were working at the radio station in his home town. They were amateurs, sure, but they were playing some amazing, exciting music; and look: they were drinking and drugging before, during and after their shifts. And they had a real amateur doing production, some Ozone Guy or something. It was a non-stop, semi-full-on party, with great music, and it was all happening a few blocks from his mother's trailer…

He might have been in Gilroy when he walked out of his mom's trailer, but when he climbed those stairs, he was climbing up to heaven.

No one remembers hiring Frisco, and no one remembers when or how he started working there. The why is obvious. Marty Manning had done the production work for KFAT when he was there with the rest of the Texans, but Chris doesn't remember hiring anyone to do it. Of course, the work had to get done or no ads would be broadcast, and no income would come in. The Ozone Ranger had fallen into doing Production, but he did it in the haphazard, lo-tech way that most things got done at KFAT, but there wasn't a lot of it, and no one thought too much about it.

Even though he'd had little experience in production, Mark Taylor was the most experienced jock on the staff at this point, so he was the Production Director. After working all night from one a.m. until Gordy came on at 6:00, Mark then left the DJ gig behind and went into the Production Room to do *that* work. Of course, this being KFAT, when an ad was sold, amid the jubilation of a sale came a sense of urgency to get the spot on the air as soon as possible. Today, of course! As soon as the ad ran, they could collect their money, so they had to get the ad to the station immediately, and a driver was dispatched twice a day from Los Gatos to Gilroy with papers and ad copy. Jeremy wanted the spot on the air that evening, which meant that the Production Director had to be there to produce the spot. Uh-oh.

Like most of the jocks, Mark couldn't afford to have a telephone at home, and if Jeremy wanted him to do production at a moment's notice, he'd have to be reachable by phone. Mark knew that, what with Jeremy being Jeremy, he was going to get the call during business hours, when he was asleep, and demand something instantly, so if he had to have a phone to be reached at, Jeremy was gonna have to pay for it. Jeremy hated phone bills, but he was over a barrel, and he agreed to it.

At this point, most of the office work, including sales, was being done on Laura Ellen's dining room table in Los Gatos. Paying the bills was haphazard, and the phones were sometimes connected, sometimes not, so a driver was a necessity. Ads were always an urgency. Someone had to drive it down, and someone had to produce the spot when it got there. Which meant a call to the Production Director.

And, as the Ozoner knew they would, the calls came in and woke him up.

So OZ wasn't all that unhappy to have this new guy show up and start doing production.

Frisco would show up as the evening approached and stay until well after dawn, about the time the office staff showed up. He'd disappear without saying good-bye, and sometimes he came back soon, sometimes longer, sometimes not. Sometimes he'd been to the bakery in town, where the Mexican ladies gave him bread to take with him. God only knows what they thought of this specter.

Sometimes the bread was soaked in garlic butter, sometimes not. But he always shared it with whoever was at the station when he got back.

The jocks could see him from the studio through the glass window into the Production Room. They'd see him sitting, staring at the papers with the ad copy that had been driven down, and then getting to work. Sometimes he'd sit for a few minutes, sometimes he'd sit, immobile, for so long that the jocks were afraid that he'd passed out, and sometimes he had. But he always got to work, and the work was always first-rate.

And he always knew another trick or technique. Always. The staff learned how to do production from Frisco. It was a remarkable coincidence that a production man of his caliber and experience would be living in Gilroy and available, at such low wages, for employment. Imagine wanting to do a home video of your daughter's third birthday party, and a neighbor just moved in next door, and he'd love to shoot it for you and help you edit it, and his name was Martin "call me Marty" Scorsese. KFAT was unbelievably lucky to get Frisco. And Frisco was unbelievably lucky to find KFAT. But Frisco had no brakes and both his feet were on the accelerator.

"He was a mentor to me," remembers the Ozone Ranger, "he taught me so much about production; he was a great human being. But God, what a character!" And consider who's saying that.

The earliest anyone remembers Frisco being there was sometime after Cuzin Al first got there in February of'76, and by March or April, Frisco was helping to set up some microphones for a live broadcast from the Production Room for Al's bluegrass show. Al remembers him helping with the set-up that night, and after that, he just... stayed. But Al does not remember who played that night or when, so the date is lost.

Frisco was drawn to KFAT like a lost man is drawn to a distant light. You may insert your own metaphor here. I won't mind. Take your time. Here's some Frisco stuff.

Chris Feder swears that he didn't hire Frisco. Neither Laura Ellen nor Jeremy has any memory of hiring him and Larry Yurdin was gone before Frisco showed up. All the staff loved Frisco and remember that his work was always great. They also remember that he was drunk a lot at first, but then it got worse. At first it was booze, but it got to be anything.

Uppers, downers, pot, booze, speed, whattya got? But that was later, and the production work that he did when he got there was just exactly what KFAT needed.

Not only did Frisco do all the production, he willingly taught others to do it. Sully remembers picking his brain and loving it. She was amazed at the tricks he showed her. All the KFAT jocks were new to production, and Frisco had the willing ear of all of them except Tiny, who wanted nothing to do with production work. And Buffalo Bob, who wanted nothing to do with any kind of work.

Tiny had two active kids to take care of, and locking herself in a room where quiet, concentration, and a lack of interruptions were essential wouldn't work for her. Instead, Jeremy gave her the job of cleaning the station with her off-air hours. She did that passably well, but some of the staff were casual about the clutter they generated, and she never quite caught up with the mess.

The Ozoner and Frisco worked the same hours, and they got along well. Frisco was easy to get along with when he was coherent, sullen and obstreperous when he was not. OZ let him work in the Production Room undisturbed.

KFAT's front door was always open, and listeners wandered in frequently. Most of the staff remember a sign that Frisco, who hated unnecessary interruptions from strangers, put on the outside of the Production Room door: "This may look like fun to you, but these people take this job very seriously. They are working very hard, they're trying to put together some show. Please make your visit short."

Sister Tiny also remembers the nights when her boyfriend, Ken, the crazed, abusive Viet Nam vet came to harass her. Everyone was rightfully afraid of Ken, because he was invariably drunk and crazy, and he had beaten up Tiny and other drunks. There were times when Frisco had to hide her in the Production Room, and despite Ken's screaming, would not allow him in. But there was one night when that would not work, and Ken was completely and violently out of his mind and determined to get at Tiny. Ken was drunker and crazier than ever, and in his rage he wanted to beat her, and there seemed to be no stopping him. But Frisco was drunker and crazier than Ken, and with Tiny behind him, it was he who stood up to Ken and backed him off when everyone else was scared helpless. Tiny knows she'll never forget that moment in her life.

She also remembers Frisco's kind heart, and his concern for the other jocks. In his early months at KFAT, back when he was sometimes sober and had thoughts of others, he was always a member of the Fat family. Tiny remembers, "Well, I was so broken hearted all the time. One Christmas I had a broken heart, and I was so bummed, and my kids were there, and I was working because I always worked Christmas, and Frisco took my kids in the back room with the Christmas tree and the presents, and he made sure they had a fun day."

He was noble of heart, was Frisco, but weak of spirit, and he liked to get fucked up.

Chapter 40: The Big Box

Jeremy came to the station late at night to tinker with stuff. He was always tinkering somewhere on something. Late at night was a good time for tinkering, and the Ozoner, who saw him regularly in the early days, remembers, "Yeah, tinkering, doing strange engineering things. I was also on the night that it first went to high power. I was on the air when it first went from a low power Gilroy thing to… Loma Prieta," and people started calling in from far corners of the state.

Now it was almost a week after the transmitter move, Jeremy was working at the station late into the morning, and the Ozoner was on the air. Jeremy came into the studio and talked with OZ. They wondered how far the station was reaching now, and they wondered aloud how they could find out. They knocked a few ideas around, the usual preposterous stuff and a few decent ideas, when the Ozoner remembered the big box in Room A, and said, "What about the big box?"

It was a huge cardboard box, about three feet high, with all of the 45's that were sent to the station. KFAT played no 45's, and anything that was put out on a 45 rpm record probably wasn't Fat, anyway. Unless it was something like an early Sun recording of Elvis, or Johnny Cash or Carl Perkins or Jerry Lee Lewis, and all of that stuff was on albums.

Jeremy asked Ozone to announce that we'll give the box away to the farthest caller. Already experienced with a Jeremy Lansman Operation, the Ozoner reminded Jeremy that he would have to pay for the postage, and Jeremy agreed.

Fee Waybill of the theatrical rock band The Tubes called as his character Quay Lewd, but he was only calling from Marin County. Calls came in from Sonoma and Mendocino counties, and some towns south of Monterey that no one had ever heard of and could not verify. After Jeremy and Mark decided to disqualify the guy who said he was calling from Cleveland, watching the Cuyahoga River burn, the winning call sent the big box out to a guy in Yosemite Park, about 200 miles away.

People were listening from even farther away than Jeremy had hoped. This was a really good sign. And people were taking their jobs loosely seriously.

Chapter 41: Unkle Doll

Sherman and the Ozoner did the overnights. Ozone would do Mondays through Thursdays, and Sherman would do Fridays, Saturdays and Sundays.

Everyone knew that Sherman Caughman liked his dirty songs. In fact, until last year he was still playing them. He was on KPIG radio, which used to be owned by Laura Ellen, and on Saturday nights they turned Sherman loose. Sherman has always been hard to handle, as he was on these nights from eight until ten. Then, from ten until midnight, he did his "Dirty Boogie" show. It was as you would expect.

At KFAT, he never felt the need for a stage name, so he was just "Sherman," until 1976, when a company put out a funny doll for adults named "Uncle Sherman." The doll was of a seedy, middle-aged man who wore a raincoat that, when opened, revealed the doll to be (ahem) "anatomically correct."

It was marketed as a "Flasher" doll, and it was briefly popular among a small, sick segment of America. People like you. Of course, someone sent one to KFAT, and the Ozone Ranger immediately started calling Sherman "Uncle Sherman." It was perfect, and it stuck. Even Sherman started calling himself that, changing the spelling from "Uncle" to "Unkle."

And so it goes to this very day.

Go online. Look it up: Unkle Sherman.

It was a perfect conflation of fantasy, reality, a toy and a real-life dirty old man.

Chapter 42: WHO STOLE MY SCREWDRIVERS?

It is one of the most enduring memories of the KFAT era, and was the object of much impassioned speculation: who stole Jeremy's screwdrivers?

If you get a close look at a good copy of the KFAT bumper sticker or some of the T-shirts, there is writing around the outside of the sticker. It says: EAT MORE GARLICK... WE WISH WE WERE COWBOYS... HIGH CHOLESTEROL, LOW PROTEIN RADIO... WE'RE IN IT FOR THE $$... SADDLE SWEAT MUSIC... WHO STOLE MY SCREWDRIVERS?

It was the Ozone Ranger's shift, and it was late at night, maybe three or four a.m. Jeremy was working, tweaking the station, as always. At first, he was working in the studio, and then he was working in the Production Room, and when he came back to the studio, his screwdrivers were gone. *What the WHAT???* He'd set his box of tools in the hallway just outside the studio, right under the window looking into the studio, then he'd gone back into the Production Room. He had to make an adjustment, and only been there for a few goddam minutes... *And now his screwdrivers were gone!*

If that doesn't mean much to you, then you, my friend, have either had insufficient exposure to the professional building, electrical or maintenance trades, or you have been insufficiently involved in home repair to have an opinion in the matter. Ask any pro who uses tools on the job, and you will find a level of agreement that will surprise you. You will learn how attached a pro becomes to his or her tools.

Anyone who uses tools professionally takes pride in them, keeps them clean, operational, and, most importantly at a critical moment, knows where they are. Jeremy, you will have guessed, was among those with an extensive collection of tools, and he used them as would a surgeon. Jeremy lived by his tools, and one night his collection of screwdrivers disappeared from the hallway in the station, and to this day he has never seen them again. They were sitting in their case, with the lid open, in the hallway, next to his toolbox, on the floor, right under the window that looked into the studio. Then they weren't there anymore.

This being KFAT, several things could have happened in those few minutes. People were always coming in and wandering around; the Ozoner enjoyed the company, and as long as he did his work, Jeremy was not about to interfere. But there were also junkies in the vicinity, and he'd had experience with that lot back in St. Louis. He'd lost a couple of turntables back there, and some microphones and assorted small objects had also disappeared. Maybe a tape deck now and again, or a lamp. But, Jesus- *his screwdrivers?*

It could have been a junkie. One could have wandered in, taken the screwdrivers and split, and that could have happened while he was in the Production Room, and the Ozoner might not have seen him. It could also have been any one of the disreputable characters that hung around the station, but no one had been at the station except him and the Ozone Ranger when he went into the Production Room to make the adjustment. Standing, looking at the carpet where his screwdrivers had been, he and the Ozoner were alone at the station.

Bob Cassidy was asleep in Room B, two doors away in the room farthest away from the control room, both doors shut against the noise and possible intrusion. But who knew who might have wandered in while the tools were in the hallway and immediately absconded with the screwdrivers, in a hurry to keep them for himself, sell them, or exchange them for a fix? Who knew who had been there during those few minutes?

Jeremy was in a rage. This was serious, disrespectful shit and it would not stand! *This was bad shit* and he was *pissed!* He was already *way too stressed out* with the way the station was running; he felt helpless and lost in this town and with these people at the station, and with the people outside the station. He was enraged. And helpless, *which made him even madder.*

But he shouldn't be helpless! He owned the damn station and he wanted his tools back! And he expected someone to return them to him! Right now! *He was pissed!*

He went to a closet where he kept his tape and wire and solder and diodes and such, he got a can of red spray paint and came back into the hallway and stared again at where his tools had been. He angrily shook the hell out of the can then he painted the wall across from the window that looked into the studio, in foot-high letters, in capitals so it could not be missed:

WHO STOLE MY

SCREWDRIVERS?

And no one ever erased that message. As long as KFAT stayed in that building, that anguished cry was on the wall.

By the time I talked to Jeremy about this in April of 2004, he still had no idea who had taken his screwdrivers. But I knew. Investigative journalism *fucking rules,* man!

It seems that Jeremy had been right in that anyone could have come into the hallway outside the studio during those few minutes, and someone did. In his back room, Bob Cassidy had woken up and wanted a drink of water, and he shuffled out into the hallway.

Everyone at KFAT knew Jeremy's toolbox- he frequently had it with him and we'd always joked about it. Bob now says that when he saw them, he thought that the case of screwdrivers sitting next to Jeremy's toolbox might have belonged to "someone not necessarily connected to KFAT." He says he thought that someone might have come to visit the station to drink some beer, request a song or two, and hang out with the jock, and this "someone" might have brought their own screwdrivers, and left it, open, on the floor in the hallway outside the studio, looking like what *everyone at the station knew* was Jeremy's toolbox. Uh huh…

In any event, Buffalo Bob took the screwdrivers, put them in his truck in the back lot, and went back to his room and shut the door, ignoring the yelling that ensued shortly thereafter.

Of course, this was in the very wee hours; neither man had seen Bob, but both men knew that Bob had alcohol and guns behind his closed door. They knew he was often drunk or passed-out from drinking. He was mean when sober, cruel when drunk and downright nasty when wakened. And he had those *guns* back there, so no right-thinking person was going to knock on his door and disturb him that night.

When questioned the next morning if he knew where the tools were, or if he had seen anyone at the station the night before, he told them that he had been asleep and missed everything. He said he thought he heard some yelling sometime in the night, but that happened a lot, and as long as it stopped, he ignored it and went back to sleep. He was sorry, he said, he couldn't help in the investigation.

As investigation it was. Jeremy was really pissed at the betrayal of trust as much as the expense of a new set of screwdrivers. The message on the wall had been painted by the owner, and no one saw fit to paint over it. Until Jeremy did something about it, it would remain. It was the first thing that visitors to KFAT saw, and it let everyone know immediately that this was a strange, special place.

It remained on that wall for years, a spectral, anguished cry from the past.

WHO STOLE MY SCREWDRIVERS?

Twenty-eight years later, I told Jeremy that it was Bob Cassidy who had taken his screwdrivers. He laughed, but I swear the laugh had an edge to it.

Chapter 43: I Stole The Screwdrivers

Sully was usually on time for her shift, and sometimes she came in early. She was like that. She had been studying production work with Frisco, who had a particular fondness for Sully, and he'd been spending a lot of time with her, showing her production tricks. The only times to see him were long after her shift, when he got there, or before it, while he was still there, so sometimes she got there early to work with Frisco. One such morning she arrived at the station and Frisco was standing in the hall, looking at the wall. When she approached, she saw the message, and she stared, unable to believe it at first. But it was KFAT, so… She and Frisco laughed about it. Jeremy was forever losing things and asking people if they had seen them. This was going a bit further, but... this was KFAT. Sully said, "You know, let's do something that's kind of funny, but we might get fired."

And Frisco said "Okay."

Sully remembers, "We typed up a suicide note, saying 'I stole the screwdrivers and I thought I would be able to live with myself. My mother needs an operation, but now I find I can't live with myself and I'm sorry. Goodbye.'"

"And then we take this… this is mean, I wouldn't have done this if I was a mother… we took Elsie's favorite doll, and Frisco made a little noose and he, I remember him crooking the doll's head just in the right, particular way, like a broken neck…"

They hung it from the light in the hall, outside the studio. It hung in front of the message on the wall, and they pinned the note to its chest.

Sully said, "You know, either he'll laugh or we'll get fired." Jeremy laughed.

And so did Awest, who came by to show Jeremy his design for the station's new logo. He saw the new sign and said, "I love this guy! Where is he?" An inveterate collector of happenstance art, Awest saw the message on the wall, loved it unreservedly as both art and lore, and felt that that one line best described the insanity that he so loved about the station. He knew immediately: he would put the line on the bumper sticker, so he added it to the rest, and put them on the outside of the logo, where you can see it in the photo section of this book.

Chapter 44: The New Crew, An Angel And A Cop

There was an electric excitement at the station. It was a new format, a new sound, and everyone there was just as new at their jobs, just as fresh and excited and happy and enthused as they could be. They all knew this was their job-of-a-lifetime, and they were all running with it. For some, it was an introduction to pride.

Of course, all the personal issues that each of the new staff would bring to the job would sooner or later show up and jeopardize the whole thing, but no one thought of that back in the summer of 1976.

The gig was exciting as hell. It was fresh and it was real, and they were playing the most exciting music in America. Every one of them knew their rock, and knew that it had gotten stale, vapid, lazy. None of this crew liked what was rock 'n' roll back then. Even Buffalo Bob, whose show rocked the hardest, was rocking the country cuts. He'd discovered Hank Williams, Jr., and it had opened his ears to a whole new slate of country-flavored artists that he'd never heard or had stubbornly refused to hear. Hank Jr. had opened Bob right up, and now even Bob couldn't be immune to the excitement at the station, and even he was doing the work, looking for the country rockers. That was his thing, and there were a bunch of gems in the stacks, and all of them were Fat.

The rockers weren't everybody's thing, and the other jocks were finding their way and looking for new music, too. The listener lines were always busy. There was no bell on the phones, but when someone called, the light on the phone buttons would blink. Terrell Lynn remembers that he never had to look at the phone to see if someone was calling; someone was always calling. Or two or three callers were waiting, and that was as many lines as they had, when they had any phone lines at all. Jeremy couldn't always pay the phone bill, and sometimes Continental Phone turned them off until... well, this was a Jeremy Lansman Operation.

There'd been talk among the staff that Bob's shifts were too rocking, and others were definitely more country, and some were more traditional country, Patsy Cline and Loretta Lynn and George Jones and... well, you know: KFAT was too hard to classify easily into any category, so you have to go to a list, and it's too many to name. Bob's stuff was heavier on ZZ Top and Lynyrd Skynyrd and... well, whoever rocked but was Fat. Some Rolling Stones songs were, some weren't. George Thorogood. Jerry Lee Lewis, Joe Ely, Delbert McClinton. I wish I could remember more. There were lots.

And, as he was now MD, why, who'd bitch about his shows? Laura Ellen might, but Bob wasn't listening to her. As MD, Bob was getting to listen to most of the records first. He made some comments on a few at first, then he didn't, but he'd leave a bunch of records out for the DJs to sort through. Of course, rank has its privileges, and Bob got his pick of any record he wanted, to keep or sell. Fortunately, the records were coming in twos and threes. Here's a test:

The Premise: Bob is making about $65.00 a week and taking home a lot of records.

The Question: What did Bob do with all those records?

Choose One:

A. He kept them

B. He sold them

C. He did neither

D. He did both

(The answer appears at the end of the chapter.)

Bob kept some of the records and sold some, on one occasion unloading 8,500 records to one store. Then, I imagine, he bought some booze or he went to dinner or a bar, or bought something, or spent it in some way, maybe on rent or gas or food or in a bar or something else, like a bar, but we've lost the thread and must get back to the excitement at the station.

All these people were the new staff of a new radio station, with a format that hadn't been heard before and wasn't being heard anywhere else on the planet, and the listeners were crazy about it and calling constantly and everyone loved it, and in Santa Cruz, everyone was listening to it.

Any bar or restaurant worth going to in Santa Cruz had KFAT coming out of the speakers. Bars had jukeboxes that were unplugged. Restaurants that featured a Sunday Brunch tuned in to Sister Tiny's Gospel Hour, and then stayed there when the regular Fat music came back on. Of course, for a while at 8:30, when the Fat Dork came on, you could walk the streets of downtown Santa Cruz and suddenly hear a wide variety of musical styles coming out of the different establishments where before there had been a pleasing uniformity of sound.

People were listening, people were calling, and the response was almost always favorable, even though some of the songs challenged what a few of the callers had considered to be reasonable bounds of civility. But almost everyone loved it. They'd never heard anything like it before, and all these people had been through the rock thing and were burned out on the rock thing, and especially what the rock thing had become after it had lost it's integrity and become a corporate thing, and they'd been, without knowing it, hoping, waiting for something new and exciting, as rock had been only ten years before, because they missed the excitement and the joy that rock used to have, and now they had it back in a way that no one had even thought of before that crazy bastard Larry Yurdin thought of it, and now here was his vision, tweaked, modified and done to its illogical extension, and all these years later it was a success, and Larry wasn't there to see it, but he was elsewhere, hearing about it, and sort of following its progress while working on other stuff, and the people who had hired him at KFAT had been forced into hiring a bunch of inexperienced lunatics who were learning the music just a step ahead of the listeners, and in fact oftentimes learned the new music *by being listeners*, and it was their job to excite themselves and the other people on the staff, and the people at the Green Hut down the street, and the listeners, and that's about all they cared about.

I know that sounds like a lot of fun and all, but there was also a lot of pressure on these people. They were new at their jobs and new to the music, and the library had so many albums that no one outside of a musicologist would have heard of, and there was the stuff that *some* of the staff knew about and played. And then there was all the new stuff coming in. Even the old stuff was coming in. Record companies were happy to empty out their back catalogues for the only station playing their older stuff. Old and new, it was all coming in the mail every day.

KFAT happened at a time when there was a niche for an alternative to rock. The stuff KFAT was playing had to come from somewhere. Sure, there had always been folksingers with a guitar and a song, but there was also so much hipper stuff coming in. Joe Ely was amazing. He was really a Texas guy, but he *rocked.* Not just Buffalo Bob, but anyone could play Joe Ely on their shift and he fit right in. And all the others! And Willie!

Willie Nelson had moved to Austin when Nashville had no interest in him. (I was about to say that Willie couldn't get arrested in Nashville, but then I remembered who I was talking about, and wrote the other thing.)

Willie had written some great songs in Nashville, but his career was going nowhere. After a bitter divorce and his house burned down, he moved back to Texas where he was amazed to find a town full of long-haired, pot-smoking people who LOVED his music. He was a hero in Austin, but his music still wasn't being heard anywhere else. Now there was a serious music scene in Austin, and Waylon was over in Lubbock, and Ely was just up the road...

Tiny: "We broke Willie Nelson in California, I swear to God. I never heard Willie Nelson. He was not on the air anyplace in California until we played him, I kid you not. He was in Austin, he was not in California."

Terrell Lynn: "Don't you remember going to, like, the first concert, and nobody was there? He had the booth with the Hells Angels. You could put a blanket anyplace."

Jerry Jeff Walker was a drunken, stoned-out hippie redneck good-ole-boy who'd done the Nashville thing, done the LA thing, and had gone to Texas and kept playing his thing, his way. But in Texas he found a home.

And then came KFAT, the first people to play Willie in California, and the doors began to open to concerts by the Texas delegation. Willie begat Jerry Jeff, who begat Joe Ely and Delbert McClinton, who begat Guy Clark and Tom Russell and Van Dyke Parks and Tompall Glaser and Jesse Colter and Townes Van Zant and Michael Martin Murphy, the original "Cosmic Cowboy." And more all the time.

Listeners had been calling in asking, no, begging for KFAT stuff, like hats, T-shirts and bumper stickers. Awest had designed a logo, printed up some T-Shirts, and they were selling like hotcakes; Jeremy had promised that if he designed a bumper sticker, Jeremy would get it printed. Awest didn't want the old stand-by rectangular bumper sticker that looked like every other bumper sticker and said something catchy or stupid about the station, giving its name and call numbers. Jeremy wanted something less conventional, too, and that was Awest's specialty. He had been looking for a logo for the station, and he'd found it in the library on an old Bob Wills record whose cover featured a bumper sticker with a cowboy with blue and yellow rays behind his head. Awest took that image and made it Fat. That would be the logo he'd put on the T-shirts and it should be the bumper sticker, too. For a bumper sticker it was a weird shape, unlike any other station's stickers, a shape that would need to be die-cut, but Jeremy liked it.

He told Jeremy that the stickers needed to be made from special stock so they wouldn't chip, peel, or fade in the sun, and that and the die-cut would all cost more money. But Jeremy wanted to get as much impact from as little money as he could, and Awest told him that this was the way to go. Jeremy agreed, approved the funds, and they were ordered. The staff couldn't wait, and a week later, they were back from the printer.

Willie Nelson was doing a show at the Santa Cruz Civic Auditorium that night, put on by the Hells Angels, and KFAT was promoting it. Terrell Lynn, wearing a new cowboy shirt for the occasion and his big white straw cowboy hat, went over the hill with me, to the show, and to hand out the new stickers after it. I was hoping to get a station ID from Willie, so I brought my tape recorder. We also brought a KFAT T-shirt with us to give to Willie, and we tried to get backstage to do that. But the backstage was guarded by a mountainous man wearing Hells Angels colors, and seeing that we'd have to get past him to see Willie, I approached the giant and told him that we were with KFAT and wanted to see Willie to give him this T-shirt, and I held out the shirt to show the Angel.

The Angel wouldn't look at me or the shirt, or acknowledge me in any way. He just didn't respond. Statue-like, he stared straight ahead and said nothing. Terrell Lynn sized up the situation, turned, and started walking away. At about 6'5", the Hells Angel was almost a full foot taller than me, so I had to look up when I said, "Excuse me, maybe you didn't hear me. I'm with KFAT radio, and we're helping to put on this show, and we play Willie a lot and we love him, and my pal and I would like to see him and thank him and give him this T-shirt," and I held out the T-shirt again.

Again, no reaction. Terrell Lynn was getting anxious, as he knew that this was no way to talk to a Hells Angel. The Angels' last security gig had been at Altamont Speedway for the Rolling Stones, and everyone knew how well *that* had turned out. Terrell Lynn was ready to bag it and go out front to wait for the show to start, but I looked up at the Angel and asked: "I don't suppose there's any way I could *muscle* you into it…?"

The Angel looked down at me, puzzled, like he wasn't sure what he'd heard, while Terrell Lynn was backing up several feet to avoid getting any blood on his new shirt. The Angel looked down at me as if he was trying to figure out if he'd heard that right, then he turned away and stared straight ahead again and ignored the punk in front of him. He never said a word, but both of us got the message, and we left and went to see the show.

We didn't try to go backstage after the show was over, either. We had work to do.

After the concert, Terrell Lynn and I stood outside the Auditorium as the concert- goers left the building, asking people if they wanted a KFAT sticker. People were confused at first, not being sure what was being asked of them, but as we held out the stickers and offered them to be taken, for free, people got it, and immediately began taking them. And talking to us.

Soon there was a backup at the door as people were prevented from getting outside by the crush of people outside the door, taking the stickers and talking to the two guys who actually *had something to do with KFAT*. We had to back up further toward the street to allow more people out of the auditorium. The crowding got more dense as more people tried to leave and were stuck inside the hall by the back-up outside. Someone in the back of the crowd shouted, "Hey! What's going on?"

Someone yelled back that they were giving out KFAT stickers, and that was the end of the grumbling, but then there was a surge as we had to back up all the way to the curb to allow more people out of the auditorium; people were asking for two: "one for a friend." We'd brought two thousand stickers to an 800-person show, and they were gone long before the venue had cleared out.

On the way back to Gilroy, a California Highway Patrolman stopped us and told me that one of my brake lights was out, and started to write me a ticket. I lied, saying that I hadn't known about it, that I was sorry, and that it must have happened tonight, at the Willie Nelson show, where I'd just come from, 'cause, you see, I worked at KFAT radio, and I'd "just been to see Willie Nelson, and we'd been giving out our new bumper stickers, you see, like these here, and…"

The cop stopped writing, put down his ticket book and asked, "You work at KFAT?" To which I modestly replied, as if surprised that the cop knew what that was, "Yes, yes I do. And we were handing out our new bumper stickers in Santa Cruz tonight, the ones that just came in from the printers today. Would you like one?" The cop asked for two and put away his ticket book.

I told the cop that I did "Chewin' the Fat" for KFAT, and as I had my recorder with me, would the cop like to do a station ID for us? Thus, I skated on the ticket, and KFAT had an ID that went, "Hello, this is Officer ______ of the California Highway Patrol, and you're listening to KFAT in Gilroy."

The music scene in Texas was growing and recording, it was spreading to California, and KFAT was about the only station playing them. Genial Johnny, who'd been with the first crew at KFAT, was back in Santa Cruz doing his once-a-week show, "Road Apple Radio," on listener-supported KUSP, and he was playing them, too. KUSP got as far as the outskirts of Santa Cruz, and KFAT had been going into San Jose and little towns south of San Francisco, and of course Santa Cruz and Salinas and Watsonville and Monterey. But that was with the old transmitter on Mt. Madonna. It was a new game, now.

The first staff had been waiting eagerly for the transmitter move, until they got tired of the wait, the bouncing checks, the smell, the town, the equipment and the bullshit. The transmitter move was going to earn them a living wage and lots of perks, and it was worth waiting for. Until it wasn't worth waiting for any more, and they left.

The new staff wanted it to happen, too, but they were all pretty damn happy to have the jobs, and they wanted the signal to boost, but they'd never ask Jeremy about it. Except when they asked him about it. But Jeremy had also been waiting, just like everyone else.

And then the change got approved, and the transmitter was now on the tallest site in the Bay Area. There was a new game in town. Laissez bon ton roulette: let the good times roll.

The Answer To The Test: D) Bob did both.

Chapter 45: The Staff Hangs Out

Tom Diamant was a record rep, but he was also a pal. He represented artists whose music was being played at KFAT, which alone would have made the station stand out in Tom's esteem. Trying to get the albums he repped played on radio was rough work, as some non-comms had one- or two-hour shows once a week that played his stuff, but no one in the world was playing it constantly--and commercially—except for KFAT. So Tom loved the station. No, I really mean that: he *loved* the station.

He listened to it whenever he was where he could get it. He was a fan, and because he lived in the East Bay, he could go to KFAT personally. His budget from those small, independent labels was pretty thin, so there was no way he was being flown to, say, Idaho to meet with a PD from some small-town college non-comm station. But KFAT was, you might have heard, different. They played his music and all the rest, and he loved all of it. Okay, he'd heard enough of "Moose Turd Pie," but the rest was awesome, and he came down and he liked the staff and he kept coming back.

I'll get back to Tom soon, but when he and I first talked about KFAT, the first thing that he remembered was that the staff all hung out at the station. He's been to over a hundred radio stations, and he says that in all of them except KFAT, the jocks did their shifts, put their records away, took care of whatever station work they had to do, and left. AT KFAT, there were always jocks hanging out. It was the only social scene they had outside of Augie's or the Green Hut.

And there was almost always one or two of them living there. Yeah, living there.

Buffalo Bob lived there a few times, in Room B, finally being tossed out by Laura Ellen because he had yet again left his hotplate on and a small fire had started, the smoke billowing out from under the door while Bob was gone somewhere unknown with the only key. The door was easily knocked down and the fire, detected early, was smothered. It was the carpet burning, and as you would have guessed, there was a sufficiency of greasy build-up on the carpet to inhibit any flames from spreading quickly, while producing a surplus of smoke. Eventually, the smoke cleared and Bob cleared out, but the odor lingered for months. But by then, it was Tiny's problem. One of several.

She never forgot the embarrassment of having a newspaper reporter and photographer showing up to write an article on the "wacky new station," and they found and photographed Tiny cooking dinner for her kids. All she could afford were vegetables, and she was cooking them on a Hibachi on the deck just off the back staircase. It was a great photo for their paper, a great visual, but Tiny was mortified, and she was afraid that other mothers—or the authorities—might think that was *all* that her kids were getting for dinner, that that was all she could afford to feed them, and that she wasn't a good mother. The first two were true.

That was her paranoia talking, but she was deeply embarrassed. Afterward, no outraged mothers called, no children's welfare agency got in touch, and the embarrassment passed. But Tiny's maternal instincts never wavered, and she was always aware that she was in some way fucking up her kids. She always had an eye towards how she could do well by them, whenever she could. But it was tough. It was just damn hard, living at KFAT with two kids. Also, Tiny liked to party…

Down the block, Hans "Tex" Teufel, the cook at the Harvest Time was a Fathead, and he sent over grilled garlic, butter mushrooms and other treats whenever he could.

Probably everyone slept somewhere in the station at some point, some for extended periods, some for a nap or an overnight. Laura Ellen took naps there, as did others, and "sleeping it off" happened often enough to no longer be of interest, or worthy of comment. It was just life, going on at KFAT.

I kept a sleeping bag in my car, in case I couldn't finish my work in time to get back to San Rafael to take a job as a substitute for that day. If there was a teacher out for several days and I had committed to those days, I always kept my commitment. I needed to complete my work for my next show and get back to San Rafael for my real-world job, a hundred miles away. I couldn't blow off the teaching gig, as that was what was paying the bills. There was a couch in the Meeting Room that I got to know well.

Several days a month I would work through the night and, as dawn came and I had to be in my car heading north at 5:30, and for a few months as I was leaving, I would go into a back room and wake Gordy, who needed the half-hour to wake up, go out for coffee, get back and be on the air at 6:00. During those months, Gordy was living on a red vinyl Barca Lounger that had appeared in the Meeting Room from somewhere.

The Barca Lounger was in slight disrepair, but it was, without a doubt, the most comfortable chair at KFAT. But as a bed? Gordy thought not, but for three months he called it home.

There were several reasons for hanging out at the station as opposed, say, to going somewhere else. The lack of money, of course, was a prime factor in staying put. Also, anything happening worth seeing or doing was in Santa Cruz, but that was almost an hour away, over the hill. Getting there was one thing, but sucking up all the free drinks and then trying to get back, was something else.

Also, no one liked to talk about it, but Sully remembers a persistent problem that kept her and others from socializing: "I introduced a show with Leo Kottke once, and I remember going there and just being very conscious of the fact that my clothes and hair and everything smelled like garlic."

When Tiny or Sully or Sherman or anyone else got it together, they moved into or near Gilroy. There weren't a lot of apartments in Gilroy, but that wasn't a real problem for the staff. Outside the town of Gilroy now are a lot of fancy McMansions, but in the mid-70s, there were small houses and cottages on larger properties and medium-sized farms. There weren't a lot of them, but there wasn't a lot of demand for them, either, in Gilroy, and the staff found what was there for the taking in their price range. Clean but slightly broke-down was at the upper edge of their price range, and that usually meant remote, which was fine by the jocks.

They almost always shared places with other staff. They moved around a bit, but the staff always stayed in close contact. Except for unusual circumstances, like a family event out of town or something, they had nowhere else to go but the station, or whatever event the station was involved in, or could score free tickets to.

They ate at the same places because they all had the same access to the same trade-outs. When the smorgasbord place in Gilroy started a trade-out for one meal a day per approved person in exchange for advertising with KFAT, the staff celebrated and started eating there daily.

They ate together and they hung out together. It was Gilroy, for God's sake, and there was nowhere else to go. Not during the day, and usually not at night. So they hung out at the station. Like a family.

Wherever they went, they carried the smell of Gilroy with them. It took a couple of washings to get the garlic smell, or whatever smell was currently wafting through Gilroy, out of their clothes, so they never caught up with it, and it followed them like a shadow. The worst was the cauliflower. They carried it with them, and people smelled it when the staff went out. People made jokes about it, ha ha, and the DJ's laughed, too, ha ha. Sometimes, it just made sense to stay in Gilroy, maybe hang around the station.

There was always someone doing something strange at the station, and there was always some gossip to spend some time with, with whoever was there. Whoever was on the air wasn't doing the kind of work that precluded him or her from chatting, or from putting on a long cut and going to another room to hang out and have a beer or a joint.

And, of course, on some afternoons, and most evenings, there were people who stopped by, often bringing stuff to share. The front door was open, and there were the regulars who came by and became friends, and there were the folks who were fans, but hadn't been there before. They came to party, and even the regulars came by to see what was going on. It was a social scene and there was almost always something going on, and it was almost always different and almost always the same.

Cigarettes were smoked, beer was consumed, joints were passed around, and sometimes some coke or crank would show up. There were empty Jack Daniel's bottles, beer bottles, beer cans, full ashtrays, and new cigarette scars on the console. By the time three or four or five a.m. came around and the party started fizzling out, the poor jock seldom had either the energy or the inclination to put away his records, much less clean and tidy the place up, empty the ashtrays and the trash.

It pissed off the morning jock, and it usually fell to Gordy to cajole the night guy to help clean up. That was the key: if Gordy didn't help, it never got done.

But the jocks soon found that there needed to be some standard of order, in order to function. It was cute at first to get paid to play music and party all through your shift, but you couldn't keep pissing everybody off by being a slob. This was increasingly a family, and if you pissed someone off, you heard about it pretty quickly. Everyone was living in close quarters, and like any newly-formed society, they quickly learned that they needed a few rules.

And like any society, they quickly learned that a large portion of the society were going to be helpful partners in the new endeavor, and some were going to be… resistant… to promising to live by the rules, or even to try playing well with others. Or to give a fuck what others thought or said. That would be Bob Cassidy.

Frisco would be the other notable failure in this regard, but Frisco *tried* to live by the rules, and there's your difference. Bob tried when it suited him, which meant that he was happy to perform his functions, as he understood them, but outside of those functions, there were no absolutes that he would agree to. Foremost among them was sobriety. But his shows rocked and were increasingly popular, so again: fuck you and leave him alone.

As for the rest of the staff, they all understood that living in close quarters meant some give and take, and for the most part, it worked very well. You might think of these people as amiable fuck-ups, but you would be wrong.

Fuck-ups they might be, in a societal overview sort of way. But consider the standards.

There were people on the KFAT staff that were not fuck-ups at all, but they just couldn't be graded by the conventional standards.

The 60s were an aberration. It was a brief burst of youthful exuberance, of freedom, of experimentation, of joy and enthusiasm and can-do-anything attitude of such a magnitudinous proportion that had never happened before in any culture in the history of the planet. But that rare, pure force was petering out and giving way to the rising tide of an increasingly homogeneous, formulaic culture, where convention was the rule, not freedom, not experimentation, not anarchy. The searching, questioning spirit of the 60s was all but gone, and a restless productivity emerged, and commercial success became valued, rather than joy and excitement. The Bay Area had been one of the centers of the revolution, but what was left was now the last pocket of that rare, rebellious, exuberant air. Virtually an anachronism at birth, KFAT breathed that rare, exuberant air and burned with passion at both ends.

KFAT was the last gasp of the Sixties. Though it started in 1975, that which gave KFAT its freedom and permissiveness (close but different), the chemical indulgence, the general indulgence, the promiscuity, the joy, the carelessness, the enthusiasm, the willingness to experiment, the openness and so much more, represented a period that had largely passed by the time of KFAT's inception. Jeremy and Laura Ellen were of that period, and they had hired people of that period and that inclination. Culturally, the hippie movement was waning; people were growing up and getting jobs. The Revolution was over.

But there was still a core of believers, and KFAT hired them, almost exclusively. No, it's not that they were fuck-ups, although, certainly there were *some*, but most of them simply did not fit into the growing paradigm of steady employment, marriage and various and sundry responsibilities. In a word: respectability.

Respectability? It's just not for everyone.

I know that respectability is an admirable thing, and I normally recommend it highly, but some people just do not have the aptitude for it. And some of those people found their way to KFAT.

And they huddled together because they were the only ones who were doing what they were doing. It was fun, it was exciting, it was revolutionary, and there was nowhere else to go.

It was Gilroy. Have I mentioned the constant smell of garlic?

Chapter 46: Morning Gordy

Gordy was still taking the bus down to Gilroy every morning. He'd always lived in San Jose, and on his way to KSND, he'd pick up the racing forms, take the bus to Gilroy, deliver the forms to the Green Hut, and go to the station. He didn't have a lot of what you'd call friends, and no one he knew knew or cared what he was doing. Now all these new people were there, the music was hot, the station was hot, and for the first time in his life Gordy Broshear was a winner, and he liked it fine. The KFAT staff was a family and he was a legitimate part of it, and he liked that a lot. He did the Morning Drive slot for the station, and that was a very important shift. He was good at it, and the listeners liked him. Sherman and the Ozoner were always glad to see him when he got there, Karl Hess would be there soon to do the news, and the day began anew.

In the early days the station kept going off the air, and Gordy remembered that Jeremy was never in a real rush to do anything, and the hour it took Jeremy to get to the transmitter from Los Gatos gave him and Karl time to go out for sandwiches. It happened regularly, at first.

For a while, KFAT had the "Fat Ass-Trological Report" along with the news. Then it was just the news. Then Karl Hess was gone. Rumor had it that his wife, the "Beauty Queen," hated the freaks at the station and wanted a real career for her husband. Karl never brought his wife around, and some think it was because she didn't want to be there, while others think that he didn't want her to see how crappy and unprofessional the station was. The truth was that the station was a broke-down, cheap, piece-of-shit small-town radio setup, full of what could only be described as disreputable types with no clear future outside of an arraignment, and his wife wanted better for him. It's called "normal."

The staff called her "the Beauty Queen" from Karl's having brought her to one KFAT event, and that was the staff's impression of her. Nothing is known of either her reaction to Karl's new work-mates, or their conversation on the way home from that event, but she never came with him to any more KFAT functions during his brief tenure in professional radio.

In any event, Karl was soon in a new career, and he has been the only person I've spoken to for this book who refused to talk to me about the station. He told me that he had moved on with his life and didn't want to look back on his time at KFAT. But everyone at KFAT loved Karl J, and remembers him fondly. Some think he was the funniest guy they'd ever met.

Fortunately for KFAT, this wasn't where its listeners went for their news, anyway, and the station didn't really suffer from the lack of a news guy. Gordy missed the banter, and so did the listeners, but it wasn't a deal-killer for anyone. Besides, there was still that fifteen-minute talk show, Chewin' the Fat, at 8:30, and that was usually interesting.

Laura Ellen was in charge of the office. She'd be in one of the offices, either in Los Gatos or in Gilroy, handling the traffic and the correspondence and phone calls, and Jeremy would handle the equipment. With Larry and Chris gone, Laura Ellen wanted to tinker with the sound of the station. Keeping the Texas music, the progressive country cuts and the hardcore country cuts, she wanted to add more progressive country, more bluegrass, more blues, folk and traditional music, more comedy and more out-there stuff. Laura Ellen knew that more and more great music was coming in, that no other station was playing it, and their audience would eat it up. The jocks listened to her because besides being management, Laura Ellen was also family, and they liked Laura Ellen's suggestions, saw that she was in tune with the staff and the listeners, and except for Bob, they increasingly took her advice and added more of what she wanted on the air. And it was working: the calls with good feedback increased again.

Laura Ellen was exerting more control, adding albums to the "Hot Box" and the DJs started looking there for new cuts. She knew the audience wanted new stuff and KFAT was never going to be about playing anything all the time. Other stations had limited playlists, and she never wanted KFAT to be like that. Whatever came in that worked was Fat. Laura Ellen said yes to the wide-open playlist, the staff said yes, and the listeners called in to say they agreed.

Gordy Broshear remembered working under Larry and under Chris, and said that it was under Laura Ellen's direction that the sound came together. It was the time when the jocks were always reaching for new music, trying it, talking about it with the other jocks, and the excitement meant that the jocks were taking pride in their jobs, some of them for the first time in their lives.

GB: "Laura Ellen wanted to get some more roots in there. Bluegrass, blues stuff from the KDNA library. That's when the KFAT sound started developing."

Laura Ellen tried to hold weekly meetings, trying to corral the staff to show up. This was difficult because of their different schedules. Why would Buffalo Bob show up for a meeting at noon when (a) he wasn't on until four and (b) this would ruin his whole day of drinking, and (c) he wasn't about to take direction from Laura Ellen anyway?

Sherman and Ozone were asleep during those hours and were excused from the meetings.

But she held the meetings, and as many people came as would, and she would hold up an album and talk about it, and she started a primitive color-coding chart to guide the jocks on what to play. She put a swatch of colored tape around the bottom corner of each album, and the color stood for a type of music. The reds, blues, blacks, greens, oranges, whites, grays and yellows were for roots-y stuff, the blues records, the classic country, rockabilly, oldies, gospel, the progressive country, bluegrass, swing, comedy or out-there stuff, at which KFAT had excelled in collecting.

KFAT wasn't paying any attention to the charts. KFAT was literally off the charts.

Gordy's only real limitation on what he could play was a matter of decorum. When he took over from the overnight guy, he kept getting requests from the leftover overnight listeners, and they wanted the raunchy stuff, which Gordy couldn't play in the mornings. Instead, he got his kicks by playing the strange stuff. He had a special feature that he called his "Just Before Eight O'clock Early Morning Thing."

He featured some comedy, his childhood Spike Jones stuff, some Disney stuff. "One day I was late, and didn't start the segment until after 8:00 and I got about 30 calls from listeners who I had made late for work. Those people all said that they had to sit in their cars until after the segment played." It was Gordy's first realization of the impact he and the station were having.

I once asked Gordy if he ever stole someone else's segue. "All the time," he said. "You heard a good one in the evening, and it would be repeated the next morning or afternoon."

Sometimes Gordy was asked to introduce an act from the stage of the Catalyst, Santa Cruz's biggest live-music nightclub venue. He didn't get paid for it, but it was a thrill, and he did it. He was very insecure and very overweight, had never gotten much attention, and this was his first chance to be a celebrity, to meet the public, bask in some attention, and maybe get laid. The getting laid part wasn't working out, but he had fun and he did get a lot of drinks bought for him, and he came in to work hung over a lot in those days. On the air, he'd complain about his hangover. One night he was in a bar in San Francisco, and when he ordered his drink, a guy next to him said, "I know you! You're Gordy!" then he told Gordy that he remembered him once saying that he was so hung over that he woke up that morning to "the roar of my avocado ripening on top of my refrigerator." People were listening.

Besides the hangover issue, Gordy had a lot of health issues, and they were increasing. Serious health issues. Everyone knew when Gordy had used the bathroom, and knew to avoid it for at least an hour until it was safe to go in there again. Was it ever really safe to go in there?

His stomach was always fucked up, and he was often in real distress. Why he insisted on telling his listeners about these illnesses, in graphic detail, remains a mystery to many of the staff. Several of them spoke to him about cutting back on talking about his illnesses, but he did it, anyway. Tiny remembers the women who called Gordy to commiserate, or recommend doctors or tell him about cures. Tiny thought he liked the attention, and he continued to do this until his illnesses forced him off the air.

Gordy also delighted in regaling listeners in a chatty manner about the intimate comings and goings of the jocks, and Tiny was frequently pissed at him for telling everyone listening who she had gone home with at night or come back with in the morning. She was pissed off at him for this a lot, because she went home with a lot of different fellers back then, and if Gordy knew about it, he talked about it. It was good for Tiny's image as a wild woman, but bad for her self-esteem, and really bad for her parenting career if any Social Services people were listening.

Laura Ellen remembers, "Yeah, he did that. Monday mornings, one of his traditions was to go through the trash and try to figure out what kind of party they had had over the weekend. And I remember getting calls from Sister Tiny, 'If Gordy tells who I went home with one more time on the air I'm just gonna quit, I'm not coming back!'"

Like everyone else, Gordy also had money problems in addition to his love problems and health problems. He knew others had it easier. Terrell Lynn was getting laid regularly, and the Ozoner was getting blowjobs on *his* shifts. Don't even *mention* what Tiny had going on. He knew they were partying, but the parties always happened after he was gone. He was insecure and jealous. So he told the stories to his friends on the air, and several of the people at the station suggested again that he cut back on the personal gossip, both about staff indulgences and about his poor health.

Laura Ellen: "He was just a gossipy old lady. He'd tell everything. He'd whine on the air about his weight..." He was in love with a woman named Marjorie. It was obvious to everyone that she was a) married, and b) uninterested in a romantic liaison with Gordy. Gordy moped over her for years.

LEH: "He'd pine for Marjorie." She once told him, "Always pick a married woman to pine for because it was safe, you probably won't get what you wished for."

Once the transmitter moved in the fall of 1976, Gordy realized that taking the bus down was more trouble than it was worth, so he bought an old beater of a car, and he and Terrell Lynn rented a house in San Juan Batista, about a half hour south of Gilroy. The drive to San Jose was a half hour, but the bus took an hour. This was a good move for both men, but the booking action out of the Green Hut had to find a new source for the racing forms.

He remembered his Monday schedule: "I'd get up at four a.m. to do my show, do production work in the afternoon and then drive up to Palo Alto (to the Fat Fry- Chapter 82) and start drinking before the doors opened, drink through the show until it closed at two a.m., then drive to Gilroy, stop at the Longboat for coffee and then do my show." But his health issues were stacking up.

Chapter 47: B-Side Players

Subs, fill-ins, wannabes and used-to-be's. Part-timers, any-timers, old-timers and first-timers. They are all part of the radio landscape. Karl J went from college radio to KFAT in one easy leap, and then he leaped into the real world. He stayed at KFAT for a few months and then he got a real job. They came, they went.

Dalton Hursh, aka Reno X. Nevada came through and plied his trade on KFAT for a few weeks, until Chris Feder fired him because Reno no longer cared.

Some of the people who came through made names for themselves in radio, most did not. For many, their tenure at KFAT was all they'd ever know of professional radio. For some, it was a stopping point to oblivion, for a few it was a stopping point on the way to somewhere bigger.

Bill Goldsmith tried out for KFAT, and was told he didn't have it. He came back a few years later and found steady employment. Which changed, him or the station? A little of both, I think. Now he does incredible radio on the internet.

Someone who's coming later in this book will predict that the jocks who found steady employment on KFAT should enjoy their ride because it would be over for them when they left KFAT. And for the most part, he was right.

But come through they did, and come through for the station they also did. Cosi Fabian was an English girl who made the club scene in London when London was the swingin' center of the world. She got famous in London for being famous in London, and was known in the scene for being known in the scene. She came to America to get out of town and try something new, and she came west. She got caught up in the revolution that was going on in San Francisco, and stayed there, falling into a variety of situations, and one of them was KFAT. She did the news for a while, calling it "News From A Broad," and she was sharp, incisive, clever, and she was a success.

She lasted a few months while Travus T. Hipp was off somewhere, and then disappeared from radio. She's now an activist for the reform of prostitute's rights. Look up Cosi Fabian on the internet.

Steve and Helene Kane called themselves "The Citizens Kane" and did a self- penned interview/variety show that lasted a few months. A few people remember them, most don't.

There were others, and others remember that they were there, but the details get fuzzy and mix and morph into other memories, other details, confusing all who try to remember what went on, on the periphery of lives that were suddenly moving too fast to take in the peripherals. The jocks were hanging out and hanging on and being pulled along. Someone new would come in at the end of your shift and announce that they were your replacement, that the last guy who did it wasn't coming in tonight, and they were on next. So it went.

People came, people went. Sometimes it seemed as if there were a lot of them, then not so many. Every jock remembers people coming tentatively into the hallway outside the studio and looking in at them. They were looking for Laura Ellen; they were looking for jobs. Some got a tryout, others didn't.

Sometimes at first it seemed as if anyone with an ego could take a swing at it. Terrell Lynn took a swing as his alter ego, Skip Towne, doing a shift as someone else, and that was fun, once.

Several jocks remember Jeffrey Kramer, but none of them can tell me anything about him. Not what he was like, or his taste in music, or how long he lasted. No one can give any details about him and some told me only that he was tragic, but his name keeps coming up. I never knew him, and one jock now tells me he heard that he died. Rest in peace and all, Jeffrey, but who were you?

The jocks all recall any number of people coming up the stairs and staring at them through the glass window from the hall into the studio, across from "WHO STOLE MY SCREWDRIVERS?" They'd see that, and see that it had been there a long time, and they'd wonder if this was real radio. Was this what it was like?

But if they had to ask that question, they were not going to last. KFAT was for the people who looked at that wall and said, "cool." Some of them could stay. Some for a while, some for a week, some for a day.

There were also some really good jocks who came and went; some of them, indeed, very talented. Singer-songwriter Tom Russell jocked at KFAT for a few months, as did singer and producer Hewlett Crist, and both are worth looking up in their own right. Well known within radio engineering circles, the "Barefoot Wonder Boy" Don Mussell jocked at KFAT, and his wife, Rachel Goodman was "Calamity Jane" at Fat for a few months. Singer and Fat fave Mary McCaslin did some shifts.

Too many people came through to put into the book, but they all contributed. Some of those contributions were embarrassing, some merely fruitless, some were rewarding.

Thanks to all of them.

Chapter 48: Jeremy's Composite Clipper

In the summer of 1976, Robin Marks was looking for something to do. One of those techie wizards, he had worked for several radio stations back east, and had just come out to the west coast to visit his brother, who had an apartment above San Francisco's famous Haight Street. The apartment was pretty high up in the hills, and Robin, with nothing much to do one day, began fiddling around with the radio dial to see what the radio scene was like out there. He "liked to look between the nooks and crannies on the dial," looking for the weird, the unusual, the interesting. When he found a strong, clean signal, he stopped dialing, and he listened.

It wasn't just a strong signal, it was clean. And clear. To a radio technician, it was a beautiful signal. And it was playing "all this really nice music."

He had always wanted to like Country & Western, but he never could, "because they played too much shit, and every now and again they'd play a good song. But I wanted to like it, and here was a station I could like. It was right in between [listener-supported Pacifica station] KPFA (94.1) and [rocker] KSAN (94.9)."

Intrigued by the cleanliness of the signal, he called KFAT and spoke with someone who turned out to be Jeremy, who knew right away that Robin was an engineer because they weren't request questions, they weren't what-was-that-song questions (KFAT got those constantly). These were techie questions, and Jeremy and Robin struck up a friendship that has lasted to this day.

Unknown to Robin at the time, the conversation with Jeremy was circling the vortex, and when he accepted Jeremy's invitation to come to dinner, he was pulled right into it.

It was during one of the periods when the phone company had taken the phones out again for unpaid bills. Jeremy had routed the calls to the only phone he had that still worked, at their home in Los Gatos, and that was where Robin reached Jeremy, so Robin went first to Los Gatos, where he fell into like with both Jeremy and Laura Ellen. He was new in town, he didn't know anyone other than his brother, and these people were friendly and were very much like him. Unusual, for sure, but that was good.

Laura Ellen made dinner, and Jeremy and Robin stayed up all night working on a cigarette pack-sized device that would display the phone number of the incoming call. This was long before pagers and cell phones, and caller-ID was a technology that had yet to be invented. But Jeremy and Robin did it that night, just screwin' around with spare parts.

By using a technology that they invented as they went, and using KFAT's sub-carrier (don't ask), they proved that it worked. Robin was in the vortex now, aboard for the ride, and he says, "it proved to be a pretty good way to waste the next three years of my life, until I got a real job."

Over those three years, Robin would be an engineer at KFAT, a fill-in DJ, a sounding board and a fellow experimentalist for Jeremy. That was big for both men, but perhaps especially for Jeremy, who needed a friend badly. Friend? I could say 'co-conspirator' but that comes later in the story.

Showing Jeremy's appreciation for a fellow technician, Robin earned $175 a week, and I doubt that anyone at KFAT knew that. The jocks were still earning $3.25 an hour and struggling mightily. They knew they were the Makers of the Sound, but landlords didn't care, and they couldn't pay for squat.

Robin also remembers cashing his paychecks if he "went to the bank first," and remembers more than once when he would take his paycheck to the bank and, without bothering to look at it to read the amount, "the teller would hand you the check back because there was no money in the account." It didn't affect his relationship with Jeremy, he "just thought it was part of the territory."

One of Robin's first jobs at KFAT was to build their phone system. Jeremy didn't want to pay phone bills. He thought they were a needless expense, and that drove him crazy. Phone calls were just electrical signals that you could throw around, and paying somebody else to do what he could do for free pissed him off. Robin built a system that used KFAT's own transmission lines. It was a highly unusual system in the outside world, but it was the *Modus Operendi* in a Jeremy Lansman Operation.

KFAT's studio was two doors away from Continental Telephone, from whom they got their service, when they got service, until the unpaid bills forced Continental to cut the station off. But Jeremy was a techie, and as he was so close to Continental, one day he met one of their techie's in the parking lot, and they got to talking. They talked tech talk. They discussed carrier lines, sub-carrier lines, and other technical techie tech stuff, and Jeremy mentioned his need to get phone service cheaply from Gilroy to San Jose and San Francisco. Back then the service from Gilroy to almost anywhere was very expensive.

KFAT couldn't get phone service in Gilroy without paying Continental all of his past-due bills, so that was out. Jeremy got a phone number in San Jose, where he wasn't known and his credit was still good, and he asked for three lines. We're gonna get technical for just a bit, but if I don't do it, then

I'll have to waste a lot of space explaining it, and to do that, first I'd have to understand it, and that's not going to happen, so here's what Robin said:

"The transmitter was in the San Jose dialing area, so Jeremy ordered three lines installed at the transmitter, and we figured out a way to transmit them on the main station carrier so that they'd go from up there down to the studio, and ring in the studio. And then the response from the phone in the studio would go use the station-to-transmitter link."

The new phone system worked quite well and cost Jeremy nary a cent, which made him happy until the day the phones stopped working again.

It was an hour drive from the station to the top of Loma Prieta, and when Jeremy got to the shack that housed the transmitter, the door of the shack had a piece of paper taped to it. It was from the FBI. It was a search warrant.

After all this was over, Jeremy filed for his FBI file through the Freedom Of Information Act, and found out that a guy who used to hang out at the station, and had a thing for Laura Ellen, had a grudge against Jeremy and knew about the weird phone hook-up, and he'd called every nearby police department, finding no one interested enough to investigate. Eventually, he called the FBI.

The Feds had recently busted a guy who called himself "Captain Crunch"* who was making and selling "blue boxes" that allowed people to get phone calls for free. They had gotten a lot of publicity for the bust, they'd tasted PR blood and they wanted more, so when a disgruntled guy called and promised to tell them about another such case, they listened. And, of course, the FBI already knew and had a file on Jeremy, so they got excited.

They got the search warrant and went up to the transmitter site, broke into the shack, and disassembled Jeremy's private phone service. Jeremy's response to this will be discussed in another chapter, but as you're reading *this* chapter, we'll finish this story.

The FBI had gone to the transmitter site and found Jeremy's "weird little black box" and didn't know what it was, but knew that it was fishy. Several other radio stations had their antennae there, but this was the only one of these. But if they didn't know what it was, they couldn't charge him with anything, and that was their years-old, fervent desire. They had to know what it was in order to charge him. Then they could arrest him. At last!

The FBI went to the Continental Phone Company and found the guy who knew Jeremy and they interrogated him about what Jeremy had done. Jeremy's FBI file showed that "they wanted to know how this guy Lansman was making free long distance calls from Gilroy to San Jose, and they wanted to know what kind of equipment he was using, and which of Continental's equipment Lansman was using and how he was abusing their phone lines."

The engineer at Continental told the FBI that Jeremy wasn't abusing their

---

*Captain Crunch cereal used to contain a prize inside. One prize was a whistle with the same frequency that the phone company used to send a signal past a station that charged the user. By blowing the whistle into the receiver, the user by-passed the usual lines and went straight to their destination without being charged. One man duplicated the whistle and sold the device until the FBI caught him and sent him to jail. I hear Steve Jobs was making these too.

phone lines because they didn't have any lines into KFAT. The FBI couldn't find any laws that Jeremy had broken. Of course, the FBI already had a file on Jeremy, and they didn't like what they read in it. They wanted something to pin on him and they didn't believe the Continental guy, so they continued their investigation, which eventually led them to another of Jeremy's techie buddies. The Feds' method was to come into a place unannounced and demand to conduct an interrogation, but to first obtain a promise that the interrogatee would not alert the subject of the interrogation of the investigation.

The friend they found, John Higdon, was in his office at San Jose rock station KOME when the FBI showed up and asked Higdon for an interview regarding a wiretapping charge against Jeremy Lansman. Higdon said, "Well, you've got him cold." They perked up when he told them he knew all about Jeremy's little phone system. The FBI agents grinned at each other and said, "Okay, tell us more."

Higdon told them "that they had Jeremy 'cold' on a tariff violation." The violation was that Jeremy had connected a wire directly to the pay phone circuit, and that was a Fair Trade Act violation, and, if they completed their investigation, filed all the papers carefully and correctly, and prosecuted Jeremy to the maximum extent that the law would allow, he would probably be found guilty and fined twenty-five dollars. He also warned them that Jeremy would never pay the fine, so they'd have to take him to court to collect it.

Higdon repeated that they'd better fill out all the forms carefully and submit them properly and by the book, because he knew his friend Jeremy would be reading their filings, and Jeremy knew the laws as well as the people who had written them, and he'd be looking for mistakes.

Jeremy was an avid reader of all things technical, and it turned out that he had taken advantage of a poorly written law. The law was written so broadly that one could interpret the language to mean that when I call you and you answer the phone, you are tapping the line. You will have guessed that no prosecution followed the exposure of this heinous act of civil disobedience, but it did force Jeremy to get even more creative in his pursuit of free telephone service.

Jeremy was never questioned by the FBI about this issue, and he never heard about it again. But this isn't a chapter about the phones, this is a chapter about Jeremy's Composite Clipper. Unless you know what they are and have one at home, please read on.

Robin (remember Robin?) had been cruising around the dial and stopped at KFAT because it sounded so good. I had never been aware that different radio stations *sounded different* from each other. I always figured they used the same equipment and had the same signal, and the only difference was what they broadcast. Not even close, big-time journalist!

As a techie, Robin knew immediately that he was listening to a station that sounded good, and that was why he called and spoke to Jeremy about the signal. Jeremy had an amazingly simple and elegant—and cheap—way of making his station compete with the biggies.

I'd like to state here that I've been having a good time researching and writing this book, but I have my limitations, and explaining technological complexities is not my strong suit. So here's Robin again, and I hope you have as much luck as I did understanding most of it: "What Jeremy did… the problem with a low power station is that you at least have to make it sound loud, because people aren't going to be getting much signal wherever they are, and a weak signal should be as loud as possible.

"The trouble with loudness is that you're allowed to broadcast only so much of it. The FCC limits the deviation of the signal. How to make it sound loud to make it stay within the limits…

"The whole industry has grown… since Jeremy did this. It's called Composite Clipping. You take the entire signal, stereo and everything, and run it so that it's basically going through a diode or a diode bridge, and that in effect cuts it off at a certain point, above which would result in over-modulation of the transmitter. He figured out a way of doing that without making it sound bad."

Robin went on to explain that "Before KFAT, to my knowledge, no one had ever done Composite Clipping before, but afterward it became the basis of all the audio processors."

The soul of what Jeremy put together with a couple of diodes, some resistors and an operational amp, all for under ten bucks, is now in virtually every radio station in America. Ladies and gentlemen: Jeremy Lansman!

That's impressive and all, but Robin doesn't like what they've done with it since Jeremy put the first one together. He says that this technology has been pushed to its extremes, and the effect is that all the radio stations push their signals as much as they can, but they push the signal beyond its limits, with sole concern about loudness, and no concern about the quality of the sound. With the trained, specialized ear of a radio engineer, Robin can't stand to listen to radio these days. He says it sounds "awful." It's his job to know these things.

Jeremy's job at KFAT was Chief Engineer. His mandate (to himself) was to do two things: to make the transmitter put out as strong a signal as he could, and give it as good an *audio signal* as he could. As an engineer, he was committed to making it sound good from a professional standpoint.

Robin had another friend who was an engineer, and for Robin, Harkey's word was the last word in pronouncements of an electro-audio nature. Harkey thought KFAT was the best-sounding station on the dial. He lived at the wrong end of one of San Francisco's hollows, and to get KFAT, he had to wire his tuner to a lighting fixture in his hall. Some big electro-brain, you might think, but he got KFAT at home and his neighbors didn't.

More from Robin on the Composite Clipper: "The secret of KFAT's good sound and audio processing was some circuitry that he'd wired in behind the board without stabilizing, and there was a cold solder joint on an op[erational] amp, so sensitive to the touch that if anything nearby it got touched, and the station would almost go off the air. So you learned that you'd better not mess around with anything back there, because the things were just hanging together back there with snot and baling wire."

One thing that the jocks would disagree with Robin about was when he said "almost go off the air." All the jocks remember the station going off the air regularly at first. Sometimes it was an excess of heat at the transmitter shack, or there was too much of a temperature differential between the shack and the studio, or a sudden gust of wind that might sway... just a bit... the antenna that was *relatively* secure on the roof, or it might be that someone hit the wall or stomped their foot too hard near the Composite Clipper and the mess inside the wall got jiggled *just a bit,* and that might be why the station went off the air. But those are just some guesses.

People tried to remember to walk softly by the record stacks, and to try not to knock the wall between the stacks and the studio.

Sully remembers "sometimes the station would go off the air and you had to jump up and down in a certain place and it would go back on the air. If the station went off the air, it sometimes meant something was loose back there..."

Sully remembers Robin eventually coming to fix it, opening up the back where the wires were and going, "Aaarrgghhh!"

Sister Tiny: "It was like this huge spaghetti thing of briars."

Sully: "It was all alligator clips and twisted together..." she remembers. "It was a mess, and "something was loose back there, so you'd jump on a certain part of the floor, and it would come back on."

Robin: "This had happened hundreds of times. Jeremy didn't want to get in there and neaten things up. I didn't want to neaten it up, either. I just re-soldered all the wires and the problem went away."

But that repair was still a ways into the future, and now we must get back to the station.

Oh, yeah, Robin has something else to say: "...so finding a station like KFAT, which sounded so pure and musical, was quite nice. And an amazing thing when you went there to see it, and it looked like such shit."

Chapter 49: Open Door Policy

Try it today. Just try it today.

You say you wanna go to your favorite radio station and talk to the jock? You wanna go there and hang out with the jock, right? Go ahead and do it, brother or sister, but bring your lawyer's business card and some money for bail if you're going to insist on seeing that jock, because you ain't gettin' in, pal, and that is the that in that.

But this is KFAT, so come on in.

There was no sign on the door or above the door. No neon, no paint, no poster, no placard, no logo. Just four dirty three-inch high brass numbers, 7, 4, 5, 9 above the door. But on that glass door, right in the middle at about chest height, was a KFAT bumper sticker, all five inches high and three and a half across, and to any Fathead who had gotten as far as Gilroy and managed to find its main street, that could mean only one thing: KFAT inside.

Silent from the street, you could hear the music as soon as you opened the door. Twenty-three steps up, and you were on the small landing at the top. From there you could go straight, to the Production Room or the bathroom (God help you), or to the right to the studio, the offices and back rooms. Unless bursting of bladder upon arrival, a visitor would always be drawn to the source of the music, the studio. And so they were.

They were almost always polite and respectful, and usually they brought some goodies with them. The Fatheads knew that the KFAT jocks liked to party and liked to party some more, and they came prepared to contribute to the party. Wouldn't you? If you were going to a party at a friend's house, like, say, a barbeque or a dinner, don't you bring a little something? A six-pack? A bottle of wine? You know what I mean.

People showed up, came upstairs and hung out. They usually brought a little something, like beer, some coke or a joint, but even if they didn't, it was always cool. Terrell Lynn remembers "People wandered in a lot. Sometimes it was the Gilroy police.

"Sometimes I'd be standing there with a joint in my hand and turn around, and there were the police, and they wanted to request a song. They'd be driving around listening to us. It happened a few times a month at first." He remembers the cops that came regularly.

TLT: "Two cops. I turn around, and I got the joint in my mouth, and I'm kinda looking at this album cover, and I got the cover in one hand, the other cover in the other hand, and I turn around and he goes, 'Hey, how you doing?' They were requesting a song. So I put the album over the joint. I thought they were just going to take me away. Didn't say a word."

Then, one night a cop who had been coming to visit Sister Tiny regularly told her that he couldn't come up there anymore. He said that all the cops knew what was going on up there, and he couldn't come back.

ST: "He was so nice, and finally he told me, 'you know, Tiny, I just can't come up anymore because I'm afraid if somebody finds out, I'll lose my job.'" He was genuinely sad about it, and he said he could overlook all the illegal stuff personally, but he had a career and a family to think of, and...

More people started showing up at the station at odd hours. Odd people showed up because they were too odd to work, and they had their afternoons free. Odder people showed up at night, and there were more of them.

Unless the station was on Ken Alert for Tiny, the door was unlocked and people came in. Once in a while someone was looking for a room to rent for the night, or a hooker. In the wee hours, it was the only unlocked door on Monterey Street, and some of the winos or junkies were just trolling, checking doors as they cruised or stumbled down the street. This one's open? Where's it go? I'll jush... go in there... an' sheck id out...

More often, they knew where they were going, and they wanted to party. Usually they Brought Their Own. BYO was for parties with their pals, but Bring Something For The Jock was what you did when you went to KFAT. Terrell Lynn still laughs when he remembers the overnight shift when the Muffin Tin Twins, a couple of 19- year olds from the East Bay, up near Oakland, came down with a muffin tin covered with aluminum foil.

Terrell Lynn logically surmised that they'd brought a bunch of muffins, which was fine with him, because he liked muffins and he was getting a mite peckish. But when they took off the foil, each concavity of the tin had a different treat in it. In one were some capsules, in another, tabs of LSD; in a few were different buds of what looked like high-grade pot. In another was a chunk of hashish, a particular favorite of Terrell Lynn's.

Then there were the pills, the powders, and a chunk of something that Terrell didn't recognize. The Muffin Tin Twins were young, and they had stamina. The Twins and Terrell Lynn stayed up all night and used it all up. Everything. That might sound like heroic consumption to you, and it was, but let me remind the reader that (a) Terrell Lynn was in shape for this, and (b) it took them all night. Kids: don't try this at home. And stay in school!

Tiny had a cute guy visit her once, and then send her a package. In it was a big rock of cocaine, about the size of a golf ball. Reports at the time indicated that it was of excellent quality.

Sully was sent a package from a guy at Stanford University, just up the road apiece. It was a very fine, beige powder in a small, laboratory-looking little bottle with thick glass, and it came with a letter. The letter told her that everyone he knew at Stanford was listening, and the powder was MDA, a rare and powerful psychedelic, and "If this isn't your thing, please excuse me and toss this away."

As her mother will be happy to learn, Sully "wasn't about to take something from somebody I don't know." Ever adventurous, Terrell Lynn said, "I'll take it."

The guy sounded like "he knows what he's talking about. He was so polite."

Sully: "He was very polite. It was a typed letter…" Terrell Lynn has fond memories of the contents of the package, and spent many happy hours thereafter wandering in the Black Rock Desert in Nevada.

Again, thanks, listeners, and keep those cards, letters and packages comin' in.

Sully remembers: "But it was different with people who came from Gilroy. The people in Gilroy didn't know there was anything about us."

Outside of the police and the usual suspects, no one in Gilroy had any idea that something unusual was happening in their town. The local newspaper, *The Gilroy Dispatch*, never mentioned KFAT, and after KFAT's initial attempt at having a relationship with the town by hosting a pancake breakfast that only Gary Steinmetz of the Green Hut attended…

After that, KFAT was on its own in Gilroy.

Laura Ellen remembered, in those early days, going to a broadcaster's meeting in San Francisco, and finding someone who lived in Gilroy. They were talking about a restaurant in town when a man came up to them and asked her if she was Laura Ellen from KFAT. She was, she acknowledged, and the man went wild in front of them both, ranting and raving about how great the station was, and how popular it was in San Francisco and how everyone in the City was talking about it and how amazing it was, and how honored he was to meet her. He went on and on, praising the station, and repeating that everyone in San Francisco was talking about it. He was visibly impressed by meeting her.

The man from Gilroy stood there in open-mouthed astonishment: he had no idea that anyone from out of town even *knew* there was a radio station in Gilroy. He'd thought it was a little local station, just like it had always been.

The FCC demanded that all announcers, i.e. jocks, should have licenses. Jeremy sorta wanted the jocks to get them, and he mentioned it periodically. Tiny had some friends keep her up for three days studying for her third try at the test, and she passed that one. Terrell Lynn also took the test three times before passing. Bob promised to get his, and dared you to ask about it. Sully got hers, Gordy already had one, and Sherman remembers when his license, acquired back in Broadcasting School, saved his butt when he first got there:

"I was signing on the station one Saturday morning at six a.m., because at the time they didn't have a full-time staff, and they were shutting down at midnight, hiding the key under the garbage can in the back."

"I went on the air at six and got there at 5:30, and the key wasn't there. In the studio there was a hole there where an air conditioner had been, and it had a piece of plywood there. And I thought, if I don't get the station on the air, someone's going to be pissed. So I moved the plywood and was crawling in the hole, and I got about three quarters of the way in the hole and I feel somebody grabbing on my leg, saying, "Hey, hey!" and I turned around and it was the Gilroy cops, and they thought I was breaking into the building."

"But sitting outside by the door was a stack of records that I brought to play. They had followed me as I drove into town... I guess I looked suspicious... and they followed me and were watching me, and my license was on the wall and I showed them my driver's license and they saw the other license, and said OK and left."

But when the station was on the air, that front door was unlocked.

Cuzin Al: "I used to have regular visitors. There was this schoolteacher that lived in Gonzales, and I'm not the only guy he visited," he says, "but on Sunday nights he'd be coming back from San Francisco with a load... and he used to bring by these beautiful big Thai Sticks, oh man, and he'd leave 3 or 4 of them there with me, and he'd smoke one with me..."

Sully's shift was over and she was hanging out with Frisco in the Production Room one afternoon while Bob had visitors in the studio. Old friends from his past, if you know what I mean, and the friends left just as the LSD was coming on, and Sully remembers, "he started laughing. He was on the air and a commercial or something struck him as funny and he started laughing, and then I ran in and took over, and then he thought that was funnier. That was *so funny!* He kept laughing, and he was, like, [he] ended up sitting on the floor of the studio, with me doing his shift, and he laughed for about 35 minutes. Every time I opened the mic you'd hear "*ho ho ho...*"

Dr. John stopped by one night with Aaron Neville, and told Sherman that he'd been listening, and said Sherman was "the best chair actor" he'd ever heard, then asked Sherman to move "that pile of records." then dumped a pile of cocaine on the counter where the records had been. Neville had laid an album cover flat over the Hot Box and was rolling Thai Stick joints, and thus went the night.

Sherman also remembers the two cowboys who came in one night, carrying a half keg of beer. One was tall and thin, the other was average height, a little chubby, and they looked like typical redneck guys. They said they were from Gilroy, they'd been listening, and decided to come over and drink some beer with the jock. It was late at night, the door was open, and they walked in with the beer. They said they liked the music and their friends all listened, and they'd like to stay and drink some beer with Sherman, who was at least as happy about the beer as he was about the company, and the two cowboys pulled up a couple of chairs and settled in for a visit.

They favored the traditional country stuff, and Sherman played a long set of it for them. They smoked their cigarettes, drank the beer, and they laughed and talked about this and that. They talked about the music and the station, and their work as cow-hands, and then some more about the music. They were getting comfortable with Sherman, and Sherman was having a good time with them, and the two visitors stretched out their legs and leaned back, and then the tall one pointed his thumb at his friend and said that them and their friends liked the station a lot, and listened to it a lot, but there was one thing they all thought, and that was that the station played "just too damned much nigger music."

Shocked, Sherman's head jerked up and he bellowed, "*What did you say?*"

The cowboy said that "them and their friends" liked his station and listened to it a lot, but them and their friends felt "they played too much of the niggers… Especially that nigger, John Lee Hooker. D'ya have to play so much of the niggers?"

Sherman had heard enough and was furious, and told them that they had just crossed a line. He told them that they couldn't talk like that at KFAT, and besides, John Lee Hooker, who lived in Gilroy—in fact, just three blocks away—was a friend of his, and they couldn't say that about a friend of his, and they had to leave. *Now!*

"But we jes'…"

"No, man, *get the fuck out now*!"

Surprised but calm, they stood up and politely said they'd leave. They packed up the beer and the cups they'd used, and the tall one handed the half keg off to his pal. "But before we go," he said, reaching behind him for his wallet and taking out a business card, he left it by the turntable, the one nearest the door, "we jes' want you to know who's listenin'." Sherman said nothing, just stared hard at them as they carried the beer out the door, with both of them, saying, "good night, y'all." Sherman looked away and finished the beer.

Still steamed, Sherman dumped the last of his beer and, now in a foul mood, went to the stacks and pulled out a few records by John Lee, cued them up, and played a few cuts. As the second song was playing and he was alone, he picked up the card. They were from the Ku Klux Klan. The card is in the next photo section.

Chapter 50: This Way

Working late through a Saturday night, I was editing an upcoming "Chewin' the Fat" in the Production Room when Bob told me that Laura Ellen wanted to talk to me on the phone. Laura Ellen told me that Tiny was unable to make it for her Sunday morning shift, The Gospel Show, and could I do it? I thought.... Huh?

"Well, uhhh... you know, us Jews don't get a lot of gospel in our services... uhh, I don't know much about..."

"Well, look, Gilbert, I've got no one else who can come in, and you're already there... "

"Well, yeah, sure, but you know I don't know a lot about gospel... "

"Doesn't matter. Do your best, have a good time, and don't let the records run out. Talk if you've got something to say. You're good with talking."

Buffalo Bob was doing the overnight that night, and if he was not in the mood to hang out and do the Gospel Show, then there was no fucking way that he was going to stick around for a few more hours and do the Gospel Show. And so it fell to me. Laura Ellen had taken several minutes, and given me clear and explicit instructions as to which turntable and microphone controls worked which function, and in which direction they either turned, flipped, switched or pushed. I wrote it all down. I wanted badly not to make any mistakes. This was my first time DJing, and this was live. Listeners would be listening. Jocks would be listening. Laura Ellen would be listening.

I finished my work by 2:30 that morning, looked through the library, picked out a bunch of what looked like Gospel albums, then went back into the Production Room and spent a half hour playing some of the music that I'd found, making notes. At 3:30, I went into the studio to ask Bob to wake me at 6:30. "Sure," he said. And he did. He woke me at 6:30 with a foul oath of indeterminate intention, and I drove the four blocks to The Longboat and got some coffee to go. By 6:50 I had all those records out and I had the piece of paper with Laura Ellen's instructions laid out on the board in front of me. I had about an hour of music planned out, I'd take my first vocal break after about twenty minutes, and I knew what I wanted to say.

Bob was antsy and already mostly out of there, and I said, "Watch me," and went through the controls according to the instructions Laura Ellen had dictated, and Bob said, "No, no, no! That's fucking wrong!" and he proceeded to tell me that the mic switch worked the exact opposite of what Laura Ellen had told me. Not up! Down! I didn't know what to think. Could I have written the instructions wrong? No, that couldn't be: I'd been very sure of that. Could Laura Ellen be wrong? Well, I fucking doubted it. But she had been on the phone, and Bob was right there. I told him what Laura Ellen had told me, and he glared at me, and snarled, "You gonna believe her… or *me* ?"

God, he was scary, so I said, "You, Bob." He said, "rrrraammahhmrr," and walked out the door, hooking his thumb over his shoulder, saying, "play that last record" and he was down the stairs and gone.

I looked around, the place was empty. I'd never seen it like this before. It was seven o'clock in the morning, people were listening and I was alone and in charge. And it was Gospel Time. I played Bob's last song, opened the mic saying, "This is K-F-A-T, Gilroy" at the top of the hour. I didn't hear it through the headphones, but then I wasn't sure if I was supposed to hear it or if I hadn't turned on the headphones, or what had happened, but it was only that brief ID, and I didn't worry about it. I had other things to worry about.

What I worried about was whether the song would play and if it went out over the air. Then I hit the play button, and heard the song over the monitors. This was a good sign, and the needles on the meters that measured volume were jumping around. This was also a good sign, so I played out that record, and then the next, and the next, and it seemed to be working now, and after the first few, I had a few more picked out, and I was ready for my first-ever live vocal break on KFAT. I had a song cued up on my right, and when the song on my left finished, I took a sip a water, a deep breath, flipped the switch down, reminded myself to speak slowly, and said, "Hi, this is the Gospel Show on KFAT. It may not sound like your typical Tiny. In fact, it's *not* your typical Tiny. It's your topical Gilbert, and I'll be sitting in for Sister Tiny today. Here's something to think about on a Sunday morning," and I hit the play button on the right turntable, and heard the song over the monitors. I flipped up the switch, and took the headphones off. Then I wondered: had I *heard* my voice through the headphones? I'd been so nervous that I hadn't noticed, and now I didn't know or remember if I'd heard myself through the headphones or not. It didn't matter. It was over, the show was on the air, and I knew I could do it. Then the phone rang, and Laura Ellen wanted to know why there was a half-minute of dead air just then?

That fucking Bob!

I told Laura Ellen what Bob had told me, and she said, "Between me and Bob, and you listened to *BOB*?" She was right, of course. That fucking Bob! When I reminded her of this years later, we thought that it could have been that Bob made a simple error. Or maybe not. That fucking Bob!

Another time, and for this one I know the date, and it was February 17th, 1979. Don't ask how I know the date, because it involves a guy I knew being shot by the police in a hail of gunfire, but that is the date. Laura Ellen needed someone for a Sunday night after Demento, and I was usually there on Sunday afternoons, so she asked if I could do the shift. I had a week's notice for this one, so I said yes, definitely. I also asked, and I wouldn't say no, anyway, but instead of the normal Fat stuff, could I bring in my records from home and play the stuff I like? I said, "it won't be Fat, but it'll be Fat." Laura Ellen had no problem with that, and I announced on my show that I'd be sitting in as DJ that Sunday night. I called it "Irregular Fat," and I invited Sherman and Terrell Lynn in to hang out, smoke something, snort something else, and overlook my shit. Done, done, done and done.

I was going to tell you who I played, but instead I put it on the website. Yes, the music, too. I started with "Magical Mystery Tour," and went into Blondie's "Fade Away And Radiate." I had a ball, and thirty-seven years later I still listen to it, and it still rocks me. It comes from a time when people *listened* to music, a time that has sadly all but passed. Thirty-seven years later I still consider it an outstanding set of music.

Hell yeah, listen to Irregular Fat!

Chapter 51: The Chapter Before Chapter 52

You'll think that the title of this chapter is me being a wise-guy. Well I am, but it's not.

Chapter 52 gave me conniptions, and after I wrote it I never rewrote it, which I do for everything, and then I avoided it like the plague. Any plague. I had given myself the responsibility of coming up with a chapter that defined, that quantified and qualified, that explained the excitement of KFAT: the music, the jocks, the attitude and the listeners. Everyone knew the station was special, but they knew that by listening to it, not by reading about it, so how was I going to convey the buzz, the joy, the excitement, the pulse of KFAT to a reader who wasn't there? KFAT was a rare, serendipitous conjoining of the right people at the right place and the right time. The right time? How would I explain that? How would I explain the station? How would I explain a single artist or a single song that someone reading the book hadn't heard? How could I explain the experience of hearing Red Steagall when the reader would most likely have never heard of, much less heard, Red Steagall?

Why focus on Red Steagall? I don't know, he just popped into my mind when I needed an example of a great artist who no one had heard of outside of KFAT. Red just popped in there once, and I used him whenever that need arose, but that wasn't going to help me help the reader understand the station or the excitement.

And then one day, while rambling through the dusty halls of the internet, I came across a couple of sites that didn't feature KFAT, but referred to it.

I saw that journalist Paul Lambert, in trying to explain KFAT, once wrote that, "…after the rebel outsider excitement of early Fifties rock 'n' roll, shown by Jerry Lee Lewis, and Chuck Berry, rock had been grossly overtaken until the Beatles and psychedelia had restored its outsider status. But after the deaths of Janis, Jimi and Morrison, rock had become the bloated glam rock of Foreigner and Frampton. At KFAT, the old country honky-tonk shit-kickers parties with the new outlaws of Willie, Waylon and Kristofferson and a new hybrid was born, Americana, that remembered that rock 'n' roll or country rock was the expression of the outsider, the don't give me no shit strain of

America, the let me smoke my joint and drink my booze in peace.

"The excitement was the birth of that new hybrid; rock was no longer the city white boy's toy and country was no longer just for the redneck. This music spoke to people as no other rock had since the early days. It was alive, free, in your face, take no prisoners, goddam I'm glad to be alive and don't give me no shit about it.

"Now we knew why our parents had warned us against these people. We have seen the enemy and they are us."

Lambert went on about the music later in the piece: "These guys were not great just because you could get high and enjoy them, they were a representation of a strain of American thinking and feeling that needs to be let out of its cage every generation or so. In fact, the world needs them again, now!"

So, after reading that, I didn't re-write the next chapter. All I want to do now is remind you that rock, in the form of Elvis Presley, was a hybrid of country, blues, swing, R&B, gospel and more. It hybridized, and Elvis may not have been its first voice, but his was the voice heard around the world. Then it became bloated and insincere and profitable with the Frankie's and Bobby's and it became a business thing. The early doo-woppers were replaced by factory-trained acts, and became less intense, less passionate.

The Beatles and Stones etc. liberated it, gave it passion and drive, and then the passion fell off and it fell to corporate acts like the above-mentioned Foreigner, Loverboy, Toto, etc. By the early 70s the British Invasion was over, the culture, and its radio, was bereft of passion, and in this sterile atmosphere, Disco thrived.

KFAT was exactly what Paul Lambert said it was: a hybrid. It was new and exciting and it was like you were first hearing early rock, or first discovering the Beatles: something to get excited about. And knowing that you were in on the excitement, but it only went as far north as Sonoma and as far south as Salinas: it was a special thing, it was limited in its geographical reach, and it was OURS and fuck the corporate asshole acts and fuck the disco shit and get me another Longneck, darlin', and let's get to rockin'!

The music and the jocks: that was something to get excited about!

So now I'll send you to the chapter that I dreaded. I loved writing it. Had fun writing it, but it wasn't what I wanted it to be, and I knew I'd want to largely re-write it. But I didn't re-write it, because I think Mr. Lambert made my point.

The music and the jocks: that was something to get excited about!

Oh, yeah, and don't forget to check out Red Steagall.

Chapter 52: The Music & Emmylou

It was fun, it was exciting. The jocks all had jobs now, which was a new thing for most of them, and they were still getting used to that. 'Imagine that,' they must have thought, 'I have a job.' Yes, they did, and Jeremy and Laura Ellen had a working radio station, and now KFAT had a more powerful signal from a new site. Excitement was in the air, and on the air.

But it was always all about the music. The jocks were new at it and getting attention for the first times in their lives, but if it wasn't for the music, they wouldn't be there. If Larry or Chris or Jeremy or Laura Ellen or Lorenzo had wanted a normal radio station, they'd have gotten normal radio people to work there. Jeremy, Laura Ellen and Lorenzo all wanted something to stir things up. That was why they'd been the only people in America willing to hire a guy like Larry Yurdin.

Larry was on to something bold and new and exciting, and he knew that there was enough of this new music, if you spread it out with folk and blues and country and gospel and western swing and bluegrass and comedy, enough to be a format.

And he'd been right, but now it was even better. Larry knew there was enough music out there, but now there was even more being made, more being recorded, and it was coming into KFAT in records and tapes. And it was *real* music, not in the sense that other music wasn't real, but in the sense that the songs dealt with a reality that people could relate to. Often humorously, frequently tragically, but always... real. I mean, whoever was listening knew exactly what the singer was singing about. So much of rock music had lyrics that just made no sense when you could hear them. Disco was huge everywhere else in America, but it was bullshit. The singer-songwriters like the Beatles, the Stones and the Kinks weren't making the music in 1976 that they had in 1968, and even 1972. But now there were the Eagles, Jackson Browne, James Taylor, and they were all Fat. And Tom Waits, who would never fit into any category other than that he was Fat. Who else played Roy Buchanan?

The Band had blown people's minds with their American folkloric influences; their first two albums opened minds and changed attitudes, and you could argue that "Music From Big Pink" was the first Americana album (others might cite Dylan's "John Wesley Harding"). But this was before anyone was calling it that. Guys like Bruce Springsteen were selling because they made music that people understood. And that's what all the music was like on KFAT. There were breakup songs, and you understood that. There were prison songs and you understood that. Travelin' songs, drinking songs, bustin'-up homes songs. Sad, funny, true or false, fast, slow, talking or singing, these songs spoke to the listeners. More than any other station out there, KFAT spoke to the Fatheads, and the Fatheads got it.

And the jocks were such obvious amateurs, the Fatheads got them, too, as they would a friend. And my God, the music they played that you'd never heard before. I don't know how many people in California had ever heard of Rodney Crowell before KFAT, but the guy was a monster.

He could sing and play guitar like a fucking champion on fire, and he was writing some absolutely incredible music. Where had this guy been? His albums were totally solid, and he was only the leader of *someone else's band.* Of course, that other band was Emmylou's.

Emmylou Harris had been a small-time local folksinger in Maryland in the early 70s, who years ago had charmingly written a letter to "Broadside" magazine, the folk music scene's flagship publication, asking for advice on how to learn to be a singer, and was now performing in small clubs whenever she could. At the time, Gram Parsons was looking to record a solo album after his departure from The Byrds. Parsons had taken over the guitarist slot in the formerly folk-rock Byrds, and taken them in a country direction. The only album he was a member for, "Sweetheart of the Rodeo," is still a beautiful collection of music and widely and justifiably considered a classic.

But Parsons wanted more, and he struck off on his own, forming the Flying Burrito Brothers, an all-country rock outfit. Or was it an all-rock country outfit?

When he wanted a backup singer for his solo album, friends told him about Emmylou, and he went to see her sing in a small club in Baltimore. Impressed, they worked out a few arrangements, and Gram asked Emmylou to fly to Los Angeles *right now* to sing backup on the album. She had been struggling to make $100 a week as a singer, and Gram would put her up at the Chateau Marmont and give her $500. It was her first professional gig outside of Maryland and she was blown away by the offer, but she took it and flew to Los Angeles.

She has since said that Gram taught her to sing country, but her talent had never been in doubt. As it turned out, during the recording, Gram and Emmylou became lovers, and the already-in-progress separation of Gram and his wife was now assured. Then Gram died of an overdose.* When the album, "Grievous Angel" came out after Gram's death, his wife, as she was still

*Gram's friends knew that he wanted to be cremated at his beloved Joshua Tree National Park. But he was still legally married, and his wife wanted him buried with her people in a cemetery in New Orleans, which Gram did not want. So his manager and another friend snuck into LAX, where Gram's body was awaiting air transport back to New Orleans, stole his body, took it to Joshua Tree and burned it.

legally his wife, had control over the album, and allowed Emmylou only one credit—as a backup vocalist—and eliminated all photos and other references to her from the album covers.

Since then, Emmylou Harris has worked non-stop and proven herself to be a brilliant singer as well as a songwriter,** with a voice and inflection that draws you in like into a bath of warm honey. There is sweetness and sadness in her voice, and it came across the airwaves and into the hearts of the jocks and the listeners. Emmylou Harris was the queen of the new music.

I've mentioned Joe Ely a couple of times, and as a reference he is unsurpassed, but I've only mentioned Delbert McClinton once. Shame on me. The guy is a giant. Delbert sang like his heart depended on it, and every cut burned with the joy or the sadness of the moment. Both of these guys rocked the blues and rocked the country. They were a revelation to the people listening, who were starting to tell their friends about KFAT and play it at work, so even more people heard the music and the weirdness and the jive. And there were so many other great artists that no one had ever heard of.

But where were these artists coming from? Why didn't America know about them? And my God- there were so many of them! And any of the following artists could be next on KFAT.

Who today remembers Alice Stewart and her band, Snake? Duster Bennett? Anyone?

Ellen McIlwaine? Tom Russell? Anybody? Sleepy La Beef? I *know* you know Shel Silverstein. How about Dave Dudley or country queen Kitty Wells? Karla Bonoff? Gary Stewart? Marty Stuart? Yeah, Dwight Yoakum rocked! Elizabeth Cotton and Jean Ritchie.

Does anyone know who Louis Jordan was? You should look him up. Randy Newman? C'mon!

DougKershawFreddyFenderEttaJamesCarlaThomasLauraNyro JimmyReedMuddyWatersLeonRedboneOzarkMountainDaredevils TomTHallJuiceNewtonConwayTwittyDaveVanRonkLightningHopkins CarolHesterRickySkaggsCharliePrideBuckOwensBessieSmithHoytAxton BennyGoodmanFatsDominoBobDylanEverlyBrothersGeneAutryJohn FogertyMartyRobbinsMaRaineyJoyofCooking,DavidBrombergSonnyTer ryandBrownieMcGheeDireStraitsTheBlastersDanHicksandtheHotLicks JoyofCookingDixieDregsNarvelFeltsQueenIdaSpikeJonesStrayCatsJona-thanEdwardsBruceSpringsteenJudyCollins. Pinetop Perkins. John Hartford, Dave Edmunds and Nick Lowe. James Brown. The Andrews Sisters. Steppenwolf. And there wasn't a good doo-wop song that *wasn't* Fat.

We played Bob Marley, Peter Tosh, Jimmy Cliff and anyone from "the Harder They Fall" album, and then we found lots of other reggae guys, and of course *any* Clapton, including the John Mayall stuff. Les Paul and Mary

**Bereaved by Gram's death and too broke to get back to LA, Linda Ronstadt flew her out, and the canyons above LA were on fire. Emmylou saw the beautiful canyons and hills, and the sky that appeared aflame, and saw a connection to her feelings about Gram, and wrote the song "Boulder To Birmingham" on the spot. Listen- you'll hear it..

Ford (how high *was* the moon?).

Mac Wiseman had a career that as of this writing has spanned 55 years, most of it distinguished and significant in the canon of American music. He's made influential musical contributions over that period, but have you seen any of his music in your iTunes selection?

Chip Taylor was a major KFAT favorite. True, he is probably best known as a footnote, if at all, to the rest of the world for writing "Wild Thing," the hit by the Troggs and later immortalized by Jimi Hendrix, and "Angel of the Morning" the hit by Merilee Rush, or for being the brother of actor Jon Voight. But for KFAT listeners, his "(I Want) The Real Thing" and "I Wasn't Born In Tennessee" were regular requests, and his tunes and his voice were perennial KFAT favorites. You'd have to love this guy. His album, "Somebody Shoot Out The Jukebox" is on the list of any Fathead who goes looking through old record bins and garage sales, and if you find one, put it up on eBay and see what happens. But listen to it first.

You know- we haven't even mentioned Willie Nelson yet. I'm so sorry, Mr. Nelson.

Now, I might not be the best reference for an artist like Willie Nelson, but even as a youth on Long Island, I'd been drawn to some of the new country bands, liking Poco and Rick Nelson's country outfit, the Stone Canyon Band, the core of whom later formed the Eagles. I didn't know most of the artists that KFAT was playing when I first got there, and as the talk show host it wasn't my job to know them, but I loved all of the music when I heard it in my car or at the station. I couldn't get KFAT at home in San Rafael, but I always had it on in my car. So, I'd been exposed to a lot of Fat music, but never paid all that much attention to Willie Nelson. I'd heard him, of course, and knew that there was a big buzz about him at the station, but there was such an overwhelming wave of new music that I was being suddenly exposed to, as were most of the listeners, and Willie was great and all, but there was so much other music, too. It was exciting, yeah, but it was overwhelming.

And then, soon after I'd gotten to KFAT, maybe a few months, Willie was playing in Oakland, right across the Bay from me, and Laura Ellen got me a couple of tickets. And Willie, as they say, blew the poor boy's mind.

I'd been wrong. Willie wasn't great. No, WILLIE FUCKING NELSON WAS *SO* FUCKING AMAZING!! He was incredible! He was an incredibly gifted and natural singer, and his voice and his singing style were unlike anyone I'd ever heard. You felt comfortably woven into a Willie Nelson song.

And he was, by far, the coolest fucking guy I had ever seen.

He was looser, and yet he had more control, and was having more fun, than any singer I had ever seen. He was *easily* the coolest guy I'd ever seen. His songs were fantastic. And he wrote them! Everyone in the crowd knew every song by heart, and I was looking around me, wide-eyed, at an arena full of people who knew something amazing that I had not known. I had known that Willie was popular. KFAT told me that, and all these people told me that, but WHY WASN'T THE WHOLE FUCKING WORLD IN AWE OF THIS GUY?

I was, in a word: blown away.

The next day I called Laura Ellen to thank her for the tickets, but before I got to the "thanks," I screamed at her that WILLIE WAS FUCKING AMAZING AND HE WAS REMARKABLE AND WHY DOESN'T EVERYBODY IN AMERICA KNOW ABOUT THIS GUY?

And have I mentioned? KFAT was the only station in California or anywhere else playing him. And he had his pal Jerry Jeff Walker, who was a hoot and totally cool. Jerry Jeff was really totally cool. "Pissin' In The Wind." Come on! "I Heard You Been Layin' My Old Lady," "Desperados Waiting For A Train." Come on! And he's that guy that wrote "Mr. Bojangles."

And don't forget Waylon Jennings. How much time can I spend on Waylon? Okay, you're on your own with Waylon Jennings. Please look into him. You'll be glad you did. Start with "Are You Sure Hank Did It This Way?"

Ray Campi and all the Rockabilly that KFAT played- and nobody else in America was playing Rockabilly. And Ragtime, too. And Buffy Sainte-Marie. And all the southern rockers like the Allman Brothers, Lynyrd Skynyrd, Marshall Tucker, Black Oak Arkansas, ZZ Top and the Eagles and Poco and Rick *and* Ricky Nelson and others, like Tom Petty. Roy Orbison, Roy Orbison, Roy Orbison. Marshall Crenshaw and *of course* Flatt & Scruggs. Steve Martin, Eddy Rabbit, Gordon Lightfoot, Freddy Fender, Carole King. Wet Willie. We played *anything* by Van Morrison. The Velvet Underground and Jesse Ed Davis. Sandy Bull and Bob Seger.

The Amazing Rhythm Aces were amazing rhythm aces, and some of their songs will blow you away. Ever hear "Burning The Ballroom Down"? Try to, please. But give yourself a few minutes to recover. "The End Is Not In Sight" is one of the best songs ever written in any language in any category at any time in any place in the history of the world, and if you don't believe that, then you haven't heard it, and that's all I'm going to say about that song.

Fuck it, that goes for *both* of those songs.

Ry Cooder. Billy Joe Shaver. *Any* Cash or Carter! Wanda Jackson, Marcia Ball. When Tracy Nelson sang "Down So Low," a listener who had been driving called to tell us that she had to pull over to the side of the road to listen.

Many listeners knew *about* Patsy Cline, but when they heard her now, they were blown away that she wasn't being played everywhere else. Jesus Christ! That was Patsy Cline and she'd had *hit records,* for God's sake, and why wasn't she being played anywhere else? Her death had stopped the flow of new product, and country stations had moved on without her. And one time when Sully played Patsy Cline and followed her with Aretha, a lady called to tell her that she'd been driving, and she'd just missed hitting something when she heard that segue. Oh, they were listening.

U. Utah Phillips had "Moose Turd Pie," which was, without question, the most requested song on KFAT, and a true station phenomenon, and I guaran-fucking-tee you that this was the only station in the world that played that song. This song gets a whole chapter later.

Please- somebody tell me you remember Commander Cody and His Lost Planet Airmen.

Ledward Kaapana and any of the Hawaiian slack-key players, Tanya Tucker, Vassar Clements and Bryan Bowers and the Beatles in the same set? Try 'em in the same set with The Red Clay Ramblers. David Bromberg, Lonnie Mack, Dolly Parton, Linda Ronstadt, and yes, that's right, Ella Fitzgerald *before* she was a star; both Reilly and Maloney, Mary Kay Place, Townes Van Zandt, Maria Muldaur, Stevie Ray Vaughn, Crystal Gayle, Johnny Paycheck, the Boswell Sisters, the Holy Modal Rounders, Even Stevens. B.W. Stevenson. Elizabeth Cotton. Lord Buckley. The Blues Brothers. John Fahey. Flaco Jimenez. Grateful Dead! Janis Joplin!

But where else would the New Riders of the Purple Sage be played if not on KFAT? Or the Rowan Brothers or Alvin Crow or Asleep At The Wheel or Pure Prairie League or Bobby Bare or Jessi Colter or Kinky Freidman and his Texas Jewboys, or Leon Russell or Danny Gatton or Danny O'Keefe or Jonathan Edwards or Tim Hardin or Dan Hicks & The Hot Licks or Canned Heat or Mary McCaslin or Dave Van Ronk or Lacy J. Dalton or Jill Croston (wink) and Ricky Skaggs and Rosalie Sorrells and I'll stop for a while to catch my breath. (The Persuasions.)

Okay... And KFAT played the guys who were known, but were from the Golden Age of country. These guys weren't fitting in to the current crop of country hits, so they weren't getting much play—or none—on traditional "country radio:" Buck Owens and Mel Tillis and Johnny Paycheck and I know I mentioned Patsy Cline, but here she is again, and so is Merle Haggard and George Jones and Tammy Wynette and Loretta Lynn and Jerry Reed and Eddy Arnold, and because I can, I'm going to throw in Jerry Lee Lewis and Fats Domino again, and Ray Stevens and I'm probably about to lose you, so I'll stop.

Wait! Can I say two words here? Little Feat. Thank you. Oh, and Leonard Cohen.

Johnny Cash has gotten so much attention of late, and he is now so revered, and deservedly so, that it must be hard to imagine that country music had played his hits back in time, but time had passed and *no one* was playing his stuff anymore outside of KFAT. Probably anything that Johnny Cash recorded was Fat, and that was true of so many artists.

Neil Young- everything from Buffalo Springfield to Crazy Horse and solo stuff. And most of Stephen Stills.

Besides, they were all Fat, and whatever was Fat was played. Monty Python was Fat. Cheech & Chong and the National Lampoon were Fat. The High Wire Radio Choir, Hudson & Landry, Steve Martin and the Firesign Theater. Remember Lester "Roadhog" Moran and his Cadillac Cowboys? That was the Statler Brothers. And yes, we played Mel Blanc. Whatever was funny, and while you might have heard of Justifyin' Justin Wilson, you've probably never heard of Jerry Clower, and his records were fucking hilarious. Today, Chris Rock would be Fat, Martin Lawrence never.

Does anyone remember that before Martin Mull was an actor he was a singer and a comedian, had put out comedy records, and toured the country as a comedian? He once toured with his living room furniture, naming the tour "Martin Mull and His Traveling Furniture." Funny, very funny. And as you would imagine, a very intelligent funny. Smart and funny was Fat.

There was so much out there that was Fat, and it kept coming out and coming in. So much of it was new and virtually unheard until it got played on KFAT. Wherever else was Jesse Winchester going to be heard? And he was fantastic. Same with Randy Newman and David Lindley and El Rayo-X and Mary McCaslin again and Larry Hosford and a lot of people that you wouldn't have heard of without KFAT, and if you don't know who they are as you read their names, we have the lack of KFAT to lament.

All of those people in the last paragraphs are outstanding artists, and if you don't know them, I invite you to look for their recordings. You'll find many of them somewhere, but a lot of what KFAT was known for was the stuff that you almost certainly won't find today, not in stores, not on iTunes and not on websites, but it was just amazing music and a revelation for listeners when they heard them. That's why the KFAT library was so revered by the traveling musicians who stopped by.

That's why people today still trade KFAT airchecks online. Also, how many radio stations that went off the air more than 30 years ago still have people trading *every day* for memorabilia on eBay? It was the music and the memories of the music and the times. And that's why old recordings of KFAT airchecks are on the net right now, at www.kfat.com. It was the music, man. Or go to http://audio.lns.com/more_airchecks/kfat/1982_12/ and find 58 45-minute KFAT airchecks. I hope whoever was/is doing that, I hope they they're still there and put up more.

If you think that I infer that much of the KFAT library was too obscure *for you*, and obviously *not for me,* and if you think that I am somehow inferring that I am in some way better than you, well, without knowing the depth of your knowledge about Fat music or your true feelings on this issue, allow me to insert a preventative disclaimer, so that you may read it and see that the aforementioned premise, if you *had* entertained such a notion, would have been fallacious, so that we may move on.

About that, I have this to say: While it is possible that I may know more of the music discussed herein than most people, first let me say that I will gladly stipulate that I understand that this was largely due to a fortuitously happenstance occurrence by which I was thrust into a position whereby I would, with the merest act of consciousness whilst at the station, have picked up an unusually large, statistically speaking, exposure to, and hopefully subsequent familiarity and concomitant retention of the music under discussion, and that, secondarily, subsequent to my travels and discussions with many people on this subject, I want to state clearly and for the record that in the course of those travels, I have met several people who know considerably more than I do about the music presently under discussion, and, their having demonstrated said knowledge to me, I have, in those aforementioned circumstances, graciously and willingly acknowledged that superior breadth of knowledge, and therefore hopefully I will have established that I see nothing to be gained by espousing an attitude of superiority to anyone else whom I do not know, but who might very well have the greater knowledge than me of such obscure and arcane music as we refer to in this chapter, and I am both resolved to that as well as in admiration of that knowledge, and so I propose that we consider the issue settled, and we move on, okay?

I believe my point, before I was interrupted by myself, was that while you might find some of the artists mentioned above in stores or on the internet, there was much of the stuff that KFAT played that just won't be found. Assuredly, collectors have collections, and I have met a few people who went to the sale at the Gilroy Public Library after KFAT had expired, where all that was left of the station, about half of the remaining records from the library, were put outside for sale, and people came from all over California and other western states to see what was there. Several people have told me over the years which KFAT records they have, and what personally written comment from which jock is on the back cover. Most of them bear the stamp that was imprinted on them by Laura Ellen as she put them into the stacks, "**STOLEN FROM KFAT.**" But that is at the end of the story, and for now, the records were just now coming in that will not be heard again.

The Dusty Chaps? "Don't Haul Bricks On 66" is great. *Really* great. The Cleanliness & Godliness Skiffle Band. Remember "Don't Bogart That Joint?" That was Fraternity Of Man.

Blues? You think we didn't have *all* the blues guys? We played Robert Johnson, Blind Lemon Jefferson Bukka White, all the Kings- BB, Albert and Freddy, and then there was Muddy Waters and Howlin Wolf, Son House,

Leadbelly, and John Lee Hooker was a friend of ours and lived three blocks from us. We played Mississippi Fred McDowell and Blind Willie McTell, Little Walter, and any blues artist from the golden age that ever recorded.

What the hell, we also played Taj Mahal and Billie Holiday and Professor Longhair and Ray Charles and Dr. John and the Stanley Brothers and Ralph Stanley and the Louvin Brothers and Lefty Frizzell, John Hartford, Lyle Lovett, Bill Monroe, early Elvis, Joe Cocker, Jim Ringer, Norton Buffalo, Roy Rogers, Gene Autry, the Lovin' Spoonful, Cowboy Jazz, Riders in the Sky, The Kinks, Carol King, Joni Mitchell, Arlo Guthrie, Pete Seeger, Chuck Berry, Creedence Clearwater and the Blue Ridge Rangers, JJ Cale and Eric Clapton, John Mayall, John Hiatt and Fats Waller and the Flying Burrito Brothers and Tom Rush and Phil Ochs. Koerner, Ray & Glover. Eric Andersen and maybe Judy Collins. And Joan Baez, of course.

I don't think a day went by without ol' Doc Watson gettin' a spin, and yes we played Bobby Bare right before Dave Edmunds and Nick Lowe, and I bet you've never heard of Bob Hall, but as they say here in Mexico, *¡Qué pianista!*.

We played the Jim Kweskin Jug Band and the Memphis Jug Band. R. Crumb & His Cheap Suit Serenaders was the band fronted by the famous cartoonist, Robert Crumb. Remember the cover to Janis Joplin's first big-time album release with Big Brother & The Holding Company, "Cheap Thrills"? That was Robert Crumb and his distinctive cartoon style, but also distinctive were his old-timey musical stylings, and those fit right in on KFAT.

Chuck Wagon & The Wheels was a perennial Fat favorite, with "How Can I Love You If You Won't Lay Down?" and "Ice Cold Beer, Red Hot Women" and of course "Disco Sucks." Your brain will hurt when you hear Sol Hoopii's "Hawaiian Cowboy." And *I know* you've never heard of Robb Strandlund, much less heard his version of the Bee Gees' "Stayin' Alive." Strandlund recorded the song, but never put it on an album. The only recording of it ever to leave his possession was sent to KFAT, and that was the only place in the world where you could hear this amazing song that totally rocks. That song may be Bluegrass or Newgrass, or I don't know what, but it *so totally rocks* that I can't help but dance when I hear it, and I actually have a friend that I met when he emailed to ask me if I knew that song. That song! That song that almost nobody knows! Go online and listen to "Stayin' Alive." Do it now!

And that guy that I met because of that song? He's still a friend, and I know that because he knows he's going to have to buy his own copy of this book, and he still likes me.

Well, he says he likes me. Why? What's he told you?

The Siegel-Schwall Band was hot, but try to find the Cache Valley Drifters, or the Coon Elder Band or Chuck McDermott & Wheatstraw, or Free Beer or California Zephyr or James Talley or Dakota Dave Hull & Sean Blackburn, or Roly Salley, who still plays bass for Chris Isaac, but maybe you've heard his amazing song, "Killing The Blues, "and I say find it by Happy & Artie Traum, because it will become your favorite song in the world for at least a day, and when was the last time that happened to you? KFAT played it. Robert Plant and Allison Krause covered it, and they are deservedly well-respected artists, but the Traum's version kicks its ass!

The Wild Tchoupitoulas were a wild bunch of Louisiana Bayou Indians backed up by the Neville Brothers, and if you can find their record, grab it. Go "Meet De Boys On De Battle Front." That was a Buffalo Bob song, for sure.

J.D. Crowe & The New South were a hot band, but where are they now? Patrick Sky, The Goldcoast Singers, Pork & The Havana Ducks? The Light Crust Doughboys? Randy Travis? Son House? Memphis Minnie?

You'll still find Commander Cody & The Lost Planet Airmen, the Seldom Scene, Old & In The Way, or Kate Wolf or Paul Siebel or Danny O'Keefe or that outlaw biker rebel ex-con-says-he-was asshole David Allen Coe or My God, Bob Wills! Ladies and gentlemen, Bob Wills! Remember Merle Haggard's song, "In Texas, Where Bob Wills Is Still The King?"

Bob Wills is *Willie's* idol, for God's sake! Outside of a couple of small stations in Texas, you couldn't hear him anywhere, but KFAT played him.

Slim Whitman, Danny Gatton, Ray Wylie Hubbard, Dave Dudley. You might even find Little Roger & The Goosebumps somewhere on the net. Ever hear of them? Maybe I'll tell you later about their mash-up recording of the theme from "Gilligan's Island" sung to the music of "Stairway To Heaven." That song sent a team of lawyers from Warner Brothers down to KFAT's studio in Gilroy to demand an immediate cessation of playation. (I just made that word up.) It seemed clearly understood to the lawyers when they left Gilroy that afternoon with "both" copies of the song, that KFAT had agreed to a complete and total cessation of the use of that recording, but after their departure, and once they would be halfway back to San Francisco and stuck in traffic and unwilling to turn around and drive over an hour back to Gilroy, they were made aware that the cessation had indeed been complete and total, but, sadly for them, temporary, as another copy had been hidden when the lawyers called and said they were coming down to confiscate the record on behalf of their client. Fuck with KFAT, will you? Should I tell you about it later?

I know I mentioned the Amazing Rhythm Aces, but I know you can find them, and I urge you to do so. Do it now!

But you'll never find the Boswell Sisters, or George Gritzbach or Sid Linard, but if you can find anything by Fraser & DeBolt, I'd buy it from you. Rosalie Sorrels' stuff might be out there. Should be out there. Danny Gatton and Butch Hancock should be out there, Mickey Newbury is kind of a footnote, but that might interest the completist, and he was good, but just good, but since when isn't good good enough? Let's lighten up out there. That reminds me of Tom Rush and Eric Andersen. Also- name a blues guy- any blues guy- we played 'em. Good Luck finding the Lost Gonzo Band, but that was Jerry Jeff's band and worth some attention. Jerry Riopelle? Kate Wolf has to be available out there somewhere, and she was marvelous and a big Fat favorite. Hank Thompson. Hoyt Axton. Ann Peebles.

Did you know about Michael Nesmith? That's right- one of The Monkees. Nesmith was known as the only guy in the group who could play an instrument, both before the group was formed and after. Some say Peter Tork could play, but… nah. The Monkees were over, and Nesmith was making great, innovative music and innovative videos in the early days of that medium, and doing it at his hideaway near Monterey. Check out his "Rio" some time, or "Shelly's Blues" or "Silver Moon." He was a total Fathead, and listened constantly and called regularly. His "Elephant Parts" is widely considered to be among the first music videos ever made.

And before I lose you entirely, let me mention the musicians you'll know and find, like Bonnie Raitt and Jimmy Buffett; I think you'll like 'em best if you'll look for the early stuff. Ry Cooder's stuff is out there, and probably *anything* he's done would be excellent. Look into Ramblin' Jack Elliott and Jerry Jeff Walker and David Bromberg and Guy Clark and Taj Mahal and John Prine and Albert Lee (not the guy from "Tens Years After"- that's Alvin Lee) and Little Feat and Emmylou and Ely and McClinton and Peter Rowan and George Jones and Tompall Glaser and I don't know- isn't there somewhere you can look this stuff up?

In closing: If any of this discourse interested you, I found a couple of sites on the web that might also interest you. An avid Fathead has posted www.barovelli.com/tunes/kfat.htm, where you will find a list of 2,709 Fat songs that the author of the site identifies as the "cream of the cream" of KFAT music. Elsewhere, at www.barovelli.com/tunes/allmp3s.htm (and I'm just guessing, but I think it's the same guy) the author lists some 16,343 Fat songs. So you see, it's not just me.

And remember: I wasn't a music guy at KFAT, and any of the KFAT jocks would have made this chapter six times longer than it is, and maybe you're grateful it's almost over.

One more thing: this list of artists here isn't the history of what KFAT played over the years; this was the playlist *every day* at KFAT, and you might hear any of these artists at any moment. I don't know of anyone else had a playlist like this.

So Larry had been right, and it was a shame that he wasn't there now to see his baby grow and change and evolve into whatever it evolved into. It wasn't going to change too much, because the music was the music and the attitude was the attitude, but the jocks would get better, a sales staff might be hired who could generate some money, and people could start showing up on time for their shifts, and some of them could try a little harder to stay sober through an entire shift.***

It was a stable staff of unstable people, and it was going to get more interesting, and both Jeremy and Laura Ellen knew it. But neither expected that it would get more interesting *almost every day*. Crises were cheap and plentiful when you're poor and ragged, and every day a new issue arose. Most were dealt with by a phone call or a meal or a word, a bathroom break or some baby-sitting help. Other crises were more demanding.

The station was dysfunctionally functional, people were doing their jobs, they knew they were having an impact from all the callers and the people they'd meet, but they knew they were something special when Emmylou Harris came to Gilroy to meet them.

*** Don't say it! Don't say it! Don't say "…and pigs will fly out of my butt."

Chapter 53: Word Spread

By the fall of 1976, word about KFAT was spreading throughout the Texas community of hipsters and musicians. Then KFAT was making itself known in New Orleans and Los Angeles and Chicago and Muscle Shoals and in Nashville. No one else was playing this music, and the people being played were talking to their friends about it. Then the suits in the record business knew about it.

The folk music scene and the emerging singer-songwriters in New York and Los Angeles knew about KFAT. The rural folk in their hollers might still be out of the loop, but the labels that recorded their music were aware of KFAT. Today, the music is called Americana, but back in 1976 there was no name for this hybridized integration of styles. Back in '67 and '68, when music was first breaking out of its old AM restrictions, the new free-form FM radio was called Progressive Radio, and Laura Ellen was calling this format "Progressive Country," and that's about as good a title as you'll find, but Americana is what it goes by now, and that's pretty good, too

After a show in Salinas, Emmylou Harris and Rodney Crowell came to Gilroy to see what KFAT was about, and to meet the jocks who were playing her stuff along with all the other great music. She'd heard about the station, but couldn't believe it when the stories she'd heard were all true- more than true. Some crazy guy named Frisco set up some mics, and she and Rodney played and sang a few songs in that little room they called the Production Room. Then she went into the studio with the DJ, picking records out of the KFAT stacks, in ecstatic indecision. Play this? Play that? *You have that*? Play *what?* You have *that?* the jocks made suggestions that blew her away. Songs she hadn't heard in years, songs she'd only heard about. The station was playing some incredible music that wasn't being played anywhere else. This was a watershed moment in American radio, and Emmylou wanted to see it first-hand.

Then she DJ'd for three hours, and smiled the entire time. Ask her.

It was exactly as she'd hoped it would be: real people playing real music and no one was on an ego trip. No one was better than anyone else. They all wore jeans and T-shirts and flannel and sneakers and flip-flops and boots. They all drove beat-up cars and trucks, and they were all the kind of people she'd grown up with. Real people.

Chapter 54: What Impact?

After all the years have passed since KFAT has gone silent, fans still talk and wonder about it. And when I meet them, there is a great similarity at what they wonder the most. What they all ask is this: Did the jocks know they were making an impact? Did they know how special they were?

Fortunately, this is an easy question to answer: No, certainly not at first. With virtually every jock that I have spoken to about KFAT, I have asked this question, and the answer is always a variation on "no."

Kathy Roddy, a.k.a. Sully: "We were so isolated for a long time. You know, we were down there doing this goofy stuff, but we didn't really realize we were having... that other people were listening to us. It was becoming a good thing, like a big thing, but I didn't know that. I don't think we were aware of that."

Terrell Lynn agrees: "We weren't."

The jocks knew that the listeners were on top of every cut they played, and had a lot of emotion invested in the station. The listeners called and let them know that with smart requests, observant comments and provocative questions. Of course there were jerks who called, but they were remarkably few, and KFAT's audience seemed to want to be educated as well as entertained. They wanted to know which session players played on which tracks, who wrote the original, how old was that song, or where was that track recorded? Then there were the requests for "Moose Turd Pie," and I will discuss that soon.

The jocks knew the listeners were *right there* with them a lot of the time. They may have called in one caller at a time, but the lines were almost always lit up.

Were the jocks stars?

Where? In Gilroy? Give me a break...

In San Francisco? They never went there.

In San Jose? Why would they go to San Jose?

In Santa Cruz? Well, yeah, at The Catalyst and a few other places, they were known...

But they lived and worked in Gilroy, and on a daily basis, they were stuck, isolated in Gilroy. They had Augie's, and some went to the Green Hut. There was the Swedish smorgasbord place where the station had a trade-out for a meal per person per day for a while, and they might run into each other there, but it wasn't a place to hang out.

A few lived in town, most lived just outside of town. With everyone working such different hours, it was tough to hang at home with people, but it happened. It happened sometimes, but there was always the station, and that was always the best bet to see someone. Someone was always there, and there was always something going on. You could listen to a new album, see what came in the mail, hang with the jock, say hello in the office, check out what was going on in the Production Room, talk to a caller, hang out some more with the jock. There was probably some beer left from yesterday, or someone might have a joint. There was almost always something to smoke or drink.

Gilroy isolated the staff, and being a KFAT DJ became an us-and-them thing. No one disliked "them" or made anyone from outside unwelcome. Indeed, new people were always welcome at KFAT, they were just... "them" is all.

Sales people and record reps came by. A lot of the times when a jock knew a visitor was a label rep, they'd put on the headphones so they couldn't hear them. No one in the music business would disturb a jock with his headphones on. If they got ignored, they went away.

Listeners came by, and they were usually very polite, and very welcome. They came and they were respectful, and rarely did a jock have a problem with a visiting Fathead. Some who came in were surprised that KFAT looked exactly like it sounded: messy, rowdy, a bit out of control. It was a party that they had invited themselves to and they behaved accordingly, and they brought gifts as they would to a party. They brought a little something for the jock, and the jocks always said thanks, man.

But it was still Gilroy, and a shift is a long time to play records, especially when they did it every day, day after day, and then the next day and then the day after that. The gig had its strains and its pressures, and the Flow must never cease. The pressure was constant, and the longer the shift lasted, the longer the last of it lasted. They couldn't play what they'd played yesterday, and they couldn't play what someone else had played already. I'm not asking for *sympathy* for these people, I'm just saying… It was a great job and all, but…

But Gilroy was still a small, dusty, shit-hole of a town, and the staff of KFAT were aliens to the folks in town, and it made them into an island, their own nation. Remember the time Laura Ellen was at a convention in San Francisco, talking to another Gilroyal, when a man came up and gushed at meeting her, and was worshipful about KFAT. It blew the local guy's mind that someone from outside of Gilroy, much less from San Francisco, knew, much less revered, the station. No one in Gilroy cared about KFAT.

GK: But you knew you had a hot thing, right?

Sully: "Not really, 'cause we were *doing* it. You were aware that your friends thought you were doing a cool thing. We had no sense of the ratings."

What she meant was that KFAT had virtually no ratings. Jeremy would not pay for "the Book," and what they heard from people in radio who had "the Book" was that KFAT's listenership was too small to measure.

Tiny: "It didn't matter, I was having so much fun."

Stars? Not fucking hardly. Not in Gilroy. Not at $3.25 an hour.

Elsie was in nursery school several blocks away, and at some point, everyone on the staff, and sometimes friends of the staff, if they were there at the right time, went there to pick her up. Laura Ellen was always busy, and the staff was family, so she sent whoever was around to bring Elsie back to the station. The school was shocked to see a never-ending stream of different hippies and freaks come by to fetch Elsie, but Laura Ellen would call and let them know someone else was coming and they got used to it, but they were scared and always treated whoever came by for Elsie with cautious reserve.

The KFAT staff were freaks to the Gilroyals, and the locals kept their distance. The staff kept to themselves, ate among themselves and hung out together. Did they have any idea that they were having an impact? If they were, ha ha, where's the money? Where's the flashy threads and new cars, man?

No, these people just went to work every day and did their jobs, and knew that it was a fun gig. They were all happy to be there, but were they important to anyone? Did it all mean anything?

Huh? What do you mean? What's that mean? The jocks didn't even have business cards.

Tiny: "But you asked me when we knew that our station was a success, [and] I started getting an inkling when this real radio dork who I had listened to as a child, Gene Nelson..."

Gene Nelson, also known as "Emperor Gene Nelson," was an institution in Bay Area radio. He'd been on the air for decades, everyone knew him, and his show was incredibly popular among the older listeners.

Tiny: "And he came up to the station and he was a real creep. He was real dorky and he had white shoes and a white belt, and he was like, "Oh, this is such a piece of shit..." you know, making fun of us, and I just looked at him and I said, "I used to listen to you on KYA or whatever when I was a kid."

"It puffed him up even worse. But I'd thought he was, like, *it*, and here he was, the most dorky guy I'd ever seen."

Another indication that the station was having an impact was when Bonnie Simmons, formerly of KSAN and one of the few important women in San Francisco radio, came to Gilroy. Tiny says, "She came up to the station, I remember. She wanted to get a look at me."

In 1978, a magazine devoted to Bay area music, named BAM (for Bay Area Music) began publishing. They were going to have an annual Bay Area Music Awards show in San Francisco, and award BAMMIES to deserving acts. For their first show, they invited me to be a presenter in the Folk category. People in the Industry were listening.

Cuzin Al: "But the other thing was, somebody knew the engineers at KNBR in San Francisco, who was that? Was it Robin? Because on Sunday night they'd come down there and... the guy was the chief engineer at KNBR in San Francisco, and he said that Frank Dill and all those cats listen to our station."

KNBR was a Bay Area radio powerhouse, and every year they sponsored a Bathtub Regatta, held in the estuary at Jack London Square in Oakland. Two weeks before the race, a listener called KFAT during Terrell Lynn's shift. She and her husband had made a bathtub entry, calling it "Ernest Tub," and wanted to know if he would like to be her partner in the event. Terrell Lynn jumped at the chance, and—of all things—they won. The event got television coverage on every channel in the Bay Area, and that was when Cuzin Al felt that "*everybody* in San Francisco knew about the Fat."

That event got so much coverage for the station, in fact, that the next day Lori Nelson made a call, arranged a trade-out, and three days later she handed the jocks their first KFAT business cards.

The business cards told us we may be having an impact on our own office, but here's George Frayne of Commander Cody talking about us.

The Ozoner never thought much about having an impact; he was just trying to get through his shift, one shift at a time. But he knew something was afoot when two reporters showed up from Japan. They'd heard about the station, and wanted to talk to the "Ozonah Alainjuh." They took some photos, and one of them drew a caricature of the Ozoner, promising to send him a copy of the magazine when it came out, but they never did.

One day a book showed up for me in the mail called "Honky-Tonkin'." It was a guide written for the European traveler in America, and it listed all the funky bars, blues clubs and honky-tonks in America, and KFAT was featured as the radio station playing the most interesting music in America.

And right up the road in Palo Alto, there was a garage where the first Apple computer was taking shape, and in that garage KFAT- and *only* KFAT- was heard. Of course I couldn't know that until 2011, when Steve Wozniak was quoted in *The Gilroy Dispatch*, saying, "Music is a very important part of my life, and I credit most of that to what I discovered when I discovered KFAT, back when I was designing the early Apple Computers. To this day I constantly tell people how my music 'center' came from those days. I constantly search out the sort of music KFAT played. It was so different, yet so good, musically and lyrically and comically."

These were all interesting indications that KFAT was having an impact, but the indication that I am most impressed by was when Bill Graham, the undisputed 600- pound gorilla of concert producers in the Bay Area called up Laura Ellen and asked her what acts he should book. That might not have much of an impact on you, you hopelessly clueless reader, but those who know what that meant were mighty impressed.

Sully may have thought that they were "down there doing this goofy stuff" in the early days of the gig, and they never got swelled egos about it, but soon enough they were serious about the work. They knew people were listening, and they cared what they put on the air, music, comedy, vocals, whatever. They were having an impact, but it was nothing to get a swelled head about. It was all really cool and all, but rent was due this week, and they didn't have the money. Again.

Chapter 55: Tom Diamant

Terrell Lynn says he slapped on the headphones when a record label rep came by. He wasn't insulting him, but he was ignoring him. If the rep got a little insulted, well, fuck him.

Reps should know better than to disturb a jock on the air.

Other jocks brushed them off, excused themselves, but they were working, ignored them or snapped at them to get away (guess who?). Reps came by to sell their stuff, and wanted to talk to someone on the music side, not the sales side. At every other radio station, the reps would go to the receptionist and ask to speak to the Music Director. Or the Program Director. You know: someone in management.

At KFAT, there was no receptionist. Oftentimes, Laura Ellen was out of the office, and therefore, so was the Music Director, the Program Director, the Station Manager, the General Manager and the owner, as she was all of the above. But the DJ was always there, and the reps thought they could speak to them. Fat chance.

But Tom Diamant was different. He was a record rep, but he was also a friend, a fellow enthusiast for the new music. He worked with the new music, but as a rep. The labels that he represented were the small independent labels where KFAT got most of its music. Tom lived in Berkeley, and he used to come down to KFAT, and he was always welcome. He was the only record rep that hung out at the station, partied with the DJs and their friends, and went to events with the staff, both KFAT events and others. Tom came to KFAT whenever he could, and he remembers a lot of it.

TD: "In the mid 1970s, I had a record company called Kaleidoscope Records. We had just started it, but I was also in record distribution, and one day two guys and a woman came up to get some promo copies. I was involved with distributing Rounder Records and Flying Fish and County, and just a lot of the traditional roots-filled music labels of that era… I think it was Buffalo Bob… and it might have been Jeremy, and I don't remember who the woman was, although I do remember not being introduced and asking who she was, and she said 'Oh, I'm just his girlfriend,' pointing to Bob, and, it being Berkeley, I thought it was rather politically incorrect, even though that term didn't exist back then.

"So they came and they explained the concept of the station to me, and I gave them a bunch of promos. And then I was supplying them with promos, and this was a time when the independent folk music world was really exploding, in record labels.

"Originally in the 1960s, there were a handful of independent labels. Arhoolie was one of them, the outlaw movement, all that kind of stuff was just starting to happen. Asleep At The Wheel was down here playing for Smithsonian Folkways, County Records, Yazoo; prior to that there were not that many independent labels dealing with non-hit kind of music. So I was involved with distributing this whole flourish of new labels that were happening. Also happening at the same time was a large alternative country scene that was just starting to happen.

"That would be early 70s, maybe… I just know that that would be when all these record labels that were recording, and all these people who figured out that with a little money they could put out their own record, and they could distribute it. And get it out into stores. And it was also the time when there were a whole bunch of record stores that were expanding. Tower Records was expanding, and there was a company out of Santa Cruz called Odyssey Records that was just opening these full catalog stores, and nobody had opened full catalog stores before…

GK: Could you explain "full catalog stores?"

TD: "It just meant that instead of carrying the Billboard Top Two Hundred, and a bunch of classical music, your basic store was the hits, some classical, maybe a little bit of jazz, but there weren't that many labels, except maybe some blues. Chess Records, Savoy Records… so these stores existed that the companies like Rounder Records could find a place for their product. Where people would walk in and see this whole diverse collection of music. Now, the problem was that there was nobody playing it too much. There was the end of the alternative rock radio, the KSAN kind of thing, where they would play all sorts of stuff that was still vaguely around back then; there were still remnants of Free Form rock radio, but it was mostly college stations.

"So when KFAT came along, they had this wonderful collection of music to choose from. Not just on the major labels with the country stuff that KFAT was playing, but the folk music labels, the Philo labels with 'Moose Turd Pie' that got played to death on KFAT. And they would make it a hit, 'cause I was distributing Philo back then, I was taking it into the stores, and when KFAT played 'Moose Turd Pie', which had the word 'shit' in it, which you can't say on the radio anymore, it would sell. When Mary McCaslin… when her albums first came out, it would never have happened without KFAT, I don't think, as strongly as it did.

GK: Nationally or locally?

TD: "Locally. But it always helps. Your fan base starts locally, and then the word gets out. And certainly, people came from out of town and they found out about KFAT, and I'm sure they went back from wherever, Arizona, or wherever they came from, and said, "Man, there's this wild station, and they play whatever they want, and they're actually a commercial station, and that was the difference. I've been on the radio since 1976, on KPFA, playing traditional bluegrass and old-time country music with my show "Panhandle Country," but it's KPFA, it's a local station, it is 59,000 watts and simulcast in Fresno, so I do have a huge listenership, but that's very rare, and it's usually small, peanut-whistle stations that played that kind of alternative music. But KFAT was a commercial station, and they sold commercials. They had that guy who's a classic... sleazy radio...?

GK: Bob McLain?

TD: "Bob McLain. I knew him. He got that rat-trap of a car that Sully drove around, in exchange for some air time; it was just this wheeling-and-dealing kind of thing...

"So it was a commercial station, and that's what made it different. I don't remember when I started going down there. Odyssey Records was this huge explosion of stores, and they were based in Santa Cruz, and so I started going down to Santa Cruz, and I remember my route was I'd go to Santa Cruz, and I'd do my business, I'd go to the main office and I'd spend a few hours selling records, and then I'd go over to (Santa Cruz non-comm) KUSP and hang out with those guys for a while, and then I went to go and have fun.

"I drove over the hill, and I went to Gilroy. And I usually had a bunch of promo records, so they were always glad to see me, and they were glad to see me because I wasn't the major labels... I was representing all these other labels, as well as my own label, and so it was easy... it was the kind of thing where you'd walk in, and you'd have these records in your hand, and you didn't... there was no receptionist, there were dogs all over the place, sleeping on the floor, there were people sleeping on the floor, children running around, Laura Ellen's kid was running around, Tiny's kids, and then you didn't talk to the Music Director, you walked in to the DJ, and he took the LP out of your hand and slapped it on the air! Just slapped it down on the turntable and said, "What cut should I play?" or "Let's try this." And so it was just immediate."

GK: Had that happened at any other radio station?

TD: "I knew DJ's well, and sometimes I'd visit them when I knew they were on the air, but the other stations were college stations, or other non-profit stations, so they had the Latin Music Show, or the local call-in talk show..."

GK: Block programming...

TD: "Block programming. So, depending on what time you were there, maybe, you could time it so the bluegrass guy was on the air, or the folk guy was on the air, but at KFAT it didn't matter, because they were happy to see you.

"Yeah, and I remember the first time I went down there. I was supposed to... I was in charge of putting together an ad for something, and this was the first time I met Sully, and I walked in there and I expected to meet somebody at an appointed time, which was absolutely ridiculous in retrospect, particularly with Frisco. And I showed up and Sully's on the air, and I walked in and there's no one around, or if they are, they're not paying attention to you, there was no one there, and there was this person on the air, and since I had already had my radio experience, I was a little nervous about interrupting someone on the air, but I walked in and said that I was supposed to meet somebody named Frisco here to do this ad, and Sully said, 'Oh, Frisco's not here. He doesn't show up this early.'

"So she called him, and I hung out for a while with Sully, and then Frisco showed up, and that was the first time I came to KFAT, and it was a pretty comfortable place. And it was an outrageous place.

"Yeah... and I hung out there. It was too much fun. In fact, everybody hung out there. Because you know, when you go to work, you go to work and then you go home. At KFAT, you didn't do that. More likely than not, there was the on-air person, and there was 2 or 3 or 5 of the other on-air people, just hanging out.

"They had this big old chair... an overstuffed chair that you could sit on the arms of, and I just remember more than a few times, like 5 of us... somebody sitting in the chair, somebody sitting on an arm, somebody sitting on the floor, somebody sitting on the other arm, just hanging out and having a good time, and just talking about music or life or whatever. And it paid off. I remember hanging out there once, and it was during the Salinas Rodeo, and Laura Ellen or somebody came running in and said 'Rodney Crowell is doing a solo act!'

"And this was when Rodney was part of Emmylou's band, and he was just trying to break out on his own. He was doing a solo act at the Salinas Rodeo and Emmylou was going to be singing harmony with him, is going to be backing him up. And KFAT had tickets, so we all just got in somebody's car, drove down there, he gets out there and starts singing. It's just him and his guitar, and then Emmylou comes out and starts singing, and Emmylou was still... she was a country star, but a lot of people still didn't know who she was, and somebody from the audience yelled out, 'Who's the woman singing with you?'

"And I remember her, almost shyly saying, "Oh, I'm Emmylou Harris." And then the place exploded, because she had had country hits at the time, but I guess she wasn't visually known as well. But moments like that, if you were at the station at the right time, you often just got whisked off into either a party or somebody doing something, or a show like that, that just... I mean, that was a remarkable musical experience, seeing Rodney Crowell like that. I see him today, and I think he's marvelous today, but that kind of thing, where… they were the central… if you had something like that going on, if you had a show you wanted to promote, you called KFAT. And KFAT of course had their Fat Frys.

A Note From Gilbert To The Reader: Well, Tom's chapter is certainly having an easy time writing itself, so why don't I just let him write the rest of it? Here, in no particular order, are some highlights of our talk:

- "KFAT certainly had an influence on the industry in terms of ticket sales and album sales. Definitely."
- "One of the exciting things about that era and about that station… was that the DJs were just *so* into the music that they… and so into their programs, and into the whole concept of the station… that they all wanted to find something better than the other guys. They wanted to find that track that no one else was playing, that they would play first, and then everybody else would start playing. They had those stickers on the albums that people wrote all over, and everybody just… these days, so much comes out that it's called the "three second needle,"* and I don't like that…I don't like that…"
- "They spent time listening to that stuff, and often times they spent time listening to it on the air, so sometimes some stuff went out there that never got played again, because the DJ was on the air and the thing was new, and they said, 'OK, let's see what this thing sounds like.' But they were really into the music, and when the Fat Frys happened, they came to the shows. Everybody was young, and everybody had a lot of energy. Everybody could stay up 'til four in the morning, everybody could get drunk and go to work the next morning. And there was a love and an energy for the music, and so when you brought albums to them, you knew that they would have a chance."
- "It was just a wild station. I remember there was this Don Reno banjo track that was ten minutes long, and there were at least one or two DJs that loved that track because it gave them enough time… there was a liquor store a block away, and it gave them enough time to play the track, leave the station, sometimes with no one there, walk down to the liquor store, and get a six-pack of beer and walk back. I think there were a bunch of tracks like that, and everybody had their favorite…"
- "The freedom was really rare. And it was a really rare… a really good grouping of DJs that were able to entertain and play the music that the people were interested in hearing, and turn the people on to the music. Certainly there are some whose musical taste I liked better than others, but I can't think of a weak DJ there."

---

*When a DJ, MD or PD auditions a track and gives it just three seconds to make an impression.

- "A lot of people, because of the musicians coming through, knew about KFAT back in other parts of the country. How much influence, it's hard to say. It was a huge influence here..."
- "You can take the music and play it, but you can't take the whole package. The zaniness of the DJs, those wacko phony commercials, the Fat Grams, (chuckles) there are memories I'm not saying into a tape recorder, but... there are all these little elements, and they all came together for that number of years, and you could take elements, and you could recreate something, but you could never recreate that scene. You can't find Tiny and Terrell Lynn and Sully, Gordy, Sherman and that whole group of people, and the ongoing craziness of Jeremy, who was absolutely nutty, and whose style of management (chuckles) was not the greatest, heh heh, to say the least, and just all the dynamics of what was going on there, it was a unique thing.
- And when stuff like that goes on and people always say, 'Let's start a new one...' and you really have to start your own thing. You can be inspired by it and take elements of it, but you can't recreate it. To this day people come up to me and say, 'Oh, you know, so-and-so's gonna do this great thing, and it's gonna be like KFAT,' or 'Gosh, I wish somebody would start something like KFAT,' or whatever... It never happens."
- "At KFAT, it was night after night after night after night, and you didn't care if you got only two hours sleep, and I don't care if I get only two hours of sleep tonight, but I care if I get two hours of sleep tonight and the next night and the next night. 'Cause I'm older now, and I may still be able to drink 'til the break of dawn, but I can't do it repeatedly without getting sleep."
- "Oh, people love the music, that's no question. The one thing that "O Brother, Where Art Thou" proved, if you give people good music that they can't get anywhere else, and make it popular enough so that they become conscious of it, they'll buy it. That album was not played on any radio station except for alternative stations. It sold like 14 million copies."
- "Look at music festivals. Look at the Strawberry Music Festival, look at the Grass Valley Bluegrass Festival, look at the world music festivals, the reggae festivals, all this music that is not commercially played on radio, not often seen on TV, hardly found in a modern record store, and these festivals sell out, are packed with people. You walk around Strawberry, around Grass Valley, there are hundreds, if not thousands, of people playing musical instruments. There is so much music going on right now, it's amazing. At the Grass Valley Music Festival, there are kids, five years old, playing fiddle, and there's people like me, 59

years old, playing guitar, and there's guys 80 years old playing banjo, and we're all picking together.

- "And we're all staying up 'til ridiculously late… And you know who's making money in the music industry today? At least in the bluegrass music industry, and in all industry, it's Gibson Guitars, its GHS strings, it's Martin Guitars, it's Fender Guitar, those are the people making money. Across the street [from here] is the Guitar Center, and anytime you go in there, it's packed. "And the guys I know who dreamed of being professional musicians, if they still don't have a real day job, they're giving lessons, they're making their living teaching music.

- GK: If there's all this interest in acoustic music and Americana music, why isn't there more of it on radio?

- TD: "Because monopolies own radio stations. It won't draw the sponsors. It won't get the ratings. They won't get enough ratings. You want to get really high ratings, that's all that matters. And if you're talking about a fraction of a point difference, they'll scrap it for something else. And NPR is going to all talk radio, because that's where the money is."

- "People say they can get XM radio, and 'I can listen to bluegrass and Alternative, and Americana all day long,' or 'I can listen to it on the internet,' and I'll tell you something: I don't like machines programming music at all. I hate it. I love bluegrass music, and I can't listen to the streaming bluegrass shows that are not DJ'd. The DJ puts together a segue; the best radio show that you can listen to is the radio show that they're playing a song, and it makes you think of a song, and the DJ plays that song that you thought of next. There's nothing better than that, and that's because the DJ knows what song should come next. In the alternative rock world, and at KFAT, there used to be classic segues, you always almost play the same song. And then there's the DJ coming in and saying something as simple as 'that was so-and-so from Boulder, Colorado,' or 'so-and-so is coming to town,' or 'I was listening to this record the other day and I couldn't stand this band, and for some reason, I put this record on and I found that track seven… and you gotta hear this.'
- "And you know this DJ, and you've gotta hear that track. And automated programming sucks. I have a bunch of stuff on my computer, and I put my computer on shuffle and I enjoy it because I chose all those tunes. But when I'm listening to a radio show, I want guidance. I want somebody to hold my hand, and lead me through it. And I want somebody who… if you're driving in your car and listening to that radio show,

and you get to your destination, and you don't want to get out of your car. That never happens with automated radio. Because the thing that comes next has nothing to do with what came before it. A good DJ just nails you down and forces you to listen to it, and you don't want to leave, and that's what good radio is about. KFAT on a bad day did that, and KFAT on a good day did that brilliantly."

Hi, it's me again. After I left our talk, I regretted not thinking to ask Tom if he had a favorite KFAT experience. My remorse lasted only a second, until I remembered his face when he told me that story about going to the nearby Salinas Rodeo to see Rodney Crowell and Emmylou Harris.

The KFAT DJ's were going there to enjoy the show, and were also hoping to meet those two, and as it turned out, Rodney and Emmylou were hoping to meet them. But you already know that story, so move along.

Chapter 56: The Buffalo Bob Memorial Walk-Up Window

Gilroy was a hot place in August. Well, when I say "hot" of course I mean that over the course of the day, either the mean temperature, or the peak temperature or the average temperature was hot. Whatever you called it, it didn't really matter, because Gilroy was just goddam hot in August. And dry. And hot.

Of course, there would have to be a peak temperature on any given day, and every day had a time at which that occurred. And that time was in the late afternoon, and the town would cool down after the peak at about the same slow pace as it had heated up. Which meant several hours when it was a little harder to breathe. And the peak temperature time in August is... oh, let's say about four o'clock, which was when Buffalo Bob went on the air.

The heat and the odors have been covered elsewhere in the book and should need no reinforcement, but this was Buffalo Bob, and how his shift went was highly dependent on his mood, and I think we've covered the frequency with which Bob might be in a foul mood or on the way to one. Who knows who or what he was pissed off at on this night?

It might have been because of a run-in with Laura Ellen, who might have had a suggestion for him. Laura Ellen was more than frustrated with Cassidy, that asshole; she hated him. Their relationship was... chilly. That this was the only thing that was chilly at the station at the moment did not improve his mood.

He didn't like Laura Ellen and he never listened to her, and even when he did listen to her, he still did what he had been doing all along, anyway. Everyone at KFAT knew that Buffalo Bob "owned" Afternoon Drive and Laura Ellen couldn't afford to lose him, and he knew it, too. But he just kept pissing people off. His attitude had gotten him fired a few times, and it was the drinking- always coupled with the attitude... but his shows were popular, and because he was family, even though he was still a cantankerous, belligerent, macho drunken jerk, he kept getting re-hired. He was family. With that voice...

There had been so many confrontations and nothing had changed, so Laura Ellen found that the best way to deal with Bob was to leave before he got there, and that seemed to work. She left him notes, knowing that half of what he did was going to piss her off, anyway.

Laura Ellen had been trying to influence Bob on his choice of music, but to little avail. One of her immutable rules was that the blue material, anything with suggestive sexual references, all the way to songs of outright filth, could only be played after nine o'clock at night. Sherman's playing of dirty songs left a hangover of late-night people still requesting the raunchy stuff when it became Gordy's shift, but he wouldn't play them, as the morning people who were just getting up, it was family time, and they and the sponsors wouldn't like it.

Bob was working that night until 9:00, and that was still a time for whatever would pass for decorum at KFAT. Even Bob saw the wisdom of that, and stayed away from the dirtier songs during his shift. There was plenty of stuff to play, plenty of stuff to experiment with and keep him interested. It had been a long shift, what with the heat and the lack of any air moving through hot, sticky, smelly Gilroy. There was a window with a bunch of panes, and one of those panes you could crank open with a handle as far as it went, but if there was no air moving through Gilroy, there was no air moving through the studio.

A long day, a lot of sweating and smelling and drinking and playing records and drinking, and now it was almost over, and it had been a *really* long day and he had one last record to go. It was 8:57, and there was just one more record on this shift. Bob was not in a good mood, and for his last song he wanted to hear "Humping In The Back Seat Of A '55 Ford," by Chinga Chavin.

Chavin called his music "country porn," his lyrics were blatant and all about sex, frequently of a disgusting nature. His guitar was made from a toilet seat. But most of his tracks rocked, and while Bob couldn't play him during this shift, it was just three minutes before nine o'clock, when the rules would allow the song, and this being KFAT, he played the top-of-the-hour ID and put the song on, never imagining for a second that he was doing anything wrong.

Laura Ellen was listening. I do not know which of Laura Ellen's demons was up her butt at that moment, but she called Bob and screamed that it was *too early for that song!*

She yelled, Bob yelled back, slammed down the phone, took the record off the turntable, opened the mic and broke the record in half. Then, muttering a few drunken, unintelligible imprecations into the mic, he went to the window with all the panes of glass, and he just... hauled off and SMACKED it! *SMASH!* and walked out, leaving the next jock to start their first song, which had already been cued up by the next jock. Then, down the stairs in a huff, a right turn and five doors down to the Green Hut. This called for a drink. Of course, everyone at the Green Hut had been listening, so Steimie poured two whiskeys, and put them on the bar. The Green Hut knew Bob was on his way.

The next day someone taped a piece of cardboard into the space where the glass had been. Someone had written on the cardboard:

The Buffalo Bob
Memorial
Walk-Up Window

Bob was fired that day. He was fired from KFAT several times, but he remembers this one. He came back the next day, and nothing was said about the firing. That's how it worked.

I asked Bob why he put his fist through the window that night. He said, "It was what was hanging from my arm at the time."

But this is a happy chapter.

A week went by, and one day a man came up the back stairs with an air conditioner and installed it in the Buffalo Bob Memorial Walk-Up Window. Jeremy had bought a cheap, used, low-power air conditioner. But it worked, and it was a luxury. The KFAT staff has evermore maintained its steadfast gratitude to Buffalo Bob for his sacrifice.

Some might ask, 'What sacrifice?' as by then, Bob was long since back at work.

Others were fired on occasion; Sherman was fired a lot more than the others, but Bob still had the lead.

Chapter 57: Annie Oakley and John Lee

The first staff was gone, except for Gordy, and a new staff was in place. But the station was on 24/7 now, and the new staff covered the weekdays, but there was always a scramble to cover the weekend shifts.

Tiny needed the money and was taking any shifts they'd give her, but she had two toddlers and she only had so much time to give to spinning records. She picked up some extra cash for light cleaning around the station, but it wasn't much cleaning and it wasn't much cash. She was broke all the time, and her mothering and her partying were clashing. One day her son, Finnian, came into the studio while Tiny had been on the air and she hadn't known that he had gone outside. He showed her his little tin bucket, and in it there were a couple of singles and some coins. She asked him where he got them, and he told her that he'd gone down to the street and asked for it. Their lack of money was talked about often, and Finny wanted to help. So Tiny could only take the shifts that allowed her to keep a closer eye on Finny and her daughter, Amber.

KFAT needed jocks for fill-in and weekend spots. Annie Mitchell was available, and the timing couldn't have been better.

Born and raised in Dallas, Annie had dropped out of her Master's program in Texas and become a VISTA volunteer, working in the poor neighborhoods of St. Louis. She knew about KDNA, and admired the open policy which allowed for any viewpoint, and she liked the wide selection of music and talk. As a VISTA volunteer, her job was training people to run their own community radio station, and was asked if she wanted a spot on KBDY in St. Louis, a 10-Watt station started by VISTA.

As an avid listener, she was influenced by KDNA, and had become one of KBDY's managers. Then Jeremy moved to California, and you might have read something about that.

But now Jeremy was in trouble with KFAT, and he needed a staff for the weekends and fill-ins. Fill-ins are who you call when someone can't make it to their normal shift, and you will have surmised that the list of fill-in names was never far from Laura Ellen's reach.

Annie was looking to start over in a new location right about the time that Jeremy remembered her and her Texas accent, and called her in St. Louis. She was open to coming to Santa Cruz to check it out. She knew Jeremy's mother, Elizabeth, from Elizabeth's visits to St. Louis, and she was going to stay with her while she visited Santa Cruz and Gilroy.

Jeremy couldn't pay her anywhere near a living wage, but Annie quickly fell in love with Santa Cruz and wanted to stay, working out a deal with Jeremy: she would do two weekend shifts, and be a saleswoman for KFAT during the week. This was a win/win for Jeremy and a win/we'll see for Annie.

Her first visit to the station blew her away. Coming into San Francisco from St. Louis, she went directly to Santa Cruz, where she hung out for a day with Jeremy's mother, and then went to Los Gatos to meet Laura Ellen, and they drove to Gilroy, listening all the way. She loved the music she heard, and she'd just come from the slums of St. Louis, so Santa Cruz was a dream come true, and Gilroy itself wasn't all that depressing, but the studio was a mess. She'd never been to KDNA, but she'd heard it was dirty and unruly there. She'd only been to KBDY, which was a poor, low-wattage, black-oriented, community-run radio station, and no one from *Smart Urban Living* magazine was ever going to do a photo spread on KBDY. You can use your imagination here.

But she remembered KBDY "was like a palace compared to KFAT. There weren't dogs and children living at KBDY. Jeremy couldn't... afford to pay the staff a living wage, so they lived at the station. And he would take what he could get."

She remembers "the dogs and the children and the mess and the people hanging out and the old-fashioned equipment. But the music..."

She knew some of the music because "you don't grow up in Texas without it being in your background" so she set to work to learn the rest of the new format. Notes on the album covers and talking about the music with the other jocks "was really the only way that you learned to do things there. Everybody was very supportive and really helped you to learn the equipment and the music... there was a HUGE record library, and there were so many choices to make. You had to do it speedily, and a lot of jocks would help you and pick some music for you, and tell you this was good and that was good. It was all albums and LPs and there was a lot of space (for notes) and everybody would write that 'I love this' and 'this is terrible.'"

Annie had a silver tray that she'd brought with her from Texas. It was a small corner of the Good Life, and she brought it with her to Gilroy for every shift. She also brought a bottle of wine, which she placed on the tray, along with a delicate, long-stemmed wine glass, and one joint. A thin bud vase with a single flower completed her morning air-shift ritual. She was on Saturday and Sunday mornings, seven a.m. until noon.

She did some overnights when needed, and those were the times when she learned the most about the music, but she usually followed Sherman or the Ozone Ranger, both of whom were doing overnights on Friday and Saturday nights. She was Annie Oakley.

Annie stayed with Jeremy's mother for two months, living in a closet in Elizabeth's house until she could afford an apartment. She remembers that after the closet, her 400 square foot apartment seemed like a mansion to her.

She would go to Los Gatos, to Laura Ellen and Jeremy's house, and play with Elsie while waiting for them to wake up. Laura Ellen and Jeremy were keeping vampire hours as well as trying to handle the daytime needs of the station, and sleep was a necessary luxury for them. They were always trying to plug some hole or placate some advertiser for what the jocks had done with their ads, or fixing some technical breakdown, or filling in themselves when no one else was available. They needed whatever sleep they could steal. And Elsie was two, and fully awake right about dawn.

Annie would come to Los Gatos and play with Elsie until Laura Ellen got up and made coffee. It was where Annie learned about good coffee, and she is forever grateful to Laura Ellen for that. Then they'd drive to Gilroy, listening to the station (they were *always* listening to the station), and talking. Annie remembers: "That was all we talked about. What were the segues, what was the latest, who was the hot artist, who was good, [that was] the whole thing that everybody talked about."

It was quite an adjustment for Annie, coming to Santa Cruz from a St. Louis slum. During the week, she'd try to sell ads for the station, and she found it easy to get in the door with clients, but hard to close the sale. There was no way to define the station, which advertisers needed, and the sales staff was run by Jim Gordon, who Annie remembers as a friendly, companionable Santa Cruz kind of fuck-up. KFAT had no ratings to speak of, and that made it harder.

Annie's other problem with sales was that the territory was so huge. Her territory was San Francisco, all the way down to Santa Cruz, and it was all cold calls up and down the coast that got her nowhere. She remembers that she wasn't that good as a sales person, but they were lucky to have her because "there weren't that many people foolish enough to try to go out and try to sell it in those days." She was deeply frustrated by endless futile attempts at selling ads for the station, and now that she'd moved out to California, she was worried about not being able to pay her bills.

So it was more cold calls, as there were simply no established accounts to call on or maintain. But on the air, she was a queen. The phones never stopped ringing, and a lot of what drove her show were the requests. "In ways, it would be built around the requests, and a lot of what I think led to the popularity of the station was that we played what the people wanted to hear."

"It was crazy how many people wanted to hear "Moose Turd Pie." It was probably the most requested song at the station." A whole chapter on this song follows.

"Then there were all the raucous songs, and by the time I came on, Sherman had been playing all these dirty songs, and people wanted to hear them, and I couldn't play them because it was the morning. I had fun mixing up raunchy songs with gospel songs. I was really big on Texas Swing. But I played what people wanted to hear, and if I couldn't play what they wanted, I played something close to it. I think that was a reason for the success of the station, the open-ness. We invited people to come by, we invited them to call, we tried to be as responsive as possible."

And callers were waiting twenty, thirty, and forty-five minutes to talk to her.

On some Sunday mornings, John Lee Hooker would show up at the station. He lived in Gilroy, just three blocks from KFAT, and he liked to come up and hang out. He was embarrassed about his stuttering, so he'd rarely chat on the air, but he did love to chat with Annie, and talking about the music was always a pleasure for both of them. He'd pick out songs for Annie to play, and they'd hang, just Annie, John Lee, and his driver. Annie and John Lee became friends and Sunday morning radio pals. She remembers that he was a good guy, very friendly, and very interested in the music. And he knew with one listen to KFAT that he had a friend in Gilroy. Somehow, Sherman or the Ozoner would know in advance when John Lee might show up. Annie doesn't know how they knew this, but she suspects it had something to do with the Green Hut.

Annie was also a member of the Green Hut Visitation Crew, and she wasn't above putting on a long cut and slipping out for a beer now and again. She remembers that among all the talk of the music among the jocks, one of the favorite subjects was the identification of which cuts were the long ones. That was useful knowledge when you were thirsty, and DJ work was thirsty work.

Annie says that one of the reasons for going to the Green Hut was that the station was so dirty and grungy that, during our interview years later, she still shudders and will not discuss some of the stories of the filth at KFAT. So when she wanted a beer, she wouldn't drink any of the unopened bottles that she found there. "I mean, you wanted something that was a little bit antiseptic at that point."

Annie is something of a lady. She is a genuine Texas lady who is at home in tea rooms and Honky-Tonks. She is small, and some would say she was dainty. Others might say that she was prissy, but they would say that at their own peril. She is quite a lady.

Starting her shift at seven a.m. meant that she often ran into Frisco, who was usually winding up his night of… whatever it was that he'd been doing in there. Annie remembers Frisco as "a genius. A crazy alcoholic." He'd been in there all night, and when Annie got to the station, he'd have her voice an ad or two, and then stumble off to sleep, either in some back room, if one was free of other people crashing there, or in the Production Room.

Frisco talked a crazy game, but he got the work done. He knew he was worth a lot more than Jeremy was paying him, and he always grumbled about it, but he had nowhere else to go where he could drink and work, so he drank and worked. To save time, Frisco did them simultaneously. Today it's called "multi-tasking."

Annie: "Having come from the ghetto in St. Louis… Frisco was not a problem. But he was a problem to a few white country girls."

"He didn't do anything too bad, he just talked wild stuff. He had fanciful ideas about how this could be done or that, and because he was usually a little screwed up on this or that, he would raise a few eyebrows." The problem wasn't really a problem when he was a little screwed up. The problem was a problem when he was a *lot* screwed up, which was a lot of the time and getting more frequent.

His memory was getting worse, his utterances increasingly more bizarre, his grip on reality was shaky, and anyone could see that he was heading downhill fast, and no one knew what to do about it.

Chapter 58: The Floor

At first, Laura Ellen thought she wanted KFAT to be more of a straight country station to compete for ratings in the market, so she listened to KTOM, the straight country station in Salinas, down the coast from Santa Cruz. She soon realized that, "even if we played the same music in the same order, we couldn't be a country station because our attitude was so out there. It was a real learning experience- that we're an attitude station, a lifestyle station."

Laura Ellen had begun to realize that KFAT was taking on a life of its own, a life far and away from whatever had been envisioned for it a long year ago. But that was how it worked in a Jeremy Lansman Operation.

Laura Ellen and Jeremy had made Bob Cassidy MD, but he wasn't doing anything with the title. Oh, sure, he'd made a couple of calls to the labels at first, but now the records were coming in of their own accord, and Bob wasn't doing any of the follow-up work that the job needed. There were record reps to schmooze that he wouldn't talk to, and reports to make to the trade that he never got around to reporting. Bob's two main assets to KFAT were that he made an occasional comment on a newly arrived album, and he showed up on time for his shift.*

But he had *that voice*. And his Afternoon Drive was immensely popular.

Everyone always remembered that the checks might bounce on any given payday. At first they bounced about half the time. Sherman remembers driving up to Los Gatos on some station business, and they gave him the checks to give out to the staff. They told him to cash it right away, while there was still money in the account. He proceeded forthwith to the bank in Los Gatos and cashed his check, but the others had to wait for him to get back to Gilroy with the checks, and by the time he got there with them, it was too late. All of the current supply of available money at KFAT was gone. *Again!*

The jocks were earning $3.25 an hour when they got paid, and everyone had a problem with that. Terrell Lynn had lost his teamster job delivering beer two days before he came to the station with Jim Shugart to play guitar on a commercial, and was hired on the spot by Laura Ellen. But he was living mostly on his unemployment checks and not declaring any of the $60.00 he was raking in from KFAT. He knew that without his unemployment checks,

---

*Years later, Bob asked Laura Ellen for a recommendation for another job in radio. Remembering all the years of struggling with Bob's bullshit, Laura Ellen's comments were limited to "He always shows up on time."

there was no way he could still work there.

Then Shugart was gone, replaced by another Santa Cruz salesman/hustler, Jim Gordon. But the money was still not flowing into Jeremy's bank account. It was the damnedest thing.

Bob Cassidy had been living in a back room at the station for a couple of months, but he'd left his hotplate on again and it had caused a small fire, so Laura told him he had to move out. Again. This was good news for Tiny, who'd been living in Santa Cruz, and her car was a piece of crap. She had trouble almost every day getting over the hill into Gilroy, and she was late consistently. She was on from eight p.m. until one a.m., and she'd bring her kids with her, put them to bed in a back room, do her shift, go to sleep in a back room, then take them home the next morning. Now that it was open, she just packed up all their stuff and moved her family into Bob's old room, and she cooked their food on a hibachi on the landing out the back door.

Sully remembers that it took a month or six weeks to get paid the first time. No one had money for luxuries like clothes or restaurants. With Jeremy's approval, Jim Gordon negotiated a trade-out at a smorgasbord restaurant in Gilroy, and the staff got a meal a day there. A trade-out is where you put an ad on the radio, then instead of paying cash for it, you give someone a service. In this way the full-time staff was fed, other ads got their cars repaired, and household necessities were procured.

Terrell Lynn lived in his truck, parked out back, for a while.

Gordy once lived in a red vinyl Lazy Boy chair in the Meeting Room for three months.

Bob lived in Santa Cruz with his girlfriend, Lindsay, on weekends, spending some nights at the station, and some nights at a cabin on nearby Coyote Lake. The cabin was owned by Gary Steinmetz, whose brother owned the Green Hut, where Gary worked and Bob hung out, drinking for free.

Early on, Jeremy had a small problem: he had to accede to the FCC's requirements, and put a religious show on his air. He knew that Tiny was having a hard time feeding her kids on $65 a week. He wanted religious programming, but with *these* people, who at KFAT would do it, and what form would it take? If he wanted a show about fucking up, he'd have a long list of people he could ask, but a religious show? With this godless group? Tiny was always asking for extra work, so he asked her if she'd like to do a gospel show on Sunday mornings, and she said Yes! She remembers, "I didn't know anything about it. So I got the gospel stuff I knew, which was bluegrass, and a few old corny country gospel songs, and put that together and called it The Gospel Show, and then any song that moved me really spiritually, I started throwing in. Once I played 'Keep On Trying' by Poco, and someone called and complained that it wasn't gospel, it doesn't talk about God. I said it did to me. I got into it. I liked it. I used to get high doing the Gospel Show."

With the extra money, Tiny was able to feed her kids if she shopped carefully. She remembers cooking a lot of carrots and potatoes on the Hibachi out back. Cheap and filling.

It fed the kids and it was healthy and all, but it depressed Tiny, who remembers, "There was a Chinese restaurant on down the street by Augie's... and I had to buy one dinner and split it between me and my kids. I couldn't afford to feed my kids. Sometimes I'd feed 'em McDonald's because that was the only thing that I didn't eat."

Tiny lived at the station with her kids twice: once for three months and once for almost a year. When Bob was living at the station, he'd drink beer during his shift and then wander over to the Green Hut for the serious drinking. Tiny remembers him coming back from the Green Hut, seriously drunk, coming into the studio and telling wild stories until he'd "sort of drip down the wall and be asleep on the floor and I couldn't move him. He was... there." So she'd do her shift around the sleeping figure, hoping that his snoring wouldn't interfere with her commercial breaks.

Terrell Lynn remembers "we all slept on that floor at some point."

Sherman remembers fondly the times he had a dry throat. Being a professional, Sherman knew that a dry throat was hindering his ability to do good radio, and he knew that Steimie over at the Green Hut knew it, too, and would appreciate his dilemma. Sherman would call over there for a screwdriver, and Steimie's girlfriend would show up with a screwdriver. A big one. Steimie would take a bottle half full of orange juice, put in some ice cubes, fill the rest with vodka, and send his girlfriend over with it. She was a nurse, and was hanging out at the Hut after work, still in uniform, and to this day Sherman has warm feelings at the sight of a nurse.

Sister Tiny "fell asleep a couple of times. One time Gary from the Green Hut came ran up the stairs, yelling, 'Tiny! Tiny! Are you OK?' "Sad Eyed Lady of the Lowlands" ended ten minutes ago and the station had been going ffffttt... ffffttt... for like, ten minutes."

"The nurses called me and said to come over," Steimie remembered.

TLT: "He came up a couple of times that I remember, when I'd passed out."

Of course, as any inveterate barfly will tell you, drinking is good and all, but sometimes you just need some human companionship. That, apparently, was what the long cuts on records were for. The Allman Brothers had a couple of songs that would work, and the Grateful Dead had a few. Paul Butterfield and Stephen Stills had songs that lasted a whole album side. These were professionals, so leaving just any album to track through all the cuts was unprofessional and unacceptable. No, it had to be a long cut. Then they'd scoot on over to the Green Hut.

Usually, Steimie knew what it meant when "Whipping Post" by the Allman's came on, or Stills' version of "Season Of The Witch." It meant that a visit from the jock was imminent, so he just got a drink together, put it on the bar, and waited—not too long—for the jock to slide through the door. With KFAT always on at the Green Hut, and Steimie being something of a professional in his field, he knew in advance who was on the air and what their refreshment pleasure was, and he'd have it waiting for them.

Sometimes a jock might get caught up in a drink or a conversation while there, and it might be inconvenient to have to go back to the station just to start a record, so one afternoon Terrell Lynn waited for Laura Ellen to leave, and called Steimie in to show him how to work the controls. Now the jock could be left to his drinking, and he'd have Steimie go over to play the next record. Steimie liked doing it, and it was a symbiotic relationship that worked for everyone.

Well, not everyone. The guys had the Green Hut, but Sully and Laura Ellen worked daytime, and at night Tiny wouldn't leave her kids alone at the station. Sherman went over, but more often just called in his order. Terrell Lynn liked the Hut, but also liked being at the station. He liked the people and the music and he liked the process of finding the next cut. His tastes were perfect for the afternoons.

Buffalo Bob was at the Green Hut a lot. A lot. There were times when Steimie had to call the station to ask someone to come over and get Bob so he could get back to work.

He was drinking before his show and drinking during his show, and then drinking after his show. There was one advantage to this: if Bob was Missing In Action, they knew where to find him. The disadvantages were more plentiful.

The guys at the Green Hut had a game they'd play sometimes. They'd get Bob fucked up and dare him to do something crazy. One time, they found an old 2X4 board that had been lying behind the bar for emergencies like brawls and attempted robberies. They waited for Bob's drinking to get him past a certain point, and then they bet him that he couldn't break the board with his fist. Sufficiently greased, Bob felt up to the challenge.

That he showed up for work with a broken hand was bad news, of course, but the good news was that the pain sobered him enough to work. Laura Ellen told him—again!—that his friends down the street weren't his friends, that they just wanted to make fun of him, but Bob was having a good time and drinking for free, so nothing much changed. Laura Ellen was deeply frustrated. Bob was incorrigible. She'd never met a man so resistant to helping himself as Bob Cassidy.

But the Bison Boogie was a successful show, Bob owned Afternoon Drive in Santa Cruz and San Jose, and people constantly told Laura Ellen so, wherever she went.

And he had that voice, deep, resonant and captivating, and his show was a hit. It was damned frustrating. Laura Ellen wanted the voice and the music, but the total package was giving her an ulcer, and she'd have to do something about it. Very soon, she found the answer. Bob was due on the air at 4:00, and he'd show up about 3:45, so Laura Ellen started leaving at 3:30. And it worked. Problem Avoidance, you say? Perhaps, but it worked. For Laura Ellen, for Buffalo Bob, and for the station.

Photo Section One

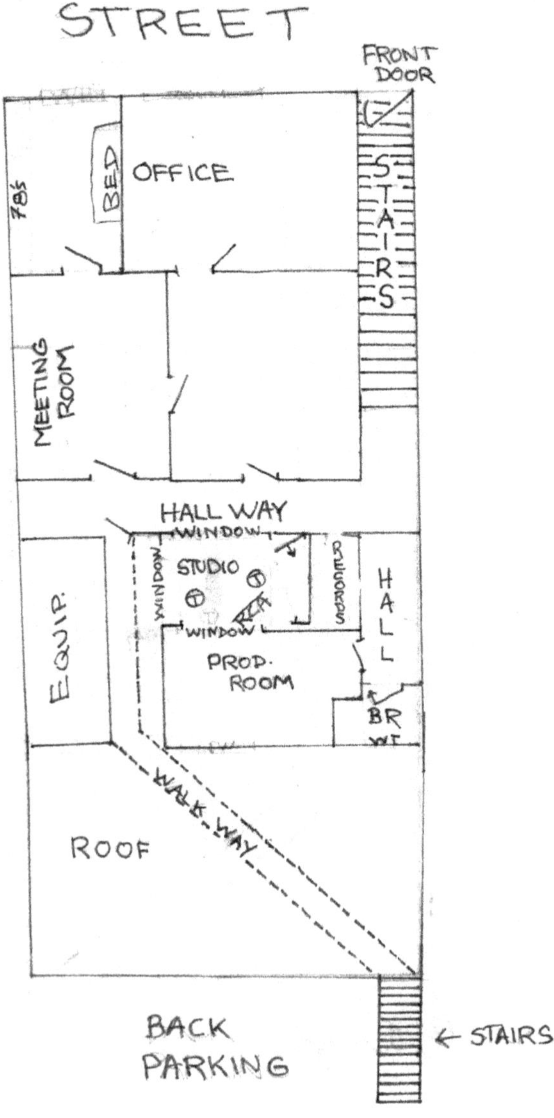

Front of building - Our door was on the left

Rear Parking - Many a battery and tire went MIA here.

Sully Does the logs

Laura Ellen Hopper - Early Days

Laura Ellen with Elsbeth

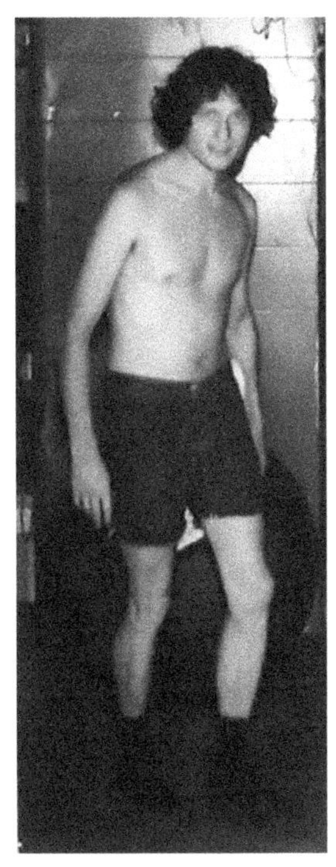

Jeremy Lansman - Early Days

Elsie and Bernice

Chris Feder

KFAT Staff: Sully, Ozone Mark Taylor (OZ), Sister Tiny, Gordy, Buffalo Bob

Terrell Lynn Thomas at the board

Gordy at the Library

Sherman Caughman "Head Shot".
Note the open beer and roach clip

KFAT Studio

KFAT BUMPER STICKER

Buffalo Bob

Ozone Ranger (Mark Taylor)

Sherman

Cuzin Al

Awest

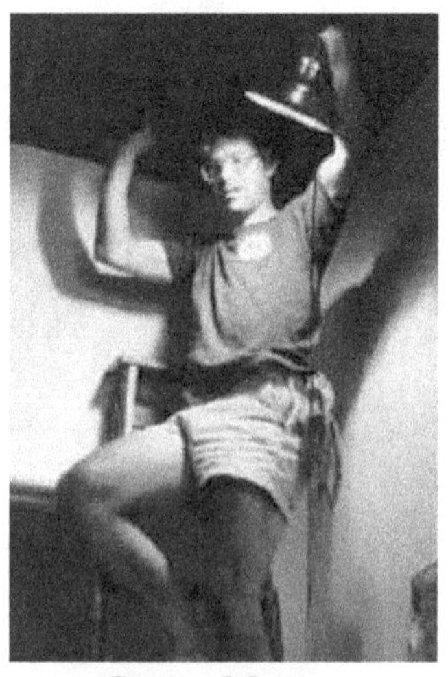
Robin Marks

KFAT Staff at ballfield: L-R KFAT Fan, Sister Tiny, Sherman, Amber, Gordy, Terrell Lynn, Mark Taylor, Bob Cassidy, Leaning: Gilbert, Kneeling with beer: Jeremy Lansman

Gilbert with Chris's girlfriend, Linda, who heard I was going to do a talk show on KFAT, and said: "Chewin' the Fat!"

Cuzin Al, Sister Tiny and Frisco in the production room

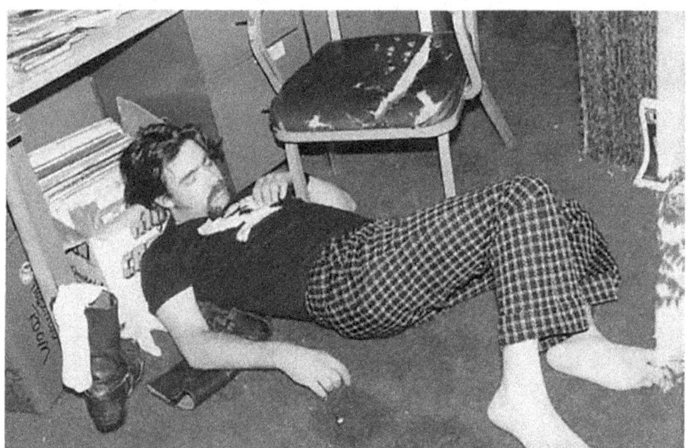

Frisco passed out in the production room

Awest and Wanda at the Hells Angels booth,
Laura Ellen and Terrell Lynn seated behind

Invisible Empire
Knights of the Ku Klux Klan

WHILE AMERICA IS BEING DESTROYED SILENCE IS NOT GOLDEN: IT'S TREASON! WHEN WILL YOU FIGHT?

For Free Information write:
P. O. Box 700, Denham Springs, La. 70726 USA

KKK Card

KFAT Beer label

Chapter 59: The Santa Cruz Sales office

The station was getting a good response in its new coverage area. People were listening, they were talking about it, and they were calling. Ads should be selling, but advertising sales was another area where Laura Ellen and Jeremy had no experience. Hippies are a trusting lot, but the creditors that kept sending those bills were not. They wanted to get paid, and therefore KFAT needed a real sales staff.

The Sales Office had been in Los Gatos, in Jeremy and Laura Ellen's dining room. Bob Cassidy had been selling for them, briefly, and that was intermittently successful, but he was now on the air, and once Jeremy and Laura Ellen got a look at his drinking habits up close, they abandoned any thoughts of asking him to keep selling in his off-hours. Jeremy had asked Sherman to sell for them, but that wasn't working out.

With the sale of KDNA in St. Louis, Laura Ellen and Jeremy had bought a bunch of antiques, and brought them west with them. Now they were having to sell them one by one and in lots, just to pay the bills. Checks were bouncing and people were getting tense again. That hadn't worked out with the first crew at KFAT, it would only work again briefly with this inexperienced lot, and they knew that something had to be done. Jim Shugart wasn't getting them any money, so he was fired.

They hired a salesman named Jim Gordon, and that started to work. He knew a couple of other people, and thus a "sales force" was now on board. But the sales people knew that Santa Cruz was the station's home base, and that was where they needed to be. They were increasingly reluctant to drive to Los Gatos, a twisting half-hour drive up Highway 17 from Santa Cruz, and Gordon begged Jeremy for an office in Santa Cruz.

Jeremy OK'd it, and they found an office on Walnut Street, and moved in. It was a big old Victorian that housed five or six offices, and it was on a quiet, tree-lined street about ten blocks outside of the main nexus of downtown Santa Cruz. There was much work to be done, both in hiring a better sales staff, and in getting people to buy KFAT air time. KFAT was popular and all in Santa Cruz, but not nearly enough people wanted to advertise on it.

Laura Ellen remembers that "we have this mystique about us," and potential advertisers were always telling the sales people, "Well, I listen, to KFAT, but my customers don't."

Jim Gordon would ask, "Well, aren't your customers kind of like you? 'Well, yeah, but... people different from me don't really listen. Okay, my surfer crowd listens, but no one else does. Or my such-and-such crowd listens...'"

It was a tough sell, and as would prove to be the template for the sales office at KFAT, sales people came, and sales people went. Some made sales, but not enough, and no one at KFAT was getting rich. But some of the salesmen were doing all right.

Bob Cassidy has a good memory for details of this sort, and he remembers Jim Gordon getting paid in cocaine. Another, Chuck Scardino was a major rip-off artist. He was a New Yorker, a hustler, and he was so sleazy and doing so much cocaine that he was getting freaky. The coke made him crazy, but coke being coke, he always needed more coke. Sully remembers:

"Once when he came with this woman, and he was just mehmehmehmeh, and he came in and opens my shirt, he pulls my shirt open, and he was so whacked out, and this woman was just standing there, and I said, 'Listen, Chuck, you know you gotta leave. I really like you, but you gotta leave.' And he goes, 'Well I don't really like *you!*' And I go, 'Fine, Chuck, I'm calling the cops.' He was just coked to the gills."

Years later, Laura Ellen still wouldn't talk about Chuck Scardino. Bob remembers that Scardino would get records and concert tickets from the station and sell them. Worst of all, he'd sell ads, slip the spots onto the schedule himself without putting it on the traffic logs in the office, get the money from the advertisers, and keep it.

Bob avoided Scardino, dealing with him only when he had to, until a Willie Nelson concert where Bob was in charge of backstage security, Scardino wanted to get to Willie, and Bob wouldn't let him in. Scardino wouldn't speak to Bob after that, and Bob considered that a blessing.

Bob also remembers that concert because it was put on by the Hells Angels, and one of his all-time enemies was there. He remembers: "This was a guy I knew from high school, and he and I have always been enemies. And now he was an Angel. And he thought he was some kind of big deal, and we had a confrontation backstage where he got to kiss my mag light."

GK: You hit him with a mag light?

BC: "I certainly did."

GK: Well, why aren't you dead?

BC: "I don't know. I guess everybody thought it was OK. We were having private words, and I made him go away, bleeding."

Hey, I told you... Bob is a tough guy.

Scardino was getting crazier and scarier, and he was finally fired. But that didn't stop him from making sales for KFAT; it just stopped him from giving the money from those sales to the station. He was so crazy that bar, restaurant and shop owners were calling the station to complain about the crazy guy that was there trying to sell them ads for KFAT and wouldn't leave, and would KFAT please tell him to leave them alone. He was out of control, and although he was offering excellent rates for the ads, he was scaring the people, and they wanted him to go away.

Laura Ellen told these people that he wasn't working for them, and that they should tell him they knew that. She told them to call the police. Those were the easy calls about Scardino. Others were more alarming, but it wasn't worth the discomfort of having to see Scardino in person, of confronting him, of having to tell him to stop, so the management at KFAT apologized when they got the calls, and hoped that he would go away. And one day, mercifully, he disappeared.

But advertisers were still calling KFAT to ask why their ads weren't running, and the sales office, now Scardino-less, had no idea that the ads had been sold. "Not sold?" the advertisers asked. "Sold and paid for!" they insisted, and they had the receipts. It was a Scardino deal, it was news to Jim Gordon, and Scardino was in the wind. KFAT ran the ads with no compensation for themselves, and blood pressure levels ran a little higher in Gilroy and Santa Cruz for a while.

The good news was that the ads were working. The local sponsors reported getting a good response from their spots, and they were loyal and supportive of KFAT throughout its career. Restaurants and bars had especially good response to their ads. Jim Gordon arranged a trade-out for the staff at The Crepe Place in Santa Cruz, and it became a years-long loyal advertiser. For years the staff ate many of their meals there for free, thanks to Gordon. There was such a comfortable relationship with the station and its advertisers that the jocks would put a menu from The Crepe Place in the folder with the ad copy, under "C," right in front of the mic, and when the logs said it was time for an ad from The Crepe Place, the DJs would just reach for the menu and start talking off the top of their heads about some of their favorite items, or what they were going to try next. While The Crepe Place had paid for a 60-second spot, the jock often winged it and the ad could end up being several minutes, depending on how talkative the jock was. All that evened out, as some spots lasted for less than a half-minute, but no one was counting.

In fact, all of the spots on KFAT were approximately a minute or a half-minute long. This was hippie radio, and the Ozoner, who started out by doing production back then, felt that the phrase "one-minute" was more of a suggestion than an actual rule.

Laura Ellen was now in charge, and she said to keep it close to the time required, but no one really needed to consult any clock back there in the Production Room. If it went on the air, if it was about a minute, it was close enough for KFAT.

Chapter 60: Laura Ellen Steps Up/In

During the Chris crisis, Laura had been coming to Gilroy daily to work in the office. Enid had been there when it was KSND and she was a bit overwhelmed by all these freaks who now came in and seemed to create a whirlwind of anxiety with everything they said and did. KSND had been a small-town station, but it was what Enid knew as normal. Now, normal was so fucking weird that she didn't quite know what to say. Her job kept her in the office area of the station, and at least she had her usual functions to keep her mind straight. And now she also had Laura Ellen working with her, helping with the books and the traffic, acting as a buffer. And she drank a bit.

Of course, there was also Elsie, learning to walk and toddling all over the place and opening drawers and pulling things off desks and tables and falling down and crying. And where there was Elsie, there was Bernice, their Weimaraner dog, sleeping or moving somewhere else to sleep, always within sight of Elsie.

Enid's drinking hadn't affected her ability to keep the billing straight, but there'd been so few ads up until now, that it wasn't all that much of a challenge. She'd been a closet drinker for years, and she ran the office with what passed for professionalism, so some of the staff never knew that she was drinking. There were others, however, who knew, but they were drinkers themselves, and weren't about to blow the whistle. What's good for the gander is good for the goose. Right, Bob?

With Laura Ellen helping out with the traffic, the office was running smoothly.

Laura Ellen told Bob she wanted more country stuff, and that was about all she told him. But the other staff members, still insecure in their new gigs, especially after Chris' firing, looked to please the owner's Old Lady. With Chris gone, Laura Ellen was now the boss, she was there more often, and she wanted to hear more country music. Not the straight country stuff. The offbeat stuff. That's what Laura Ellen wanted to hear. And she was always listening.

Jeremy's idea had been that Larry Yurdin would run the station: hire and manage the air staff, hire and manage the sales staff, get the records, pay the bills and distribute the profits. This would allow Jeremy to work on the equipment, get some money and ignore all the aforementioned nuisances. As they used to say in the back rooms at KFAT: fat chance.

Isolating Jeremy from people was a good idea, and Laura Ellen knew it just as everyone else knew it: Jeremy just wasn't good with humans.

Jeremy went along and pretended it was all good until he ran out of capital and he couldn't pretend any more. He knew a lot had to change. Then Jeremy freaked out when Laura Ellen forgot to schedule an ad run from their biggest client, and the client wouldn't pay for spots that hadn't run. Already stressed to the max because nothing was working out as he had hoped for this station, Jeremy flipped out at the loss of income, screamed at Laura Ellen, and Laura Ellen said she wouldn't do the traffic any more.

Larry was gone, Chris Feder had come and gone, they needed a Station Manager they could trust, and Laura Ellen wanted the job. She was fed up with the domestic routine she'd felt stuck in, and she loved the new format. Jeremy knew the station needed a Station Manager, he knew *he* didn't want to do it, and he knew he couldn't pay anyone to do it, so Laura Ellen was now the Station Manager.

Laura Ellen was willing and able to step into the role, but some of the work had to be delegated. She'd made Buffalo Bob the Music Director, handing him a list of record label phone numbers with one simple instruction: call collect. The MD position came with no bump in pay and Bob made a couple of calls, then stopped. Laura Ellen picked up the ball, made the calls to the labels, and the records were coming in the mail. Daily.

But KFAT was getting two or three of every album the companies were putting out and hundreds of records were coming in the mail. Bob had his way with what he wanted, and put the rest out for the DJs to look over. Bob told me his sole position on giving musical direction was: "You search, you figure it out, you do it."

Without instruction to do so, the jocks had taken to writing comments about songs on the album covers. They'd find a song they liked or hated and write a brief note about it and sign their names. This was one of the ways the jocks communicated. Meetings were infrequent and brief, with children running around and screaming, and the jocks not much interested in someone else's idea of how things should run. Sherman remembers meetings with Jeremy: "Sometimes hash brownies and wine, sometimes a bottle of Jack Daniels, or an ounce of weed that he'd throw on the floor, two 6-packs of 16-oz Budweiser's. Sometimes he'd pay and we'd go down to the liquor store and get something. No ritual, all different. Sometimes we'd sit there for an hour and come out of the meetings going, 'What…?'"

Not much got done in staff meetings: the jocks' schedules were all over the place and no one wanted to show up several hours before or after their shifts, so attendance was low. Laura Ellen was getting discouraged on the rocky path from hope to frustration, and ultimately not much changed, so the meetings were rarely called any more.

For their education, the jocks were listening to everyone else's shifts, always noting new songs and interesting segues. The KFAT library had gotten so big, and contained so many types of music, they all had to talk to each other about it or they couldn't keep up with the learning curve. There was no other way.

Laura Ellen was now in the office in Gilroy every day, bringing Elsie and Bernice. Elsie, a beautiful child with a cherubic face and curly blonde hair, was walking now, and Bernice was never far from Elsie, who liked to wander. Naked or in a diaper, Elsie was all over the station, all day. Bob remembers Elsie running around naked in Los Gatos, too, but no one ever bothered her because Bernice was always with her.

Tiny was bringing in Finny and Amber, and the three kids liked to play together. That was the good news, and the bad news. Three kids, aged 2-8, with no activities to distract them, will find activities to distract them.

Sherman: " [Sometimes] I worked during the daytime and Elsie and Sister Tiny's kids would take these boxes and ride them down the stairs, and the dogs were barking and chasing them down the stairs, and I went SHIT! 'cause I'm trying to do a show…"

Terrell Lynn: "I'd close the doors and put on the headphones."

Laura Ellen and Tiny taught their kids that no sound could be made while the red light was on. That light meant that the mic was open, and all noise had to stop.

Sometimes Jeremy would take the kids in a back room and make them scream as loud and as long as they could. The deal was they could scream now, but when the red light was on, they had to play "Quiet."

LEH: "That was probably Jeremy's way of dealing with it. 'We're going in the back room and we're going to scream, and then we're not going to scream any more after that.' That would be a very Jeremy way of handling it. Probably why my daughter is screwed up today."

I asked Laura Ellen if that worked.

"Yeah. It worked pretty much, pretty well. When the red light went on, everything had to stop, and they had to be quiet. I remember them being thrilled when the red light burned out. I said, 'hey, we can always get a new red light bulb.'

Increasingly, Laura Ellen had her hands full with the office duties and trying to give the jocks some structure when no one had much experience working within a structure. Also, she suspected that the sales staff was having their way with the station's money. She'd ask them about the ads sold, the ads played, and the money collected. She didn't think the figures were matching up, and now she had to deal with that, too, because Enid had no clue about a lot of what was happening.

Bob remembers that his paychecks were different every week, despite putting in the same hours every week. He asked Enid about it a few times and never got more of an answer than, "Gee, I thought that was right..."

Laura Ellen was trying to make the station functional, make the sound more consistent, have all the DJs play the same sort of music. She wanted the jocks to play more country music and less of the weird stuff. Always leave in the weird stuff, but play more country. Except for Bob, the staff tried to please Laura Ellen, but halfway through their shifts, they'd go in their own direction again. Tiny played the slower, sadder stuff, Bob the rockers, Gordy got a bit chirpy in the mornings, Sherman liked the dirty songs, and Sully and Terrell Lynn liked the more classic country stuff.

Laura Ellen began asserting herself, and she was an owner, so the jocks, except for Bob, began to listen to her, and things were becoming a little less hectic. Gordy thinks this was when the sound of the station jelled. Everyone got more familiar with the stuff that Laura Ellen liked, and the new stuff coming in was terrific, and everyone, including Laura Ellen, liked it. While everyone had their favorites, they were all mixing it up more, playing more of the stuff they were hearing from the other jocks. The shows were blending into a stable, consistent sound, and the station was sounding solid. The staff was stable, the station was sounding more consistent, ads were being sold and put on the air, Jeremy was off working on his equipment, and Enid was starting to get the hang of the new scene. She had more work to do, she was doing her best, and the new staff was crazy, but she was starting to enjoy them and the job. So of course, Enid had to go.

Chapter 61: More Laura Ellen, Less Enid

There are some parts of this story that have no explanation or logic behind the events. There are a few gaps in this story: black holes of fact. Two of them occur at this point in the story.

One is that Enid Wingate left KFAT within a year of the new ownership, and the other is when Richard "Frisco" Bo appeared at the station. Today, no one seems to know when, why or how Enid left KFAT, and no one seems to know who hired Frisco, or when. You know about Frisco, and we'll get back to him again.

Laura Ellen wanted to be more hands-on at the station, and her workday became busier and busier. Working with Enid on the books and traffic, Laura Ellen had a station to manage, a barely manageable air staff, a suspicious sales staff, a toddler and a dog to watch, and the office work was getting more backed-up as more ads were being sold. Jeremy was flipping out because the station wasn't generating enough money to be self-sustaining, and his plan for KFAT had depended on that. He no longer had any reserves of cash. The finances were becoming more complex, and yet they weren't making sense, and Enid tried, but had no idea where the money was going. Laura Ellen needed help. Enid was no help. Enid had to go.

Chapter 62: Less Enid, More Greg

There had been a continuous stream of sales people through KFAT's offices, first in Los Gatos, then at the Santa Cruz office. They came, they struggled, they were replaced.

More and more people were listening, and the sales were starting to reflect that. The number of spots per hour increased, and this made the jocks a bit busier. But they were all getting used to their new jobs by now, and with each shift they learned more about the music that was hidden in the stacks, and new music was always coming in the mail. The jocks listened to each other, and they were always talking about the music. They noticed each other's segues and commented on those. They'd joke about the missed segues, the dead air, the bad mixes and brag about the good ones. They all got to know an astonishing range of music, and they knew they were the only people qualified to play this wide-open new format.

They were more and more comfortable in the DJ chair, and they weren't as panicked every time they had to think of the next record; now, they were starting to think two and three songs ahead. They thought of a song they wanted to play, and they started thinking of what songs to play to build up to it. They were becoming comfortable at just the right time, as KFAT had more ads now, and those would take more of their time on the air and in production. That hadn't been the case when they sat down at the board for the first time nine months ago, which seemed like only weeks ago. Now, there were more entries in the logs, more PSA's, more commercials, more promos and the Fat Grams, and that all took more concentration.

With more ads appearing more frequently on the schedule, they also had to have a new consciousness about their sets. They had to end the set at a good place to start the ads, and they had to know which songs would sound good opening a set, and which would not. Not every song is a set-starter, but they were ever more prepared. They were getting into a groove at their jobs.

It also meant that there was more traffic coming into the station. That was more work for Enid, as well as the others. No one remembers much about Enid. This much was known: she had a husband who had killed himself, she was a closet drinker, she lived in a converted Wells Fargo wagon-stop, and it was a strange, ramshackle place.

Someone earlier may have told you that she stumbled about getting her job done, and that may be true, and I will try to find out who said that. In any event: stumbling, maybe; drinking, almost certainly. But that she was culturally different from the new jocks at the station is indisputable. She understood the people at KSND, the freaks at KFAT not so much. The one thing she was sure of at KFAT was that she was from the home planet, and these new people were the visitors.

The DJ's had little interaction with Enid. She had been a bit flustered at the sudden influx of crazy people, but she handled it well, and the staff generally liked her. But the workload was getting heavier than she ever had to manage when the station had been KSND, and she was getting over her head. Enid was gone.

No one remembers firing her. No one remembers her quitting. She just disappeared, from KFAT and from this book. I won't mention Enid Wingate again.

In 1972, Greg Arrufat graduated from UCLA film school and moved to Boulder Creek, a beautiful, rustic town up in the mountains above Santa Cruz. In 1976, he became involved with the Boulder Creek Theater, which showed second- and third-run movies to the small group of hipsters and woodsmen living up there. Fresh from film school and Big City ways, Greg wanted to help increase business at his new venue; he wanted to be innovative, and he started thinking about what films to show and what alternative ways there were to attract business to this small theater. Hip films or hip live music at the theater was the answer.

Arrufat knew that advertising was the key to making his project a success, and he asked everyone about where he should advertise. The overwhelming answer was KFAT. They were playing the music that he wanted to book into the theater, and he knew that Fatheads were exactly the hip people he wanted to know about his shows. So he went to the Santa Cruz sales office and spoke with Sales Manager Jim Gordon.

Arrufat wanted the right film for a midnight show, and he and Gordon settled on the hip western, "Zachariah." They ran the ads and the show sold out, a first for the theater.

Greg knew he wanted to do more shows with KFAT, and he was fascinated by Jeremy's suggestion that he "stop thinking theater, and start thinking theater of the mind." Greg was intrigued by this guy, and he knew that he was talking to a brilliant man. Jeremy and KFAT were exciting, and Greg was ready to be part of something exciting.

Arrufat started a regular ad run on KFAT, the midnight series was a huge success, and he knew that everyone coming was a Fathead. This was in September of 1976. Greg Arrufat now had his foot in the local media market and he didn't know it yet, but he was up to his knees in the vortex. He started planning a series of live shows featuring Fat bands and hip films. KFAT had great rates and the Boulder Creek Theater was amazed at how many people were coming. The spots cost $3.75 for a 30-second spot, and soon they cost even less because he started working at KFAT, and Jeremy gave him a discounted rate.

Greg owed Jeremy money for the spots, and rather than pay cash, he and Jeremy worked out a trade. Greg could have the ads, and he would come down to the station and work off the debt. Frisco had just gone away to dry out, and KFAT needed production help. UCLA had taught him how to edit tape and work a microphone, and his natural talent as a copy writer closed the deal. Greg Arrufat started showing up daily at KFAT to do the production work, but they also needed help with the traffic. By this time, someone who I said I would no longer mention had left the station, and Greg Arrufat became the new Traffic Manager. He remembers that sometimes he spent sixteen hours there at a time. He had to. He was now in the vortex.

The "office" was still in two places: the back rooms of the station and in the office in Santa Cruz. Jeremy had hooked up an old Teletype machine to a pay phone at the back of the station. The paperwork was being generated and sent to a couple of early computer geeks up in nearby Mountain View, who would put together logs of when the ads should run, and send them on a teletype down to Gilroy. This was a curious overlapping of the newest, most cutting-edge technology working in tandem with some of the oldest technology still in use anywhere. A teletype? If you can find one now, it's in a museum.

A pay phone? The other, regular lines were sometimes shut off for lack of payment, and the only phone that Continental Phone Company would let KFAT have was one that paid for itself. Jeremy accepted the indignity of this because someone always needed to be in touch with the station from either San Jose, Los Gatos or Santa Cruz. Besides, Jeremy could have his way with these machines, and make them do his bidding.

The Mountain View Boys were college kids, early computer geeks, working for a pittance, and they'd try to get the logs to the station by midnight, where Greg would convert them to a schedule that he could put in front of the DJs to run the ads, but when they were late, Greg had to wait for them.

Overnights, virtually the only sponsor was Cal Stereo, and if a jock knew that an ad was supposed to run, but the logs hadn't come in yet and they didn't know when or which spot to play, they'd just reach over to the 7-Up crate and pick out the most recent Cal Stereo spot and play it.

Greg had to be there because he had to hand-feed the punch-tape through the teletype machine. Greg remembers getting there at nine or ten a.m., during

Gordy's shift, writing copy and doing production work, staying for Sully's shift, Terrell Lynn's shift at 1:00, Bob's shift at 4:00, and then Tiny would come in, and some of her friends would usually help him out with stimulants to complete his work, and he'd still be there when the Ozone Ranger wandered in around one a.m.

By this time, and for obvious reasons, Greg had given up living in Boulder Creek, over an hour away, and he'd moved into a house in nearby Morgan Hill with the Ozoner.

Greg had been at the station when Emmylou Harris and Rodney Crowell showed up and he helped run the board in the Production Room so they could play. Having dried out again, Frisco was back by then and set up the mics, and taught Greg to cross-mic in a way that gave stereo sound from one microphone. Frisco knew all the tricks. They did the live mix from the Production Room, Sully was in the studio, and it was Greg and Emmylou and Rodney in the tiny room, with Emmylou and Rodney singing and Greg watching the meters. He recalls it as "an orgasm for the ears."

Then, Emmylou and Rodney went into the studio and DJ'ed for about three hours. It was an astonishing, rewarding day that Greg will never forget, but he also remembers that it screwed up his production schedule for the whole day, and he had to make up that work well into the Ozoner's shift, and he was still there, working, when Gordy showed up again.

Chapter 63: Gary Faller Finds KFAT

Gary Faller will always remember coming to KFAT for the first time. Living in Santa Cruz, he'd been listening to the station for about a year, and hearing about it all over town, and now in the summer of 1977, he was going there to talk to the owner about an engineering job. His friend Greg Arrufat told him to go there and speak to Jeremy about a gig that Jeremy had in mind.

Gary ran a small recording service in Santa Cruz, and he could set up, wire or repair anything that had to do with sound reproduction. He wanted to impress the owner of the station, so he got dressed in his best chinos and his cleanest shirt, and drove over the hill.

He'd discovered the station, as had all his friends, when the station's signal suddenly got strong enough to come everywhere in and around Santa Cruz. Like everyone he knew in town, he was a fan of the station, his car radio was always tuned to 94.5, and he'd been listening to Sister Tiny all the way to Gilroy.

It had been cool back there in Santa Cruz, almost chilly, because Santa Cruz was on the ocean, but over the hill it was hot. Real hot. And it smelled of garlic. And it was really hot.

Like all the guys he knew, he was already smitten with Sister Tiny. He knew that she was a rowdy, lusty, honky-tonk gal; everyone in Santa Cruz knew that. Everyone knew about the woman who had opened up the mic after a Lefty Frizell song and said, "Man, that song makes my nipples hard." He knew other stories about her; hell, *everybody* in Santa Cruz knew stories about her, and he wanted to meet her, and now he was going to meet her, and he was nervous about it. After all, this was *Sister Tiny*.

It was his first time in Gilroy, but it wasn't hard to find the main street in this small town. He parked his van on Monterey Street in front of the optometrist's office, put some quarters in the meter and walked up to the glass door with the KFAT sticker on it. There was no other writing on the door, no sign above it. Just 7459 above the door and that sticker. It was a job interview and he was nervous as he opened the door, heard the music and went up the stairs to the landing at the top.

The first thing he saw at the top of the stairs was a precariously stacked pyramid of empty beer cans, a sleeping dog, and three nearly naked children running and yelling in the hallway. He looked to his right at where the music was the loudest, and he assumed that would be where the studio would be. He knew that's where she'd be, and he turned toward the source of the music.

The song was a rollicking, bar-brawling, breakin'-up-with-some-ole-boy tune, and he passed the pyramid of empty beer cans and the children running around and screaming. He stepped over the sleeping dog, and there, in front of him was the control room, the studio, and there she was, in side-view. She was a slender, attractive brunette, and she was naked except for her headphones and her red panties. Later, with an engineer's eye, he would notice that the control board was practically an antique, and the rest of the gear was…but for the moment, all he could see was an attractive woman who was naked except for the headphones and those panties. Those small, red panties…

She was facing the board, reading an album cover, gyrating, rockin' to the beat, and she was so into the music that she didn't notice he was there.

It was hot, the music was chooglin', the speakers were blasting, and she was dancing at a party of one, keeping it going with the records she was playing. Bare-breasted, she was swaying and rockin' in her chair, singing along with the song and she didn't hear him or see him, but he got a good look at her. Both of her round, firm, bouncing breasts. Great breasts!

He might have thought, "So this is Sister Tiny. They're wrong about that!"

Chapter 64: Live Fat

Greg Arrufat was now spending almost all his time at KFAT, and still trying to book films and bands into the Boulder Creek Theater. The first film, "Zachariah," had been such a success that both he and KFAT had done more shows. KFAT's Sales Manager, Jim Gordon, took the theater as his account and wanted to run with it.

Jeremy was a dreamer, and his plans went all over the place, and sometimes they might even work. The success in Boulder Creek got them all to thinking about a live broadcast from the Boulder Creek Theater, but Greg had too much on his plate already, so he brought in Gary Faller, who he'd worked with at the theater to help work out the logistics for the live show.

Gary was a hippie with a sound truck, and had worked for Greg at the theater, doing set-ups and sound engineering. Greg knew Gary was insane in just the right way. He'd pulled off some crazy, cobbled-together gigs, and little would faze him, so he was a perfect foil for Jeremy, who met him that day at the station and hired him on the spot.

Jeremy wanted to broadcast a show from the Boulder Creek Theater, and they took a couple of Jeremy's Marti's (remote transmitters) to Boulder Creek, and Jeremy tried aiming them where the theater had a show scheduled with Ramblin' Jack Elliott, and Gary Marti's at Loma Prieta to get the signal there, but it didn't work.

Then, between Jeremy, Greg, Gary and Jim Gordon, someone came up with the idea of renting out the unused movie theater in Gilroy, the Strand, and broadcasting a show from there. They already knew that the signal from Gilroy to Loma Prieta would work, but they tested the Marti's, and they performed like champs, so KFAT booked Rosalie Sorrels and Utah Phillips for the first KFAT live show. Three dollars in advance, three-fifty at the door. At the door? Excuse me? It sold out in a day.

There was much excitement at the station about <u>the first live show</u>. Laura Ellen tapped Sully to be the emcee, and everyone from the station was going. Many of them had never been out in public before as a celebrity. They wore clean jeans and whatever was new. They were nervous, excited and happy. So were the listeners.

It was a huge success, giving Fatheads their first chance to come to Gilroy in their best western wear and meet other Fatheads. But most of all that night, what everyone remembers was the magical moment when Utah Phillips was singing "Daddy, What's A Train?" and paused just as a train came rumbling through Gilroy and blew its whistle. The tracks were right behind the theater, and the rumbling of the train could be felt by all, and as the whistle blew and the sound reverberated inside the walls of the theater, everyone sort of held their breath, let it out and looked around them in wonder. Thaaaat... was... cooool. The whole audience smiled, and they knew they were all sharing a remarkable moment. And the listeners heard it on their radios, too.

And you know, who knows? All the engineers and trainmen up and down the line knew about KFAT and called in with requests, knowing that the jocks always gave their requests first priority because they'd be out of range soon. That an engineer on a train passing through Gilroy was aware of KFAT is a given; no one else played the train songs that KFAT played. If we know for certain that the engineers all listened to KFAT, why not assume this engineer was listening as he went through town? And the conductor was almost certainly listening to a song about trains while passing through Gilroy, why wouldn't he blow his whistle? Y'know... let the people know he was there? Why not, if he was listening...

Oh, he was listening.

So maybe there is an old, retired railroad engineer out there, sitting in the shade, I hope, in a beach chair in Arizona or Florida, remembering the night he listened to KFAT, when he heard the story about trains and blew his whistle. Would he have heard his whistle on the radio over the sound of it blowing just outside his cabin? Maybe, maybe not, but I wonder if he knows that everyone listening to KFAT that night heard it, in the theater and at home, and in their cars and in bars, and that he created a special moment for thousands of people.

Sir or madam, if you are reading this, and it was you: yes, they heard it. And they all knew how special that moment was. I still hear about it from Fatheads. Thank you.

After the show at the Strand, Ms. Sorrells, and Mr. Phillips came back to the station, where Frisco set up the mics in the Production Room, and the show went on until almost four a.m.

The show was well received by the listeners, many calling the Fat Gram, thanking the station and asking for more shows. It also created a lot of excitement at the station, and it evolved into a regularly- scheduled broadcast of Fat music, to be known as the Fat Fry. Fat Frys would become a staple of KFAT programming on Monday nights for many years, with Gary Faller setting up the soundchecks inside, and then sitting outside in his truck with his gear, sending the signal to Loma Prieta, and thus to the radios of Fatheads, wherever they were.

Chapter 65: The Trains Kept A- Rollin'

One of my favorite films has a line in it that the three most romantic sounds in the world are a ship's horn, an airplane's engine, and a train's whistle. Not many felt that way at KFAT.

You know the phrase about being from the "wrong side of the tracks?" The tracks were just a couple of blocks from the station, and both sides of it were pretty shabby. Trains rolled through Gilroy at all hours of the night and day, but especially at night.

Consider how much fun we've had recounting the zany antics of our DJs, but consider also that DJing was a long, lonely job, especially the overnights. There were usually people who came by to visit or party, but that was not every night, and the job went on for months, and then years. Many hours were racked up alone, with some beer, some pot, or whatever, or with nothing, just those records. And those records kept coming in. And those records wouldn't play themselves.

When it was late at night, the smallest hours, and you'd been on the air, alone, since when-the-fuck-ever, but it seemed like all night, what were you gonna do? Leave? Where to? When the Green Hut closed there was nothing open but the 7-11 several blocks away, and the Longboat diner. It was a punk-ass town, and it was damned quiet once the junkies started looking for shelter and left the streets empty. There was nothing else going on in Gilroy except for the overnight processing of fruit and vegetables, the Mexican ladies at the bakery, and the overnight guy on KFAT.

Visitors usually made it all fly a lot faster, but there were also a lot of hours and long nights logged at on the air when no one showed up, and it was just the DJ and the records in the library.

And every once in a while, a train would roll through Gilroy. The conductors had to blow the whistle, anyway: they were coming through a town with cars and people crossing the tracks. They'd pull on the whistle because they had to, but they'd also whistle for KFAT. They knew we could hear them.

This was, after all, a radio station, and it had been outfitted so that street noises would not interfere with an open mic. During the day there wasn't all that much traffic, but at night the air was still, no one was moving about, and that whistle had a low, plaintive tone, that wistful, lonesome, minor note that made you stop and listen, and you let it sear into your soul. That sound just gets to you, gets to you inside.

Many listeners have commented over the years about the trains. They remember listening to the jock talk when the train came through town and whistled, and it came over the mic, and they all heard it. Listeners knew that the jocks had had no professional training, and were Fatheads just like they were. But the jocks were *the* Fatheads, and they ran the show. They were talking to you like a friend when the train came through Gilroy, and many listeners have told me that the trains used to make them feel like they were in the room there with Terrell Lynn or Sherman or the Ozoner. They felt like they were all in the room, and it was late, and they felt like they were all in this together with the jock, whatever this was.

KFAT played train songs. No one else played train songs. Sometimes a lot of train songs. Sometimes *hours* of train songs. Working late one night, Jeremy and the Ozoner got into a riff of train songs. They took turns finding the songs and playing them, and Jeremy surprised Ozoner by crawling into the stacks and pulling out albums that he had never played, and they played every known train song at the station. It took several hours, and it exhausted them.

Terrell Lynn says, "Did it with me, too. Did it more than once. But one night at midnight he was there, and we got into railroad songs, and we played railroad songs from midnight until about 5, 5:30. Just Jeremy and me and my friend, Scott. But he worked with me all night on his own. He had work to do, but he did this. We were getting all sorts of calls from all the train yards in the area, making requests and suggestions. They were all listening and calling."

Spending hours with Jeremy in a convivial, recreational way was a rarity for the jocks at KFAT, and I asked the Ozoner about hanging for hours with Jeremy, and he remembered, "Jeremy was a strange bird. No, not easy to talk to, but my strategy was to be direct with him when I needed to be." Jeremy did not seek out excessive contact with the jocks, but when in conversation with them, he was rarely less than fully engaged and sincere. He cared about everyone there, and everyone knew that. And he knew strange stuff, including those train songs.

Sherman remembers the night Jeremy was working on the board, and had disconnected the microphone. All he needed was a constant signal, so he wanted a record just to track, and he suggested to Sherman an album of train songs. They played it, then flipped it over and played the other side.

Chapter 66: Fun With Words and Numbers

The shifts were stable. It was the people who might not be stable, but the music kept coming out of the speakers and sales were drifting in. Laura Ellen was piecing it together in increments, and she knew some changes had to be made.

As Sales manager, Jim Gordon was a bit of a thief, but he could stay. They had no one else, anyway. They'd had Chuck Scardino, who was a thieving, coke-fueled, scamming scoundrel, but he was finally gone. Now, new sales people were hired and the ads were drifting in. More ads meant more traffic. The traffic is the scheduling of ads to be written, recorded and run in a certain order. The schedule of ads had to be put into the logs that were set in a three-ring binder set on a slanted board between the turntables, right in front of the jock. This way, a jock could look up and see which spots were scheduled to run, in what order, and at what time.

The cassettes that contained the recorded spots were stored in two 7-Up cases nearby, to the right of the turntables. They were all labeled on their sides, for quick identification. The cassette decks that Jeremy had gotten on the cheap were in use at KFAT and nowhere else in America. They were located above the control board, about even with, and just to the left of the slanted wood board that held the log book and the binder of written copy that the jock would have to read from if the ad called for a live reading.

Also, the station was required by the FCC to schedule the reading of PSA's: Public Service Announcements, which were for free public events like lectures, school events, meetings, library functions and other free public services. Laura Ellen knew that these were of a more serious nature, and she wanted the jocks to read those at least once off the air before reading them on the air, so they wouldn't get the information wrong or stumble on the copy and sound foolish.

The Crepe Place in Santa Cruz had a perpetual trade-out with KFAT and all the jocks ate there whenever they could, and knew the menu. The jocks would just take the menu and do an impromptu riff on the item. Spots logged for one minute often ran on for two or more minutes, but no one complained. Until the FCC stopped by.

That happened right after Gordy had just put on a really long cut and walked out the door. All the DJs knew the longest cuts. Every DJ, when listening, would hear one of those cuts come on and know that the jock had just gone out to the Green Hut or (if it was Gordy) the donut shop, and this time Gordy had just left for Augie's Mexican restaurant, a couple of blocks away. Sherman was in the Production Room, working on an ad after his all-night shift, and he looked through the window into the studio, and saw two men in suits looking around in there.

Men in suits? Well, that wasn't Fat, so he went in to see what was what.

They were from the FCC and they had just been up to the transmitter site again, and they had a subpoena and a few issues to discuss with Jeremy, and at least one was about his phone system, and was he there? No? Well, they also they had been listening to KFAT on their way down, and they thought they'd heard some mighty long commercials a while ago. They wanted to make sure that the commercials were the length that they'd been logged for. There were laws and stuff about this stuff. Laws, they said, not guidelines. They were serious. They were men in suits.

They asked who the DJ was, and Sherman knew better than to say that the jock had just gone out for breakfast, so he said that he was the DJ.

They wanted to hear a specific recorded ad that they'd made a note about after hearing it on the way to Gilroy, and Sherman got it for them. With Gordy's long cut still playing, he turned down the monitor, they got out a stopwatch, and Sherman played the ad. Then they asked to see the logs, and noted that the spot had been scheduled for 60 seconds, but had run 63 seconds. Which was close enough for KFAT, but…

They tut-tutted that this would never do, but they were polite and asked to speak to the Station Manager. It was only eight a.m. and no one else was there, so they told Sherman to tell Jeremy that he would be fined. They took the office number from Sherman and went on their way. But that wasn't the worst of it.

They had just come from the transmitter site and they were pissed about that, too. They had pages of things to talk to Jeremy about.

Jeremy, of course, had done everything on the cheap and of questionable legality. When he was about to move the transmitter, he got an ancient RCA transmitter and took it apart. He hand-polished the induction coils and all the silver-plated stuff until their efficiency far exceeded the efficiency of most transmitters that are installed and then left to perform without cleaning. And, as your mother or your drill sergeant will tell you: clean means efficient. The FCC didn't ask owners to constantly maintain their equipment, so they calculated that the power would decrease over time, and they allowed the lower percentage of 62% of capacity as an average.

Jeremy's power was at something like 85%, and KFAT was putting out 1,800 watts on a license that permitted 1,250 watts. The FCC wanted to speak to Jeremy about that, but sadly, he was unavailable and not near any phone whenever they called. When he finally responded to them in a letter, he told the FCC that he fixed the system himself, that he had no maintenance or repair records, that he wasn't worried about not having the records, as he was the one who the records were meant for, and he would be happy to take them up there to show them that he was in compliance.

The FCC wrote back to say that he owed three thousand dollars in fines, and Jeremy responded promptly in a letter, saying, "In the interim period since your inspection, I've had the use of my broadcast license and become affluent and successful in the radio business, and therefore I am willing to make the following offer: I read in your notices that you are constantly licensing non-profit minority radio stations. I am willing to donate three times the amount of this fine to any one of them that you designate."

Perhaps the reader has forgotten the "Petition Against God," but Jeremy hadn't, and he was still pissed off about that. The FCC hadn't forgotten it, or Jeremy, either. Eventually, the FCC dropped the issue, the fine for the longer-than-scheduled ads were folded into the fines for the supposed transmitter transgression, and nothing was ever paid. But the FCC would remember Jeremy Lansman and KFAT. In their bureaucratic hearts, they knew they'd be back.

But at the station, the point had been made: KFAT production people had to tighten up the spots and make them 30- and 60-seconds exactly. And they did, for a while. Fun with numbers. And that was the end of that phone system, but Jeremy was already at work on the next one.

Meanwhile, Laura Ellen kept reminding the jocks to "read the copy to yourselves before you read it live on the air," and sometimes they did. Listeners who had, without realizing it, come to expect a certain professionalism from their DJ's, were surprised and charmed to hear real people make real mistakes while reading an ad. Buffalo Bob took his time, but slurred his words rather than stumbled over them. Laura Ellen wasn't all that charmed, but again, she and Jeremy had expected to let this station develop into whatever direction it developed in, and it seemed to be working. She mentioned again to all about reading the copy in advance, and everyone nodded in agreement that it was a good idea, and sometimes they did it and sometimes they didn't. And sometimes the jocks still stumbled on the air with the ad copy, but they didn't mind. That was Fat.

I remember having some fun with the copy. I usually recorded the promos for my shows, transferred them onto cassettes and put them in the 7-Up case under "Chew-Promo," with a date. Someone in the office put the promos into the logs to run at various times every day before the show that it advertised. Sometimes, however, the promo was copy, and it needed to be read by the jock. I'd been having fun with the jocks, using tricky, multi-syllabic words that I knew were unfamiliar to them, and they were words designed to stumble over. Sure, I could use "magic," in my written copy, but "prestidigitation" was more fun. I knew that Laura Ellen had been riding the staff about reading the copy before they read it on the air, and I knew that they often didn't. And everybody enjoyed a good joke.

One time I left a note for Greg in traffic that my copy needed to be read live, and the copy that I left in the book read: "This Tuesday on Chewin' The Fat, Gilbert's guest will be a man who has researched the Antarctic, looking for a new species of flightless birds and if you don't get your hands off my tits I'm going to grab your dick and yank it off."

I wrote the copy and left it there, waiting to hear what fun it generated.

Buffalo Bob swears that Tiny read it exactly as it had been written all the way through until she got to "yank." Tiny swears it never happened with her, but thinks it caught someone else. Sherman remembers knowing about the copy in advance, having been told about it by the other jocks, but Laura Ellen hadn't heard about it yet, and he read it exactly as I'd written it, as he had been instructed. Fun with words. Fun with management.

Laura Ellen heard about it, took it out of the book and off the schedule, and insisted again that this showed that the jocks needed to read the copy to themselves before they read it live (heh heh). And they did, sometimes. She also insisted that spots should run the correct length of time. And usually they did.

Tiny swears that thing with the Chew promo copy never happened, but Tiny's memories often disagree with those of others. Like about the time she met Travus T. Hipp.

Chapter 67: Travus T. Hipp

Sister Tiny swears that she did not read my promo about getting "your hands off my tits or I'll yank your dick off." Bob says he heard it. She also swears that she never said "that song makes my nipples hard." Gordy told me he heard her say it. Others did, too.

Others remember a new ad for workers at the Safeway, the grocery chain, which was then on strike, and anyone who went across the strike line would be a scab. The first time Tiny played the spot, it got halfway through before she tore the cassette out of the deck, broke the cassette open in front of the live mic, and ranted for five minutes on labor rights, then played a long set of pro-labor songs. Others remember that, Tiny does not. Travus remembers coming to KFAT.

TTH: "One day I went to Los Gatos and met Jeremy and Laura Ellen, and sitting around and being crazy, and didn't even get a handshake deal. Jeremy said, 'yeah, sure, start next week.' And [I] started feeding the news down there, and it was a couple of years before I met anyone."

Tiny's memory may be faulty with what she doesn't remember, but she does remember seeing Travus for the first time.

Waa-aa-aay back in this book I said that if I had the time and the inclination, Lorenzo Milam would deserve a book to himself. Or I might have said that about Sherman Caughman, I don't remember. Either way, it's true about both of them, and it's also true of Travus T. Hipp, one of the first serious newsmen in the counter culture revolution.

ST: "When Travus came into the station he was wearing this big burly cowboy leather jacket, and he had this rifle, he had this hat on and he hangs up all his stuff, and he walked in. I just watched him and immediately had this... he was like a hero to me, because he was this big radical news guy."

TTH: "I got there at eight p.m. and checked out Gilroy, and drove up and down the street looking for the station, and came in, and Sister Tiny was on the air, wearing a long velveteen gray dress and I said 'Hi, I'm Travus T. Hipp,' and she said 'Wonderful!' and threw her arms around me. And I thought, 'I'm gonna like the hell out of this station.'"

"That's such a lie!" says Tiny. But Tiny also remembers Terrell Lynn hanging out at the station before he worked there, and that isn't possible, as he was hired his first time there.

Born Chandler Laughlin, he had been influenced by radio legend Jean Shepherd. The words he chose and the way that he strung them together mesmerized young Chandler.

TTH: "I had come back from the Navy, and back to Berkeley. I had become at least borderline political. I had been to many places where there were dictatorships that richly deserved to be overthrown, Cuba included. I saw people starving, and their sisters were whoring for the Navy because there wasn't anything else to do for money. I saw a lot of that shit, Dominican Republic under Trujillo, and I came back to Berkeley, and was striving to be a beatnik at that point. I went to Monterey Peninsula College and [was] elected student body president. This is 1960-62. [I was] thrown out for my son Sean, who was still a large bulge in his mother's belly, and she claimed that I had sent her to San Francisco for an abortion, and the women's dean was out for my ass. The cops were pulling up, and it was United Nations Day, and I was Student Body President, so I put on my little jacket. And me and the guy who was organizing the hustle were planting an olive tree on campus to symbolize youth's commitment to peace, and as I was planting this tree with the silver spade in my hand, I looked over the President's shoulder and saw the cops pulling up at the administration building. So thanking everybody, I said good-bye, walked into the student union and sold my books, and walked out the back door. I drove off in my MG with a girl I knew, and drove to San Francisco."

Having been politicized, Chandler saw the world in a mode of change. The Cultural Revolution of the 60s was raging, expanding, mutating, and he wanted to know what made the machine run, so he pursued a career in the news. Earlier, Larry and Chandler both worked at KMET in Los Angeles, and it was Larry's slovenliness in the studio that upset Chandler so much that he got kicked out of that station. Excellent reporting had followed at several stations, and now he was at KFAT.

TTH: "I went to Yemen, which is at the end of the world… I mean nobody even knows where it is, and checked into the U.S. Information Agency with Peter Lauffer from NBC, and he has his NBC ID card and I had my KFAT ID card, and the guy (from the Agency) looked at me and said, "You work for Jeremy Lansman!"

"You could subscribe to a service where you got sermons, and the 'Jeremy Lansman and Lorenzo Milam: Jewish Disciples of the Devil Trying to Drive Jesus Off the Airwaves' was one of the hot sermons that got syndicated to every Baptist [church] in the entire south.* There was never a Sunday, according to the guy at the U.S. Information Agency in Yemen, in which at least 3 or 4 people were not denouncing Jeremy and Lorenzo from the pulpit, and asking their flock to send letters to the FCC to keep God on the airwaves."

"He'd spent a year and a half at the FCC doing nothing but answering mail

*These sermons were in reaction to what became known as "The Petition Against God" in Chapter Two.

about it, telling people that they weren't trying to drive God off the airwaves, that Jeremy Lansman and Lorenzo Milam were not truly tools of the devil. It paralyzed the FCC for damn near ten years. I had heard of Jeremy because of that, and I had heard of KFAT from people in the radio game who knew about it."

Travus had reported for alternative news shows all over America, most notably on KSAN in San Francisco during its height as a revolutionary central clearinghouse for music, news and cultural reportage. Travus was in the middle of "the good fight," and became known among the few dedicated, serious news reporters covering "the Revolution."

But the '60s were over, the Revolution had fizzled out, and Travus found that no one wanted his firebrand style of reporting. He was a dedicated newsman, and would brook no compromise: there was the truth and little else in Travus' world. The corporatization of the media left him out in the cold, and no one wanted him until he heard from Jeremy, who wanted news and commentary from an alternative source... something that listeners were not getting anywhere else.

TTH: "I had been on for a year and a half at this point, and I had a big beef with a woman, and I walked out, and living in the boats (in Sausalito's famous Gate Five), and somehow I got this offer from Jeremy to come down and work live full time. I dragged my ass down..."

TLT: "I remember doing a show with Travus... and somehow the price of gold came up, and I leaned over and said 'what does that matter?' and he berated me on the air for like ten minutes."

TTH: "My little lecture on gold."

TLT: "I felt like I wanted to crawl under the table, but I thank you for years for bringing me into the news world, for what affects you and what doesn't. And ever since that morning, I have been a news junkie."

Travus remembers: "The most significant thing for me about Jeremy was that he was totally open. If you had any idea at all, he'd say 'go ahead on.'"

Terrell Lynn pointed to Travus and remembered: "I got more stoned with you than with any other DJ when you were on live."

As I write this, it's late May, 2012 and Travus passed away last week. Travus was an anachronism. He was a blustery, self-important blow-hard with serious tolerance issues, but he was also smart, caring, insightful, passionate and brave. He fought the Good Fight for you, for all of us, and for himself. He never followed the path to wealth, and he maintained his passion about his search to find and announce the truth. Rest In Peace, Travus.

Chapter 68: The New Sales Office

With the air staff stable, or at least showing up on time for almost every shift, the sound of KFAT was now consistent and nonstop. That was a good thing. Now to make this bastard pay for itself.

Jim Gordon was the Sales Manager, and then he wasn't. Jeremy needed a serious, professional go-getter to head up a sales force of go-getters. This was Jeremy's first commercial station, and he and Laura Ellen had signed on for a course in making a profit. Other commercial stations were doing it, and now KFAT had to do it, whatever it took.

A year ago, Laura Ellen thought she wanted to be more commercial, like the country station in Salinas, KTOM, but soon realized that KFAT was an "attitude station." She recognized that KFAT's success came as much from the jocks as from the music, and she knew that even if KFAT played the same songs as KTOM, they'd still never be the same. It was the attitude.

But the jocks had a bad attitude about getting paid $3.25, and Laura Ellen and Jeremy had talked a lot about how to sell ads on KFAT, and they realized that they had to go into uncharted waters to do it. Neither of them had ever had much truck with the slick, tie-wearing, corporate hustlers who made their livings in radio ad sales, and neither of them wanted to have much to do with that sort, personally. But if KFAT was going to pay for itself, much less make a profit, they had to play the game. The radio sales game. They knew they had to get into bed with the devil, but they were hoping to keep the sheets clean. They'd hire a real sales manager and keep him under control. They'd have a staff of real sales people, and they'd control them, too. They were both very sensitive about the ads on their station, even though they both knew they had to play the game to stay in business. It was their station, their vision, so they would hire a guy who would generate more sales without selling out KFAT's soul.

They wanted KFAT to stay the same, but they needed to hire a strong sales guy with a strong background in radio sales. They wanted to commute to Mammon, but not live there.

Jeremy thought that radio sales people were snakes, but if he was going to be in the snake pit, he wanted the biggest, meanest snake in the pit working for him. His name was Bob McLain.

Terrell Lynn remembers: "I couldn't believe it. White belt, white shoes, the whole fucking thing!"

Sister Tiny: "My hackles were up the minute I met him!"

McLain had been in radio sales for decades, and he was the consummate pro. Matching shoes and belt, polyester salesman suits, a pinky ring, a hearty handshake and a 100-watt smile. Jeremy liked him like he liked poison ivy, but it wasn't friendship that Jeremy needed. He needed a guy who could generate sales to the big national accounts and get some real money coming in. Jeremy was freaking out that the station had been on for nearly two years and more money was still going out than coming in, and Bob McLain told Jeremy and Laura Ellen that they had a real winner on their hands. He told them it "might look like a dark cloud now," but he could "make it rain money." Jeremy and Laura Ellen scrunched up their noses and signed the contract.

First, he needed to be a presence in San Jose, and he needed access to San Francisco, where the majors were. McLain needed to hire a staff of pros, and to attract them, he'd need an office in San Jose. Santa Cruz was a small-time burg, and all that town would ever produce for KFAT was small-time ads from small-time players. McLain wanted to be in San Jose and San Francisco, and he knew he'd never find anybody to sell for him in Santa Cruz, where the available pool of talent were hippies, surfers, drug dealers and welfare losers. McLain wanted pros on board, and Jeremy and Laura Ellen agreed. Jim Gordon was gone and so was the Santa Cruz sales office.

The new sales office opened in San Jose in early 1978.

When he got there, the sales staff had had a trade-out with the Bacchus Massage Parlor, which was later busted for prostitution. That was the first thing that changed with Bob McLain. But McLain knew that that little problem was small potatoes, and to put his stamp on this mess, to make some money there, he knew what he had to do: he'd have to sell Jeremy and Laura Ellen on it—and it wouldn't be easy—but that air staff had to go!

Chapter 69: Bob McLain Gets A Staff

Bob McLain was an old hand in radio, and he knew the territory. He wanted a staff of sales pros to sell this weird, unclassifiable station to the money. Money came from the big-name, big-budget sponsors, and those small-time local shops and restaurants that were advertising on KFAT now would never provide a living for a man of McLain's needs. Running an ad for a *gas station?* Or the Gilroy VW dealership: "Come to Jack Lewis Volkswagen, where less is more, more or less." A single car dealership, not a car company? Fuck that small-time shit. McLain had a nice house in the San Jose suburbs and a good life that he had earned with his ability to sell ideas to people. Convincing them that they needed what he had was his gig, and radio was his world and his language.

One could say that McLain had only one problem standing in the way of his making the kind of money he wanted, but one would be wrong: McLain had several problems facing him as he prepared to put a staff together to sell this bastard. The owners were hippies and had never worked in commercial radio before. They had ideals, and that was nice, whatever that was, but there was no place for such shit in his world. His world was all about the ratings, the coverage area, the demographics, the ad rates and the deals. It was all about rates and ratings: there was nothing else.

Oh, sure, good will was important, but he'd found that even that was negotiable. If he'd had to fuck some guy at a station a few years ago, all would be well again if he came calling from a station with good numbers. Jeremy had thought Bob McLain was a snake, but he was wrong: he was a shark. A snake may attack, but once it feeds, it stops; it stays where it is to digest. A shark keeps moving, keeps processing, relentlessly alert for those minute signals of blood in the water...

McLain was always alert for an opportunity, and that was his nature. The opportunity might be a new salesperson in the marketplace, or it might be a hint of a change in management at some station somewhere. Changes in management became a feeding frenzy for guys like McLain: a manager who had nixed your sale might be replaced by a manager you'd helped look good at some previous station. Salesmen would circle, and the first one in had the first shot at a sale, but all would want their chance at a pitch.

Even with unknown new managers- what a challenge! Get in there and be his friend! Now! And now McLain had a new station, with a completely new sound, and you'd think he'd be happy. Back to his problems:

Okay, the owners were hippies and he could probably push them around, but they just didn't understand the basics, and McLain tried to explain it to them: Numbers. Money and numbers. Why were these people talking about feelings? Feelings? What the fuck did fucking *feelings* have the fuck to do with anything?

And those feelings were about the staff; about...*that staff*? Those fucking lunatics down there did just whatever the hell they wanted, and they changed spots around, like when he first got the gig, and he went to the easiest account in the world, the Army, and he got them to buy time on KFAT, and *some of those bastards refused to play the spots!* Christ, how do you deal with something like that? And no one fired them?

Travus T. Hipp remembers: "KFAT owned 101 from San Louis Obispo north. The truckers were ours. And we had been trying to sell ads to Safeway for years, and getting nowhere, and when that wildcat local from the East Bay called a strike at that warehouse up in Richmond, Safeway bought a schedule to recruit scab truckers, and I guess they played the spot in the afternoon, and when it got to Sister Tiny's show, and halfway through the spot, she stopped the spot, opened the mic and pulled out the cart and broke it, and did about five minutes on the rights of labor, and played a two hour set of labor songs. And lost us the contract from Safeway."

They canceled the contract? Christ, McLain thought, that was *unheard of;* how do you deal with something like that? He'd deal with it, all right. And in doing so, Bob McLain became the enemy of the air staff, and the seed of its destruction. He let Jeremy and Laura Ellen know that things like that attitude down there had to change. This was the conversation he'd had with them about money and why weren't they getting it? Everyone had to know in advance what was expected of them and what they expected in return. McLain had been clear about that when he negotiated with KFAT. It would cost *this* much money to set up an office and get it running with a staff, and he would need *that* much money to keep it going. He expected to earn *this* much money for KFAT, and he expected to earn *that* much money for himself.

And he told them again that they just *had* to control their monsters down there, or no serious agency would allow their national clients to associate with KFAT. How could he sell to Ford, if the jock would tell listeners that "he drove a Chevy and liked them better," or called your client's beer product "piss water" on the air? Jeremy or Laura Ellen would *have* to control them!

Even if they tried to read a spot right, and stumbled because they hadn't read it to themselves before reading it live, that sounded unprofessional, and a serious client wouldn't like it, or pay for it. And that was when they were *trying to get it right!* No one ever simply *refused* to run a spot… but *these* bastards did that!

Yeah, that was a problem for McLain. But not his biggest.

Getting a sales staff for this one was going to take some work. The station had no ratings, and the owner had no real funds for promotional stuff and wouldn't pay to subscribe to the Arbitron Book, which gave the station ratings, so even if the station showed up on the charts, as a non-subscriber, McLain wasn't allowed to use the figures in a sales pitch. The sales staff would have to be very good, very convincing, for this indefinable hybrid. They were still new in the market, and the station needed a big boost of market awareness, some PR, some exposure.

McLain needed sales people who were… pliable. Money might be made, but it would take some time. His sales staff would have to be good, but they'd have to be patient. And that was a problem. But even *that* wasn't his biggest problem.

Remember that line from a few sentences ago about the station being an "indefinable hybrid?" THAT, my friend, was the big problem.

What the hell *was* KFAT? At the time, there were several definable formats in American radio. There was Top 40, of course, and rock, and there was Classic Rock and Oldies and Jazz and Sports and News/Talk and Country and maybe a few more. Today, of course, there are dozens of categories. There are a half-dozen categories just within the dance genre, and the same for Hip Hop, then called rap. But before it was rap, it was called Urban Contemporary.

During the period of the Sixties that changed American radio, even then, underground rock added only that one category to the available mix. And soon enough that changed into Classic Rock and there it stayed.

But McLain's problem was that whatever category was out there as an option for advertiser's, KFAT was none of them. What the hell was it? It wasn't… anything. And yet, there it was.

When a national company hired an agency to place its ads, the agency had to know where the ad buys would have the greatest impact; that was their job. Companies bought an agency's ability to create effective spots, and counted on them to place the ads strategically, to best benefit their company. And these agencies all had charts by which they divided up their ad budgets for the period. X% would go to Classic Rock, Y% to Oldies, Z% to Sports. Like that.

Products for older people did better on smooth jazz, easy listening, news/talk and classical stations, beer ads sold best to young men. You get it. Stations were categorized and budgets were divided among them. They were the categories, everyone knew them and knew which demographics were attached to which categories. But what was KFAT?

Oh, yeah, McLain knew this was going to be a problem. But he was a pro, and he could work with anything. If he knew what to call it.

He knew all of this going in, but he still saw KFAT as an opportunity for himself, and if some of the charges against him made by the staff are right, about him helping himself to KFAT's sales revenues, then he was he right about that, too.

But there was one more problem that he wasn't expecting. Despite his years in the business, and all the dirty tricks he'd had to resort to, he wasn't ready for the things his rivals at other stations said to his clients about *this* station. My God, it was almost like they wanted to destroy him! Or destroy KFAT. Sales people from other stations called his clients and told them stories about KFAT, of the outrageous things that a jock had played or said.

McLain got a call from a client who, the day before, had signed a contract, and he was told that his wife had heard he was about to advertise on KFAT, and his wife had objected strongly, telling him about what some jock had said last week that she'd heard about, and this wasn't the first time she'd heard about some profane, offensive act at that station. The client was sorry, but he had to cancel the contract. Yeah, he was sorry, and Bob McLain was furious! That same story was happening to almost all the sales staff, repeatedly, and McLain was really pissed.

Jeremy would have to handle that, *or HE would!* Either way, it had to get handled.

Thirty miles down the road from San Jose, Sister Tiny felt a chill shiver down her back.

Chapter 70: Jeremy's Phone-System-In-A-Cup

Jeremy was a genius. An angry, frustrated genius. His station was costing him money, not bringing it in, and there was only so much he could do about it. But what he could do, he would, and that meant electronics. Predictable, controllable electronics. Sometimes imaginary, sometimes crazy electronics, he'd give it a try. Fun with circuits and diodes.

The FBI had just been to see him again, and torn up the little piggy-back phone system that he'd invented and installed up at the transmitter site, and he had neither the money nor the inclination to pay the Continental Phone Company their outrageous fees.

Jeremy had two weapons at his disposal: himself and Robin Marks. This is what they did:

He had no phone lines, but he had a working knowledge of sending signals from one place to another. But I do not profess to understand what, exactly, Jeremy did. There is a reason why I have not put further effort into describing any of Jeremy's rigged-up deals, and here's a story that will help you understand Jeremy's approach to electronics.

The constant commute south from San Rafael to the City to do interviews, and then back to San Rafael again, and then a few hours later back down south again to Gilroy to work on the shows, was wearing out my car, my relationship with Susan, and myself. To ease the problem, Jeremy gave me an old reel-to-reel deck, so I could work on my shows at home. The Revox A-77 was the workhorse of the industry: no fancy effects or multi-tracking, just two good channels of stereo and legendary reliability. It was a standard, straightforward piece of equipment, until it must have had a problem sometime in the past, and Jeremy fixed it, and sometime later it came into my hands. It worked flawlessly for about a year, and it meant that I could spend more time working at home, making my life much better, until it developed the interesting quirk of suddenly erasing everything when I hit Rewind.

The first time it happened, it erased an hour of my work, and about the same amount the next time. I called Jeremy to get it fixed, but he was out of town, and wouldn't be back until the next week. If I had to wait for Jeremy to get back, and then wait until I drove it down to Gilroy, then wait for him to get it, then wait for Jeremy to get around to fixing it, and then wait until he brought it back to Gilroy, and then wait until I was back in Gilroy to pick it up, it might set me back a few weeks, and I could not afford that. But I had another solution in mind: I knew a guy.

The Grateful Dead were famous for many things, like their legendary jamming, the drugs, the fans, and also that they were always on the cutting edge of electronics for their concert sound amplification. They were well known for having the first serious concert sound systems, the first to develop delay systems for really large shows, and audio innovations of any sort in this area interested them greatly. To this end, they had a staff of engineers and sound technicians that traveled with them, and when they weren't on the road, this crew spent its time experimenting with electronics in their secret warehouse in an industrial area in San Rafael.

I lived in San Rafael, and as I needed a constant stream of interview subjects, I had been to the band's warehouse and interviewed the chief engineer for the Dead. We got along well, and he said that if I ever needed any help, to call. He was widely regarded as being an electronics wizard and innovator; he'd been written up in newspapers and magazines for the systems he had developed for the band. I called him now, and told him about my problem with the Revox, and he said to bring it over. I was there in ten minutes.

He was very accommodating, and we took the unit to a workbench in the rear of the warehouse, and he opened up the back, lit his flashlight, and took a look inside. He looked, and then, using his finger, he prodded with a confused expression. Then he took out a pen to push some wires around to look further inside. He shut off the light and looked puzzled for a moment, then he put the light back on and took another look inside. He poked and prodded and pushed at the wires, and he followed this wire and then that one to… whatever, and followed that wire to… what the…?

Then he shut off the light again, closed up the back of the unit and told me that he didn't understand what was going on in there. He knew someone outside of the factory had re-wired this deck, and he had no idea what was going on in there. He'd never seen anything wired like that before, and that all he could do was tell me to bring it back to whoever had worked on it.

And that is why I have no shame in not understanding what Jeremy did after the FBI caught him wiring the pay phone at the transmitter site so that the local calls to a San Jose number went to Gilroy with no charge. The FBI shut that system down, and what Jeremy came up with makes so little sense to me that I will let Terrell Lynn do his best to explain it:

TLT: "…the two coffee cups with the phone receiver in it, with the microphones in the bottom of the coffee cups, take the phone, stick it in the coffee cups, and [Jeremy] had it on a solenoid, so when the phone would ring, it lifted the receiver in the coffee cup and that's what you'd talk into. And that's why the request line sounded muffled. "

GK: "You mean, like you were talking into a coffee cup?"

TLT: "Yes."

Okay, then, there it is.

Gary Faller remembers his introduction to yet another of Jeremy's new phone systems. No one (Jeremy) told anyone (Gary) that someone (Jeremy) had devised yet another (probably illegal) way to make and receive phone calls. Gary found this out when he got a phone call from a jock telling him that listeners had been calling all morning, saying that the station sounded bad, and asked him to come over to Gilroy to find out why. He tuned to KFAT, and it was true: it sounded horrible.

Out on the back deck there was a small shack that housed some of Jeremy's improvised electronic innovations, and it was the first place Gary looked. He told me, "the Mosley STL (Studio Transmitter Link, which took the signal from the studio and locked it into the transmitter at Loma Prieta) out in the shed outside the studio up on the roof that looked like something out of M*A*S*H, and periodically this one particular Mosley would kind of freak out and KFAT would sound like it was motorboating (raspberry sound). Way in the background you might hear the sound of a song, but it would be way, way, way in the background… and I'd tell the DJ to keep playing records, or try to go out and reset it, and I'd….. It was very hot… it was in a… it was un-air-conditioned, like the size of an outhouse, this shack on the roof. And eventually what tied in with this, was that… I had never been informed about the phone system, and I went there where the STL was, in this kind of shack up there on the roof, and there were all these telephone modules, encoders, and decoding modules and telephone hybrid modules and I went "Whoa, what's all this stuff?"

"Jeremy had gotten them, like, a week before, and hooked them up, and was actually using the downlink, the extra frequencies that would go up the mountain on the STL link, there was actually a couple extras that could share to and from that weren't being used, but the more the phones got used by calling the sales office during business hours, the worse the records sounded. The records sounded like there was a lot of dust on the needle. It had like ten or fifteen phone lines going… and all the phones were lit and I thought, '15 lines? I didn't remember this,' and within a week there was this new equipment there and we had our own phone company."

"Jeremy had installed something new that I wasn't privy to, and when the STL tripped, I went there and said "Wow, look at all this stuff." And I called Jeremy and asked 'What's all this stuff?' and he said 'That's my new business lines to the San Jose office.'"

GK: I presume that you had never seen anything like that before?

GF: "No, I hadn't… he said they worked, and I explained to him that the more phones were used, the worse the records sounded. And he went "Ohhh, ….(disappointment)."

"It was a brilliant stroke of genius in a way, but the main reason for people listening … was the music, so I asked if we couldn't limit the number of lines to boost the air quality?"

"He said, 'mumble, mumble, yeah.' And soon Continental Telephone knew about it, as Jeremy had gone over there and boasted that he was going to offer their ten biggest customers free service to San Jose, and I think they wondered 'How could that be?' "

"And the amazing thing—if you know anything about telephones—if you dial a number, it goes through the telephone switchbox, so the calls in San Jose would go into the switchbox, and they would go to the transmitter and then the phone company would lose track of them. 'Cause they were going out through the air down to KFAT through the downlink of the STL. And vice versa: the calls from the studio would go in, and then there would be no trace of them."

GK: Did Continental ever catch wind of what Jeremy had done?

GF: "I think they got wind and alerted the FCC, and somewhere Jeremy has a framed plaque… he got a $10,000 fine for having his own phone system. But once again, the FCC's initial reaction was that they didn't believe that anyone could actually do that. Jeremy had some cutting-edge ideas… maybe not implemented in the most correct or legal manner… but they worked." Concerned readers will be relieved to know that Jeremy never paid this fine.

GK: When you first came to KFAT and looked at the equipment, what was your reaction as an engineer?

GF: "The equipment was really old, and there were some band-aids on the equipment like twisted wire. But it was how it was connected… I didn't know if I wanted to go back there. The whole thing was: 'What did I get myself into?'"

Some of the lines went into the offices at KFAT, some went to the studio, and a couple of lines were routed to the Production Room. They were regular phones like you see anywhere, but for a while, in the studio, they went into a coffee cup near the DJ, and that's why I don't know what the deal was with Jeremy's phone-system-in-a-cup. It was such a Jeremy Lansman Operation. But the Gilroy and the San Jose Sales Offices were calling each other, the listeners were calling the station, and Jeremy was calling the station. The phones worked at KFAT, but they were off the grid of the Continental Phone Company, which was two doors up the street.

Chapter 71: The Angels Listened In (And Thank God Laura Ellen Didn't)

You know you can make God laugh by telling Him your plans, right? Gary Faller planned on catching up on some maintenance work.

Gary was working late that night at KFAT. The equipment was old, funky, freaky and mostly cobbled together by Jeremy from spare parts, inspiration and cheapness. Once he had a sense of how Jeremy's mind worked with all things electronic, once he had a sense for Jeremy's electrical logic, he found he could repair and maintain the equipment, and he was proud of his part in making the KFAT airwaves vibrate with the musical madness that the jocks so reveled and specialized in.

It was almost midnight, and the station was calm. On weeknights after midnight, there were fewer visitors, and when there were, they generally stayed in the studio with the jock, and Gary would be working in the Production Room that night, so he was counting on getting his work done in undisturbed peace. Then Laura Ellen called the station and asked for him. The jock was about to go off the air and Laura Ellen told Gary that the jock that was due was:

1. too drunk to show up, or
2. too stoned to show up, or
3. too sick to show up, or
4. had car trouble and couldn't show up, or
5. had not been informed that he'd been scheduled, or
6. forgot he'd been scheduled for the shift.

Could Gary stay and fill in? He still needed to fix the gear, but that could wait until tomorrow, then he'd get paid for that work, plus get paid for this shift, too. He didn't know all that many records, but he knew a bunch, and it would be late and he could experiment, and no one was likely to be looking over his shoulder. He liked the thought of being alone at the station. And being a KFAT jock. And he could use the money. So he said, "sure," and took over at one o'clock.

It was a slow shift for a weekday, and not many people were calling in with requests. He liked requests, as they were usually for songs and artists he didn't know, and he was glad for the suggestions. It made these fill-in shifts more fun, and they went by faster. Going faster was important when you worked from one to six a.m.

So he was rockin' and boppin' and doin' his breaks, and the calls all told him he had what it takes, and it was about half-past two in the morning, and he was relaxed and he had it all under control… until he heard the thunder.

Startled, he looked out the window in the control room, but the moon was out and the sky was clear. The thunder was constant and getting louder, and then it just… stopped. It just shut off. It seemed to be coming from Monterey Street, so he went to the office window that overlooked the street and saw about two dozen big-ass Harleys, and an equal number of scruffy, tough-looking bikers in Hells Angels colors dismounting, lighting up cigarettes and joking with each other.

"Oh, shit," he thought.

The Angels had come by before, several times, he knew, but never while he was there. He knew they liked KFAT and considered it their official radio station, and that they stopped by when they were out on one of their runs, and the bars had closed. They carried with them the booze and pot and crank that they liked to party on, and Gary knew that they meant no harm to anyone at the station. But these were still the Hells Angels, known outlaws and killers, and he hoped that the visit would go well. He just didn't need this, is all.

Coming up the stairs in a noisy, clomping cluster of muscle, fat, grease sweat, and engine fumes, they were friendly and glad to be there. They all said Hi. They clearly knew their way around the various rooms of the station, and they brought out what they'd brought with them; they spread out around the station and proceeded to get comfortable.

It wasn't like there was a surplus of comfortable seating there in the control room, and if they all brought in whatever chairs and crates might be available, it still wouldn't be enough to seat everyone comfortably, so they spread out. No one would try to take the DJ's chair, and once they all settled in and he felt no sense of threat, Gary settled in, too, and enjoyed a few neighborly tokes of this and a few sips of that. It was cool. It was alright.

And it was alright. He liked them. They were great guys, they just wanted to hang out for a while, and this was where they liked to hang. They spread out into the rooms with the unlocked doors, sitting where they could, drinking, laughing and talking amiably. It was going well, so far. Everyone was in a good mood, and everyone was passing around this or that, and it was, in general, a convivial evening. They'd request a song now and again, and he'd play it, and the morning went by a song and a segue at a time.

A couple of hours in, one of the Angels asked for a song by Patrick Sky that Gary didn't know. This wasn't a problem, as Gary had been putting on requests all night from these guys and from callers for songs he didn't know, and they had all been good songs. KFAT's audience was pretty sharp about their music, and he'd observed that their taste was usually right on the money.

He didn't know this song, "Fight For Liberation," but he knew that Patrick Sky had been a folkie and he'd briefly had a middling career. When the Folk Music boom ended, he was one of the guys who didn't make the cut into mainstream music, and in frustration, he'd recorded an album of tunes that were so disgusting, so unhinged, so politically incorrect, that he'd sealed his fate for the second time. But Gary didn't know about *that* album. That would be the album "Songs That Made America Famous." That would be the album with the song "Fight For Liberation" on it. That would also be the album that featured a song called "Bake Dat Chicken Pie" whereby Sky is in the recording studio, trying to record a song that began with:

"Ya wants to make a nigger feel good?
Tells ya whut to do,
Steal a chicken from the neighbor's yard,
And bring it on home with you.
Nigger, nigger, nigger..."

Then the engineer breaks into the recording, and over the studio monitors you hear him tell Sky that he can't say that on a record, so Sky offers to try again with,

"Ya wants to make a darkie feel good,
Tells ya whut to do..."

Then the engineer breaks in again to tell Sky he can't say that, either. Sky said he'd try something else.

"Ya wants to make a Polack feel good...?
Tells ya whut to do..."

But Gary didn't know about that. Gary knew that Sky had been a folkie, but this album was unknown to him. "Fight For Liberation" had to be a folkie protest song, right? He didn't know the song, but frankly, any song about fighting for liberation was a song that Gary could probably get behind. And if a Hells Angel wanted to hear it... So he went to the record stacks, found it, checked the ending so he'd know how it ended, cued it up, and when the next song ended, he pushed the "Play" button and he sat back and listened to how the song started, what was the pace, so he could pick out the next record.

The Angels were all in a good mood, and there was a rollicking good spirit at the station, with everyone just like at cocktail parties everywhere: laughing, talking, passing this and sipping that, snorting a bit of this and smoking some of that. It was a party for the Hells Angels and KFAT was throwing it and it was a successful party. Gary was the host, and he was keeping it all going by spinning the music that propelled the good cheer.

Like any good DJ, Gary found another song he wanted to hear, maybe something about freedom, and he put it on the next turntable and cued it up so it would be ready when the Patrick Sky song was over. Then, like all conscientious DJs, Gary leaned back to listen out of the corner of his ear to hear the song, and to listen for its end, so he could play the next song. This was radio. The segue was key. It had to flow.

Sitting back, it all seemed good to Gary. It was a good party, this was KFAT, and those were the by-God Hells Angels and he was running the show. "Fight, fight, fight for liberation," sang Patrick Sky, Gary thought, "right on!" and all seemed right with the world.

"Fight, fight, fight for liberation," Sky sang. Then someone asked him a question, and when he turned his attention back to the song, Gary heard Sky sing something about "pumping LSD up Jackie Kennedy's ass, and make her fuck the working class."

WHAT? WHAT WAS THAT?

JACKIE KENNEDY? FUCKING *WHO??*

There was more. It was about hanging the Pope and machine-gunning all the nuns and destroying the Sistine Chapel and "giving money to the poor, or they'll turn your daughter into a whore." There was more after that, but that was all he heard that night, what with him being in shock and all.

And this was still radio, and there were still licenses to be kept—like his!—and this unholy shit had just gone out over the air on *his* watch and *Jesus Christ if there weren't gonna to be some repercussions about this !*

And...and...*shit!* Just when he was really enjoying working at KFAT...

But what to do? His choices were to take the record off and save his job, or leave it on and maybe lose his job *and* his license. Or he could view it this way: leave it on and maybe lose his job or take it off and maybe get a severe beating from men who had reputedly rarely restrained themselves from the administration of a well-deserved beating.

The job? The beating? He was frozen into inactivity.

Well, it was late, it was the Hells Angels, and it was KFAT. So the song stayed on, and after it ended he played the next song, and all went on as before. Gary gradually calmed down, the Angels stayed for another couple of hours until the first light of dawn and they went off on their thunderous way. No one ever complained, and he never heard a word about it. It was another night at KFAT.

Better pick up the empties, Gary, straighten the place up. Gordy would be there soon.

Chapter 72: What The Hell...?

The sales staff was tearing its hair out trying to make sales, and then the DJ's went and fucked with the ads, and that was fucking with their incomes. Sales people were being told by the clients that they had heard of the station, and *heard about it.* Uh oh...

One salesman told me about going on a sales call to a drive-in theater, where he offered to play KFAT between movies. The manager suggested putting the station on the loudspeakers right then, so everyone could hear it, see how it goes. The salesman knew that this business would only run at night, when the dirty songs got played, so he swallowed, agreed, and listened as every outside speaker, and every speaker inside every car suddenly found themselves listening to a song about a blow job.

Who knew that an off-hand, honest remark by a jock would have consequences for someone else? Who cared that a DJ said, "Man, that song makes my nipples hard"? Who cared about that shit?

Uhh... the clients and the sales people, that's who.

But it wasn't just what the jocks said, it was what they did. I don't know how the rumors got started, but the jocks were being accused by the sales staff of altering the spots to make the product look silly.

The sales staff constantly complained that the jocks were *fucking around* with the ads, changing them, and I swear I don't know how that rumor started.

Chapter 73: The Tom Campbell Shitstorm

I don't know why the sales people accused the air staff of fucking around with the ads, but by the time Greg Arrufat started doing the traffic and other office chores, Tom Campbell was on KFAT and the jocks began to fuck around with his ads, and it blew up into a real shit-storm.

The air staff had been there for over a year now, and the routine had settled in. It wasn't what many would call a normal routine, but it was *their* routine, and it was working. KFAT had become a bastion of rebellion and anarchy for themselves and for their listeners, and they knew what was Fat because they decided what was Fat. And Tom "Motor Mouth" Campbell was definitely not Fat.

Not running the Army ads was one thing: it was a personal and a cultural statement; but just a couple of the jocks had refused to run them, and those ad buys were only going to run for a few weeks, anyway. These were different. This contract would run for a year! A full frontal attack on the Fat sensibility, these spots featured a loud, hyped-up raving motor-mouth selling Cal Stereo corporate shit at a hundred miles an hour, and if you tuned in to KFAT anytime after 10:00 at night, that was almost the only ad you heard.

But this chapter is really more about Greg Arrufat and the rest of the staff—who, with the listeners—were up in arms about Campbell's Cal Stereo spots, as it was soon clear that the ads were now a permanent part of KFAT's programming. The staff knew why the ads were there, but that didn't mean they liked it, and the calls to the Fat Gram line protesting Campbell's ads were beyond counting.

Sully had been learning production tricks from Frisco and, taking some of her newly acquired skills, she and Greg set to work on a Cal Stereo spot. Sully had learned how to speed up or slow down a voice by putting a straw over the capstan of a reel-to-reel deck. Now adept at editing, they took an ad where he said "low, low prices" and they spliced in four more "low's," all six progressively slower and lower in pitch.

Nobody said anything about it, so when the next ad came in, where he chuckled, they spliced in more chuckles, and made him sound like he was choking. This ad ran for a week before Campbell called the new sales manager, Bob McLain, who called Jeremy and told him that his staff wasn't doing him any good, and Jeremy had better do something about it.

Nothing much got done, as Jeremy had some hard choices to make. On one hand, McLain was right, of course, but on the other hand, the station was somewhat of a success, and he couldn't afford to alienate a crew that knew the format and was working for $3.25 an hour.

Tom Campbell was an unabashed huckster in the world of radio advertising. He was just like the guy lampooned by every comedian who'd ever lampooned fast-talking hucksters. He reminded some of Dan Aykroyd's ads for Bass-O-Matic on Saturday Night Live. He spoke incredibly fast and had such a jarring "AM" sound that Fatheads were shocked to hear him on KFAT. And Cal Stereo was usually KFAT's only overnight account!

No one else wanted to buy ads during those hours, and Jeremy, frantic as always about not being able to pay the bills, was happy to have him. Hell, he was happy to have *anybody* who would pay for those hours and whose checks never bounced. Jeremy had invited the Nazis and the Communists and the John Birchers to talk on KDNA, and he remembered the reaction to them, so he was pretty used to controversy, but even *he* was surprised at the listener's immediate shockwave of anger and unrelenting vehemence that these ads generated. This was a true dilemma for Jeremy and Laura Ellen, and it caused them true pain.

Listeners kept up a steady barrage of calls to the Fat Gram in record numbers to voice their disapproval of Tom Campbell, and Terrell Lynn put those calls on the air. Jeremy's dilemma was whether he could afford to give up this unpopular shill, or to ignore the wails of protest against him and cash the checks. Someone, somehow, had to diffuse the controversy.

Jeremy was familiar with controversy, had created some himself in his career, so he sought out a creative approach to the problem. But this was Jeremy trying to deal with humans, and luckily for everyone, where there was Jeremy, there was Laura Ellen. A consultation was held, and it was decided that I would be dispatched to Tom Campbell's house to do an interview with him for my show, Chewin' the Fat. If he'd agree.

Campbell knew that his presence at KFAT was a shit-storm that wasn't dying out with time, and only engendered increasingly fervent demands for his removal the longer he was on the KFAT airwaves. He wanted the account, he licked his chops at what he was going to do to this kid, Gilbert, and he agreed to the interview.

Campbell's voice and his ambition had made him a very wealthy man; he lived in a twenty- thousand square foot home in the hills above Los Gatos, and Campbell welcomed me warmly and showed me around his well-appointed house, including his various collections, his model-beautiful wife, his rings, his Rolexes, his toys, his art, and his car collection, which included two Rolls Royces.

Campbell was a warm and gracious host, and made me feel welcome as he showed me around, knowing I'd be awed, which I was, and once I set up my recording gear, Campbell took over. He was a seasoned pro and I was a neophyte, as impressed as I was insecure in the presence of such wealth and professionalism, and Campbell pretty much conducted the interview as he wanted. He said that the listeners should know that his ads had been solicited by KFAT, and if the management had a problem with them, they could call him and ask him to change to whatever they wanted to suggest. But they had not made the call, and Tom was sure his ads were helpful to people because they alerted them to great sales from a great company, and then he went on to advertise for Cal Stereo a bit, and in general he had his way with me.

But, unlike in his ads, during the interview he spoke at a normal rate of speed; he was intelligent, charming and had much to say about the history of radio and radio ad sales. We talked about the changing tastes in American media, and radio in particular, and we talked about American culture. He was knowledgeable and articulate and a real charmer. I got enough material for two shows, and we aired the promos for the first of them. It was early in the first of the two shows that we discussed the ads and the angry calls that they generated. In my outro, my wrap-up at the end of the show, I told the listeners that I knew that Campbell's presence was a controversy for KFAT, and I asked them to call in and vote on whether or not they wanted me to run the second part of the interview. The calls poured in, Gordy kept the score, and the vote went overwhelmingly in favor of running the second half.

But Tom Campbell understood the problem, and he was a sport about it. If it was gonna be a problem, why not tackle it head on, and be part of the joke? And KFAT was always ready for a good joke. Campbell went to Gilroy and used the Production Room there to cut a spot that ran on April 1st, 1978. In it, he spoke in his usual frantic, rapid-fire voice, announcing a change! Effective today! Right Now! In the *new* KFAT format: from now on it was going to be ALL TOM CAMPBELL!! ALL THE TIME!!

Fatheads didn't get it at first, and stunned callers flooded the lines. But the music never changed, the listeners got it, ha ha, April Fools! and things stayed the same. What also stayed the same was Campbell's presence on the overnight air.

Other than Cal Stereo, local sponsors would always be KFAT's mainstay, but they were a tough sell, too, until Courtesy Chevrolet was promoting a big sale one weekend, and they advertised it all over the South Bay on several radio stations. When the sale was over and the manager looked at his sales reports, his salesmen had reported a significant majority of the cars sold had gone to people who gave KFAT as the place they had heard about the sale. Thereafter,

KFAT ruled at Courtesy Chevrolet, and they became regular sponsors of Fat events. Fatheads would have to put up with Tom Campbell on the overnights, but both staff and listeners would still listen.

Also, as much as the jocks hated Tom Campbell, once Bob McLain came around, and the staff got a good look at *him*, the Tom Campbell shock wore off and they all felt they could live with Campbell. It could be worse. It already was. The air staff knew that McLain hated them and wanted to get rid of them, and they hated, loathed and abhorred Bob McLain.

Another problem was that the staff, for whatever reasons, was pretty unhappy about making $3.25 an hour. Like the first staff, they'd been looking forward to moving the transmitter so they could make better wages, and now that had happened, but the wages hadn't increased. Then a year went by, the ad load increased, and still no raise in pay. The staff loved the gig, but fucking hated the pay. And now they had to listen to Tom Campbell pollute their station, and for what? They knew there was good money in those ads, and none of it was reaching them.

Someone in the sales office might be raking it in, but the air staff sure wasn't seeing it. The jocks had heard that the salespeople were making $40,000 a year—which was HUGE in 1978—while they were told how tough it was out there, and that they were lucky to get their $7,000. Okay- *almost* $7,000, but still…

Jeremy was still a Man of the People, and in populist fashion, he wanted to limit the amount of ads that ran on KFAT. The staff was behind that, but if they were going to sell out, they wanted to be paid for it.

As Traffic Manager, Greg was dealing with both Jeremy and Laura Ellen. Laura Ellen was Program Director, Station Manager. Buffalo Bob was still called the Music Director, but he was doing nothing except taking records home, and thus Laura Ellen was the de facto Music Director. She was the one to cull through the records that came in, after Bob had his way with them and took what he wanted, and it was Laura Ellen who told the staff what to add and what not to play, and put some stuff in the "Hot Box." Usually, people listened.

Greg was doing a lot of production when he got there, because Frisco was off somewhere drying out. Greg liked the kids playing in the hall and the dogs and the hanging out, he felt it was a real family that he'd found. Many people had this feeling about working at KFAT, and it was clear that the listeners felt that way to an extraordinary degree for a radio station. The listeners came to know the jocks and relate to them as friends. They certainly weren't snooty, pretentious radio people who were better than the listeners. Hell, the music was so new and exciting, *the jocks were listeners themselves,* and the Fatheads knew that.

Okay, so this chapter wasn't *all that much* about Greg Arrufat. Let me tell you this about him: he got a license plate for his car that said "FM KFAT."

Chapter 74: Promotions

Your typical radio station relies on "promotions" to make friends and increase listenership for the station. Profitable stations gave away vacations, concert tickets, hats, T-shirts, bumper stickers, magnets, buttons, coffee mugs, posters and other memorabilia. But those were for profitable stations.

Jeremy got a good deal on jalapeno-flavored lollipops, so KFAT gave those away. And then there was the dirt bike climb, where several of the jocks were put on tiny bikes with tiny motors, given no instruction other than "this is the gas, this is the clutch," and pointed up a ridiculously steep hill, where we each got about thirty feet up the hill and fell off, some more ingloriously than others, to the great amusement of the crowd, and Tiny still swears that her knees have never been the same since that day.

Then there was the time that Jeremy felt it would be cheaper to re-label cheap beer than making his own. Remember: this station was so promotions-deprived that it wasn't until after Terrell Lynn entered and won the big Bathtub Regatta up in Oakland, getting KFAT unexpected notoriety from big-time Bay Area media that Lori Nelson handed out their first business cards to the staff.

There was the Hotel Utah Experience (coming), but other than that, KFAT's only adventure into the world of promotions consisted of bribing the listeners. Jeremy agreed that, legal or not, the station would admit that it needed listeners, and it needed to show up in the ratings book. Therefore, if it was necessary to pay people to listen, we announced in a frequently scheduled promo that Jeremy was willing to pay. It would be bribery KFAT-style.

At the time, a first-class postage stamp cost 13 cents. If any listener sent in a stamped, self-addressed envelope, KFAT would send them a check for 25 cents. Listeners had to pay 26 cents for the original and the return stamps to get the check for the 25 cents.

Jeremy figured that if putting KFAT on the air wasn't profitable for him, and if he was going to have to pay people to listen, then those people should just suck it up and take a loss on the deal, like he was.

Thousands of people sent those envelopes in, and there were so many of them that they had to be delivered in mail sacks. Someone had to open them all, sort them out, print the checks, seal them in the envelopes and mail them out. Jocks, friends, whoever was nearby groaned and got them all out eventually and hoped Jeremy would come up with a better idea next time.

Chapter 75: Chewin' the Fat

There were some differences between me and the rest of my new pals at KFAT. The reader will appreciate how hard it is for me to write about myself objectively, so let's get it done and behind us.

First and foremost, I was from New York, and being from New York meant that I was… assertive. Not domineering, just assertive. I was playful and mellow, but also very verbal. I had a better vocabulary than most people, and I enjoyed using it. It wasn't that I would arbitrarily throw out multi-syllabic words that others didn't know to impress anyone, and I wasn't an aloof intellectual. I wasn't even all that intellectual, except that I loved to read and learn, and I had an impressive vocabulary, although one generally eschews self-aggrandizement.

I am Jewish, but not that much. I have a Master's Degree in something; I am articulate and sincere, and I loved the new music and the new people in my life. I lived with my girlfriend, Susan, in San Rafael, where I was a substitute teacher, and alone among the staff, I drove a sports car, one of those Volvo P-1800's that Roger Moore drove in "The Saint" series on television. I fit into the flow of the station because I'd always sought to avoid the normal, I'd had a varied career path, and I was no stranger to various illegal substances or the trafficking thereof. Not necessarily the smartest guy there, I was the college-guy intellectual of the staff, and I came down at odd hours to do my work.

I had been making belt buckles out of attractive stone slabs, using a friend to do the leather work, and we sold the belts to the tourists at San Francisco's Fisherman's Wharf. I used to hang out sometimes with Chris Feder, and I stopped all that lapidary stuff when he hired me for KFAT, but I went back to that leatherworking friend now with a design for a case to hold my tape recorder, microphones and accessories. I like custom-made stuff. Still do.

Armed with a cool case and a small but growing list of publicists willing to set me up with interviews, I also kept my eye on the news, features and book reviews. Any item that happened in the Bay Area was fodder for "Chewin' the Fat."

"Chewin' the Fat" was on weekday mornings from 8:30 until 8:45. It was an unusual idea for an interview show, as all the other talk shows were either one or two minutes, or one or two hours in length. If they were of the one- or two-hour variety, they were buried in the early hours when no one was listening and there were few ads. A fifteen-minute talk show at the peak of the important Morning Drive slot was unheard of. Neither Jeremy nor Laura Ellen remembers who thought of this innovation; Larry Yurdin has no memory of it, and neither does Chris Feder, but that was what Feder told me they wanted, so that was what I prepared. Soon after I went on the air, someone on the sales staff sold my show to The Sportsman's Shop, a sporting goods store on the mall on Pacific Avenue, which was the main shopping district in downtown Santa Cruz. It paid seventy-five dollars a week. For five shows. Woo hoo!

Putting five shows a week on the air meant a lot of work. I had to find people to contact about arranging interviews, contact them, arrange a mutually agreeable time, which was often difficult, as I was working most days as a substitute teacher. It meant working at school until 3:00, driving south over the Golden Gate Bridge to the City, doing the interview, then going back up north to San Rafael to do what I had to do there, and doing that a few times a week. That also meant that twice a week I had to leave San Rafael at ten o'clock at night, get to Gilroy at midnight, write my intro, outro and commercial, transfer the interview from a cassette tape onto a reel-to-reel deck, edit the whole interview, record the intro, outro and commercial and edit them all onto one reel-to-reel tape, then I would have to write, record and produce my promos, transfer them from reel-to-reel to cassette. Then I had to put everything away, clean up the Production Room and put the tapes in their respective slots, "CHEW-show" or "CHEW- promo" in the 7-Up case in the studio. The shows were on 4¾-inch reels, in 5-inch boxes. The promos were on cassettes.

I had to be done and in my car by 5:30 to get home by 7:30, so I could shower, shave, change clothes, be back in my car at 8:00, and be in class at 8:15. I used to love doing my promos, but by the time I got to them, I was always too exhausted to think, and when I hear the promos now, I can hear the cleverness in there, but sadly, the exhaustion prevails.

When I knew I wouldn't be home in time for the call from the school district if they wanted to hire me for the day, I'd call Susan around 5:30 to tell her if I needed to stay, so she'd know whether or not to accept the gig for me if the School District called, usually between 7:00 and 7:30. Because of my KFAT work, there were many days when, as much as we needed the money, I simply could not work that day. For those mornings when I couldn't make it back to San Rafael awake, I kept a sleeping bag in my car. Sometimes I slept at the station, sometimes I'd sleep on the way north.

If I knew a nap was likely, I'd take Route 1, along the ocean, and sleep on any number of small, out-of-the-way beaches that I knew would be deserted at that hour. I used to like spreading my sleeping bag out on the sand and going to sleep, listening to the waves. On the faster 280 north, I knew every exit and where it led, and where I could pull over to sleep, hidden, for a couple of hours.

Susan and I were always broke in those days, and I had car troubles, just like everyone else at KFAT. On the days when I had to get to Gilroy, but had no car to get there, I'd hitch-hike down from Marin County, through San Francisco, and somehow down to Gilroy. During my years at KFAT, I hitch-hiked out of every entrance and been dropped off at every exit on Highway 280.

One Sunday while hitching through San Francisco, I got picked up by a couple in a yellow Volkswagen bug. They picked me up at the south end of San Francisco, at the entrance to 101 going south. They were chatty, friendly people, and when I asked them where they were going, they looked at each other, smiled, and said they didn't know, they were just out for a Sunday drive. When I asked them what kind of work they did, they looked at each other again, smiled, and shut up.

"No, really, what?" I asked. They smiled again, shrugged, and the man told me that he was the manager of an adult sex shop, and she was his girlfriend and employee. As I was *always* frantic for material for my show, and seeing an opportunity, I told them what I did, where I was going and what I was going to do there. Then I asked them if they'd like to be interviewed by me, at the station, for my show. They looked at each other and smiled. They knew KFAT, and said that if I didn't use their real names, they'd do it.

I was often barely on schedule in arranging interviews, editing and broadcasting my shows, and I was always desperate for more interviews, more usable material, and Bang! this one had just fallen into my lap. And a ride right to the station, too!

At the station, Frisco was hanging out, and he set up three mics in the Production Room, then left, and the interview began. I got right to the sleazy heart of the matter and my guests were unexpectedly candid. I always liked the prurient stuff. And so do you. I also liked whatever was sleazy, weird, improbable, mystical or spooky. Conspiracies were a specialty, too, but prurient was always at the top of the list.

No sports, no cooking stuff. Musicians were usually boring. What could musicians say about their latest album? It was always something they were proud of, it was always their best work, and they always loved who they were recording with and for.

It was hard to have standards when I was constantly in need of new material. I had to have something on tape for every weekday morning, and I was broke, I was becoming stressed-out, and I was always in need of sleep. To this day, I believe that KFAT has ruined my sleep patterns forever. I could miss a night of sleep now and then, but I could catch up with naps. But when I began missing a night of sleep every week and sometimes twice a week, there was no catching up. Susan was in school, too, and being a substitute teacher was paying most of the bills, but this was the work I was enthused about. There had been lay-offs in the school system, and any teachers hired in the near future would be one of those that had been laid off, so I had little hope of being hired as a full-time teacher any time soon. Besides, KFAT was a lot more fun, and might actually lead somewhere. So the question was: teaching or radio? The answer was: the show must go on. Get some sleep, Gilbert.

The Sportsman's Shop in Santa Cruz was my first sponsor, and my deal with them was that I would write and record a 60-second spot for the end of each show. If they had some special promotion to announce, or some sale, they sent the notice to me at the station, and it was put in my mailbox and I used it. If not, they'd send me their ad copy for the upcoming weekend, and I'd pick something out of that to talk about. Of course I'd mention the name of the sponsor and their address in every spot, but most of the time I'd riff on whatever aspect of the store or equipment that I felt inspired to riff on. Or I would take some liberties with the copy if the opportunity and the inspiration arrived at the same time.

Like others at KFAT, I liked taking liberties, and I remember one night…

(sound: harp strumming; visual: wavy dissolve…)

…It was nearly dawn and time was running out. I had to get out of there. Having finished editing my next show, I still had to write and record the ad, and I hadn't even looked at what the Sportsman's Shop had sent me, and hadn't chosen a subject to talk about. I'd been keeping this schedule for months, I'd worked all night, and just then I was exhausted, running on fumes. I still had to clean up the studio and get in my car for the two-hour drive home, if I could even make it home, and the ad still had to get written and recorded. Exhausted and without inspiration, again and again I stared at their upcoming advertisement, at the sports equipment and the prices, and felt no excitement, no inspiration. It was 5:15 a.m., and I looked at the pictures of the equipment again, but nothing stirred. What should I talk about? I was exhausted, I wasn't thinking creatively, I was barely thinking at all, and time was running out. The Sportsman's Shop had nothing special to say and neither did I, but it was up to me to say… something. Now I had less than fifteen minutes…

I looked at the copy again and I looked at the pictures again, and… *nothing.* I was just too tired.

Then I drifted to the bottom of the ad where I hadn't looked before, and saw the address of the store. Of course I knew the address, I'd always concluded the ads with, "…That's the Sportsman's Shop, 1225 Pacific Avenue, on the mall in Santa Cruz." I'd always given the address, of course, but I'd never really thought about it before… I kept staring at the address on the ad, then had an idea. But… could I do that? It was late, it was KFAT, so what the hell?

I opened up the mic, and rather than introduce an item from the Sportsman's Shop as I normally did, and in a conversational tone I said that I thought the listeners should hear about something that had just happened in the area, and that they might not know about. In my normal conversational tone, I went on that although the story was being suppressed by all the major media, a local source told me just yesterday about a man who had gone fishing at his favorite hole near Felton, in the mountains above Santa Cruz. He dropped his line into the hole and almost immediately caught something of a weight and strength that he knew was unusual for this fishing spot.

He had a lot of difficulty reeling it in, as it fought hard all the way, and when he finally got it out of the water, he was astonished to see that what he had caught wasn't a fish at all, but a two-foot long miniature submarine with Russian Navy markings painted on it, and he was further astonished as the hatch opened and several tiny men, estimated at about 2 inches in height, all in tiny Russian Navy uniforms, fell out of the hatch and onto the ground. The fisherman pulled his knife out and, shouting and pointing his knife at the tiny men, he forced them into his fishing creel, locked them inside, and brought them and the submarine to the sheriff in Felton, who called the FBI, who came and collected the men and their vessel.

My source further told me that, on further investigation, it has been learned that the vessel was one of a fleet of miniaturized spy submarines deployed by the Russians into the system of underground rivers that flow beneath the continental United States, where they were conducting surveillance missions under our most sensitive areas, from which virtually no critical sites were protected.

Now slowly potting up the music behind me, I went on, saying that The FBI and several teams of U.S. Navy frogmen were even now rounding up the network of miniaturized subs, and it had just been learned that the sub fleet's headquarters, the main base of the miniaturized fleet was in a large underground cavern located just below… the Sportsman's Shop at 1225 Pacific Avenue in Santa Cruz.

I potted the music up a bit more and added: "That's the Sportsman's Shop, 1225 Pacific Avenue, on the mall in Santa Cruz," and slowly potted the music down so Gordy could start another set, which he'd do in three hours and fifteen minutes.

I turned off the mic, spliced that bit of tape onto the end of the show I'd just completed, put away my stuff, put the tape in the "CHEW-show" slot, and left. If someone didn't like it, Laura Ellen could deal with it. I was going home.

In the end, of course, this was KFAT and nothing was said. But a listener did send in a drawing of the captured Russian submarine.

I was proud of my reputation among publicists for being polite and well-prepared, but not everybody liked me. Maya Angelou was either in a really bad mood or just didn't like me. I left her with a clear sense that, despite her having completed the interview, she hadn't enjoyed doing it. I listened to the tape afterwards, and still couldn't find what seemed to piss her off. Sorry, Ms. Angelou.

Charles Schulz, creator of the famed "Peanuts" cartoon strip, didn't like me, either. Again, no idea why.

I know why Fred Willard didn't like me. Fred is a gifted comedian and actor, known for his improvisational skills, and you've seen him in a thousand parts. He's brilliant, and to this day I'm a fan. At the time, he was on a nightly TV show, a comic take on talk shows called "Fernwood 2 Night." Fred played Jerry Hubbard, the clueless sidekick to Martin Mull's acerbic host. It was hilarious, and I suggest you check into it on YouTube.

In an effort to express my comedic talents while reducing my need for one interview a week, I began writing and producing a comedic news show based on the "Weekend Update" segment of NBC's famously hip and wildly successful "Saturday Night Live."

I had a connection to the show, and he sent me a copy of the teletype sound effect used in the background on "Weekend Update," and I put the show on Thursdays whenever I had one together. These took a lot of time, which I did not have, so Thursday's had become "Potpourri Chew" days, and sometimes anything went. More on that later. I called my news show "The True Blue Chew News."

By the time I met Fred Willard, I had a couple of those comedic extravaganzas under my belt, and thought of myself as something of a wit. I showed up at Fred's hotel room with no questions, but with the outline of a script I wanted to improvise with Fred. Fred was dismayed at the idea, and didn't want to do it, but I had no questions prepared. I apologized, and Mr. Willard was a gentleman and a professional, and agreed to try it and see how it went. He was great, and helped with some obvious set-ups that he worked into the skit, and the show went on the air. The show was a success, but I'll bet he told the publicist that he never wanted to talk to *that guy* again.

Then there was the stripper at the Mitchell Brothers Theater.

The Mitchell Brothers, Artie and Jim, started an empire built on pornography, and were among the first to flourish in the newly expanding legality in personal freedoms. Sex was everywhere in San Francisco, and Jim and Artie bought a run-down movie theater in a seedy section of town, renovated it into a mini-mall, and reopened it as a one-stop pleasure palace of porn. There was a movie theater that held about a hundred people, there were individual booths and there were shared booths, where the Susan B. Anthony dollars you bought at the front desk opened a shade in your booth, and you could see what was happening on the tiny stage in front of you, and so could any of the other windows that looked onto the same stage, if they were open. Your dollar bought you a minute, and when the minute expired, the shade would drop again.

Also inside the emporium was a store containing all manner of pornographic magazines and films. Another store inside the complex had all the sex toys you could think of, and some that you'd never thought of. Toys for all manner of tastes. No judgment here. It was all good.

The Mitchell Brothers Theater was the most extensive and blatant of the new sex venues, and got lots of publicity, especially internationally, and every day, all day, buses would unload tourists who would get out at the theater's entrance chittering nervously, then go inside. Busloads of Japanese and German and Australian tourists would unload at the Mitchell Brothers Theater several times a day. The Mitchell Brothers Theater was the most famous and successful of the new porn venues, but they still had to compete with several other sex venues and strip clubs. To do this, they periodically brought in a famous sex star to appear at their theater for a week, and they'd want publicity for her appearance. Remember, this was before porn was everywhere.

When you think about it, short of fucking on stage, there isn't much a sex star can do that wouldn't be simply crude—and illegal—and that material could never be covered in the "legitimate media." The Mitchell brothers wanted exposure in the mainstream media, so these stars had to come up with an act. Marilyn Chambers came around once, and because she was so well known for being in one of the most successful of the early porn flicks, "Behind the Green Door," the Mitchell Brothers had to rent a larger movie theater for her performance to accommodate the interest from the press. I think I remember that she lifted barbells or something.

But the woman of whom I speak was a lesser- known porn star, and she was in town to strip or whatever to excite the men of San Francisco. Not as well known as Marilyn Chambers, she had what the press release called an act, and she was booked for a week's run at the Mitchell Brothers Theater, and the Mitchell's needed publicity.

The press was invited for a special preview of her show, and it was expected of them that if they saw the show, they would conduct an interview with her afterwards and write or talk about it to give her presence in town the much-needed publicity. Being the intrepid journalist that I was, I booked a seat for the show and an interview thereafter.

Always polite and prepared, I asked her the usual questions about her career and her... actually, I have no memory of any of my questions except one: I asked her if seeing men behave as they did during her show- acting out in such a crude, immature and disrespectful manner- did she ever lose respect for the men? For men in general?

She said that she loved doing what she was doing, and loved that it brought men pleasure, and she was proud to be providing pleasure to her fans, and she loved every one of them.

The next day I got a call from a very upset Mitchell Brothers publicist, asking what the hell I had said to the woman. The stripper had been calm the entire time we'd been together, but apparently as soon as I'd left, she flew into a rage about the *asshole* who just left, and she wanted to know why he'd asked her those things, and what did they know about this *asshole*, and couldn't they keep *assholes* like that away from her?

So I'm guessing that's another person who didn't like me. But one guy did something to me that was just uncalled-for.

I was—and still am—a huge fan of the Rolling Stones, and when I got a press release about an upcoming tell-all book about the band, written by a man who had traveled and hung out with them, and had been a supplier of heroin to Keith Richards, I got excited. I immediately called the local publicist for the New York-based publisher to book an interview. The publicist was an elderly woman named Joyce, with whom I'd had an excellent relationship over two years. Joyce told me that the author, Tony Sanchez, had just had major surgery in England, was recovering, and was unable to do a publicity tour behind the book. Joyce was sorry, as was the publisher, because this sensational subject matter would have guaranteed the book a lot of national and local press, and therefore sales.

This was the book that started the famous story that Keith Richards, addicted to heroin and on the verge of starting a tour, had gone to a clinic in Switzerland to have his blood removed and replaced with clean, non-addicted blood. Keith has since said that he made the whole thing up, and was just "winding Tony up" when he said it. But at the time, Keith had not said this, and the story generated an immense amount of publicity. The publisher was indeed sorry that Sanchez was too sick to tour to publicize the book.

A few weeks later, Joyce called and said that Sanchez was still recovering from the surgery, but he was better and had agreed to a limited tour. He was only going to tour for two weeks instead of the usual six, and only two or three interviews a day, rather than six or eight. And because I had been the first to call, and because she knew how much the interview meant to me, I could be one of the interviews in San Francisco.

Of course I had read the book, and was prepared for the interview.

As was customary, due to my teaching schedule and lack of an office to conduct the interview, I made the appointment for 5:00, at Sanchez's hotel. As usual, Joyce met me in the hotel lobby and escorted me upstairs to Sanchez's room. In the elevator up, Joyce reminded me that Mr. Sanchez had recently had serious surgery, and not to excite him or take too much of his time. I assured her of my sensitivity to Mr. Sanchez's condition, and we went to the room, where Joyce introduced us, said good night, and left.

Sanchez was friendly, and clearly fancied himself something of a hipster. He was pale, but all Limey's are pale. He wore a black silk disco shirt, open midway to his navel, and around his neck he wore a mid-weight gold chain with a three-inch gold straw attached. Having some familiarity with the territory, I recognized that the decorative piece on the chain was a cocaine straw. This being a certain time in my life, I happened to have something white with me and, pointing to the gold straw, I asked if that was for what I assumed it was for.

Sanchez cocked his head a bit to the side and, alert and suspicious, said that yes, it was. I asked him if he'd like a taste, and Sanchez bellowed an enthusiastic, "Yes! Thank God!"

We sat down, he did a line, then another, and then we did the interview. It took about fifteen minutes, and I started packing up my gear. Before leaving, I offered another line to Sanchez, he did it, and then I left.

It was a good interview; I knew the prurient stuff, and asked all the questions I wanted to. I had no doubt that Sanchez would report back to Joyce that I was polite, prepared, and a good guy, which was what she always heard from her authors about me, and I was proud of that.

The next day, Joyce called me and she was very upset. What had I done? Why had I done that? She was angry.

"Do what?" I asked.

"You gave that man cocaine, Gilbert! And he's still recovering from serious surgery! He's a very sick man. Why did you do that?"

It was a good relationship with Joyce, and I needed her for future interviews, and I knew I couldn't deny it, as Sanchez had obviously told her about it. I said I was sorry, and told her that the guy had been wearing that coke straw, and when I asked him about it, the guy asked me for a line. It was almost true, and it worked; the subject was dropped and never came up again.

But consider: why would that prick tell Joyce that I had given him cocaine? What could he possibly gain except an enemy in San Francisco? I now conclude that Sanchez was a rat who had written a book ratting out Keith Richards, and was now out on tour, publicizing that he was a rat, and maybe he was just keeping in shape by ratting me out. Fuck you, Tony Sanchez! No link for you!

Okay, Tony, let's let bygones be bygones. Here's Tony's link.

That was an interview where I couldn't understand the interviewee. Then there was the interviewee who couldn't understand me.

The Fairmont Hotel is one of San Francisco's oldest, most prestigious hotels. It is, in a word, a Noble Establishment. It sits at the top of Nob Hill, and there are but two buildings at the top of that hill: The Mark Hopkins Hotel, and the Fairmont Hotel, and both are equally elegant and famous. The Fairmont entrance was a small area that allowed passengers to be let off, but except for the rare exception, the driver would need to park somewhere else. I don't think it's there now, but I remember (?) that the area had a circular fountain in the middle. The entrance also had a small area off to the side that would hold three cars. The Fairmont also had a notoriously protective doorman, who of course would not let you park there.

I had established a relationship with the doorman, who liked me and recognized that I was a working stiff and would only be there for 30 minutes at the most, and then leave. Also, I clearly appreciated the favor, and he knew it. One day, I had an interview at the Fairmont with the author of a biography of Clark Gable.

I'd read the book, and had my questions ready. I had time to go home from school before leaving for the City, and I was getting my recording gear and questions together when the phone rang. It was my best friend, who wanted me to come over to his studio right away. I told him I was just leaving for an interview, but he insisted I come anyway. He had something remarkable to show me, and I had to come over before going to the City.

Wanting, as ever, to be polite, I obliged my friend, and what he showed me *was* remarkable, indeed. It was the first of a new kind of marijuana to come into this country. It's a long story and you can read about it elsewhere, but marijuana changed in this country around 1979, and this was what took over. The standard *Cannabis Sativa* was what everyone smoked, but it was about to be overtaken by *Cannabis Indica*, and you don't need to know why, only that pot changed very quickly and significantly in this country in 1979. Before then, no one had ever heard of *Cannabis Indica.* But this was 1978, and it was the first of the new genus we had ever seen, and my friend urged me to partake thereof. It looked like sticky, seedy hashish, and, knowing that fresher is always better in these matters, I knew I daren't delay, so I took a puff. And then another. But just the two! I still had to go to the City to do an interview, and this was some blinding shit. I was being responsible.

Thereafter, I do not remember leaving my friend's studio, or driving to the City, or passing over a little something called the Golden Gate Bridge. But I do have a few fragmented memories from that afternoon. I remember finding the courtyard entrance to the Fairmont Hotel. I know there is no fountain there *now*, but I remember riding around a fountain, my two left tires up on the curb, my two right tires on the pavement and driving in a circle on the sidewalk around the fountain. I remember the doorman running after me, yelling, "Hey! Hey! Hey!," and I remember thinking that I was doing something wrong. Recent research shows me that there is no fountain at the entrance to the Mark Hopkins hotel. Wow, good weed, man!

I remember being in a hotel room and fumbling while trying to assemble my microphone. I remember a lot of fumbling. While it usually took less than 30 seconds to unscrew the two halves of the microphone, insert a battery, and screw it back together, I think I might have taken a lot more than that. I remember I kept not remembering which question I had asked, and having difficulty figuring out from her response which question she was responding to.

I remember the author being annoyed and telling me that this was taking too long, and that concludes this portion of what I remember of that afternoon. When I heard the playback in the studio a few days later, I understood. I had a slight hesitation to my voice, and I sounded drunk. I had asked one question twice, and then another question three times, each time allowing her all the time she needed to answer the question thoroughly, before asking it again.

Also apparently, I *had* taken a lot of time to assemble my microphone, as she mentioned twice during the interview what time it was, and that it was much later than it should have been.

Other than those, the whole routine of arranging for and conducting interviews was generally a professional, efficient experience, with everyone on their best behavior, so there weren't a lot of opportunities for me to fuck up, and there were many more interviews that I did where everyone went away happy. So fuck you, Tony Sanchez.

The film critic for "Playboy" magazine loved our interview, as he was liberally supplied with some cocaine, and *he* never complained to anyone about it. You could hear snorting sounds on the tape when it aired.

Don Novello was a comedian and a writer, and is best known as Father Guido Sarducci. Novello lived in Marin County, as did I, and when he had a new book called "The Lazlo Letters" to promote, he came over to my apartment to do an interview.

Lazlo Toth was the name of the man who attacked Michaelangelo's famous "Pieta" statue with a hammer, knocking off Mary's nose before he could be restrained. Novello liked the name, and he began sending out letters to the business and political leaders of America and many foreign nations. The letters were barely sane, but seemingly sincere, and funny as hell to read, and when the addressee responded to the letter, Don printed the reply alongside the original letter. The letters were so clever, so enthusiastic, so nuts, that a surprising number of the letters went up the chain of command to their intended recipients, and several of them chose to respond personally.

Don and I enjoyed a little of my hospitality, and as the interview about the book drew to an end, Novello pointed to my tape deck, told me not to turn it off, and suddenly went into his phony Italian accent, asking me if I'd like to talk to Father Guido Sarducci: "He's-a right-a heah- you wanna talk-a to 'im?" We improvised an interview and it went well, both of us enjoying the exchange. After enough material on tape for a show, Novello happened to mention that the artichokes in his home town of Reggio Calabra, in Sicily, were the biggest in the world. I said wait- just down the road from Gilroy was Castroville, who claimed that *they* had the world's biggest artichokes, and, with theatricality, the volume of the argument rose and rose, and accusations flew back and forth between me and the good Father, each screaming the virtues of our local produce while demeaning that of the other. It was pretty damn funny, and I knew just how to use it.

I put our "heated exchange" at the end of the Sarducci interview, and I slowly increased the music volume as Sarducci ranted that he'd seen the artichokes from Castroville, and they looked "like-a *peanuts*" compared to the artichokes from Reggio, and I enthusiastically protested, insisting that the Castroville artichokes were bigger and better than *any* Sicilian artichokes, whereupon Sarducci indignantly demanded to know why the people in Castroville even *called* them artichokes, and so it went between us as our voices receded and the music swelled, and I came on for my outro.

I also got two Sarducci KFAT IDs that day, and that show became the most requested repeat "Chew", and was replayed two or three more times during my first tenure at KFAT, when I needed a re-run. Being always desperate for time and material for my shows meant that re-runs were needed more often, and I asked Laura Ellen if I could have original shows on Mondays, Wednesdays and Fridays, and have a repeat show on Tuesdays, and an anything-goes "Potpourri Chew" on Thursdays. The saleswoman cleared it with my sponsor, and that became my schedule.

I also liked Bill Murray, and it seemed to be mutual, until I was a jerk.

Murray had just joined the cast of "Saturday Night Live." It was the show's second season and Chevy Chase had just left for a film career, so they hired Bill Murray. I had a connection to the show, and was invited to the studio to see it and hang out in the show's offices. I asked my friend if there was anyone from the show that I could interview, a call was made, and Bill said he'd do it.

Susan and I went to NBC and got the backstage tour, and at the arranged time, we met Murray and wound up spending the next couple of hours together, recording, talking, with Bill horsing around, improvising skits for us. Bill took Susan's umbrella and did five minutes with it. He was in an ebullient mood that day, as he was suddenly the star of a very hot show, and he had just that afternoon signed the contract with his first manager. I also got Bill's KFAT ID that afternoon.

He talked about how Chevy had "taken up a lot of room on the show" and that several of the cast weren't all that sorry to see him go. He talked about his excitement, the show and the cast, the popularity and the opportunity and all of that, and it was a great interview and a popular show.

Bill had given me his home phone number, and this is where I did a jerk thing. About two weeks after I got home, and puffed-up with my importance at knowing a celebrity like Bill Murray, I phoned Bill at home and tried to do another interview with him. It was nine a.m. in California, but it was noon in New York, and I thought surely Bill would be up, and this would be a good time to reach him. Unfortunately, Bill was asleep when I called, and he was pissed at me, telling me that it wasn't cool that I'd called, much less at this hour, much less for an interview. I got the message and never called again. Sorry, Bill.

But while together in New York, I got Bill to do a phony commercial for my sponsor. It was quite good, and I'm going to tell you about it later in the book.

Dr. Timothy Leary came to Gilroy for an interview, and he rode right over me. He was a master public speaker, and he was smarter, he thought faster, he was more experienced, and probably higher than me. He had his way with the interview, which was all right with me because whatever Timothy Leary wanted to say was all right with me. Dr. Leary had an agenda and he took care of it, but in a gracious, if forceful manner. He was gentlemanly and of a pleasant demeanor.

After the show, I asked if I could get a photo with Dr. Leary. He willingly obliged and I gave my camera to the man who had driven Leary to Gilroy, who took the photo. None of that is remarkable, I know, other than that, when you see it, you will see that only one portion of the photo is distorted. In the left third of the photo, I and my background are in focus, as is the right third, which is the optometrist's office in the background, behind Dr. Leary. But the middle third, containing only Dr. Leary, is distorted. Check it out. Discuss.

Twenty-five years after KFAT went silent, I met a woman who used to work in KFAT's Sales Office in San Jose, and she gave me the artwork for an advertisement that Awest had drawn about my show. I never knew it existed, I do not know who ordered it and I do not know if it was ever used. Here it is: Chewin' the Fat ad.

In retrospect, one of the few disappointments during the course of my endeavors at KFAT, was that I came up with a tag line that I was hoping would catch on. I wanted a line that was clever and catchy, and, hoping to hear people start saying it to each other, and I began ending my shows with, "Keep practicing." Sadly, it didn't catch on, and I have never heard it used. Not once. Ever.

Keep practicing, indeed

Chapter 76: Frisco And The Angels

Yeah, Frisco liked to get fucked up. No one had ever told him that he had to be either a drunk *or* a tweaker, so he drank and he used a lot of speed, and sometimes downers, and so he knew where Gilroy's shady corners were, where shady people traded in this form of pain relief.

Frisco was from Gilroy and he knew its nooks, crannies and crevices. On any given night, he might smoke or snort or drink or swallow whatever he got his hands on, whatever got the job done. The only limits he knew were time and gravity, and when either gave out, he slept where he lay. He lived for months in a back room at the station, but spent many nights over the years on the floor of the Production Room.

He seemed to exist in the shadows of Gilroy, late at night and rarely seen. He came and went as a shadow would. He worked when he wanted, he always got his work done, and he reported to no one. He spent hours in the Production Room, sometimes with the curtain drawn so he could not be seen by the jock in the studio, but usually the curtain was open.

Jocks remember seeing him for a while in the Production Room, wearing sunglasses and headphones, with a bass guitar strapped over his shoulder and plugged into the board, playing to whatever it was he was listening to. No one knew that he could play or what he listened to in there. No one knew where the bass had come from, but when it disappeared, we knew: through one method of consumption or another, his bass had disappeared into his bloodstream.

When he went out, he was… gone. And sometimes when he came back, he had fresh, warm bread he got from the Mexican ladies at the bakery in town, and usually when he came back he was blitzed. The Ozone Ranger remembers one night Frisco came back, and he was blitzed. How blitzed was he? He was so blitzed that it did not register in whatever chemical freak show was playing in his frontal lobe that night that there were a bunch of Hells Angels motorcycles parked in front of the station. At least two dozen of them. A less blitzed person might have noticed those.

"The Hells Angels used to come down and party a lot in the middle of the night," said the Ozone Ranger, "I don't know about other people, but they used to come on my shift quite a bit."

In the event, it was a significant night for the Hells Angels: they'd all been in court earlier that day when their founder and unquestioned leader, Sonny Barger, had been convicted of murder, and sentenced to life in prison. It was a black day for the club, whose bonding and fraternity were legendary. It was a black day for the Hells Angels, and it was a black night.

That black night they gathered at their clubhouse to honor, to celebrate, if that is the right word, the life of their leader. Some time that night they got on their Harley's, with Sonny's wife, Sharon, on the back of a bike, and went on a "run." Records of such things are not reliably kept, so I have no way of knowing when they left or where they went on this run before the early hours, but it was in the early hours that morning that Frisco got back to KFAT from his solitary rounds, did not notice all those motorcycles parked outside the station, and stumbled his way up the stairs.

The Angels were in a dark, somber mood, roaming around the station as they normally did, snorting speed off of Buck knives and drinking, smoking, talking, cursing, laughing and remembering stories about Sonny, and digging the tunes. There was laughter now and again, but it was in low, moderated tones. The Ozoner was playing music for them as Frisco came up the stairs and lurched his way toward the studio.

Ozoner: "Frisco comes in, just bombed out of his mind... just blitzed... probably speeding and drinking and whatnot, and he starts... pawing Sharon Barger, and I don't know if he had a death wish, or what was going on, but this was Sonny's wife, and Sonny was in jail for murder at the time, and there were four or five Angels in the room there, and immediately they get very upset. As Danny the Reb told me later, out of deference to KFAT, they just hustled her out of there with a lot of angry words, and left the premises rather than do anything else."

Now, I am not one to make accusations, and I would never knowingly spread a rumor, but it was pretty well understood that people had... *disappeared* for acts against the Hells Angels of a much less egregious nature than this. Beaten to a bloody fucking pulp, well, sure; disappeared, I've heard some stories. This insult was so outrageous that even the Ozone Ranger, after a near personal best in consumption that night, recognized that Frisco had just done something so wrong, so out of bounds, that warning lights blazed brightly through his mind, shocking him that Frisco had done that.

But the Angels loved KFAT, and they'd leave no bodies, no bones or blood behind them here because of one asshole, so they gathered up their gear, got on their bikes, and left Gilroy in a thunderous roar. That party was over.

Ozoner: "And the next day I said, 'Frisco- do you remember, uhhh... what you were doing last night?'

'Mumble mumble, no.'

'You were grabbing ass on the wife of the head of the Hells Angels.' Frisco had no such recollection."

Chapter 77: Moose Turd Pie

It was a hit. It was a mystery. It was a KFAT phenomenon, and it was a huge pain in the ass. It's the only song that gets its own chapter in this book, and it's not even a song.

Of all the music that KFAT played, of all the *types* of music that KFAT played, there is only one absolute listener favorite, and every single jock who ever worked there knows what it is. And they all *hated* the fucking song.

U. Utah Phillips was a folksinger and ethnomusicologist. That means that he researched the songs as well as sang them. He knew folk songs from the 15th, 16th, 17th, 18th, 19th and 20th Centuries. He knew so many train songs, prison songs, love songs, farewell songs, work songs, courting songs, traveling songs, field songs, hobo songs and songs about the plague that *you would not believe*. He knew when they were written, and why they were written. He was a walking, talking, singing-and-playing national treasure. But this chapter has more to do with his talking than anything else.

"Moose Turd Pie" was, without doubt or debate, the most requested song ever at KFAT. It was a live recording of him in a small club as he began a song, then stopped and said, "I'll tell you about the worst job I ever had." Then he told the audience about when he was working on the railroad, laying tracks in the hot sun with a hundred other men. They slept packed in railcars, and ate in a rail car. The company was too cheap to hire a cook, so the foreman picked one of the work crew to do the cooking. This man had the job until someone complained about the cooking, and then *that* man became the cook, so no one wanted to complain. But Utah Phillips did, and they made him the cook.

He knew that he had to do something extreme to get out of the job, as everyone knew to hold their tongues no matter how close they came to gagging. In desperation, Phillips said he went out and found a large moose turd, brought it back and baked it into a pie. A huge man came in for dinner, and when served pie for desert, he took a big bite of it and bellowed "My God! That's moose turd pie!"

Then Phillips paused for a couple of beats, and added "It's good, though!"

Listeners called for this song continuously. They always wanted to hear it. They begged for it. Non-stop. It drove the jocks nuts.

Every jock had a policy about "Moose Turd Pie," and they told the listeners what their policy was when they called, and over the air. Terrell Lynn would only play it once a week. Sherman only played it if he wanted to go out somewhere. The listeners kept calling for it. Sully said no, and Buffalo Bob hung up on them.

Hewlett Crist (he's coming) only jocked on Sundays, but he played it every Sunday. Even though Laura Ellen told him not to, he played it anyway, "because it was Fat."

Rocket Man (coming) played it every couple of nights; it was late and sometimes it was easier to play it than to argue with someone. It was six minutes of his shift, gone. He didn't care. He didn't listen to the song anymore, and if he had someone there he wanted to talk to or go into a back room with, he played it. He played it until the night he'd had enough of the damn thing and the never-ending calls for it, and some listeners have told me that they'll never forget the night Rocket opened up the mic, started up a chain saw he'd brought with him, and sawed the record in half. Of course, there was always another copy of the record in Laura Ellen's desk.

Annie Oakley played it once a shift, then wouldn't listen to any further requests for it. She hated the damn song. Everyone did.

Because none of the jocks listened to it anymore, playing it gave them a chance to clean up the studio or put away some records for those six minutes. Besides, you could play *anything* after this song and it always sounded good, so the jocks didn't have to sweat the segue.

Every place like KFAT (*what?)* has a special phrase, meaningful only to those on the inside, and "Moose Turd Pie" provided one of those for KFAT. Many a joke or a quip or an awkward moment at the station ended with "It's good, though!"

One of KFAT's favorite ID's was the one of Utah Phillips saying, "This is Utah Phillips, and you're listening to K-F-A-T in Gilroy. It's good, though." That ID was procured by me when Phillips played at San Francisco's Stern Grove one Sunday. I approached Phillips after the show and introduced myself, said that I worked at KFAT, and I told him that he had the most requested song at the station.

Phillips knew KFAT. They played his music and he had played at the Strand Theater in Gilroy for KFAT's first-ever live remote broadcast. After it, he and co-headliner Rosalie Sorrells came back to the station and played live for a couple of hours. Phillips still remembered that "interesting feller," by whom he meant Frisco, who had set up the mics for the impromptu concert from the studio after the show. KFAT had all his albums, and played them more than anyone else in America. He loved KFAT.

Phillips' face beamed in pride when he heard that he had the station's most requested song, and he smiled and stood up a little straighter. He was well known as a folklorist, and he had written many articles for folk music and railroading publications, and performed thousands of times all over the United States. He had much to be proud of.

Puffed up with pride, he asked, "Really? One of my songs? Which one is it?"

I said, "Moose Turd Pie," and he visibly deflated. His face fell, his shoulders stooped, and his face looked a little like he had a stomach-ache. "Really? Oh… uhh, Okay. Good."

I had my recorder with me and I asked if he wouldn't mind, and that was how KFAT got its Utah Phillips ID.

I was sorry that I'd hurt this gentle man's feelings, but the karmic debt was balanced about a year after KFAT went off the air. KPFA, the listener-supported station in Berkeley, was having one of their fund-raising drives, and the KFAT staff was invited to do a day of Fat music to raise money for the station. I was on the bill, of course, but what would I do? An interview? Nah. The jocks would play Fat music, so they put me on the air to take calls from listeners, and have the former DJs talk with callers, and to raise funds for KPFA.

Every jock who was able to be there was there, and all were asked in advance by Mary Tilson, who produced the event (and who is coming in Part Three) what they wanted to do about the certain flood of requests for "Moose Turd Pie." Every one they asked said that they *did not want to play that song!* But everyone knew that callers would call all day and ask for it, and it was feared that many would withhold their contributions until it was played. So it was announced all day that the song would be played during my shift.

Yeah, they played it on my shift, and at the end of the day, the money generated by the KFAT jocks was *seven times* any other day of the drive, and I was really happy for them, and proud of KFAT. Still, that was six minutes of my life that I will never get back. It was good, though.

Okay, okay, I wasn't gonna put it in the book because we all hated the damn thing, but you've heard enough about it, and I have to admit that when I heard it for the first time in over thirty years, it was funny, God dammit. You know you want it, so here it is, friends: Moose Turd Pie.

Chapter 78: Teresa Needs A Job

In December, 1977, Teresa Cortez was slender, attractive, and best friends with Bob McLain's favorite Sales Gal, Geneva Nieto, the hated head saleswoman at KFAT, who the jocks called "the Dragon Lady."

Geneva told Teresa they needed someone to answer the phones and drive around and pick up tapes from various radio stations. Teresa wanted no part of the suddenly burgeoning Silicon Valley thing, and this radio gig sounded like fun. KFAT had just hired Bob McLain, who had just hired Geneva, and they had just opened an office in San Jose. New positions were opening, and Geneva wanted her friend there. Geneva told her the pay would be shit, but the perks were great: they could go to concerts all the time, have their pick of albums, and eat at a lot of restaurants where Geneva would arrange a trade-out, so they could eat for free.

Teresa needed the job, but if Teresa came from Geneva, no one at the station wanted to know much about her when she showed up in Gilroy with ad copy and messages from Laura Ellen in Los Gatos. If she was friendly with the sales staff, no one on the air staff wanted to know her.

Teresa: "When I started there, I made $1.98 an hour, but also [got] food and gas trades."

Teresa remembers driving with Geneva to San Francisco, to Berkeley, Palo Alto and other places for free meals. If Geneva wanted to eat somewhere, she'd arrange a trade-out, and bring Teresa along. Most of the trade-outs for food were in Santa Cruz, "which was a long way to go for dinner, but most of the music events were in Santa Cruz, so we ate then."

When she began working for KFAT, the office was still in Laura Ellen's house, so she'd go there, but Los Gatos wanted them out. The town had gotten several complaints from neighbors, angry that Laura Ellen's and Jeremy's house had such overgrown, untended grass. Their front lawn was a mess, and they refused to make it pretty. After many notices about this dereliction, Jeremy built a high fence around his house so no one could see the mess, and he thought the city would leave him alone, but then they were on him about having a business office in an area not zoned for this commercial application. This was at the same time that an office was rented in San Jose, which was just when Teresa got there, but she began each morning at Laura Ellen's dining room table.

Needing transportation to and from Gilroy, Jeremy rented her a truck. Which was nice, but after reading an article on the cheapness and reliability of gasohol, Jeremy took the truck back and converted it so that it used gasohol, which he made for it. The gasohol broke the engine. Teresa still remembers the astonished faces on the rental people when she returned it. It's a long story, and I don't know if the litigation was ever settled. The American auto industry meets a Jeremy Lansman Operation, indeed: Who's gonna pay for this?

Trying to organize the spread-out office work, Laura Ellen needed the office work to centralize in the office at Gilroy, so she moved her stuff from her dining room down there, and she started by doing the traffic, by which ads are scheduled. Teresa remembers, "I didn't know Laura Ellen well, didn't see her much. She'd come into the office to type up a playlist, then be off to Gilroy."

"Laura Ellen had very little to do with me… She wasn't good with girlfriends."

LEH: "When I was doing traffic, I screwed up a big Chevy account [for a local dealer]. That was really early, and they said, 'you're not doing a good job in traffic.' It was the only mistake I ever made, but it was a big enough one that I said, 'fine- I don't want to do it.'

GK: But you're Laura Ellen. Who was going to complain about YOU?

LEH: "Well, Jeremy. So I just didn't do that particular job anymore. I became Program Director instead."

Making Laura Ellen the PD meant they didn't have to hire someone else, and that was a blessing. The level of tension was rising steadily and everyone felt it, but it was focused on Jeremy, who wasn't used to any of this, and he had to pay for it all. Pressure was building on all fronts, Jeremy was losing sleep, getting crankier, skinnier, and he was going to snap.

Laura Ellen was stretched pretty thin, too, and when she forgot to schedule that ad, the money from which Jeremy was counting on, he exploded at her, and she vowed never to do the traffic again, which was when Allen got hired.

Allen was a tall, easy-going hippie, and he moved and talked slow and easy. He was unfazed by the constant chaos at KFAT, so everyone thought he was unflappable, and that was why they stood around, shocked, when, after just a few months there, he attacked Jeremy.

Jeremy was in a snit about something late one morning, and for some reason he wanted the garbage thrown out *right then*. Allen was standing there watching Jeremy as he ranted on and on about something that made no sense to him, when Jeremy turned to him and demanded that he, Allen, throw out his garbage. It was out of the blue, it was almost lunchtime, and Allen didn't want to do it just then, and went out to lunch. After he left, Jeremy dumped Allen's garbage pail all over Allen's desk, covering the work he had done so far that day underneath a pile of empty beer bottles, food wrappers, used napkins, chicken bones and the oozing of various liquids. When Allen came back and saw his day's efforts ruined, he jumped at Jeremy and picked him up and threw him down, then he picked him up and threw him down again. Then he pounded on Jeremy as everyone stood stock-still, gaping in shock until Tiny came in and screamed "Stop it! Stop it!" and Allen stopped hitting Jeremy. In the awkward silence that followed, Gordy took Allen by the arm and walked him out of the building, and he never came back or called for his final paycheck.

Tiny remembers, "I was on the air and I heard this scuffling, and I went out in the hallway, and Allen was wailing on Jeremy, and they went down to the end of the hall and Jeremy was just like (cowering) because he had went off one too many times."

Gordy had taken Allen outside, but everyone else was still standing around, in a somber daze, when Terrell Lynn showed up. "I asked, 'What happened?' and Jeremy goes, 'What happened? I'll show you what happened!' and he grabbed me and he hit me and he threw me against the wall! I was in shock! He pounded on my chest first, and then he grabbed me by the shirt and just threw me against the wall."

Allen was gone, and Teresa started coming to Gilroy, working down there and doing the traffic. She says, "I stayed away from the DJs. They were too out there. I'd heard about the drugs, and stories about them." She was going to avoid the DJs. Reader, select one:

1. We'll see about that, Teresa
2. Fat chance.

Chapter 79: Friends In Low Places

It's a damn shame and all, but everyone in the world is not going to be law-abiding or honest, or a faithful partner and a constant provider. Some people are going to be criminals. You have to take them away from society, away from bad influences, until their time is up and, according to the law, you release them, hopeful that they will gratefully assimilate into the culture and assume productive, tax-paying lives, but where actuarial tables are able to predict with chilling accuracy the length of time until their return to the prison community.

And the worst of these offenders are put into fortress-like prisons that are specially built to hold the worst of these criminals. One of those fortresses is called Soledad Prison, and it's about 45 miles down the highway from Gilroy. Among the few privileges granted these hard men doing hard time, was to have a radio, and it was to KFAT that those radios were invariably tuned. It was a rare and very much valued lifeline to the outside world.

Soledad's relationship with KFAT went back to its earliest days, when a late-night jock had a masturbation contest, with women calling up and masturbating on the air, and the jock went on the air, asking the guys at Soledad to be the judges. After that night, KFAT was, without question, the official station of Soledad Prison.

Who had more time on their hands and a greater need to be entertained than these hapless men? No one, that's who, and they were among the most fervent of KFAT listeners. Besides the radio, they were also granted the occasional phone call, which they used to call KFAT with requests for songs. The jocks knew when the call was coming from Soledad because the caller always started the conversation with, "I've only got one minute to talk..." and they gave the prisoners preferential treatment with these requests. The prisoners knew that the station got more requests than they could play, and the fact that KFAT played *their* songs gave them a strong feeling of solidarity with KFAT. When the KFAT jocks played a request from a prisoner, they announced it with "This song goes out to the country club," and the guys there all knew that was what the jocks called their joint. It was their station, and they were rabid fans. Calls and letters to KFAT from Soledad were a part of almost every day.

Bread & Roses is an organization formed by Mimi Fariña, Joan Baez's younger sister, and its purpose is to bring entertainment and solace to those in confinement, in hospitals and prisons. Mimi knew all about KFAT, and she proposed that the station play a softball game: maybe against the guards, maybe against the prisoners, and KFAT agreed. The negotiations took a few weeks, and when all was done, a date was set and the KFAT staff was told not to wear blue jeans and not to bring in any drugs or booze.

Some of the prisoners were allowed on the side of the field, but most watched from inside while the KFAT staff played against the guards, and got creamed. Screams and cat-calls abounded, but it was all somewhat respectful, and all in good fun. Everyone had a good time, and the calls and letters came pouring in, thanking us. Everyone wanted to do it again.

Another visit was organized, this time for a concert, and KFAT booked Zydeco legend Queen Ida. It had a rocky start, as the performance was threatened when, musicians being musicians, upon their inspection as they entered the prison grounds, an empty vodka bottle was found in a drum case, and the drummer swore he hadn't known about it. Also, one of the musicians wore jeans, and the show was delayed for three hours while he went home to get another pair of pants. By the time he got back, and after some severe tsk-tsking, the warden said he believed that the empty bottle and the offending jeans had been a mistake, the crisis was over and the show went on.

Most of the prisoners were allowed into the show and they all knew all the jocks. Tiny got the biggest cheers and the most obscene proposals, and everyone felt good about it. The connection between Soledad and KFAT had been made a long time ago, but those guys had nothing but time, and now the bond was stronger than ever. KFAT was there for the guys at Soledad, and the calls and letters of thanks poured in.

I think you also might consider that several jocks had to look around them that day and think: there but for fortune... Several of them knew that the way they lived and who some of them hung out with, they might be there themselves. In that event, it would be good to have a friend inside. KFAT and Soledad were in this together until they got a letter from a white supremacist inside, complaining about all the "nigger music" they were playing.

Mail was usually delivered during Terrell Lynn's shift, and Sherman was hanging out with him when Teresa came into the studio with the letter and said, "read this." Terrell Lynn and Sherman were righteously pissed, and they went on the air immediately, saying they wanted to talk to the guys at Soledad. They told them what was said in the letter and they said, "We don't play that game, man. We don't have color here. So, if we get any more of this shit, we just ain't playin' yer requests anymore."

Immediately, the station received a barrage of calls and letters, begging the station not to judge them by one asshole. They said they'd take care of it. Another caller said they knew who had written the letter, and they'd make sure he never wrote another one. And no one ever did.

Tiny got the most letters from Soledad, and she began to wonder if the prisoners, isolated and away from women, might fixate on her, and it might not be as safe for her and her kids as she'd like. The door from the street was always unlocked, and anyone could wander in.

Like the night a man showed up and asked if he could watch for a while. He seemed calm and safe, so Tiny said okay. He had a large paper bag that he kept on his lap and he was mostly silent during her shift. With the few pieces of conversation, she learned that he had just been released from Soledad and this was his first stop; the paper bag held everything he owned. Now his silence began to worry her.

He sat with her until dawn, and then, thanking her, he left. She never saw him again, but she began to worry about her relationship with the prisoners. She stopped answering their letters, but this was KFAT, and she always played their requests.

Chapter 80: Almost Broke My Back

In the late Fall of 1976, two marijuana growers from northern California, having recently harvested, dried and manicured their crop, were sitting in the kitchen of one of the growers, comparing, contrasting, and sampling thereof, when a third gentleman of similar horticultural endeavors drove up, perceived what was happening, and begged them to wait while he raced home and came back with *his* best efforts. Which they did, and he did.

When he returned, they spent an idyllic, if chemically altered, afternoon, discussing the this's and thats of their harvests and their trade, and as sunset fell and they all had to be elsewhere, they vowed to do this again next year. "Next year we'll do it right," they swore, "we'll get all the best local growers to enter their best, we'll have some food brought in, we'll get some independent people to be judges, and we'll have prizes." And thus was born the Second Annual Harvest Festival.

In December, 1977, there were eighteen entries, with each contestant paying a hundred dollar fee to help defer the cost of the event. A caterer was hired, and also a fellow who played hip but soft acoustic guitar. Five people were recruited to be judges, people whose experience with the herb was extensive, and all had to be in excellent health. They also had to give up a three-day weekend. The judges would convene in a beautiful, secluded house in the woods somewhere in northern California, and commence their judging on Friday afternoon, stay over that night, and begin again the next morning. The judging would continue all that day and into Sunday, when the guests would arrive. The Harvest Festival would begin around noon, and conclude about six p.m. All times approximate, of course. Each grower was required to provide one bud for display—to be seen but not smoked—and an ounce of buds from that same plant for consumption.

The attendees were met that Sunday morning in pre-arranged places, and upon presentation of the invitation, which allowed for one guest, they were admitted into a van with blacked-out windows, so that no one could call anyone from the event to say where they were, because they didn't know where they were, and neither could they tell anyone where they'd been when they got home, because they didn't know where they'd been.

Because this is a book about KFAT, there is a connection, and I'll get to it now. I won't spend any time describing the dining room with its three eight-foot tables for the entries, the banks of theatrical lighting over each table, the wooden bins that held each entry, the display bud on an antique plate, the buds for smoking in an antique bowl, the pipe, cigarette papers, matches and a lighter in each bin. I won't tell you about the ballots based on the Davis 20-Point Wine Tasting System, by which all the entries were graded on Appearance, Taste, Fragrance, Strength and Overall.

Not gonna mention the electron microscope in the den, in front of which lay eighteen 3x5 cards, numbered and mounted with a small sample of each entry, which you could view magnified, up to 500 times.

The food was great, the entertainment was perfect for a group of stoned people, and took place as the ballots were being tallied in another room, and when the tally was completed, the entertainment ended and the awards ceremony began. But this is about KFAT, and it so happened that one of the judges for this event was my best friend who, upon my request after the event, conferred with the two producers of the event, and they agreed to let him do an interview with me about the Harvest Festival, under strict rules of anonymity, of course. And so Mr. X came over to my apartment with eighteen baggies of an undisclosed herbal substance, and we proceeded to do an interview about the Second Annual Harvest Festival. And it nearly broke my back.

I was accustomed to interviewing people who were articulate, alert, knowledgeable about their subjects and used to being interviewed. Also, most of them were sober.

By the time Mr. X and I had done some preliminary, uh… conferring about the uh… subject of the interview and some other things, about an hour had somehow passed. We were well into the research portion of the afternoon, as opposed, say, to the interview portion that had yet to be done. And was done, eventually. But time and excess will take their toll, and by the time I had the batteries inserted, the mic assembled, and asked my first question, we were both pretty wasted. My first question for my friend was, "So, uhh… how many entries… were there, Mr. X?"

His response was, "...............uhh...........uhhh, there were.......... uhhh.....eigh… teen."

And thus it went. For about forty-five minutes. I had a fifteen minute show that included the intro, the outro and a one minute commercial. I had to make what is above into, "there were eighteen." And you should have heard the more complex questions, the ones that required complex thought.

In a normal interview, with sober people who were used to being asked the same questions for the several weeks of their publicity tours, I usually needed between ten and twelve edits to complete the work on the body of the interview. On a show with a lot of edits, like a True Blue Chew News, I had to do twenty or more.

On the Harvest Festival show, there had to be a hundred god-damned edits! I'm tellin' ya, it almost broke my back. But all is not woe and travail:

The producers of the Second Annual Harvest Festival heard the show, as it seemed they were Fatheads, and invited me to the Third Annual Harvest Festival. I went to the Third Annual Harvest Festival, and it was amazing! But it gets better!

A few months after the Third Annual Harvest Festival, *High Times* magazine published an article about it, written by someone who had not been there, had only heard about it from someone, and the information was inaccurate and misleading, and the producers came to see me. They knew that the story was "going to get out there," and if they couldn't prevent that, at least they wanted the story to be accurate. They wanted to invite a writer to the Fourth Annual Harvest Festival. They asked me to suggest the right outlet, and I suggested two worthy writers. Both writers got approval from their editors; a California magazine, *New West,* was going to give it 500 words with a humorous slant, and the other, a national magazine, wanted to do a 4- 5,000 word feature story with a photo spread, and was named *Rolling Stone*. See ya, *New West*.

The producers of the event also asked me to be the emcee for the event, and I accepted, and remained in that position until the demise of the event after the 20th Annual Harvest Festival. *Rolling Stone* sent Tim Cahill out, and the event was written up in the March 6th 1980 issue, with Richard Gere on the cover, and the article is "Teahouse of the Harvest Moon." My photo appears inside, from the elbow down, in my capacity as emcee, giving out a prize. You could look it up. Man, what a party! What a wonderful event that was, year after year for twenty years. But that show almost broke my back, god-damn it.

Chapter 81: Remote, But Not Remotely Normal

Jeremy loved technology, and Jeremy loved good radio. KFAT was so alive that it needed more live shows. That first show at the Strand had been a big success, and he saw that a community of listeners was forming. He thought they should have a place to gather. So much enthusiasm came from the Strand show that listeners, jocks, office staff, sales staff and management all agreed, and that had never happened before. And it never happened again. What brought them together was everyone's desire for a regular live show, and everyone wanted Jeremy and Laura Ellen to respond.

Of course, everyone knew that a surplus of people would be unwilling to drive to Gilroy, and that limited the potential audience for the show. San Jose is quite the little city these days—ninth largest in the nation, you know—and has major venues to attract the major acts. But back in those days, the concert universe of northern California had but one shining star, and that was San Francisco. The cost and availability of major acts wasn't as much of a problem for KFAT as it might be for other stations, because who among their artists were major acts that played the large venues?

Jeremy, Laura Ellen and Bob McLain wanted a presence in San Francisco, and someone on the sales staff knew a guy in Bill Graham's organization…

Bill Graham was the undisputed king of live music in northern California. He and Chet Helms had put on shows at the Fillmore Auditorium and the Avalon Ballroom, respectively, starting in 1966, and some would point to these shows as the start of the 60s. Helms eventually went out of business, and it was Graham who put his indelible imprint on the booking and staging of Rock 'n' Roll in America, and his influence is still felt even after his untimely, tragic death in 1991.

Long gone from the Fillmore, he now had other venues in many cities, among them a relatively small 250-person nightclub in San Francisco called the Old Waldorf. It was perfect. The broadcast from the Strand had meant sending the signal to Loma Prieta from Gilroy, which was already a known equation; this time the signal would come from San Francisco, so there was work to be done. Once he saw that the signal could be sent from the Old

Waldorf to Loma Prieta, the show was booked for the next available night, a Monday six weeks away.

Jeremy delighted in such challenges, and this time he had a co-conspirator in Gary Faller. Much contemplation ensued, followed by a series of discussions, followed by a period of hypothesizing, which led invariably to speculation. To Jeremy, the difference between speculation and action is motivation, and Jeremy was very motivated, so all of the possible convoluted wiring and signal-throwing options were considered, narrowed down to a list of probabilities, among which the least likely were eliminated, and what remained became a plan. And that plan was thenceforth implemented. And implementing means action!

Gary, of course, had his truck, which was a remote recording and broadcasting facility. Oh, boy, does that make it sound professional. I don't know if it was called a step-van or a box-truck or what you'd call it, but try to imagine the old kind of truck when you were a kid that used to deliver bread or laundry. It was one of those, and inside was a crazy mix-up of electronics blinking in all kinds of colors, and there were wires and cables hanging everywhere. And the outside of the truck, well… it was painted kind of weird and, overall, looking at it, it had this… *attitude* about it.

Funky as it might look, such a remote facility was rare, and at the time, ahead of its time.

The plan sent Gary behind the stage, up some stairs, through dusty, unused rooms and climbing over the roof of the Old Waldorf, hanging out of a window, stringing about 250 feet of cable up from the sound booth in the club, then over to another room, then out the window to *there*, then over the ledge, down the wall, then, secured to a pipe by duct tape, down to the ground (more tape), around the corner (tape) and along the brick wall and across the sidewalk, all the way to his truck. And all of it was buried in duct tape. Any pro would recognize instantly that it was an unusual way to run an operation, but there it was, and all those cables led into this funky old truck with the… *attitude*.

But inside, inside… Packed within the truck were all of Gary's electronics, with all those knobs and sliders and buttons and switches and faders and toggles and lights and meters, and mixed in with all that there were also a bunch of Jeremy's electronics, with all *those* knobs and sliders and faders and toggles and lights and meters, and when it was all plugged in and powered up, it was a sight to behold.

Gary Faller: "Jeremy had gotten some submarine cases, some regular navy submarine kind of military cases, and put the linear amps in them so we would have a hundred watts driving the Marti signal. Then, it was pretty cutting technology to have a Marti. It sounded great. And Mr. Marti had modified the transmitters for Jeremy for a wider bandwidth, so it kind of sounded pretty FM-ish, and you couldn't hit 'em too hard, you couldn't over-drive them too much, but they put out a pretty clear signal.

"It was a stereo signal, and a lot of people were running mono even then on remotes, and back then, people were even using telephone lines and the signal was horrible. And so it was pretty advanced stuff we were doing for a little radio station out of Gilroy."

Besides all of Jeremy's weird equipment, Gary had his board and his cables and his meters and monitors and mics and antennae, and all the mic stands and booms and headphones and gadgets and tools and rolls of tape and other essentials. And you can assume that a lot of Gary's stuff didn't look normal either, know what I mean? And it was all in that truck with the… *attitude*… that had all the cables coming into it from around the corner and up the wall and over the ledge and across the roof and into the building, and it certainly didn't have that *professional* look about it, and it was indeed a sight to behold. And you know who beheld it?

GF: "Yeah, the FCC office was right across the street, and they came into the truck and toured the truck, and they went 'Ohhhh… wow.'

"They came by, and they were looking at the Marti's, and I had run, like, 250 feet of cable up near the roof of the Old Waldorf, and then we shot it back to Loma Prieta with the Marti's, and they wanted to know how we were getting it there, and we tested the power by holding a fluorescent tube in front of it, and if it had enough power it would light the fluorescent tube as you held it in your hand."

Terrell Lynn has a photograph of him inside Gary's truck, holding a lighted fluorescent tube and smiling. He smiled until just after the photo was taken, when Jeremy leaned over and whispered, "uhhh… I don't think you want to sit there," and explained that it was an excess of radiation that was exciting the elements inside the tube, and causing them to fluoresce. Terrell Lynn thanked Jeremy, handed back the tube and bolted from the truck, which was about when the guys from the FCC showed up. One simple phone call had told them that this was a KFAT broadcast. Oh, the FCC knew about KFAT and they knew about Jeremy Lansman. Whatever was going on in that truck, it was a Jeremy Lansman Operation, and they wanted to see it.

GF:"Ha, ha… their reaction was, like, 'well, I don't know if we have anything covering this.'"

GK: You mean laws? Regulations?

GF: "Yeah, laws, regulations or anything, I mean they were looking around, and they were pretty amazed that, at the technology then, that was pretty cutting-edge technology.

"They said, 'Wow, I didn't know anyone could do this.' From a technical standpoint, it was kind of amusing, because the Marti receivers and all, were way ahead of what even the big rock stations were doing who had a lot more money, technology-wise, being able to go right into the transmitter and have a good signal. And especially from San Francisco to Loma Prieta, as the crow flies, that's sixty or eighty miles, so we were able to hit the transmitter and they (the FCC guys) were just totally impressed.

"They were aware of KFAT, and they were aware of the frequencies, and they were aware that they [we?] weren't interfering with anyone. And went 'Ohhhh… wow,' and they didn't have anything covering this… laws, regulations or *anything* …"

They knew about Jeremy. They'd heard stories about Jeremy. Everyone in the FCC knew stories about Jeremy. And now they *had* a story about Jeremy. They'd been there, inside the truck, they'd seen it, they'd talked to him.

They'd gotten to look over all the gear and inspect all the meters, and everything had to be up to code: it had better be up to code or someone was goin' down. And they'd seen it all, and it was all up to code, whichever parts of Jeremy's system might actually be governed by the code, and it was all a little shaky, but it was all up to code. Even if they didn't entirely understand it, what they did understand were their meters and their laws, and this was all Sunday-go-to-meetin' clean. With grudging respect, the representatives of the FCC knew that they had seen an unusual configuration of logic, genius, and parsimony, which is also known as cheapness.

It was unexpected, it was effective, and it was clean. It was a Jeremy Lansman Operation.

The show that night was another big success, and everyone took note.

Chapter 82: Fat Mondays In Palo Alto

There was so much to dislike about Bob McLain, and one thing to love, although the jocks hated to admit it: Bob McLain convinced an easy-to-convince Jeremy to have a weekly live show to broadcast on Monday nights, and he brought in Bob Simmons to book the shows.

Jeremy remembered how successful that live show at the Strand Theater in Gilroy had been, and he was excited about the recent show at the Old Waldorf, and wanted to do more shows, but he was resistant to committing to a weekly show and all the responsibility that would bring. He had enough on his plate and he worried about who would book it? Who would pay for it? Where would it be? Who would do the accounting? Bob McLain wanted the show to be a weekly feature on KFAT, and McLain knew Bob Simmons could do all the work and make it a success. McLain needed it to be a success because he needed KFAT to have a higher level of public awareness, so he could sell more ads, at higher fees. Jeremy was ready to be convinced that he should do it on a regular basis, and he met with Bob Simmons and agreed to the whole package.

Simmons had been involved with KSAN back when it was still KMPX, and he had booked some shows for the Avalon Ballroom back when Chet Helms was fighting Bill Graham for the best shows in San Francisco. Simmons had booked them all: The Dead, the Airplane, Quicksilver, and Janis and Big Brother. He had cred.

Living in San Francisco, Simmons rarely came to the station, but there are things one does not forget: "I seem to remember that it was funky. There were, uh, housekeeping issues. I remember I had never seen a station that used such cheap gear to try to stay on the air with. I mean, the spots ran on these tiny double speed cassette machines that Jeremy had doctored. 'Cart machines? We don't need no stinkin' cart machines!' They were tricky to run correctly. I remember the dirty staircase coming up from the street. The station was over a... what? Optometrist shop? I remember being impressed that Jeremy had one of the first home computers I had ever seen..."

Simmons did his work from home, but came down to Palo Alto for every show that he booked. He had only occasional interaction with the jocks, but met them all, and listened to the station whenever he could. He was a fan.

"I remember getting along fine with Gordy, Kathy Roddy, Laura Ellen, Terrell Lynn Thomas, and even Jeremy. I did not see them too often, though some came to the Fry. People seemed to give me some slack for some reason. Maybe it was because I was such an enthusiastic fan? I remember telling everyone how great I thought they were and what they were doing. To me, it was like a miracle, an irreverent radio station that was NOT just in it for the dough, and trying to satisfy every dumb-ass listener in the world with their 'lowest common denominator,' the kind of clowns that 'radio' thought was 'funny' at the time. Just the motto was enough... "The Wide Spot On Your Dial."

"Sully had the sly-est sense approach. It was a radio station that was busy being true to the idea of itself. It was very much like the early KMPX/KSAN, except it was playing country/folk/blues/rock/American music. And it had a fantastic sense of humor. The people were funny without being the [center] of [the] humor; sarcasm mixed with irony and dry wit."

Bob Simmons: "The only one I ever worked with was McLain. He reminded me of the quintessential small-time radio hustler. It is a type that has pretty much disappeared. Perhaps that is why I have this affection for him in my memory. He was worse even than the sales character played by Fred Willard on WKRP. He wore these shiny iridescent lime green suits. Polyester to the max. He had a high pompadour hairdo, and had big teeth. He smiled a lot and cracked terrible jokes. He wore a pinky ring. I knew him at KOME, where we had worked together before. He was about as far conceptually from the Country & Western hippies of KFAT as it is possible to get. I think he had started out as a car salesman where he learned his fake, over-hardy bravado style. He was so utterly, transparently insincere that you had to love him for it. It was like he was saying 'I don't believe a word I am saying, and if you do, then you're an idiot, and you deserve everything I do to you.' In a way isn't that the car salesman's creed?"

"But it was McLain who sold Jeremy on the idea of the Fat Fry," Simmons said, "and it was McLain who introduced me to Jeremy, and told him I was the guy to do it. McLain knew that I had been part of the production team back at the Avalon Ballroom in the late 60s, and that I had experience putting on live music, so he had confidence that I would put together good shows. I don't know much about how McLain manipulated the sales books, but he and Jeremy worked something out. I think Jeremy was dependent on Bob, since Bob could bring in clients like the big stereo stores and the big bucks they represented. It was one of those classic battles between programming and sales that happens at some level at every radio station. McLain signs a contract with Tom Campbell, the most odious radio voice in the Bay Area, and suddenly "pure KFAT" is now covered with Campbell- McLain sleaze...

and KFAT has to run the Campbell ads for Cal Stereo or whatever the name of that big shitty retailer was. The staff freaked out. But early on, it was McLain who brought the bucks into KFAT and allowed the station to survive... if not prosper."

Simmons remembers that "Jeremy was impressive with his apparent technical knowledge. However, I did not like his certainty that he was surrounded by simpletons. He was bright, but not nearly as bright as he thought he was. He had a lot of energy though, and he was right often enough that he took this as a confirmation of his brilliance.

"I think it is something that he caught from Lorenzo, who probably really was a genius of the first order. Laura Ellen always seemed like the more level headed of the two. Jeremy would make some outlandish comment or make a statement that might fly in the face of reason, or even good manners. Laura Ellen would then give you one of those, 'See what I have to live with,' looks, that somehow would make Jeremy more tolerable. Jeremy, to me at least though, seemed like a pretty decent guy... highly opinionated, but ready to do anything to make the station work.

"Jeremy hired me and just said, 'go get us some shows and we'll see how it does.' He gave me almost no money as a budget. I went to Freddy Herrera and Bobby Corona at the Keystone in Palo Alto and they jumped at the opportunity to do a live show on Monday nights... it would be like an extra 'weekend' day for them. And that could sometimes be the difference between a profit and a loss." Palo Alto was midway between San Jose and San Francisco.

"My relations with Freddy and Bobby were always good, except for the night when I booked The Crammps with Dan Hicks. We had almost 1500 people in the room, and Dan Hicks had done a great show... maybe that was one of the nights we had Dana Carvey on, or one of the other great San Francisco comedians whom we booked from time to time. Bobby was ecstatic… he was going to have a $5,000 bar night which was really good for them. Then The Crammps came on with the worst sounding music anyone had ever heard. Poison Ivy or Luxe was wearing a torn tutu and a fright wig, and the band just started making what sounded like very loud feedback with a drumtrack and screaming lyrics. Maybe it was clever... but who would know? The Crammps were unlistenable.

"The audience piled out of the hall like it was a fire drill. You never saw a place empty so fast. The Dan Hicks crowd was definitely not in the mood for The Crammps.

"When Bobby saw me he said 'I'm going to kill you! How could you have booked those two groups together?' I wasn't able to answer. I wasn't sure if I had done it out of perversity or ignorance. I knew a little of The Crammps' reputation, but I had no idea that their idea of "Punkabilly" was so far from reality or the music of amphetamine influenced punkabilly like Rank and

File or the rockabilly of Ray Campi. The Crammps were pure make-you-want-to-stick-your-fingers-in-your-ears-and-run-away noise. Of course, this appealed to some. About 3 people had come to see them. Bobby yelled at me, 'You cost us five grand!' I told him to put it on my bill and I got in my car and drove away."

The Fat Fry? Ever the populist, Jeremy wanted the shows to be cheap, so they charged only ninety-five cents to get in. (I guess they couldn't charge 94.5 cents.) Ninety-five cents? Why, in today's economy, that would be over a dollar!

Bob says, "I came up with the name. I remember one night as Jeremy looked out at a large audience and said... 'You really did it didn't you?' I was pleased that he said that to me. I took it as a nice compliment. I remember earlier, I think he had been ready to cash in the whole concept after the third week and we had had a really bad show. Robert Hunter, the lyricist and songwriter for the Grateful Dead had sold me on the idea of letting their band play. I did not realize Robert Hunter was one of the world's shittiest vocalists, and there was a reason the Dead would not let him sing with them. He made Jerry Garcia sound like Pavarotti. It was one of the most excruciating experiences…"

Bob asked the KFAT DJs to emcee the shows at first, but there was no money in the budget to pay them. Once the shows got popular enough to have a budget for an emcee, he hired from the amazing pool of San Francisco comedians, including Dana Carvey, Robin Williams, Bobby Slayton, Barry Sobel, Rick & Ruby, Darryl Henriques and others.

"But we had great shows, like when The Tubes played as "Cowboy Fee and the Prairie Pastries". They did an epic version of "El Paso" with Fee getting shot and spouting fake monkey blood. He had big wooly chaps on and looked great. Their country and western version of "Baba O'Reilly" from the Who was immortal. God, to have a tape of that…"

Well, as long as I'm letting Bob Simmons write most of this chapter, why not let him finish it by remembering Gary Faller and his home-made mobile broadcast truck?

BS: Gary was "a tall, laconic drink of water. Always ready, always congenial. Sometimes slow, but ready for anything anyone tossed him. He loved doing the shows, otherwise he wouldn't have worked for the bad money KFAT offered.

"Gary's mobile truck had a 24 track board. We only recorded a 2 track mix, but everything was double microphones, so the mixes were very good. We sent the signal back to the station via a Marti microwave which was also state of the art."

BS: "I loved the station. I wasn't crazy about Sister Tiny, but I worshipped Kathy Roddy (Sully). Gordy was good. Laura Ellen never cracked a joke, but she played the right music. Terrell Lynn was fun because he would play anything.

"I liked the mix. Red Clay Ramblers into the Beatles into Gid Tanner and the Skillet Lickers, or Grandpa Jones plays the music of Stringbean. I loved the fake spots, especially the ones I did, like for the Idi Amin Hot Tub Company.

"I thought it was a really important station. It showed you could do a 24/7 radio station that was run by a spirit much like that that guided "Saturday Night Live" or other great humor programs, and make a profit while doing it. It was what creative radio was all about. I wasn't that much into the rural mystique, but I was completely into the fact that it was an intensely creative space on the dial, with programming being done by certified, dedicated whackos."

As Bob has written the majority of this chapter, some of you might think I should pay him for it. Well, you make a good case, but… no.

Would you?

Chapter 83: George Thorogood Plays the Fat Fry

In 1977, there were probably 2,500 radio stations in America that played rock music. The definition of "rock" had splintered, and now there were several formats that played some variant of rock. Country music was a variant of rock as much as rock had been a variant of country music. But there was also soft rock and hard rock and country rock and metal and punk and "urban contemporary" was breaking out and heading towards hip-hop, and there was progressive rock, folk rock and oldies and classic rock, and fusion rock, and all were some form of rock.

There were some labels that were major players, like Warner/Electra/Asylum and Columbia and Capricorn and others. And when a major label sends out their promotional albums for new acts they want to break, they can expect the Music Directors at the stations will pay attention to these releases, to listen to them and put them on the air.

At most stations, jocks were simply told what to play by the format consultants hired by the station, and their ideas were rarely solicited. At many stations the Music Directors would listen to new stuff and decide, based on their evaluation of the song and its potential for their station. Just because a song was good didn't necessarily mean that it fit into your format. That's how it works at radio stations: when an album by a new artist on a major label arrives at the station, it will get listened to by the Music Director, as well as potentially by any of the jocks that are interested. The bigger the label, the more likely a new artist is going to be listened to.

There were thousands and thousands of records released each year that went out to the radio stations, and a release from a small label like Rounder Records in Cambridge, Massachusetts would almost always go unnoticed and unplayed. Rounder Records was a tiny company that specialized in folk music. Their roster was almost exclusively folk, and there were probably only a few dozen non-comms with a folk music show who might be interested. Only one commercial station in America played anyone from the folk roster on Rounder Records. Take a guess.

The mail usually got there just as Gordy was going off, and one morning, it came with its usual assortment of bills, albums, notices of upcoming releases, and requests for various functions to be announced as PSA's. Rather than trust the jocks with anything sharp, whoever was in the office would open the packages and hand the records off to a waiting Gordy, who looked through them eagerly. Music was the lifeblood of KFAT; it was why they were there, and that was almost all they talked about. The DJ's always looked for new records with good cuts in the pile of new stuff, and in the stacks. Laura Ellen had a box she called the "Hot Box" and she encouraged the jocks to give those a listen. New artists and new cuts were the currency they traded in. They all wrote on the albums, making recommendations on songs. They talked about the artists and the music all the time.

It was of great, if short-lived, pride to whoever was the first to find a particularly worthy artist or song that the others would play.

Gordy had an advantage in that he was on from six in the morning until ten. By the time he got off, it was still early in the day and he had plenty of energy to do his production work, and could listen to the new stuff right after it came in. That morning, he found a record from Rounder in the stack of new arrivals. It was from some guy improbably named George Thorogood and his band, also improbably named The Delaware Destroyers. Why was that name familiar? Wasn't he going to play the Fat Fry next week? Is that who this was, whoever this was?

George's cover showed a young man in a black T-shirt, leaning over his guitar in a classic rock pose. He was bent over the guitar, his face a rictus of rock guitar grimace, his feet splayed. He was clearly a rocker. Gordy recognized songs on the album by Bo Diddley, Chuck Berry and Hank Williams, and it definitely did not look like Folk Music to him. So he went straight into the production room, put it on, and George Thorogood *thoroughly fucking rocked him!*

I'm not going to say that Gordy, uh... hid, uhh... hoarded... kept that album hidden, because I don't know that he did that, but for sure, the next morning, he was the first one to play that sucker and watch the phones light up with callers. George Thorogood was the genuine article. He had his roots deep in the right place and his chops were shit-hot. Instantly, he was a big success at KFAT and all the jocks were playing him as soon as they heard him, and they heard him immediately because they all listened. And besides, the listeners were already calling for him. He was blowing everybody away. Then he played the Fat Fry and a week later, KSAN up the road in San Francisco started playing him and BAM! *Like that* he went national! He was a break-out hit all over American radio and he played to larger and larger audiences.

But that was later. This was in 1977, and George Thorogood and the Delaware Destroyers were still exclusively a KFAT item and they were going to play a Fat Fry. It was their first California gig, and George was nervous. Did anybody know who he was? Would anybody show up at this huge place? On a Monday? Would they like him? It was a big house, and George was 3,000 miles out of his element and really nervous.

Gary Faller was the engineer that night, and he was going to set up the radio gear from his truck, so it could be sent to the tower at Loma Prieta for broadcast. The opening act would go on around eight, and George would go on at nine. Gary got there at noon to oversee George's equipment loaded in, and help the band set up to soundcheck. A soundcheck is when the band sets up its gear and plays at concert levels prior to the doors opening, so the engineer can adjust the levels to accommodate the varied acoustics of the various venues that they'd play. It was (and still is) standard procedure before every show. This is also when the sound engineer for the radio broadcast could adjust his controls and settings so the show would sound good over the air.

Just after George and his two band-mates loaded in, the club manager informed them that, while he regretted not having informed them earlier, the club had also scheduled, and there would be performed, a wedding that afternoon, which would run into the evening. He said he hoped he could have everyone from the wedding party out of the club by seven, and he hoped they could work around that. The doors would open to the public at 7:30, and the opening band would still go on at eight.

He also told them that the other band, the one playing for the wedding, was just then arriving, and they needed to sound check *their* gear, because that was for the *wedding,* and he was sorry and all, but a sound check for the opening band and for Mr. Thorogood just wasn't gonna happen.

You could ask any professional musician you know: if you're gonna play a gig, either you or someone who works for you has got to do a sound check. The club manager was adamant: No sound check for you! This was an important gig for George, and with their sound check canceled, both the band and Gary were thrown into distress. This was bad ju-ju.

But God works in mysterious ways, and the wedding band was already in a snit about something as they were unpacking, and then they got *really* pissed off when they thought they were being asked to give up *their* sound check… and who had even heard of this guy, George Thoro-who? So both bands were in a snit, but it was the wedding band that took their snit to the couple getting married, who by now were in some heightened state of anxiety about the wedding and all, and then *they* got pissy at this development. Tensions rose, words were exchanged, and then the wedding couple and their band got into it in a *seriously* snit-like way, then *that* escalated, and words were spoken at a higher volume, words which could not be taken back, and speaking of back, many of those were up in that animated screamfest before the band resolved the disagreement by turning *their* backs on the wedding couple, packed up their gear, loaded it into their van, and drove off. They'd quit.

When George, who had wisely placed himself well out of earshot during this discussion, was apprised of the development, he approached the now-panicking couple and told them, "We'll play your wedding." And they did. And thus it was ordained that after a hasty sound check, the wedding would have music, and during the performance that afternoon, Gary Faller, outside in his truck, got his sound check for the broadcast that night.

Terrell Lynn was there that night. No, the Fat Fry, not the wedding. George was totally new to touring, and he was nervous and unsure of himself, worried that no one would like him. Jittery, he said to Terrell Lynn, "Well, here I go..." and hit the stage like a pro, losing himself in the music. Terrell Lynn remembers them as the loudest band he had ever heard, and they blew that crowd away.

Once the cat was out of Gordy's bag, KFAT had been playing George all week and all weekend, and the house was over half full, which was good for a Monday night, so the Coronas were happy, as was KFAT. Then, the people who were listening on their radios in their cars heard what was happening at the Keystone, and they turned their cars towards Palo Alto, and the people listening at home heard what was happening, and *they* got in their cars and headed for Palo Alto, and the house filled up and the show sold out. And the Keystone was a big house.

Tiny was there, and says, "He was bitchin', man! That was more music than I'd ever seen three guys make! He was awesome!"

And if you're looking for a happy ending, Gary Faller clearly remembers the night, and recalls his excitement when people who were still unfamiliar with Thorogood heard the show on the radio and drove right over to the Keystone. A sellout crowd turned out for the magic of that show, and the house rocked to its roots that night, and George, the Delaware Destroyers, the Corona Brothers and KFAT all felt good about it. As did the fans.

As did, one supposes, the newlyweds.

Chapter 84: The Hotel Utah Experience

It was one of KFAT's greatest moments, and it started in Augie's Mexican restaurant.

Jeremy, Laura Ellen and I were having dinner one Sunday night, and the talk, as always, was about the station. San Francisco was the big deal in those parts, and everyone wanted a place in its spotlight. KSAN was northern California's premier rocker, owning a huge percentage of the listening public in their 20s and 30s. KSAN was next to KFAT on the dial, and even though the formats were different, everyone knew that they were going for the same audience. KSAN was winning, handily, in the coverage area and in the ratings.

Because it cost tens of thousands of dollars for a station to subscribe to the ratings book, Jeremy did not subscribe. He had friends at other stations who would show him "the Book," so Jeremy, and thereafter everyone at KFAT, knew the ratings. KFAT rarely showed up in any position significant enough to threaten anyone at KSAN, but still, it was now there, in the Book. Jeremy wanted better ratings, so the sales staff could sell the ads for higher fees.

Jeremy was as good as anyone at the station in spinning wild, hypothetical schemes, but his schemes were usually more out there than most. Everyone remembers his plan to drive his van up to the transmitter, wrap it in coils of electrical cable, turning his van into a huge spool of electrical potential, then plug it all into his generator, tune it to 94.5, and jam it with juice. Jeremy was counting on creating a massive field of energy that would blast the entire listening area, freezing everyone's radio on 94.5.

Some of Jeremy's ideas were wild, some were merely impractical, but you had to watch out for what you agreed to with Jeremy, because if he could think of it, he could usually do it. There were a bunch more of Jeremy's flights of frequency-modulated fantasies, and I'm sure a few of them will come up elsewhere in the book, but...

Back at Augie's that Sunday night, Jeremy, Laura Ellen and I were talking about making ourselves better known in San Francisco. We discussed this and that and then a few how-abouts and some what-ifs, and had just gotten to the maybes, when I suggested that maybe we could set up a tent outside KSAN's studios in downtown San Francisco, have people live in it and broadcast from it. It was crazy and it was nuts, and Jeremy liked it. Ever practical and based on planet Earth, Laura Ellen threw a sour glance at her partner, and he saw it, thought a moment, then said, "No... we'll go somewhere right nearby and do it! Not from a tent... maybe a building nearby." And Laura Ellen brought it all home with: "No, I've got it! An abandoned building!"

As long as Jeremy was still sending the signal to his transmitter on Loma Prieta, it was still KFAT the listeners heard, and Jeremy knew it would be legal—if it was temporary—no matter where we were making pests of ourselves.

With an eye towards an increase in ratings and ad rates, we would do it during a ratings period! We'll do it over a weekend! We'll do it next month! And thus was born the Hotel Utah Experience.

Emissaries were dispatched to find a suitable location. Someone knew of a semi-sleazy hotel in a semi-sleazy neighborhood in San Francisco's South of Market Area, also known as SOMA, that was undergoing a renovation, and would consider talking to someone at KFAT about halting the renovation for a weekend, and letting us use the entire second floor to broadcast from. For free.

For free, eh?

A station representative was dispatched to check it out, Lori dealt with issues like accident and liability insurance and substance abuse policies, and it was quickly approved. A date was set, and plans began in earnest. There would be plenty of rooms for everyone to sleep in; it was just that there weren't any doors on those rooms. Doors? We wouldn't need doors as there weren't any walls, either. The rooms were framed in with 2X4's, the sheet-rocking had barely started when we got there, but they'd promised to sheet-rock some bathrooms, and that was done, so there'd be some privacy in there. Other than the bathrooms, you could see through the beams and headers from one end of the building to the other. Walls? We don't need no stinking walls! Exposed electrical cable was hanging everywhere, and surely someone was going to trip over... something, but it was free. The hotel promised to sweep it out and make it safe by the time we got there. It would work! We'll all camp out! It'll be fun!

This was exactly what Jeremy needed. He had become deeply frustrated by all the work and attention that KFAT demanded, and the constant money drain. That had not been his experience at KDNA in St. Louis. *Those* people never expected money from him, and all he had to do was play with the equipment. Here, he was constantly pulled in several directions by frequently desperate people and situations, and he was tired, stressed out and pissed off. Now, here was something he could sink his teeth into: figuring out the needs, logistics and equipment for broadcasting from San Francisco for a few days. It was an engaging electronic equation to solve, and Jeremy was happy again.

A promotional spot was cut and scheduled. In it, Sully said that 60% of San Franciscans know where Gilroy is, and 70% of the people of Gilroy know where San Francisco is, "and KFAT was going to correct the imbalance." KFAT was coming to San Francisco for a visit.

Jeremy raided the Production Room and took what he needed from there. That no actual production would take place over that weekend was obvious, anyway, as everyone would be at the hotel.

Jeremy had his faithful Marti's, the portable transmitters. That was the easy part. He couldn't use the turntables at KFAT, as those were set and balanced, so he called on Gary Faller to borrow two portable turntables. Done.

Mics were no problem; everyone had those. Done.

Calls were made to friends of KFAT in the City, to musicians playing in town, to anyone who would be happy to know we were there.

Air shifts were assigned. Stackable plastic milk cartons to hold the records were to be acquired from a nearby Safeway, and the midnight acquisition of those was assigned.

The staff got together the night before, and people were assigned what to take and when to get there. Laura Ellen gave them all mistake-proof directions to the Hotel Utah.

The hotel management had thoughtfully gone through the second floor of the hotel and removed all the nails, broken glass and sharp edges. The cables were twisted out of the way and the floors had been swept. With the completion of those preparations, the Hotel Utah was ready for KFAT.

Jeremy and Gary got there a day early, chose a room and set up the turntables, a cart deck, mics, lights, and cleared a space for the milk cartons with the records that would start arriving later that afternoon.

Listeners had been apprised of the upcoming event, and were excited. The jocks announced the move for Friday morning. No one I spoke to remembers the dates of this adventure, and the why of this will become clear. In any event, calls were coming in, making suggestions of where the jocks should go when they got there. Those callers weren't getting it: the jocks would be at the Hotel Utah, on the second floor and at the bar downstairs; they had no money for shows, clubs, restaurants and, for the most part, food. They weren't going anywhere. But visitors were welcome.

The morning shift belonged to Gordy, and he would kick off the party from Gilroy. He announced that he and Terrell Lynn would leave Gilroy in the morning, give the shift to Sherman, then drive to the City, stopping at a bar now and again to fortify themselves. He told the listeners what time he and Terrell Lynn would be leaving that morning, and Gordy and Terrell Lynn started getting calls from listeners to tell them where they hung out where KFAT was playing all the time, and where they'd be welcome. They were going to leave at 8 o'clock in the morning, and people called from places on the route and said they'd be there with friends. Gordy and Terrell Lynn made notes. Then the notes became a list. It should have been a ninety minute drive, and Gordy was supposed to be on the air in San Francisco at 10:00. But looking at the list, they knew they'd have to leave earlier. Gordy announced the new departure time: 7:00 a.m. Listeners called, saying that they'd be at such-and-such place, waiting for them.

Gordy had been talking a lot on the air about hitting some bars on the way up, and giving his likely route. He'd said they could take Highway 101, or the much more scenic Highway 280, but El Camino Real was a real street, with people and businesses going all the way to the City, and that would be their route. That El Camino Real was lined with bars, and the highways were not, had influenced their choice. The plan was to call in with progress reports, Sherman would put their calls on the air, then play the music again down in Gilroy.

When Gordy got to the hotel, he'd use the pay phone downstairs to call Sherman, and tell him to throw the switch that would move the broadcast from Gilroy to San Francisco, then Gordy would run upstairs and finish his shift. (Gordy running anywhere… even now, I'm laughing.)

Shift schedules were set, rides were arranged, supplies were procured, sleeping bags and mattress pads were loaded into cars, trucks and vans, and the staff took off for San Francisco.

Tiny had a funky red pickup, known as "the Red Dog," that Jeremy filled with a few bales of hay, and parked it in front of the Hotel Utah. That way San Francisco would know that KFAT was in town, and this was where they were.

Sherman was left at the station to run things until Gordy told him to throw the switch, and Sherman is *still* pissed about it. "They left me to do a 13 hour shift because they had all gone to the City, and taken all their favorite records and moved to San Francisco for the weekend, and they were trying to get it all set up, and I had 13 hours to play from a bad selection!"

Jeremy wanted Sherman at the controls, and wanted him to wait at least an hour after Gordy's call before coming up, to make sure that everything worked and stayed on the air. Everyone on the staff had had to fix the gear at the station many times, often with Jeremy yelling instructions at them over one of his telephone inventions. Sherman knew where everything was and what it was called, and how to twist, squeeze, push, poke, prod, coax and manipulate the errant piece of equipment into operation. For some reason, when it came to throwing the switch, Sherman was everyone's favorite, because it always seemed to work when *he* threw the switch. Bob Simmons insisted on Sherman doing it for the Fat Frys. No one knew why that was, but it was Sherman's gig, and Jeremy wanted Sherman there for *at least* an hour, then he could come up.

At seven, Gordy announced their departure, then he and Terrell Lynn were in the car, and on their way. Sherman did his best, playing from a shitty selection, and periodically Gordy and Terrell Lynn would call in from a bar along the way. They almost got past Morgan Hill, the next town north, before they stopped at a bar to report their progress, and Sherman put them on the air. They had a quick drink there and moved on up the road, to the next bar, in San Martin, the next town north. They had a drink and called in to Sherman, reported on their progress, then moved on.

By the time Gordy and Terrell Lynn got to the third bar, people on the route were listening, drinking, expecting them, hoping they'd show up. And when they did, a cheer went up and drinks were happily bought for them. At every stop along the way.

TLT: "We would have people stop on the street… and come up to us and hand us a beer and cut you out a line, and say, 'come on, you guys! I know you can make it to your next stop!"

And at every stop they called Sherman to report their progress. Then the time between reports also increased. (Physics is an inexact science.)

Listeners were riveted to their sets to hear the drama unfold, and it became increasingly funny to hear these two increasingly drunk men try to maintain their composure and get out of the bar and up the road, and Laura Ellen was hoping that the cops weren't listening, too. Listeners were into it, filling bars along the route, laughing and hoping that the two men made it to the hotel safely and without police interference. Now that I think of it, this was really a job for Buffalo Bob.

Then there was a brief, uncomfortable period of radio silence until, confused and stumbling, Gordy and Terrell Lynn got to the hotel. TLT: "By the time we made it to San Francisco, we were 3 hours late, and fucked up."

As in: *thoroughly fucked up*. But they'd found their way to SOMA, and to the Hotel Utah and… what a shock: A KFAT DJ was drunk! Who's gonna tell Laura Ellen that a DJ was too fucked-up to do their shift? Who's gonna tell Laura Ellen? Oh, she knew, all right; she'd been listening with everyone else. A drunk jock? Alert the media! Oh, wait. They *were* the media, and they already knew. And of course, Laura Ellen knew what to do.

Tiny! C'mere!

ST: "I didn't want to be on the air! I wanted to be downstairs partying!"

Gordy and Terrell Lynn's last call, as planned, was to the studio in Gilroy, whereupon Gordy passed out and Sister Tiny sat down at the mic, and once Sherman flipped the switch and she was on the air, the first thing she said into the mic was to tell Sherman to get up there right away. It was too exciting to make him wait. People were already there and milling about. This was exciting stuff for staff and listeners alike, and it was a radical way to run a station. Few people outside of the station knew what it meant to be in a Jeremy Lansman Operation.

This being KFAT, fans were welcome to visit, and they knew it. They were all curious as to what these people were like, these people they'd been listening to in pleasure and frequent disbelief, and they showed up in droves. The second floor wasn't what anyone would mistake for spacious; when no one was there, you could see from one end of the floor to the other through the exposed framing studs, but now the second floor was packed with KFAT staff and gear, with groups of Fatheads milling about, chatting and staring and laughing and drinking and smoking and hanging out in framed-in rooms with no walls and cheap, temporary light fixtures connected to exposed wire, hanging precariously from the ceiling beams, not necessarily attached to anything firm. Oh, this was *such* a Jeremy Lansman Operation!

People showed up bearing the usual gifts that they brought when they came to the station. Joints, booze and cocaine were happily consumed at prodigious rates, and everyone was having fun. Sherman remembers almost nothing of that weekend, only that it happened "all in a coke blur..."

It was cold, and there had not been any heat installed yet in the renovation. And it was cold. Have I mentioned that it was cold? My God, it was cold. Scarves and gloves were worn. Hats were worn. Sully remembers: "The hotel was being renovated, and it was freezing cold. I slept (with Tiny, I think) on the carpeted floor of an unfurnished room, in a sleeping bag."

There was a bar downstairs, and that was where most of the staff hung out. It was warm there, drinks were gladly bought for them, and conviviality ensued. Stories were told, friends and fans were regaled. Many a soul was enlivened, for few visitors were prepared for the candor of the KFAT staff. They were real, actual people. No airs, no inflated egos; they clearly loved their jobs, and were having a good time, but they also seemed to have no pretensions. Those who have read this far know that they had little enough money for rent, food and gasoline, they wore nothing fancier than their newest jeans, their cars were all broken down, and they were stuck in a tiny backwater, breathing in the constant sour smell of processed garlic. Pretensions? Airs? Hell, no, but buy 'em a drink.

Much drinking and much regaling occurred during those three days. Around the corner and one door down was the Trocadero Transfer, a popular gay disco, which one might think held little interest for the KFAT staff. Sully remembers: "We got comped to a gay disco, and a bunch of us went. Everyone was dressing like the Village People, so we decided the guys in cowboy hats were our constituents. I thought they were refinishing the floor, but the acrid smell turned out to be poppers, some kind of stimulant that's inhaled... someone explained it to me. We were the country cousins, for sure."

They went that one time, staying briefly, looking to the gay crowd as out of place as they felt. While sober enough to move, Bob went there once with his girlfriend, Lindsay, saying he wanted to beat up a few queers, but left at Lindsay's insistence without causing any trouble. Other than those two forays, everyone stayed in the Hotel Utah bar. They were stars there, and the drinks were free.

Of course, I would have a show that weekend, and I'd arranged an interview with Paul Krassner for Sunday morning. Krassner is an extremely interesting fellow, and had helped bring about the 60s consciousness as much as anyone. Too few of you will know of his magazine, "The Realist," that was unlike anything else in its time. It was an outlet for the hippest writers, reviewers, thinkers, artists and cartoonists of the day. No one else would publish those people, but Krassner did. "The Realist" was the bridge from the beatniks to the hippies and it tried to maintain a regular publication schedule, and occasionally succeeded. Think: KFAT in magazine form.

Krassner hung with the hippest people, and was a formidable thinker and wit himself. He always had an interesting view on things, and he could articulate it clearly, at length, and still be funny. He was an ideal guest, he lived in San Francisco, and he agreed to be on the San Francisco edition of Chewin' the Fat.

The reader will not be surprised to learn that confusion abounded at the Hotel Utah that weekend, and although it had been an excellent idea to schedule Mr. Krassner for a live interview at a specific time on Sunday morning, in the event, it didn't play out that way. A musician regularly heard on KFAT, Ramblin' Jack Elliott stopped by with a guitar, and was playing music and being interviewed when my show was scheduled. Krassner saw what was happening, and graciously offered to hang out for a while to see what happened on the air.

Then Barry Melton of "Country Joe & the Fish" dropped in. He brought a guitar and offered to play, but first he put both hands in his jacket pockets and pulled out a dozen joints in each fist, which he threw out to the crowd gathered around the microphone, announcing, "If we're gonna get Fat, LET'S GET FAT!" A rousing cheer went up as he tossed the joints to the crowd, and everyone lit up. There were now about a hundred people smoking joints in the area as he unpacked his guitar and started to play and sing.

But Krassner had an appointment later that day, so after a couple of hours hanging out while yet another Fat musician stopped by, I grabbed my recording gear and Krassner and I found a corner room with electricity and no one nearby, and we did the interview there, for later broadcast. A first for me, there would be no editing on this tape. I set up my gear, turned it on, and introduced Krassner, who said, "So this is what they mean by 'live on tape.'" We both paused, thought about that for a second, and went on with the interview. He was brilliant, as always, and the show went out, unedited, later in the day. No one remembers when, including myself, or what we talked about.

Everywhere, the milling and the chattering continued; people came and went, the jocks played music and interviewed visitors, and when some people left for the night and things settled down on the second floor of the Hotel Utah, and the bar downstairs closed, the overnight jocks came on, and Sherman and Rocket Man had visitors of the overnight persuasion, who brought friends and stimulants of the overnight persuasion, and the music never stopped.

That first night, someone stole the CB radio from Tiny's truck. Welcome to San Francisco, KFAT.

Everyone who was a part of the Hotel Utah Experience remembers it as a highlight of their careers at KFAT. Everyone lost a little sleep, but everyone had a good time.

We went in, we got out, and no one got hurt. It was a Jeremy Lansman Operation.

Chapter 85: Gordy Goes On Vacation

Gordy was due for a vacation. Well, we all were, and we all needed one, but Gordy got one. But where would he go? And with what money? Jeremy came through, trading out a two-week rail pass for Gordy in exchange for a series of spots for the railroad. He could go anywhere they went for two weeks, but where would he go? Where *could* he go? He knew no one outside of the Big Triangle of San Jose, Gilroy and San Juan Bautista, and he had no money for anything known to normal people as a vacation activity.

Gordy had grown up in San Jose and was now living in San Juan Bautista, a town of well over a hundred people. San Jose, when he went back to it now, was way too urban for his tastes: too noisy, too crowded, too many big buildings and too many people. San Jose was too big, and he avoided San Francisco like the Black Plague. Other cities had little appeal for Gordy on this trip.

Disneyland? Not alone, and not on his budget. Fancy restaurants? Not on his budget.

The Grand Canyon? He didn't think there were rail lines there, and anyway, Gordy wasn't much of an outdoorsman. With his weight and his health problems, anything with walking was pretty much out. Museums? Nope.

What, then? Well, he'd… travel. He'd hunker down and plan a trip and set a date so KFAT could arrange for a replacement. He'd go places that sounded interesting, he'd look around, stay a day and move on. He didn't know where to go in the Midwest, but St. Louis had an interesting musical history, so he could look into that. And music might be interesting in the home of the Chicago Blues. And Kansas City had legendary Barbeque, and *that* was worth checking out. But the east coast was a cipher to him: he knew no one there, and the whole concept of the east coast was overwhelming. Something important had happened in Boston, he remembered from school, what was it again? Williamsburg? No. Memphis? Where? Graceland? Nah. Well, maybe Graceland. And then he remembered that he knew someone in Vermont.

Gordy had spoken a few times with a guy from Philo, a label in Vermont, and they had gotten along well. Maybe he'd go to Vermont. Jeremy wouldn't appreciate it if Gordy called someone on the east coast, so he waited for the guy to call again, and plans were made for Gordy to visit a small town in Vermont.

He'd have to wing it for the rest of the trip, but now he had a place to go: Vermont.

Knowing that he'd be switching trains in New York City, he asked me what he should see while he was there. I asked how long he was going to be in the City, and Gordy said, "three hours."

"Three hours!" I screamed, "That's crazy! That's insane! The first thing you have to do is forget this 'three hours' bullshit, and spend some time in Manhattan."

Gordy was appalled at the thought. "No, no, I'm only going to spend three hours there, then catch my train to Vermont."

"Gordy, you have no idea what you're talking about! You know the downtown, the financial section of San Francisco? The part with the really tall buildings? Well, those buildings are *nothing* compared to the buildings in New York! And *imagine that downtown area magnified by a hundred- no- a thousand times!* and you have some idea of Manhattan. And the energy…! And the…"

"Well, I understand that, but I'm only going to be there for three hours."

"Gordy- you don't know what you're saying! You gotta stay at least *three days,* man!"

Cutting to the quick, after much three hours/three days back-and-forth, Gordy was adamant and I gave in. I told Gordy that if there was only one thing he was going to do in Manhattan, it could only be one thing, but Gordy had to know that I *insisted* on this. He had to do it- he had to promise me. If Gordy was going to do only one thing in Manhattan, he had to leave Penn Station, where he'd arrive, walk a couple of blocks to the Empire State Building, go to the top, to the observation deck, and look at the City from there. He had to promise to do that.

As Gordy was happy that he now had two plans for the east coast, he promised. The rest would come as it would, and he didn't care: he had enough money for cheap food and cheap motels, and he'd get by.

But who would KFAT get to replace their morning guy for two weeks? Morning Drive was an important shift, and the listeners had become accustomed to an extensive knowledge of this quirky format and a certain attitude from the jocks, and they knew what they expected from Gordy in the mornings. The jocks on hand were already being used elsewhere on the schedule, and there weren't many people they could turn to who could slip into the KFAT vibe. Not for Morning Drive. Not for two weeks. It may have sounded laid back to the listener, and it was, but that had been learned and earned over time. As few people as there were with the right attitude, there were fewer still who knew the music. It was a pretty exclusive club, Laura Ellen knew.

The listeners were avid listeners, they knew the music, and many could tell which jock was on by their selections alone. They were savvy about the music, and Laura Ellen couldn't put a newcomer into the Morning Drive slot. Weekenders were happy to help. Sherman was doing overnights on weekends, and he could do a week of mornings, if the Ozone Ranger would do five overnights instead of four. Sully would stretch her shift, Terrell Lynn would stretch his shift. Everyone helped out. But there was still a week missing in the schedule. No one remembers who had the idea, but it was genius.

Once Jeremy had given up the Fat Dork Show, Cuzin Al's bluegrass show went from two hours to three hours, ending at nine o'clock on Sunday nights. For the slot after Cuzin Al, Jeremy worked out a good rate to play a syndicated show by "Dr. Demento."

"The Dr. Demento Show" was a successful, nationally syndicated show out of LA, featuring Barry Hansen as the Dr., laughing, shouting, ranting maniacally as he introduced the weirdest songs he could find. Novelty and comedic records were his thing, and the songs were almost always interesting. Jeremy knew the cost of the show relative to the cost of paying a DJ, factored in how despised his "Fat Dork Show" was, and took the plunge. Dr. Demento was now a popular Sunday night feature on KFAT. If this odd hybrid of a station was going to have a syndicated show, it could only be by Dr. Demento, and so it was.

Although obviously syndicated, and with production values clearly exceeding those heard on the Fat One, it was popular, and everyone was a bit surprised that it engendered no adverse messages on the Fat Grams. He was a weird, eccentric character, shouting breathlessly as he introduced the next song or told a brief anecdote about the artist.

He had been on KFAT for almost a year, and when the problem of the missing week revealed itself, someone thought of using Demento for the slot. Once contacted, he said he knew all about KFAT, and he'd love to. But his syndicated shows were pre-recorded, and here he'd have to do it live. Uhh, he said, he'd love to.

The week approached, and Demento told KFAT that he was used to having an engineer in his studio, and he didn't want to play the records, just introduce them and talk occasionally.

How it worked was that Terrell Lynn DJ'd from six a.m. until seven, then engineered for Demento from seven until nine, then DJ'd again until ten, when Sully relieved him. The jocks would feed Demento cuts to play, and Laura Ellen wanted him to emphasize what was in the hot box. Awest brought in his 78's, and Demento brought records from his own collection. It was some of this and some of that, and jocks and listeners alike thought it was Fat.

Terrell Lynn remembers playing the records, with Demento standing behind him, doing a dozen or more jumping jacks to get breathless before he'd point to the mic, at which point Terrell Lynn would flip the switch and lean to the right. Demento would lean in over Terrell Lynn's left shoulder and gasp into the mic. Demento's jumping up and down was part of the reason why he always sounded so demented on the air: he was out of breath.

Now, that might work in his studio in Los Angeles, or at some other stations, but this was KFAT, and jumping up and down in the studio might shut off the station, so Terrell Lynn had to point out to Demento where he *could* jump and where he *couldn't.* So Demento went out into the hall to jump up and down, and then ran back into the studio, pointed to the mic, and leaned in to introduce another song.

Besides going out into the hall for his work-outs, Demento also might have needed less exercise than normal to sound out-of-breath, as he'd met a local musician named Larry Hosford, and they had quickly bonded over booze and blow in the clubs of Santa Cruz, and Demento began spending his nights out, arriving for his shift at KFAT, pretty much worn-out, having just driven back over the hill to Gilroy after another night of excess with Hosford. Demento and Hosford were instant pals, and the good Dr. made a hit out of one of Hosford's songs, "Homerun Willy."

Why was Hansen hanging out in Santa Cruz and not Gilroy? Well, this being a Jeremy Lansman Operation, plans had been made for the big star from LA to sleep on the couch in Sherman's living room. But he had little enough reason to go back to Sherman's lumpy couch in Gilroy when the party was over the hill in Santa Cruz, and back in Gilroy, on his first night there, Sherman's basset hound, Fargo, had snarled at Hansen and prevented him from getting to the bathroom, so he had to lay on the couch the rest of the night with a full bladder, terrified of the dog. After one night of Fargo's protectionism, Hansen met and started partying with Hosford and sleeping it off on a boat in Santa Cruz Harbor, and he started getting to work late, but that was Fat.

But Barry Hansen was a true musicologist, and no phony. He knew a great deal about the music and he was a real fan and scholar of a lot of obscure stuff. He knew a lot about the blues and jazz and western swing, and that counted for a lot at KFAT, and the entire staff liked him a lot. So did Larry Hosford, and so did Santa Cruz.

Gordy eventually came back, and things returned to what passed for normal at KFAT.

Shortly after he came back, I was still working in the Production Room one morning when Gordy got there for his shift, and I asked him if he'd gotten to the Empire State Building. Gordy said that he'd taken my advice. He got out of his train at Penn Station and went right to the Empire State Building, and went to the top, to the Observation Deck, as he had promised. He told me he took one look, saw what was down there, and stayed for three days.

Chapter 86: Max, Maxine, Murphy & Me

It has been well established that Frisco was spinning out of control, his orbits an outward-bound spiral. I had to deal with it, as I would arrive at the station at midnight with only five hours to complete my work before I had to get back in my car and drive home to shower and change, so I could work as a sub. I would call down to the station the day I planned on coming to Gilroy to leave a message for Frisco, letting him know that I needed the studio that night. When I wasn't using the studio, Frisco would stay in there through the night. It was his home as much as anywhere was.

Frisco had a cat named Max, who seemed to have made his peace with Bernice and Sherman's dog and OZ's dog, and any other dogs that came by the station. This was Frisco's home, and Max's home, too. Max had his ways, but everyone was eccentric at KFAT. He was a good cat, and if he liked you, he would let you near him. Everyone remembers Max tearing down the hallway toward the bathroom, making a sharp right into the Production Room, and climbing the carpeted walls up to about four feet before jumping off and sitting, giving himself a good licking, obviously pleased with himself.

One night, I got there at midnight and noticed a change in Max. He was big, very big in the middle… perhaps *suspiciously* big. I suspected that Max was pregnant. I asked Frisco if he had noticed the change in Max, and Frisco had not. I said that Max might be pregnant, and that a better name for him might be Maxine. Frisco didn't buy it.

It was one of those nights when I had too much to do and knew I couldn't make it back to Marin County to work that day, so I called my girlfriend around two, and told her not to accept any teaching spots for the day, then I went back into the Production Room and worked until I was done, and I was exhausted, nearly asleep. I went out to my car, brought in my sleeping bag, and laid it out on a couch in the Meeting Room. I got into the bag and fell asleep almost immediately, when Max jumped up on my chest, turned around a few times and settled down on my throat for a nap. I woke up and pushed Max onto my stomach, saying, "Do you mind, there, Max? I might want to breathe a little later…" then I fell back into a deep sleep.

About two hours later I woke up, hearing peeping sounds coming from my stomach. I opened my eyes and saw that Max had given birth to two kittens on my stomach, and she was licking them clean, and they were peeping. I knew that cats look for the safest place to bear their litter, usually finding closets and hidden areas, building birthing nests. It was in a cat's nature to seek out the safest place to bear their litter, and apparently, Max thought that my stomach was the safest place at KFAT. Think about *that*.

One of the two kittens was an orange Manx, with no tail. Seven weeks later I took that cat home and named him Murphy.

Chapter 87: Toyota Gets Fattened

The staff was stable, the new sales office was making sales, traffic was getting done and they were starting to broadcast live shows. KFAT was building its audience steadily, as more and more people talked about it. Now that the transmitter site had moved and the coverage area increased, they were getting calls from all over the Bay Area.

In fact, just what is the Bay Area? How far north does it go? How far south? Once the signal got over the mountains, how far east would it go? The only real border was the Pacific Ocean to the west. KFAT was hearing from people who were listening in the City, in Marin County, further north in Sonoma County, and in pockets even further north, in Mendocino County.

To the south, the signal went past Salinas and Monterey into King City, and KFAT had many new friends at Soledad Prison. To the east, they were hearing from pockets everywhere. People in the Gold Country and Yosemite Park were calling in, as they were out in Modesto and Turlock. There was so clearly nothing else like KFAT on the radio, and delighted listeners were calling in from remote areas to tell the DJs that they were having people up on the weekends to listen to KFAT.

People called and wrote to tell the station what lengths they'd had to go through to get KFAT. We heard about the new antennae that were bought and installed all over northern California to get KFAT. People let the station know that when their antennae didn't bring in the station, they installed an antenna elsewhere and wired it to their houses so they could listen. Desperate attempts at wiring miracles were reported.

One fellow called from Oakland to tell them that when he had to move, he brought his radio to the various houses he was considering, to see if KFAT came in there, and he vowed that he wouldn't move until he found a house that got KFAT.

Word spread to the staff that KFAT was now the talk of the "San Francisco cocktail circuit," but no one at the station knew what that meant. They asked me because I lived in Marin, and as the East Coast Jew with a graduate degree, I was, by default, the sophisticate of the group. I thought about it but couldn't help. What the hell did that mean? Surely it meant... *something.*

The air staff and the office staff knew that KFAT was a hot... something. But what did it mean at the end of the day? The DJs were still making $3.25 an hour and having trouble eating, paying rent and buying gas for their cars.

Tiny had no money and no food for her kids- again, and called Jeremy and said that she was desperate. Jeremy told her to come right over with her kids, and he and Laura Ellen fed them, and that crisis passed, but it would be back again tomorrow.

The staff were heroes wherever they went in Santa Cruz, but they knew that going into a bar and expecting others to buy you drinks all night could only last so long.

The staff had been waiting for the big move to Loma Prieta, and that had happened, but no one got a raise. Jeremy was counting on the new sales staff to generate more money, but while the ad load was increasing, his income didn't seem to match its steady growth.

The ads increased, but not the revenue. Where was the money going?

Well, that's a good question, and it would be wonderfully satisfying to tell you where it went. But I don't know, and anyone involved either doesn't know, or isn't talking. The air staff was convinced that a little of the money had gone into Jim Gordon's pocket, but he had been replaced by Super Salesman Bob McLain, and despite the increase in ads on the air, no new money was coming their way. So maybe McLain was, y'know...

The old sales office in Santa Cruz was an hour away from the studio, and the new office in San Jose was a half-hour and a million miles away. No one from either staff had any business at the other place, and the two staffs hardly ever met or mingled. Even at KFAT shows, they rarely spoke. This was a growing problem, as the separation of locations increased the level of distrust between the two staffs. The DJ's thought the sales staff was ripping the station off, thus ripping *them* off, and the sales staff was dealing with the blowback from the lack of professionalism of those lunatics at the station, who played dirty songs and did whatever they wanted with the ads that the sales staff had busted their humps to sell, and thereby prevented them from making a living.

Actually, that wasn't the only problem that the sales staff had with the jocks.

Bob's Bison Boogie was on from four 'til eight p.m., when Sister Tiny took over. Done with his shift, he was ready to go down the street to meet up with his crowd at the Green Hut to abandon moderation and get to the serious drinking. He'd wait for Tiny to get there, late as she usually was, get out of the chair, see that Tiny was *in* the chair, and he was gone like a wobbly shot. Sometimes Tiny was able to persuade Bob to put away his records, sometimes not, but either way, Bob was off for another night at the Green Hut.

There were long periods when Bob lived at the station, and after consuming sufficient quantities of booze down the street, he'd stagger back to the station, tell wild tales to Tiny, and then pass out right there in the studio. After an hour or two, he'd wake up enough to stagger to his room in the back and pass out again, this time, everyone hoped, for the rest of the night. This happened regularly, and Tiny was used to it. She just hoped that his snoring nearby on the floor of the studio wasn't picked up by the mic during her breaks. It happened all the time, and ordinarily, it might have passed un-noticed.

But this was a big night for KFAT.

The sales staff had long despaired, and often complained, that no major sponsor would touch KFAT. They'd tried, and failed. The station played such outrageously out-there, dirty, disrespectful music that the major corporations couldn't afford to be associated with KFAT. The jocks and the music were profane, willful and unpredictable. KFAT always did well for the local sponsors, and Fatheads supported those sponsors faithfully, but the real money was with the national sponsors. Beer makers, car companies, airlines, etc. The national sponsors were all afraid of KFAT. It frustrated the bejeesus out of the sales staff, but it was true: who knew what the jock might say? Even if they just played the spot and didn't screw it up, the big-time advertisers and their agencies didn't want to be associated with some of the music the station was playing, and from the moment he got there, McLain had been complaining bitterly about this to both Jeremy and Laura Ellen. He emphatically wanted the lunatics on the air fired, and the owners were resisting him. He showed them the numbers they were living with, then he showed them the numbers they *could be* earning, and then came a breakthrough.

It had taken months, a huge effort and a massive con job for Bob McLain to sell KFAT to Toyota. The sale represented the first major coup for the sales staff, and was their first opportunity for some real revenue for KFAT, *and* the chance for KFAT to prove itself in the national market. At last, the elusive stream of money might start pouring in. McLain, elated but cautious, showed Jeremy and Laura Ellen the contract- and warned them about who they should pick to read the spot. He'd be listening. So would they.

The rep for Toyota had called the sales office to learn what time the spot was scheduled to run, so he could hear it. He wanted to hear what KFAT would do with his account. He'd heard the stories, and McLain swore to him that all that was over, that KFAT would treat his spot respectfully. Toyota wanted the spot to run in the evening, but running the spot during Bob's shift was asking for trouble, so the spot would run on Tiny's shift.

Laura Ellen found Tiny in the Meeting Room when she heard about the ad, and read her the riot act about reading the copy beforehand, and not making any mistakes or comments. Tiny listened and agreed, but the next day Laura Ellen told her again how important this was for the station- for *everyone* at the station. Tiny swore she'd read the copy at least five times before she read it live, and would add no comments. "Read the spot five times," Laura Ellen insisted, and Tiny said, "I'll read it five times first, read the spot, then play a song." Laura Ellen said 'ten times!' and left, but even after multiple assurances of this, everyone's nerves clenched tight as Tiny opened the mic and started talking. The Toyota rep would be listening, as would the ad agency rep, and so were McLain and Jeremy and Laura Ellen, who listened with their stomachs in a knot. Laura Ellen had been in Tiny's face about the ad, and everyone knew how much this spot meant to the station, so she read it several times before going live.

McLain knew no one could get through to Bob, but it didn't matter, as he would be off the air. Yes, McLain knew that Bob would be off the air, but he didn't know that Bob would be back. He didn't know that Bob was *living there.* That night, as was his routine, Bob came back to the station from the Green Hut, outstandingly drunk, fell against the wall and slithered down to the floor, telling Tiny some of his hair-raising tales. Tiny listened, both to Bob and the last song of the set and the magic moment approached. But Bob wasn't done with his story yet.

Tiny was used to ignoring Bob, and even though he wasn't done telling his story, she was concentrating on the upcoming ad, so she told him to shut up and turned her back to him and faced the control board. At this point, she was listening to the song that was about to end, and forgot about Bob while he was still on the floor of the studio, still talking, and she continued concentrating on the end of the song while, behind her, Bob, was struggling to get off the floor, and he staggered in her direction. She had the next record cued up and half a minute to go on the current song when she'd told Bob to shut up. She put on the headphones, kept her back to Bob and faced the turntable, concentrating on the copy, waiting for the song to end, which was right about... now. She flipped the switch, opened the mic and started reading about the sale down at so-and-so Toyota dealer.

She was so nervous about doing this right, and she was concentrating so hard on waiting for the song to end, opening the mic and reading the copy without making a mistake, that she didn't notice Bob get up from the floor, stagger towards her and lean over her left shoulder. She was so intent on getting it right, and she heard only her own voice reading the copy, and she was unaware that Bob had said anything, but the clients and Bob McLain and Jeremy and Laura Ellen were listening, relieved that Tiny was close to the end of the spot, when they heard a deep, rough, resonant voice, a voice that unmistakably and alarmingly belonged to Buffalo Bob Cassidy, who leaned over Tiny's shoulder, unaware that he was speaking into the microphone and said, "...fist fucking."

WHAT??? WHAT???? FIST *WHAT*??

Tiny and everyone else knew what this spot meant to the station, to everyone; for two days everyone had been talking about it and what it meant that someone like Toyota was taking a chance on them, and they all knew that if they had any hope of making enough money to pay their bills, this had to go well, and she had been concentrating on reading the spot and she didn't know that Bob had said anything, until the first thing the next morning, when calls were made to the sales office, and then to Jeremy and Laura Ellen, who were expecting it.

Toyota had raised hell, and as shit rolls downhill, first it traveled south to McLain, thence to Jeremy, then to Laura Ellen, and it landed on Tiny. Sister Tiny had no idea that anything was amiss when Jeremy came in the next morning while she was typing up some PSA's, and he came up right behind her and started yelling. She got the picture immediately but didn't want to take the blame for it; she took as much of the blast as she could, and threw her chair back and screamed, "Okay, Jeremy- I QUIT!"

And she shouted so loud that Laura Ellen, who was down the hall in the bathroom, yelled from behind the closed door, "NOOOooo!!!"

Chapter 88: Tensions Mount Between The Staffs

The feelings between the air and sales staffs were so hateful, it was all but open warfare. The jocks resented the sales staff for their apparent affluence, reaped, they saw, from the efforts of the jocks. It was the jocks playing the music, the magic that made advertisers want to give them money. The money was for *them,* for their taste, for their experience, their ability to put it together. They had been at this for years, through financial hardships that were supposed to be long in their pasts. They'd been promised money for their work if the station got popular, and it was. Ads were coming in and being produced and broadcast every damn day, and there were more coming in all the time. But where was *their* share of the bounty?

Jesus Christ! These people had put up with enough to get a decent wage. No one on the staff wanted Corvettes or Mercedes', they wanted their cars and pickups to get a tune-up and some new tires, maybe some brakes. They wanted to buy some new clothes. The sales people, they reasoned- they were all doing all right. They had nice cars and nice clothes and they took clients out to nice restaurants, and they always had tickets to shows and movies and events, things that they got from their association with KFAT. And their cars had brakes.

And the jocks, man, they *were* KFAT, dammit!

As for the sales people, they had real problems with the air staff, those dicks. Who didn't know that one end of the business has little to do with the other? One type of person goes in for sales, while another smokes enough pot to get hired at a freak show like KFAT. You know, if these sales guys wanted to be on the radio, they'd be ON THE RADIO. They didn't have the laziness required for such a gig, and besides, they all knew there was no real money in being a DJ. The only money to be made in radio was in sales. Selling time. Selling time on radio. Selling radio time. When you've got a good station, everyone knows who you are, and they all want to buy time on your air. There was some good money in that, but those assholes down in Gilroy didn't know anything about the game. They didn't know the rules, and they acted like damn babies, crying about this and that all the time, wanting to get what just wasn't there to give.

These sales people weren't there to make DJ wages. They all knew the market, and they knew what kind of money they could expect to extract from it, and that was why they were in sales. The industry had some standards, including monetary, they were easy to understand, and every salesperson understood them. Your station had various rates for different times. You sold them, you got a percentage.

So why did it so clearly evade those dicks down in Gilroy? They fucked over *everyone* when they fucked with the ads; they ran them when they wanted, they altered the content, changed things around, and refused to play some of the ads. That was just *unheard of* at any other radio station in America! It just *wasn't done!* But it was done at KFAT.

Sales people got complaints from clients, and those complaints hung around the office and were complained about perpetually by the sales staff. When a new guy came aboard, he or she was told what to do and what to watch out for. The what to do had to do with the way Bob McLain ran the office, and the what to watch out for was about the jocks.

In all of KFAT's years, there was only one salesman who was liked by the staff, and that was Michael "Killer" Kilmartin. In his heart, he was something of a Good Ole Boy, he loved KFAT, and Kilmartin did something unusual when he first came to the San Jose office: he asked for directions to the station. No one from the sales office ever went there, and they thought he was a little strange when he said he wanted to go to the station, meet the jocks, see what was going on down there.

He got the directions, and he got to the station. He came upstairs with two six-packs and introduced himself, and offered everyone some. He was friendly, respectful and happy. He was a cool guy. THAT was a first, and then Killer hung around for a couple of hours, and everyone liked him.

He came back several times, and was always a welcome guest at the station. He'd be happy to help the jocks with tickets or with anything that he could do for them. But most of all, he was a bridge, a connection with the sales department, and a source of information about what was going on up there. There was little actual conspiracy that he reported, but it helped to have a friend on the Other Side.

"Killer" drank and partied with the jocks, and unlike the rest of the sales staff, he always hung out with them at the shows.

But with no money to pay their bills, living on leftovers and trade-outs, their cars in desperate need of repair, having a friend like Kilmartin wasn't enough to stem the incoming tide of discontent down in Gilroy. Not nearly enough. Laura Ellen remembered more about the crisis and her position as management: "I think this is a first in the industry. They have all those bonus incentives set up for sales people, and I asked 'why does programming never have incentives and bonuses?' "

GK: Wasn't it that everyone on the air staff hated Bob McLain because he was making a good living and driving a fancy car, and they were so broke?

LEH: "But that's still happening! The sales staff makes a... it's very disproportionate. I mean, who wants to go make sales calls? There's a reason... we don't wanna make sales calls, that's why we do programming. I'd hate it. Plus, those people have to be... they take rejection all day long. I mean, you go out on five sales calls a day... how many of those people are going to turn you away flat?

"I can't deal with that rejection. If I got that percentage of rejection from people calling me on the air, I'd be out of being a DJ. I couldn't take it. I'm just not that type of personality. A lot of people get off on the challenge of it, but I couldn't do it. There's a reason they make the kind of money that they do, and I don't hold that against them."

You might think it reasonable for the jocks to feel the way Laura Ellen did, but she was seeing both sides, and the jocks were not. They knew they were broke, and they knew that it was the jocks who made KFAT a saleable entity and yet it was the sales bastards who reaped the rewards. They knew what they knew, and they knew what Teresa told them.

Teresa, firmly implanted in the office in Gilroy, came from the sales office, and so was hearing stories from, and about, the sales staff, which she reported to her friend Rebecca Fenker, and thus to everyone else at KFAT. She reported that up in San Jose they were making big bucks. She said, "[The] sales staff [was] doing well, financially. They felt they were doing the footwork to get the spots, [and the] money belonged to them. I made bank deposits for $40,000 and $50,000 a day; [the] sales staff did well. Money was coming in big-time."

She said the sales people got 18-20% of sales, and they were pulling in "$40-50,000 a year for all sales staff."

The jocks were enraged. Teresa remembers that the trade-outs the sales people got for the air staff were not enough. The sales staff ate well, had nice cars and nice homes, and "the air staff complained a lot. [There were] at least ten meetings to try to equalize the pay." Terrell Lynn was often the most vocal in his outrage. Bob McLain took note.

I remember one crisis in early 1978 that prompted an immediate staff meeting to protest the pay. They were still getting $3.25 and hour and had just found out that Laura Ellen had hired someone to clean the station- and she was being paid $5.00 an hour. At that meeting, confronted by this news, Laura Ellen said that the cleaning lady got the higher pay because "she had the glamour job." This stirred up the staff, despite Laura Ellen's assurance that there was no money to pay them any more.

Up at the sales office in San Jose, Geneva Nieto, the friend who brought

Teresa to KFAT, had a bad attitude. In Gilroy, Geneva was much hated as the most arrogant, bitter of the sales people. Geneva would snipe and sneer at the air staff at every opportunity. She was known as The Dragon Lady.

Geneva flipped out at Annie Oakley. Sweet Annie? Geneva wanted a real nice bumper sticker, but Annie knew a guy who would work cheap, and they came up with the "I FOUND IT" idea and had thousands of them printed up before Geneva got wind of it, and Geneva tore into Annie.

Teresa: "The sticker was a goof on the "I Found It!" bumper stickers for Jesus freaks. [They were] a big success, everyone wanted one. Geneva hated them. She wanted control of everyone and everything. They finally fired her because she was so greedy and had a bad attitude."

Bob McLain was the Sales Manager, Teresa told me, "and you couldn't trust him. Money was disappearing, either to Jeremy with his toys and his house, or....."

Everyone in Gilroy knew the station *had* to be making money- you could see it in the logs. There were ads on the air, and someone was getting paid for them. Money was changing hands, man.

No one in Gilroy knew where the phantom money was, and neither Jeremy nor Laura Ellen was able to stand up to McLain, but Jeremy tried, and once showed up at the sales office, trying to see what was going on. Teresa remembers, "Jeremy hung around, meddled, and got thrown out of the office by Bob McClain. Jeremy took it casually."

All Jeremy learned was that he wasn't welcome in the sales office, but he'd already known that...

Laura Ellen knew that the jocks, feeling less and less like family, were getting a raw deal, but she was powerless and it was tearing her up. The jocks were angry and the family bonds were breaking apart. If the station accounts were empty, there was no money to pay them. The jocks weren't helping themselves, either, she felt. McLain wanted a "tighter" and cleaner sound from the station, and swore that if that happened, he could sell it, and they'd all make money. But the jocks wouldn't cave in and dilute the format.

Laura Ellen begged them to listen to reason. They had to change the format or there would be no raises. She asked them to think about it, that a compromise was necessary. Did they compromise? Did they calm down?

Yes, remembered Laura Ellen, the staff heard her problems, and they calmed down. But for how long? Laura Ellen remembers, "not long. Because the staff revolted, and we said, 'well, make up your mind. Do you want more money or do you want to do what you want to do, because you can't have it all. Can't have both.'"

GK: Did they understand that?

LEH: "No! They did not understand that! And that was the problem, and that was an era when I got into huge fights with the staff. Huge fights. 'Well, you don't want to do what I want, you don't want to do what this guy wants, and yet you want more money. Where are we going to come up with the money? Where are we going to get the money? We have to become more mainstream.' Just made sense to me- didn't make sense to them.

"'Well, we think we should get more money because we are who we are.' You know, and it's taken a lot of strong people, a lot of strong people. You know- [later, they] re-did the station and they had the card file- the playlist. You had to play by the numbers, and they really made the station much more consistent and much more mainstream, and the station actually made more money in those days..."

But those days were well into the future, when Laura Ellen was no longer a part of KFAT, and for now, the tensions were mounting, the jocks were pissed off and sniping at Laura Ellen and the sales staff, the sales staff was pissed and bitching to McLain about it, and McLain was pissed and calling Jeremy to complain about the jocks, and Jeremy complained to Laura Ellen. It was making them crazy. It was making *everyone* crazy. The tension seemed to be ratcheting up each week.

Jeremy was taking the calls from McLain, getting his side of the story, and he understood that. He was also aware of the other side of the story, and seeing it at the station. These people were his friends, they had worked together for two years, and he felt their pain. Both Jeremy and Laura Ellen were feeling a lot of pain in those weeks and months. The tension was getting to everyone, including the Gilroy office staff; it was getting worse for the jocks, and it was starting to leak out into their shows, and something had to be done. Well, what could be done?

The jocks had some ideas. Here was one option: pay the staff a fair wage.

Not gonna happen? Okay, here's another option: the jocks had heard that McLain was away at the moment, buying a boat in Hawaii. In Hawaii? Buying a boat? How about selling the boat and giving them half the money?

No? Okay, here was another option: get the sales staff to reduce the percentage that they received, giving the difference to the jocks. Also not gonna happen.

Then how about this? Charge so much money for ads that no one will buy *any ads*?

No? Okay, Bob McLain would show he was a reasonable man. If everyone in Gilroy thought that everyone in San Jose was ripping them off, he'd try to fix it. He told Jeremy about a woman named Lori Nelson, and he wanted her to come in and do the books. She had the credentials from the business side of radio, and she'd show those bastards down in Gilroy that *there was no money* to pay them. Leo Kesselman (coming soon, I swear) was consulted, approved the hire, and KFAT now had a bookkeeper. She'd start next week.

This brought little joy to either Laura Ellen or Jeremy, as this new hire had to be paid, which meant even more money going out, and what if she showed them that McLain was right? That no money was mysteriously evaporating, that it simply wasn't there?

Jeremy and Laura Ellen, already nearly three years into their big California adventure, were perpetually tired, stressed-out and pissed off. Some in Gilroy thought they saw small cracks in their relationship, but with those two, who the fuck knew? Stubbornly, they still thought they had a good deal here with this station, and it was exciting for Laura Ellen, and tempestuous for Jeremy, but there had to be some way to make it work. Nothing like this ever happened with listener-supported stations. Nothing like this had *ever* happened at a listener-supported station. Was this what it took to make a buck?

Jeremy had to hold onto this puppy for just over another year before he could even consider selling it. And then, it had to be a working station, or the sales price would be halved. It had to work.

Bob McLain called again, and again asked Jeremy what he was going to do about the mess the station was in. Jeremy didn't know. Bob suggested that they all—everyone on the air staff—come to his house for a barbeque with the sales staff, where they could all meet each other and talk it over. We'll all drink, relax, and eat barbeque. We'll talk to each other.

Jeremy had three thoughts in succession:

1. A barbeque?
2. At no cost to himself?
3. Excellent!

Ergo: a barbeque at Bob's it would be.

Chapter 89: Barbeque At Bob's

And so it was ordained, in March of 1978, that next Sunday a barbeque would be held at the home of Bob McLain. The air staff were all invited, except whoever was stuck on the air that afternoon. Of course, everyone at the station was talking about it, but the talk was not hopeful. There was no sense of optimism, no belief that anything would change after the barbeque: McLain was going to shovel shit on them and make them eat it with their burgers.

The general feeling at the station was that Jeremy had made a deal with the devil when he hired McLain, and they would go along with the deal as long as they got their share, too. But with that not happening, they only saw McLain's comfortable lifestyle, and the apparent comforts of the sales staff. This boat-buying thing with McLain had been the last straw, and they were going to ask him about *that*, too.

When Tiny looks back now at all that happened in the next year, it was the hiring of Bob McLain that signaled the end. Tiny remembers "it was downhill from there."

They came in carpools, and arrived at close intervals. McLain welcomed them and told everyone, as they got there, how happy he was to have them there, and "C'mon in the back!" The jocks were sullen and cautious: this was the lion's den, and they were watchful for signs of betrayal. Everyone was herded to the back yard, and few people that day spent much time inside. Their friends were all outside, the beer and the food were outside. This would be their territory.

The sales staff wanted to change the sound of KFAT- the sound that the jocks had created and they loved. Long before coming to McLain's, both sides knew their positions were unmovable, that neither side would budge. Bob McLain would yell into Jeremy's empty pocket, and the jocks would hear the echo, but they would always resist the change.

McLain wanted to dilute the format so he could sell it. The jocks revered the format and wanted it to remain unchanged. Jeremy and Laura Ellen wanted peace, but there was no peace in sight. They knew the format was wonderful and unique, and they knew that to make a living at it, they were being asked to make it less unique. Jeremy wished fervently that the whole thing would go away. And yet again: fat chance.

A couple of members from the sales staff were there and everyone was uneasy, but everyone was willing to talk. Bob introduced the air staff to the sales staff, and, getting to Terrell Lynn, said, "and this is our boat-rocker."

Terrell Lynn said, "I'm only rocking the boat 'cause I'm looking for life preservers, Bob." That was good for some uneasy chuckles.

Tiny was tense and drinking a lot of beer, and she started feeling queasy in what she knew would be a long period of illness from getting too much sun. "I remember my face got really red, and I was pissed because I knew instinctively that it was the beginning of the end. I knew it. I knew we were being sold out. Because they were trying to sell us like a cheap car salesman. And Bob McLain was a weasel, and I knew that something bad was going to happen. I knew it."

It didn't take long. Soon the subject of wages came up. The staff had been making $3.25 an hour for so long that their resentments had built until they were on the verge of erupting. Mark Taylor, introduced to the sales staff as the Ozone Ranger, remembers "arguing rather vociferously that it was time to start paying the staff four dollars an hour. And I remember going, 'Come on, you guys- it's time to pay the DJ's *four bucks an hour! Okay*?' and getting a lot of hemming and hawing about that."

McLain responded by telling the air staff that they held very special positions, and everyone in the office in San Jose appreciated that. But they should also remember that if they weren't working at KFAT, that they'd never find another job in radio- anywhere. He told them that after this KFAT gig, their careers in radio were over, and that they could enjoy themselves, sure, but they should also remember that 'radio was a business, and if we can't sell your talent, then you can't earn a living, and to help yourselves, you're all going to have to learn how to be more in the game that we play in, called 'radio sales.'

He said that everyone had to realize that they had to "tighten up" and be more "professional" if they wanted to make any more money.

The Ozoner led the charge for a while, saying again that they deserved that raise to four dollars, but nothing got resolved. Terrell Lynn spoke up with similar concerns, and so did several of the DJ's, but Mark and Terrell Lynn were the most vocal. Certainly, even those whose diatribes against McLain remained more felt than expressed, all felt that their opinions were clearly represented that afternoon. Ozoner felt that he was speaking for all when he made his views heard, and all agreed, and when Terrell Lynn spoke, all agreed. There was a clear position, and it applied to all the air staff.

But nothing happened. No promises, pledges or vows were made that day to in any way improve the lot of the jocks. Mark Taylor remembers feeling that everyone on the air staff had agreed with everyone else on the air staff that the barbeque had gone pretty much exactly as they had feared, and knew, it would.

But Bob McLain remembered that Ozone fellow as a troublemaker. And that Terrell Lynn guy, too.

Chapter 90: The Chapter With The Tease

You watch reality television. You know you do.

So you know that every week on television's reality shows, someone has to go. You know that. But how about when some regular, scripted drama or sitcom that you watch regularly teases you with a promo for the next week's show, telling you that some long-time, beloved character from the cast is going to leave the show, often in some shocking manner. That certainly doesn't happen every week in a long-running show, and they play it up and promote the hell out of it, in teases, especially during and after the show before the one with the surprise ending, and all during the next week in daily promos telling you not to miss it.

They don't tell you who's going away, of course, or how. They want you to watch the next show to see which beloved character goes, and how. And that's what this is like. This is the chapter where I tell you that in the next chapter someone is going to leave the KFAT family forever.

In the early days, the station had a large and varying cast, but that quickly morphed into this narrow cast, this small staff of KFAT, and they had become a close-knit cast, a family. And losing a member of a family is a rare and powerful event, as it was this time.

Someone is going to be fired, someone you know and love, and who has sacrificed to help make KFAT what it was. They won't tell you in advance on television, and neither will I.

I won't tell you who is getting fired. You'll have to wait. You'll have to read the next chapter. Don't bitch about it. That's how it's done. Turn the page, already.

Chapter 91: Time To Go

Back when this staff was first coming together, he did a few fill-ins, then settled permanently into the overnights. At first, he was the only person there with any production skills, and so he did that until Frisco got there. The Ozone Ranger had been there for nearly two years and it was getting old.

It must be noted that Mark Taylor was separated from his overnight duties soon after he mouthed off at Bob McLain's BBQ. He doesn't remember why he got fired, only that "I think it was probably a number of things. It was probably time to go. It wasn't missed shifts... I was pretty reliable in getting the work done."

He was getting to the station on time most nights, and by KFAT standards that was good, but something was amiss, and although no one remembers what caused his dismissal, the Ozone Ranger hung up his headphones in April of 1978, after almost two years. It was Jeremy who fired him, although neither man remembers how, or why. That's Fat.

It had been his longest-lasting job, and he was cast adrift in Radio World, finding that no one wanted to talk to him after he told them where he'd been the past two years. That bastard Bob McLain had been right.

He said, "I couldn't get arrested after KFAT, and... it was because of the reputation that the station had, or the reputation that I had, perhaps. There were a lot of... let's face it... it was not an era of moderate habits, and KFAT exemplified excess in a number of ways."

After KFAT, Mark Taylor had "a couple of rough years." He worked for two years at a garlic and onion processing plant in Gilroy. He was a card-carrying teamster at Gilroy Foods, driving a forklift. The KFAT DJ's took note. In the world of radio, they were outcasts, isolated rebels living inside a fragile bubble. The world had abandoned the hippie ethic, and had moved on without them. They could breathe the air inside the bubble, but they knew that outside of it, they couldn't survive. They got the message: Bob McLain had won an important battle, and any more missions to push the bubble would be suicide missions.

Mark Taylor found his way back into radio, but it was several years later, and it was 3,000 miles away from KFAT, in Hawaii. Not that that there's anything wrong with that.

Chapter 92: Rocket On Board

Mark DeFranco was a curious mix of influences. Having grown up in San Jose, one might not be surprised to learn of his attraction to methamphetamine and old Harley-Davidsons, but one might not have suspected his love of the blues. Mark loved the blues, and knew quite a lot about them, as a musical form and as a lifestyle. He wasn't a motorcycle club kind of guy, more of a lone wolf on his bike. He rode an old Panhead or a Shovelhead or one of those old Harleys from the early 60s with names like that.

He rode that bike everywhere, but he also had an old '55 Chevy pickup because, as Cuzin Al has observed, "if you're going to own an old Harley, you're gonna need a pickup to come get it."

I do not know much about Mark's life before KFAT, and that information is, sadly, no longer available to me. I'm not going to include an Epilogue chapter so I will tell you now that until recently, Mark lived in a shared room at a state-run convalescent facility in San Jose, his health and his grasp on reality irreparably damaged. But in his days on the air on KFAT, he was a vibrant man. He loved music and he loved radio, and the first anyone saw of Mark DeFranco in Gilroy, he was in the doorway of the studio on a Sunday night, asking Cuzin Al if he could hang out, and Al said, "'Course, brah- sit ovah theah."

Terrell Lynn remembers him hanging out during his show, too. Others saw him, and he was always polite, non-intrusive, and in a cheerful mood. No one at the station was going to be put off by a little grease on some jeans or under some fingernails. He was a good guy, a regular visitor, and soon he was fetching records for the jock, and then he was cueing them up, then he was doing the segue and spinning them. Mark DeFranco started, in true KFAT fashion, with fill-ins and overnights, and he was always ready to fill in.

Then Jeremy pulled the plug on the Ozone Ranger, and there was a need for an overnight guy. Friends, meet Rocket Man.

Why "Rocket" at all? No one knows that, either, but I suspect it had something to do with a rocket. Goes fast. Lots of speed. Speed! That's the connection! Rocket Man liked his speed. Methamphetamine was cheap and available everywhere in San Jose, and a few places in Gilroy.

Meth might not be a bad idea for a man doing an overnight shift- say, one a.m. 'til six- and that was what he did. And he partied. He had been circling this job for months, and now it was his. It was the coolest station on the planet and he was the overnight guy. And his show would be a party. Just because it was late was no reason to be slow, no reason to be sad. It was late and it should sound like a party at the station with Rocket Man, so the party should, actually happen at the station with Rocket Man!

Rocket would host a gathering of San Jose's shadier people there, most nights. If no one came by, he'd mention *that* on the air, and then hang out, rockin' the airwaves until people showed up.

Sister Tiny well remembers her times in residence at the station, when she'd wake up to use the bathroom in the small hours, and find a dozen or more strange people milling about the halls, offices and whatever room was unlocked at the station. Strange people in at least two senses. She remembers the thick smoke and acrid smell, the snorting of powders, the loud, enthusiastic discourse, the empty refrigerator, the empty bottles and cans left lying around and the bathroom being occupied.

That damn Rocket! It fucking pissed her off!

She doubts she'll forget the night she was in her room at the very back of the station, two thin doors away from whatever was going on outside her room. She was getting it on with a guy when she felt someone pulling on her leg, and, knowing that both of her lovers' hands were already in use elsewhere, she looked down to see a complete stranger in bed with them, pulling on her leg, crawling upward, looking for an opening to join in.

That damn Rocket! This was *his* crowd! The Ozoner used to eat her food and smoke her dope and borrow her car, and he used to call her in a whiny voice and say, "Oohhh, I'm so drunk… I can't make it into work… I'm in Boulder Creek…" and Tiny had to do nine-hour shifts because of him, but he was "like my little brother" and she put up with it. He was a drag, but he was family, and the extra hours meant much-needed extra bucks for her. But that was the Ozoner, and this was a whole other order of bullshit, and she needed to get herself and her kids out of that back room at KFAT. One morning, Tiny called Jeremy and Laura Ellen and cried that this was it, she had to get out of there, and they had to help her. They did, and she did.

She moved out, but the parties continued nightly. Now there was more room at the station for more people.

Of course the Hells Angels knew they were cool with Rocket on board, and when they came, the place seemed to clear out of everyone but them, but otherwise it was a pretty open party. Rocket always kept control and asked people to clean up as much as possible, and Gordy remembered a reasonably manageable mess when he got there for his morning shift. Gordy would help him clean up, and it usually wouldn't take too long. Gordy liked hearing the stories, and it gave him something to talk about on his shift. That damn Gordy!

Rocket Man loved the blues, but he was also an avid Fathead, and, now that he was in the vortex, he listened all the time, always paying attention and always learning. Though a latecomer to the now veteran staff, Rocket Man quickly found his stride, and his shows soon were of the highest caliber. As casual as KFAT was in many aspects, the quality of the music and the quality of the shows was always the first and foremost priority, and both staff and listeners were paying attention. Rocket was a quick study and soon attained that proficiency.

With an easy transition from Ozone to Rocket, Laura Ellen could stop worrying about finding someone—anyone—with knowledge of their eccentric library who would be willing to work from one until six a.m. for little pay and few benefits. Having a reliable overnight jock was a blessing, and Laura Ellen was glad to have Rocket Man aboard. Get some sleep, Laura Ellen.

I remember Rocket as always being cheerful, and he and I used to have a thing we did. I was working late one night when Rocket came in for his shift. Through the window in the studio, he saw me in the Production Room, and he came in to say hello.

He said, "Hey, Gilbert- how're ya doin'?"

I said, "Oh, I'm able to sit up and take nourishment."

For some reason, that just killed him. He fell down laughing, repeating, "sit up and take nourishment..." After that, he used it himself. Several Fat DJ's remember that he used that line, on and off the air, for years, and he never forgot that line when he saw me. Whenever we saw each other, he always asked me if I was sitting up and taking nourishment, or he'd tell me that that was his condition, if he got it in first. He always laughed when he said it, and it was our special greeting.

When I saw him in his bed that last time, in his diminished condition, my heart broke. I'd been told his room number by the staff, but when I walked into the room, I didn't recognize the man in there, and had go back out to look again at the number on the door to see if I was in the right room. When I came back in, the man in the bed looked at me, confused, obviously not recognizing me, either. I'd never seen the Rocket wearing glasses, and this guy's glasses were thick like coke bottle bottoms, so I asked, "Do you know who I am?" and he shook his head, no. I said, "Well, I might not be as thin as I used to be, and I've lost some hair, but I'm still able to sit up and take nourishment." His face brightened, he smiled and weakly, just above a whisper, he said, "Gilbert!"

The Rocket was easy to like, easy to hang with, and an easy-going guy with a lot of soul. But his use of who-knows-how-much speed and who-knows-how-much-of-what-else had taken its toll. He was damaged and homeless, and his organs were shutting down. He was drooling. Despite our shared greeting, which he still retained, he was not completely aware, and it was obvious that he would never rejoin the outside world. He had a pocket-sized cassette player, but no radio, and some cheap headphones and a few cassettes. He still listened to music, and he still liked the blues the best. Perhaps mercifully, he died a few months later, in April of 2006. I believe that he is happier and better off now than he was when I saw him in that place.

But let's remember that this memory was decades away, and at this time in our story he was still a fun-loving, vital, rockin' guy, so…

Rock on, Rocket Man!

Chapter 93: Heeere's... Leo!

Jeremy and Laura Ellen were still going crazy with the endless bills, the money that seemed to be hemorrhaging daily, weekly and monthly, and all the tensions at the station. They had a friend, and I've mentioned him before.

Some readers will remember that several times I have alluded to Leo Kesselman. I said "he's coming-" but I lied. He'd already been there. He'd been there since the beginning, in the background, lurking, keeping an eye on the station, giving advice when called upon, and now, in an increased capacity, he'd been there for several months. I know, you feel:

A. hurt
B. betrayed
C. angry
D. confused
E. curious

How about if I say "I'm sorry" and we move on?

Leo Kesselman was a lawyer and a pleasure-seeker. He was successful in his practice, had made a lot of money, and lived in a big, beautiful house in the hills above Los Gatos. His principal interest was his own gratification, in whatever form that might take.

Living in Los Gatos in the early 70s, Leo heard KTAO, got in touch with Lorenzo, and they became friends. He met Jeremy and a pregnant Laura Ellen when they came out to visit Lorenzo, and Leo liked that "Jeremy was crazy in his own way, like Lorenzo." They were "iconoclasts about everything," and they were "fun, creative, innovative, didn't pay attention to authority, the kinds of things I respect in a person."

It was Leo who found Tiny for KFAT, and he was who told Jeremy about her.

When Jeremy and Laura Ellen decided to settle in Los Gatos, Leo's realtor found them a house, and he became their lawyer and advisor, and their friendship deepened. Leo advised them when they were thinking of buying KSND, initially recommending that they not buy the station. But they did, and he advised them throughout their years with KFAT.

GK: As their attorney, did you ever tell them that you were concerned about the more raucous elements of the station?

LK: "Constantly. I was the voice of conservatism. I tried to tell them where the lines were and when they crossed over the lines, in regard to programming and conduct of personnel."

GK: How did they react to your advice?

LK: "It varied. Sometimes they'd act, sometimes not."

GK: Was that frustrating for you?

LK: "No, that was just the way it was. I cared in that my friends were involved."

GK: Did you call Jeremy and Laura Ellen to warn them about the station pushing the limits of what was permissible?

LK: "Yes, that would be a fair statement. At that point, the FCC was very conservative. There were lots of stupid rules & regs that you had to toe the mark on, or people would turn you in. He'd ask sometimes, but more often I'd hear obscenities, promotion of drugs on the air, and I'd tell him 'that's dangerous stuff. People are going to object to that. You run the risk of not getting your license renewed…' "

GK: Do you mean the DJ's or the music they were playing?

LK: "Both. The DJ's were themselves, but they were at the edge of what was permissible for radio at that point. Laura Ellen tried to keep it at a minimum. Since it survived, they were successful at it. This was in '76-'78, and then it became successful. And the more famous it became, the more money it lost."

Leo remained the "voice of conservatism," but it was a sporadic contribution. Lorenzo was out of the picture, spending what was left of his money and his remaining health in other cities, on other non-commercial stations, which he would eventually leave to those cities. Jeremy wasn't asking for Leo's help all that much the first year, and, not wanting to interfere excessively, Leo limited his comments to only the most outrageous infractions of legality that he heard.

But by early 1979, three years of frustration and poverty had passed. Losing money and not knowing why, Jeremy and Laura Ellen begged Leo to come in and look around, make some observations and help put the station on the money-making track that McLain was failing to do. They wanted the jocks to have free rein in their selections, but they were willing to make some changes if he could make the station profitable. He often listened to the station; these were his friends, and he wanted to help.

Leo is not a modest man. He thinks very highly of his abilities, and is quick to point them out to anyone willing to listen. He has the track record to back up his high level of self-regard, and, always fascinated with radio, and liking the challenge, he wanted to come in and turn the station around. At his friends' request, Leo got involved with KFAT. Leo looked around and fell into the vortex. He was now the station's General Manager.

He was a strict business guy, and he had little room for charm or sentiment. The station was in turmoil, what with the jocks' constant complaining about needing money, and his suspicions about the sales staff making off with more than their share, so he stepped in, offering to bring some business reality to a place almost completely lacking in it. He didn't trust the jocks, he didn't trust McLain, he only trusted his two friends, and then, by nature, only reluctantly.

Leo is an iconoclast and a hedonist. He does what pleases him, and he has been successful enough to indulge in those. Being conservative was not his style, and soon after he increased his involvement with the station, he had gone to a Broadcaster's Convention to represent KFAT. He did not fit in with the normal business folk, and was treated with disrespect, and he knew he'd remember that for the next meeting. "When I went, *I* was the outlaw." So for the next convention, "I sent Bob [Cassidy], and nobody could believe it. He blew them away. Ask Bob about it. He was drunk, as he was for most of the past 30 years." Bob had been invited, the drinks were free, and Bob didn't give a shit. Fun for Leo, fun for Bob.

LK: "I was there or the office every day. Mostly at the office [in San Jose]. KFAT was losing its ass, about $25-30,000/ month. I turned it around. Increased advertising sales, put more emphasis on it, and cut expenses. There was a lot of money wasted, spent on what [it] shouldn't have been spent on."

One focus became the Fat Fry, which had proven to be a popular, well-attended event every Monday night. Bob Simmons had started booking the shows there, and made it a success. Now Leo took a long look at the books, and what he saw made no sense.

"The Fat Fry was at the Keystone, losing $15,000 a month just on that. I met with the Coronas, I said 'Hi, guys, I'm the new Manager, and we're paying you $5,000 a month to use your club every Monday night, which is a dead night, you'd probably be closed anyway. We've been paying for a band to play here, and you've been kind enough to take all the gate and all the bar."

GK: They took it all?

LK: "And we paid for the band!"

Leo told them "there's going to be some changes. They said 'fuck you. We've made you what you are today.' I said, 'I don't know, but I think you've got that wrong. In terms of who made who, we're not going to do this for a variety of reasons, not the least of which is we can't afford it.' We took it to the Saddle Rack (in San Jose)."

After a month, maybe six weeks, the Fat Fry moved back to Palo Alto. "They begged us to come back."

Things like this were Leo's mandate. He made changes, and KFAT started to turn around, financially. "When we came back, they paid for the band, we got the gate, they got the bar, I don't remember the rest. We went from losing $15,000 to making $10,000 a month. Things like that. It wasn't a very business-like organization."

Fixing the Fat Fry was easy, but fixing the sales office was not. Leo saw "it wasn't very good. Good people, but not very organized. It's one thing to be loose on the air, it's another to be loose in your business operation. The people in the business end were as loose as the people on the air. Things got tightened up and sales increased.

"The station went from losing $15- 25,000 a month to making $30,000 a month. If I didn't turn the station around and stop the losses, the station would have been sold back in '78. Jeremy and Lorenzo were losing patience. It was fun, but not *that* much fun. I had an opportunity to buy it, but it was too much work."

Leo was elbow deep in the vortex, making decisions, negotiating deals, rendering opinions and shaping the station into a business. The jocks, you will have guessed, were none too happy about his increasing influence, but they might have gone along with it all- if their paychecks got a little fatter. They'd heard the promises before, and they knew Leo was about cutting costs, not raising them.

Leo had been a behind-the-scenes presence at KFAT from the start, then he'd thrown himself into the vortex, and now things were calm and constant, but the empty bank accounts of the DJ's was also constant. They weren't happy; Leo was making money for the station, but the jocks were still piteously underpaid. Their checks weren't bouncing any more, but neither were they covering all their modest expenses.

As General Manager, Leo was also in charge of the sales office, and Bob McLain didn't like it, but he saw that the new guy was about business, not sentiment, and he felt that, finally, his way would prevail. McLain felt that with Leo behind him, he could now tame this bastard of a station, get rid of those asshole jocks, and make the changes he'd been wanting to make. Yet again: fat chance.

Leo was in charge, and wouldn't get behind anyone. They'd have to get behind *him,* and he told that to McLain. But this was more of a crisis than he'd bargained for: the programming changes McLain was insisting on- and the jocks were resisting, the money that seemed to be unaccountably disappearing before it reached the bank, the jocks' constant whining about money, and all the rest of it. Jeremy had tried his hand on some of the responsibilities, but that hadn't worked out, and now Jeremy was drifting further away. Leo was a great help, but he was getting restless, wanting to get back to more pleasurable pursuits. Laura Ellen was out of her depth and needed to listen to someone other than Bob McLain, and most of her friends were now enemies, who she couldn't talk to about this stuff, anyway. She needed a smart, friendly voice with a lot of business experience. McLain had advised hiring Lori Nelson, an experienced trustworthy businesswoman, who he said would do a good job of organizing KFAT's office in Gilroy. Leo met with her and approved the hire. But to Leo, everyone at the station, except Jeremy and Laura Ellen, was suspect. Everyone.

Chapter 94: Meet Lori Nelson

Reader Alert: This is a very important chapter about a woman who was very important to our story. Please read it here.

One of Leo's important decisions was that Laura Ellen needed a real Office Manager for the office in Gilroy. At the same time, McLain wanted an ally in Gilroy. Meet Lori Nelson.

Leo liked a challenge, and this KFAT business thing was a challenge that he could get into, but he wouldn't want to stick around and do it forever. Going in, he knew he'd make some changes, enjoy the game for a bit, and get bored sooner or later. These were his friends, he wanted them to succeed with, and then without, his input. He told Jeremy and Laura Ellen that they needed a strong business presence in the office every day, and they agreed.

Lori Nelson: "Jeremy heard about me from Bob McLain. I had worked for (giant accounting firm) Arthur Anderson, in San Francisco, who put me into KSJO (San Jose), and I spun out of there when they sold that radio station. Then I went to work for a builder and landed in Gilroy. And that's where Jeremy found me, and fortunately or unfortunately, Gilroy had put a moratorium on building, so I couldn't sell any houses.

"This was… 1978… and Jeremy hired me as a controller to put his sales and his records in a computer, was basically what my first job was with him. And I felt, well, while I'm waiting for the moratorium to be over, I'll go put their work together, and then I'll go back to the builder."

GK: What were your first impressions of Jeremy?
LN: "Well, I knew that he was a hippie from Haight-Ashbury, a lovechild, I knew who he was, and that didn't matter to me. I knew he was a genius, and being a techie, I knew that he was capable of doing incredible things with the signal. I knew that emotionally he was very immature. I knew he threw little tantrums. I knew that he was very sensitive, and I had a lot of respect for Jeremy. But I also understood where he was coming from, and I understood the need he had for me to balance out his business…

"Well, when I walked into KFAT, I was in culture shock, because of the environment, and all the [financial] records were in boxes all over the place, nothing was put together right, and it was kind of a nightmare. But I did get it computerized… their financial records, their logs for their commercials, the business side of the company."

Lori remembers the office, and her "... wooden desk with about six wooden slats, none of which fit, so there were a number of holes in the top of it, and my chair was a cane chair with a hole in the cane, and a pillow where the cane was, and that's where I sat."

GK: And you worked out of that office?

LN: "I worked out of the office in Gilroy. I can remember going to the bank one day, 'cause I handled their finances, too, and the banker saying, 'You don't look like a KFATter,' and I said, 'No, I'm just wearing my costume, they're just wearing their costumes. You're just wearing your costume. Don't tell me you dress like that every day.' And I said, 'I'm the beard. I'm the gal that's holding this together.' And that's the way it was.

"But I was not into country music. I was not into the scene at all. But during the period of time I was there, I finally grew into it. I started wearing cowboy boots and started wearing cowboy hats. And I had a lot of interesting things happen. But (when Leo backed off) I became General Manager at KFAT."

Leo was backing off of his daily duties as General Manager, and both Jeremy and Laura Ellen had seen that a strong hand had helped calm the waters, and they wanted to keep things as calm as possible. Jeremy also had a secret, and it was an important reason for finding a strong, steady, but caring hand to help run the station. He knew that Leo was losing interest just when the station was about to need it the most. No one knew it yet, but Jeremy was starting to think about getting out.

Leo had been looking for another GM, and he'd been impressed with Lori. He still felt strong emotional ties with this station and its crew, God help him, and he wanted someone to feel some emotion for them, too, in his absence. It couldn't be Bob McLain, or even someone like him. He'd spoken with a few choices, but Lori was already there, she knew the operation inside out, she'd shown she could work with that lunatic Jeremy, she had a big heart, and she was the devil he knew.

LN: "And so Jeremy was looking for a GM, and Laura Ellen was still the MD, and basically running things, and I had a good relationship with Laura Ellen. I had a good relationship with all the people there. And so [Jeremy] came in and told me one day that he'd finally made the choice, and the choice was me, because he trusted me with the people, he knew that I cared about them. He knew I could handle the financial end of it, as rocky as it was.

GK: You worked in Gilroy. What was the relationship like with the sales office?

LN: "Well, the sales office was for me, so it was a very good relationship with me. But [for the sales staff] it was a lousy relationship with the DJs. Because they would continually screw up on commercials. And the sales people would have to face the clients, and they would say they weren't going to pay for a commercial that the jocks didn't do right, or didn't do at all."

GK: How would they screw up a spot?

LN: "Well, when you put a commercial on the air, you do not put two car dealers back-to-back. You pay for certain premiums for Drive Time commercials, and these DJs, hell, they didn't care when they put them on. They put them on out of order, [or] they didn't put them on at all. Or they would just have a live written statement, and they'd be so stoned or drunk that they couldn't even see it, and so they just talked it."

GK: They ad-libbed the ads?

LN: "All the time."

GK: But there was a log where they were supposed to play a certain spot at a certain time..."

LN: "And they didn't follow it. Well, they were stoned out of their minds most of the time."

GK: Did the DJs screw up ads in other ways?

LN: "Well, it all depended on their emotional frame of mind at the time they were on the air. And we were very lackadaisical about discipline. And that's why the sales office did not like the DJs, because they had to face the customers with these guys screwing up all the time. They were always letting people off the streets into the radio station at night. They were always smoking marijuana and having a really good time together."

GK: When the DJs would screw up and the sales people would have to confront their clients, who did they call? You? Laura Ellen? Who dealt with it?

LN: "Well, usually, Bob McLain, the Sales Manager, would call down and chew out the DJ- he'd go directly to the DJ that screwed up. And then explain to them that we were losing revenue because they aren't doing things right. So Laura Ellen tried to clean up their act, and I implemented a program that if they lost revenue off of a commercial, then what they lost was taken out of the DJ's paycheck. Well, I soon found out that I couldn't do that. It was against the law."

GK: Did you actually do that?

LN: "Oh, I actually did that."

GK: And how long did that last before you had to stop?

LN: "Not very long. But the DJs got the word."

GK: How long did that policy last, and how did they find out that you couldn't do that?

LN: "Laura Ellen went to work for the DJs and talked to Leo, and he said we couldn't do that. So I said, okay, I'll do something else."

GK: What did you do?

LN: "One of the things I did was when they lost the key to the radio station,

I charged them $75 to get a new key or they couldn't get in."

"So I worked closely with the customers, with the sales staff, very closely with Laura Ellen, and I was constantly in contact with Jeremy."

GK: How much contact did you have with the DJs?

LN: "Every single day."

GK: So you were like the middle person between sanity and...

LN: "I was the beard."

GK: How frustrating was that?

LN: "It wasn't too frustrating, because I understand that the DJs were the stars, and they're the ones that made us popular and they're the ones that kept us earning revenue, and I wanted them happy. And I would make arrangements for them to go to the marijuana festivals, I turned my back when they were drinking beer and smoking dope, even though it was *really* against the law in those days.

"But I respected the position that everybody had, and they all played a role in the success of that radio station. I was very involved in the negotiations of contracts with our clients, I was very involved with the T-shirt sales. I can remember Willie Nelson came to a concert over at Santa Cruz Fairgrounds, and there was a lot of other stars there, but he was really big for KFAT. I talked to all the entertainers that came through there.

"And of course Laura Ellen managed all that. And Jeremy, when he was there. But I can remember standing backstage when Willie Nelson was playing, and looking over the entire audience, and it was a maze of KFAT T-shirts. It was just incredible. I also remember when a truck driver would come driving into the radio station to buy a KFAT T-shirt, and we didn't have any- we were always sold out, and he would pay this guy in a T-shirt on the street to take the T-shirt off his back, exorbitant amounts of money to get that T-shirt.

"So I would call on the local businesses, and one of the big businesses there was the automotive. It was Al Sanchez, the Volkswagen dealership, the GM dealership, I knew 'em all. I'd call on them and they were really into television, and they didn't want anything to do with the radio station, and I says, 'you know, we're neighbors, let's be friends.' I said I could bring business to you. I will put a T-shirt in your office, in your locations, and we'll do a promotion on the air that says 'stop by Al Sanchez's VW and buy your KFAT T-shirts.'

"They were so bombarded with people to buy T-shirts that they called me up and said 'get these Fatheads out of here, because we're not selling cars, we're just selling T-shirts!' And I said, 'Well, don't our ads pull?' and they started advertising on the air. So I did a lot of that."

GK: But that's a salesman's job. Did you get a commission from that?

LN: "Hell, no, I was General Manager, I got paid very little. But the salesman took it over and kept it serviced and kept it going. I got the door open.

GM's do that, you know. And I did that with a lot of businesses in Santa Cruz. I called on everybody local that I could reach while I was there. And so, we sold radio ads to Bank of America, Mervyn's, all the biggies.

"But KFAT was also the last theme radio station where, when we had an earthquake, they played songs like 'Shake Rattle and Roll,' they played songs that pissed off the women libbers like 'Boobs A Lot.' And they were always playing songs that were appropriate to the environment at the time."

GK: Did we hear from women libbers?

LN: "All the time. They wrote us letters, they picketed us, we had an open line to anybody, and so the Ku Klux Klan called in on the Fat Gram and talked about murdering people, and we got *The Santa Cruz Sentinel* coming over and picketing us, and telling us they were going to put us off the air, and we told them they had the right to say anything on the air also..."

GK: *The Sentinel* was objecting to the...

LN: "...Ku Klux Klan talking on the air 'cause they're murderers."

GK: And the Ku Klux Klan was threatening murders on the air?

LN: "Of course, of course. But you know, someone would call in and say 'Terrell Lynn, I want to hear the mating call of a whale,' so he would put on something that sounded like the mating call of a whale. That was part of the licensing, that we gave direct access to the public. So we had these Fat Grams, and they would call in and ask for all kinds of stupid things, and the DJs would support it. And it was probably one of the most outrageous radio stations you ever heard."

"Which reminds me, Jeremy gave me a Porsche to drive while I worked there. He didn't pay me anything, but he got a trade-out for a Porsche, and I drove a Porsche. I drove a Porsche when I worked at KSJO, and he heard about it- it was one of the incentives to get me there. So Jeremy was a good guy."

GK: Did the equipment break down much?

LN: "All the time. It was a nightmare. If the equipment wasn't breaking down, the DJs were breaking down, or the salesmen were breaking down. It was like riding a bucking bronco to stay on that radio station. And when it was over, I was so glad. I said I'll never do this again. But it was exciting while it was going on.

"And Jeremy was colorful as hell. I would attend the advertising meetings with all the radio people once a month, for him. This was the media groups getting together. Every television station, every radio station, every advertising agency, everybody who was in any kind of a media position at all came together once a month and had a meeting. And one month he decided to go, and so he went across the street to Woolworth's and he bought a clean pair of pants, a clean shirt, and he bought a tie. And you know he always went barefoot, he never washed his hair, he never took a bath. He reeked, but he was so charming. Loving is the best way to describe him. And so he comes over with this tie, and wants me to help him with it, and I thought he wanted it on his shirt. No, he put it through the belt loops and tied it around his waist for a belt. And that's how he went to the San Jose advertising club. Well, it was the last time he went."

GK: What was the reception like for Jeremy?

LN: "Well, they treated him really bad. He stuck out like a sore thumb. These were suits. These are people who are professionals. He decided that this wasn't a place for him."

GK: Did he have a bad time, or did he just laugh it off, like 'screw them?'

LN: "I'm sure he laughed it off and screw them, but he didn't want to waste his time and go back there."

GK: So how did they treat you at these meetings?

LN: "They treated me like I was gold. Because I represented a radio station that was making it on nothing. Plus, I had a lot of influence from KSJO, from KNDY radio, from prior experience. And it was a good time for me. But then, I was a professional in with all these mavericks."

GK: And you spoke their language and looked the part.

LN: "Right. I made them look straight when they weren't."

GK: So you'd be there on a daily basis, and where were you living?

LN: "Santa Clara. I drove every day, it was quite a time. And during the time of the gas shortage, I caught the Greyhound bus down there. When I'd come riding into the Greyhound bus terminal in San Jose, there'd be hookers everywhere, and it was quite a time, wasn't it?"

GK: So what was it like hanging out at the station? You were doing your work, the DJs were on the air, but were there people hanging out? Smoking pot?

LN: "It was just like two different worlds. I was in the office doing financial statements, and doing all the banking, doing all the collections from customers that the sales staff didn't get done. Business as usual, it was just two different worlds. And I would go have lunch with them, I would go to the laundromat with Laura Ellen and her dog and Elsie. I did horseback riding with some of them, over at Martin's Stable."

GK: How about Elsie. Did you pick up Elsie from school?

LN: "A lot of times. But I was definitely the business manager. And I got more relaxed and more fitting in, and at one point I had a relationship with Cuzin Al, who was the bluegrass specialist on Sunday night. You know, I never saw him down there because on Sundays I stayed home, most of the time.

"But we went as a family to Hawaii, and he was over there on business 'cause he worked for Mission Clay Products, where they sold pipes for sewers, and his territory was Hawaii, Oregon, Washington and California, and we were over there. I had taken my whole family, my youngest child, my son, had just graduated out of high school and I have another daughter… and we all went, there was ten of us, and we went to celebrate and we had a wonderful time. And he was there at the same time, so he came and joined us, and I tried to get rid of him. He invited me to breakfast, and so I showed up with all ten of my family. I thought for sure that would get rid of him, right?

"But I didn't realize that he was Italian, and he just liked it. So when we came back to the States, he just kept asking me out and asking me out, and I said No, no, no, no. So I threw him a divorce shower. I invited every single woman I could find- I got his little black book, and I got all their names and numbers, and we all had a party with him so he could find himself a girlfriend.

"Well, he just kept pursuing, and in a weaker moment, I don't know what was wrong with me, we got together. And I think I was having an empty nest syndrome, where all my kids were gone, and he still had two teenagers, and I thought, 'well, I'll just help him with his kids.'

"But we're like the Odd Couple, we're like day and night. And he was into the DJ scene and I was not. And it was a pretty rocky road, and finally we agreed to disagree and we went our separate ways."

What Lori failed to mention during our interview was that when she said "we got together," what she meant was that they got married.

GK: How long did you stay together?

LN: "Probably five years."

GK: Who were the sales people? Bob McLain was in charge…

LN: "Yes, but Geneva Nieto was one of the key salesmen, and the sales office was a swinging door. What they would do is, they would hire green sales people, they would come in and establish new accounts, and once they got their new accounts established, they'd fire the sales people… 'cause their volume wasn't high enough… and Geneva Nieto would take the account over and make buck-o money off of it. And so there was many, many sales people."

GK: So Bob McLain would fire them? How long would they last?

LN: "Well, probably about three months. It was quite a game. It was pretty unethical. A lot of people hated Bob McLain because of his ethics. Now, Bob came from CBS, as a pageboy in the early days of radio, and he became very, very wealthy, and into sales in LA, married 5 times, had kids from each one of them. One day when we were working together at KSJO, all five of his wives showed up with subpoenas for back child support. He was a real slick-talking devil. But lovable.

"He liked the shock treatment. He would say to you: 'Come on in, take your clothes off, sit down, let's have a talk!' At the office! He loved to shock people. But he was an incredible salesman. He would call people up over and over again, and he'd say 'I've been thinking about you today.' And he brought in sales like you wouldn't believe.

GK: How did he land at KFAT?

LN: "Well, Jeremy found out that he worked at KSJO, was successful there, it [got] sold and he hired him, just like he hired me."

GK: A lot of the air staff was convinced that Bob McLain was skimming money...

LN: "They always thought that. Bob McLain earned every dime he made, but he did it very slick. McLain didn't skim any money anywhere, because I handled the money and I could tell you that for a fact. But they *hated* him because he *hammered* them about the lousy way they did the spots. He *hammered* them about their image, he *hammered* them about taking a bath. He hammered them about everything. They hated him."

GK: I can see the DJs at KFAT not wanting to hear it from a Sales Manager. Hear it from Laura Ellen, from Jeremy, from you, you were there every day, not from Bob McLain. That was commercialism, and that was antithetical to the hippie ethic...

LN: "That's what Jeremy hired him to be, Sales Manager, not General Manager. He felt that he had earned the right to be the GM, but Jeremy said, 'You're too mean to my people. Lori wouldn't be that way,' and he hired me instead."

GK: I remember the BBQ at McClain's house that turned into an ugly confrontation...it was supposed to be a bridge between the two staffs...

LN: "Jeremy was trying to make that happen, and when he discovered that he couldn't make it amicable, that's when he said, 'Lori, you're going to run things. 'Cause you will not treat these people bad.' "

I told Lori that I also remembered the staff meetings where the jocks complained bitterly about the money they were making.

LN: "But you know, in those days, everybody begged to be a DJ, so they didn't have to pay them anything- there was 15 people standing outside the door, after their job."

GK: Was there a lot of people wanting to be KFAT DJs?

LN: "Absolutely. People were dying to get on the air at KFAT radio. And it was that way at KSJO, and at KNDY. It was very typical of radio in those days. They were not a union, now they're union. Now they're told what to play, they put them in a can, they're not even in a radio station when it plays. But in those days, the opportunity to be a DJ was wide open, so they could have gotten DJs and not paid them anything. And so Jeremy was the one who did that, and Leo was the one that did that, paid them $3.25 and said no, they weren't going to pay them another dime. It was not Bob McLain, it had nothing to do with how much money the sales people made."

Leo was still the GM, but he knew that Lori was on top of the game, both offices seemed functional, so it was time for Leo to back off from KFAT, maybe take the occasional call from Jeremy or Laura Ellen. Lori would make it work.

Chapter 95: Living Around

While Bob, Tiny and Terrell Lynn lived in Santa Cruz when they started at KFAT, Sherman, Gordy and Cuzin Al lived in San Jose, and Sully lived in Palo Alto. Jeremy and Laura Ellen lived in Los Gatos, but moved to Morgan Hill, the next town over from Gilroy. Then the rest of the staff moved into the area.

Gordy moved to Morgan Hill for a while, then moved to San Juan Bautista, a half-hour south of Gilroy, with Terrell Lynn.

Terrell Lynn had lived in a place with Sherman in Gilroy for a while. Sully took an apartment near the 7-11 in town. Then Sherman took up with Teresa, and that was when Terrell Lynn moved to San Juan Bautista. But Sully stayed there with Terrell Lynn a lot.

Tiny moved to Gilroy, then to a couple of places in San Juan Bautista, "down at the end of the road and by the lettuce field, with the geese crossing the road and stopping traffic," then to Morgan Hill. Rocket Man would live wherever he could, when he could.

Greg Arrufat moved into a house with the Ozoner in Gilroy for a while, who also lived with several of the staff, including Sherman and Cuzin Al.

Terrell Lynn left San Juan for Sully's apartment, and Travus T. Hipp moved in with Gordy, who remembered: "I'd get off at three o'clock and drive to San Juan Bautista- find Travus at the Mexican place and get fucked up on margaritas, or one of us would get a loaf of sourdough bread, still hot, and a half-pound of butter. We'd smoke hash and play chess and eat until the bread or the butter ran out."

Good times.

And, as the station and everyone connected to it was their world, as it might, romance would also enter their world. Sully dated Bob, but they never lived together. Tiny dated Bob, and lived with him briefly. A few times.

Sully and Terrell Lynn were a couple for a while, as was Sully with Rocket Man.

Sherman and Teresa were a couple in their tiny 800 square-foot house in Gilroy.

Cuzin Al had just hooked up with Lori Nelson, and they were now married, so Al left the house he shared with Rocket Man.

Travus had ladies coming through, but will not speak about who among the staff he has slept with. Tiny had the personal best in that department, as Gordy kept everyone up to date on that, informing listeners who Tiny had left with the night before, or who was still there from last night when he came in that morning.

It pissed Tiny off. A lot.

I mean- she did that a lot, and Gordy talked about it on the air a lot, and it pissed her off a lot.

Yeah, good times.

Chapter 96: Up In Smoke

In 1978, Universal Pictures was preparing to release the first movie by the popular counterculture comedians Cheech & Chong. These two stoners were a big hit on FM radio and the college circuit; the routines from their albums were imitated, referenced and quoted by hipsters and "heads" in high schools, colleges, malls, and in front of 7-Elevens all over America. Back then, you only needed to say, "Dave's not here, man" to elicit howls of appreciative laughter. Everyone knew the routines.

This was going to be a film for the hip stoner crowd that still populated America's most desirable youth demographic. The films were cheap to produce and easy to market, as everyone knew what they'd be getting with a Cheech & Chong movie named "Up In Smoke." With the comedians under contract for sequels, Universal strongly wanted the film to succeed so they could put out more of them.

The studio decided that they'd put a "push" behind "Up In Smoke." They decided to do something more, something special for this one. They wanted this film to be a smash.

When studios put out a major release, they use all manner and means to promote it. If the film is major enough, and the studio has so much money into it that it's worth spending a lot more on its release, they sponsor "junkets."

Junkets are trips to a hotel or resort to see the film and interview the stars. The locations are frequently in Los Angeles, Las Vegas, exotic spas and other hotspots. But those junkets are only for the representatives of the major media, and only then for a big enough movie. NBC, CBS and ABC, of course, were the major media of the day. A guy like Gene Shalit, NBC's movie reviewer, would get to go, but the guy who reviewed movies for the NBC affiliate in Des Moines could either see it at a preview screening for the local media or watch Gene's review on the "The Today Show" along with everyone else.

Local TV and newspapers were local media. Are you a film reviewer in Albuquerque? The studio will set up screenings in your city so you and the other local reviewers can see it –free, and before it opens to the public. Los Angeles? Chicago? You're off to Vegas.

The studios always wanted the media to review the film and interview the stars. Ads for a film cost money. The brief film reviews they got were free and would alert the public to the film's opening, and its merits or faults, but the interview segments lasted longer on the air, and free air time was free advertising. Interviews were better than advertising because it didn't look like you were selling a product, and if your old friend Gene Shalit got to lob softball questions to your star, everyone got warm and fuzzy feelings towards the star, and by extension, the movie. And if Gene thought it was important enough to put someone on the air, then you should think so, too.

But with "Up In Smoke," the studio wanted to reach an audience that wasn't watching the network morning shows, wasn't reading the reviews in the newspapers, and the studio needed them to know that the film was out there and want to see it. Universal had the money and the motivation for an unusually large promotion campaign, but where and how should they spend the money to reach this non-mainstream crowd? How, indeed…

Someone at the studio had an innovative idea: a junket for hip radio people. Why don't they send all of the right people from all the alternative media in the western states to one place, put them up at a nice hotel for a couple of days, give them a hell of a party, show them the film, and then make Cheech & Chong available the next day for interviews? It was decided that the idea was genius, and someone set about making plans.

My friend Bill MacLeod worked for the agency in San Francisco that had the Universal account, and it was he who arranged the screenings, interviews, media buys and media access to actors, actresses and directors for the studio. Universal told Bill that they had a big budget for this one, and they had been discussing what they should do with it. They told Bill what they had in mind, and it sort of shocked him. No one had ever done this before, and they wanted to do it in his backyard. Universal Studios had targeted two weeks before the film opened, and wanted to fly all of the film reviewers, Public Affairs Directors and talk show hosts—whoever would talk or write about the film—from seven western states into Oakland and San Francisco, put them up at the beautiful Claremont Hotel in the hills above Oakland, give them a huge party with unlimited food and an open bar, then send them in buses to the theater that they'd rented to show them the movie. The next day, everyone would have access to Cheech & Chong for 10-minute interviews that were to be scheduled before they arrived at the hotel.

The night after the interviews, Saturday night, they'd be on their own, and Sunday morning the drivers would pick them up and take them back to the airports.

A junket for radio people? No one had ever heard of one before. This was very big news for all involved. Usually, these people got a poster from the movie and maybe a T-shirt. No one had ever been driven, much less flown anywhere, much less put up luxuriously and treated so extravagantly. It was going to be a ripping time.

Bill MacLeod's travel coordinator completed the arduous work of coordinating with the hotel and all the airlines and drivers, scheduling arrival times with ground transportation, and by four that Friday afternoon, all had arrived, checked in, and begun wandering around and meeting their counterparts from other stations, other states. There was a lot to talk about, and everyone was almost giddy with excitement. The cocktail party was set for six that evening, and dinner would be in the ballroom at 7:30. The buses would arrive at 8:30 and at 8:50, take them one mile down the hill to the theater; they'd be in their seats by 9:05, someone from the studio would say a few words and the film would start at 9:15. It was a great plan, it really was. What could go wrong?

These people were young, underpaid, unappreciated members of the fringe media, and they had never been given anything like this. Before this, they had barely been noticed, and who knew when- or if- it would happen again? So they dove into it, and indulged themselves greedily.

They drank like fish at the open bar. Free drinks? Why, thanks, I'll have another.

Many a person was seen to disappear in ones, twos and threes outside to the balcony for several minutes, as the odor of marijuana wafted gently back in through the open windows of the ballroom.

Hors D'oerves? Thanks, I'll have a couple and just stuff them into my face. Wait! Don't go yet; I'll just wrap a few in this napkin here and put them into my jacket pocket for later. Dinner? I'll find some room for the chicken. Is that cake? Yes, I'll have some…

They availed themselves heartily of the bounty, stuffing their mouths continuously and copiously. Just short of outright bacchanalia, the eating and drinking at this affair was an impressive demonstration of gourmandic excess.

In fact, I was shocked and disappointed to find that no single word exists to describe such a heroic intake of food. I believed that this sort of gastronomic over-indulgence was worthy of its own word, so I made one up for you.

Haven't you ever made a fool of yourself with food? Pizza? Ice cream, fries, nachos? Anybody? When you eat until you hurt yourself, explaining it is easy: you were hungry and you were stupid. It's not a colorful explanation and it bestows no glory, no sense of accomplishment on anyone. Your stomach is twenty minutes behind your mouth and it got you in trouble again. But what if you ate and ate and ate to a point of… *almost* excess? What if you ate right to where pleasure ends and discomfort begins, but you *did not cross that line?*

Wouldn't that be something to feel good about? To be proud of? Wouldn't you want to tell others about your accomplishment? But what would you call it? Such control, such discipline, such foresight should have a name. But it didn't, until now.

I call it Gastroexcessiphoria.

Eating as much as you want of whatever you want, and still feeling good afterwards is now: Gastroexcessiphoria. Think about it. You're welcome. But we were talking about the fools that ate too much and drank too much at that party thrown for them by Universal Studios. My God, they made pigs of themselves…

Then, at 8:30 sharp, it was announced that all would now board the buses for the trip to the theater. Everyone was in a jovial mood, bursting with an excess of good food, unlimited good booze and good cheer, and ready to laugh at the clown princes of pot comedy, Cheech & Chong. Okay!

Which was when it all went to hell.

There were two buses, and the driver of the bus I got on had gotten to the party with his bus before the dinner, parked behind the other bus, went inside, and, as no one was checking, had helped himself to several drinks at the open bar. He had helped himself to enough drinks to get lost on the short, direct trip down the hill to the theater.

If the driver had exited the hotel grounds and turned left (which was the only way he could go), all he had to do was to stay on that road down to the bottom of the hill. If he had stayed on that road all the way to the bottom, he would be at the theater that Universal Studios had rented for the night. Unfortunately, the besotted driver decided to take a turn where no turn was called for, turned again, and we were stuck in a cul-de-sac, unable to turn around or back up.

Being inebriated does not necessarily preclude the use of one's motor skills, and eventually, with the help of a few of the passengers who had to get out of the bus to shout things like, "Back! Back! Stop! Turn it to the left! Back! More! Forward! Back!" the driver got the bus back on a recognizably downhill route to the theater, and thus we arrived, 45 minutes later.

Of course the screening had to be delayed until we got there, and those in the theater—maybe for fear that they might go outside to smoke pot and drift away—were asked by a studio executive to sit in their seats until we showed up. In those soft, comfy seats with arm rests… after a long day of travel and serious use of food, booze and whatever else… So, after a happy, jocular few minutes, the mood in the theater declined from jovial to restless, then to sleepy... and within a half hour, many of those in the theater began to doze. After a fifteen or twenty-minute nap, the mood of the crowd was boosted significantly as the second busload staggered and trooped into the theater amid the cheers, jeers, and good-natured banter from those already there. It was pretty funny, and the energy level of the crowd looked good again. When we were all finally in our seats, someone from the studio made a few announcements, welcoming us and reminding us all of our appointments to interview the two stars the next day, telling us how excited they were to meet us, then the theater darkened and the film began.

It was now past ten o'clock at night, after a long day of travel and gluttonous consumption. They had all, with the exception of Bill MacLeod, his staff and the studio executives, been drinking heavily and eating prodigiously, then settled in a warm, dark theater, and asked to sit back, relax, and watch a movie. Within minutes after the film started, many of those in the theater were fast asleep.

I don't know this for a fact, but Bill and his crew, and the studio executives there must have thought there was a problem from the lack of response at key laugh lines, and they must have been nervous the next morning, when the interviews were set to commence. And by breakfast, Bill knew there was a problem.

Headaches and hangovers abounded, and few came down for breakfast. Then, a couple of people with an appointment to interview Cheech & Chong canceled, and then some others did, too. Too many people had fallen asleep and missed the entire film, and had no questions for the stars about the film. Everyone who called to cancel had a story, and everyone thought that they were acting alone. Soon they all knew what was going on: not everyone knew what the movie had been about, and no one wanted to go into the room and embarrass themselves in front of the stars or the Universal Studios executives. All in all, a few people thought they were acting alone and did the same thing. Bill still had a lot of interviews on his schedule, but not as many as he'd said he had in order to justify this junket. And Cheech & Chong were flying in from Los Angeles and were anxious to get it over with. Bill was frantic.

This wasn't something that you could explain to Cheech & Chong, and this clearly wasn't something you wanted to explain to the studio suits. Bill needed a few extra bodies to interview these guys, and he called me at home that morning for a favor.

Would I still be doing my interview today? Yes, of course. Then he explained his predicament, and could I bring a change of clothes, and come back a few hours later to interview them again as someone else?

Bill knew he needed to account for every interview he'd scheduled, and he was calling in favors. Local guys, of course—radio people, writers, friends—anyone with a tape recorder or a notebook who knew how to conduct an interview without embarrassing Bill. He thought it would work, and it did. For my first interview, I wore jeans, boots and a white western shirt, and then later I had different glasses, khakis, loafers, a hat and a blue broadcloth shirt and a sweater, and different questions. I wore the hat low and tried not to look at them.

For my second pass at Cheech & Chong, I tried to act in a sullen manner. I knew that I couldn't pull off a phony accent, and I wouldn't risk embarrassing

Bill by blowing it and giving the scam away. In the end, it worked. Stars all know that to publicize their movie, they are going on the road to answer the same questions repeatedly and meet a hundred people that they will never see again and don't need to remember. And it would help if they pretended they were interested in the questions.

After the stars do their work, it goes to an editor, and they don't usually see it again until it opens some months later. Generally, stars hate the publicity part of the business, but they do it because it's part of the business and they must. Usually it's in their contracts to do these tours, but no one likes it. The same stupid questions, asked every day by the same people with the same jobs everywhere in America. They just try to get through it. I'm going to imagine that Cheech & Chong might have used some form of herbal relaxant to get them through the repetitious ordeal of the interview cycle. Neither of them looked at me strangely during the second interview. They had no memory of my having been in that room a few hours earlier, asking them questions. We did the second interview. They had no idea.

To this day I don't know how many friends Bill called on for the hustle, and we never discussed it again, but I guess I pulled off my end of the scam. Or maybe it was the pot.

Chapter 97: Chuck Wagon & The Wheels

There were lots of bands that were played only on KFAT, but there was only one band that was *identified* with KFAT.

Formed in 1977, the name Little Dickie and the Geldings was considered too offensive by the media and club owners, who wouldn't book them until they appeared under another name, so they became Chuck Wagon & The Wheels. They played a hot, irreverent rock 'n' roll that was closer to country, but their attitude made it rock. They played the clubs in Tucson, but they knew they were too irreverent for a major record label. They knew the songs they played would keep them out of the larger venues, so they formed their own record label, recorded their songs, and went looking for someone to play them on the radio. "Red Hot Women And Ice Cold Beer," "How Can I Love You (When You Won't Lay Down?)," "There's Too Many Women To Love Just One," were in their show, but it was "Disco Sucks" that got them their first radio airplay.

Disco was the raging, cross-cultural phenomenon, and it was all over the radio, leaving a sizeable group of disgruntled listeners with little to listen to. "Disco Sucks" spoke for many, and stations that wouldn't play their other songs were playing it. In fact, there were exactly thirteen stations in America playing it. But just that song: no station anywhere wanted anything other than "Disco Sucks" from Chuck Wagon & the Wheels.

Chuck Maultsby wrote me: "So by God, I mass mailed out something like 200 copies to country stations, starting with the western states. KFAT just happened to be on this Billboard list I got hold of. Of that 200, only 13 played anything at all, and only Fat had us in any "rotation" at all. 'My Girl Passed Out In Her Food' was the one…

"I had never heard of KFAT 'til I guess it was Bob [Simmons] called, and told me they were playing us, and we should come up and play. I said O.K. and we did, and that's how we got sucked into the vortex. 'Disco Sucks' amplified everything! KFAT played us and almost no one else."

They drove in from Tucson, and they thought they'd stop in at KFAT to say howdy.

Chuck went on: "I will try to convey to you the astounding culture shock that me and the band experienced the first time we visited the Fat studio in Gilroy. Imagine six young men from super-conservative Arizona, who's prior experience visiting radio studios was a galaxy away from what we walked into one summer morning in Gilroy (after an all night drive)... and even before we got to the studio things were taking a turn for the bizarre because of the heavy, almost nauseating smell of garlic in the air (how could anyone live with this?), and we had tuned into the station while driving.... what the hell kinda station is this, and what in hell are they playing? Who is Jesse Jeff Winchester and what in hell is an Idi Amin Hot Tub... I felt like Willard going upriver to meet Col. Kurtz... and when we arrive in downtown Gilroy (which didn't look none too prosperous) we go up a rickety flight of stairs to a door with a Fat sticker on it (and that's all)... we knock... no answer... so we open the door to look in and the place looked like it had been fragged... no receptionist, but there was a guy sleeping soundly under a desk... we go in further and see a guy (through the glass) in the control room and he sees us and he extends a large pipe and gestures that we should come in and have some (and I'm sure he had no idea of who we were)... we introduce ourselves as Chuck Wagon & The Wheels and he spins around, removes the record that's playing and replaces it with 'My Girl Passed Out In Her Food.' We were flattered that he did it and shocked that he did it (wasn't that against the rules?). He never introduced himself (we supposed that we were supposed to know his name, so we let it ride)... about this time another guy walks in and does introduce himself as Terrell Lynn (I'd never heard a name like that… especially for a man). He was a very friendly fellow and he invites us to "go up on the roof" where he produces a pipe (here we go again) packed with some killer stuff that's out of our league and he also invites us to help ourselves to some cans of beer in an ice chest full of luke-warm water... it's ten in the morning! (do people actually drink before noon around here?)... and on this roof are a couple of sleeping dogs and a chicken! A chicken!! We knew we weren't in Arizona anymore..."

KFAT was playing the hell out of their album, as almost all of the songs were hot, fast, funny and Fat, and later that day they struggled to get out of Gilroy and up to Palo Alto for the gig that night.

Chuck wrote: "We were amazed at the crowd that was at the Keystone.... and on a Monday Night!!! We were doubly amazed at the reaction of the audience to everything we did- especially 'My Girl'... they cheered and danced and partied like it was New Years Eve!!! We knew we had found a vortex and we embraced it... we lost a lot of money on that first little Bay Area tour... but it was worth it!!! And we kept working it until the Bay Area was our most lucrative "market"... KFAT gave us strength and encouragement (not to mention those Nor-Cal chicks were ... ah... really 'fun' girls, with the best weed!!)."

Fatheads always loved Chuck Wagon & The Wheels, and their shows were always well attended. I asked Chuck if he had a favorite Fat Fry, and he wrote, "One time we are doing a Keystone show with Elvin Bishop (there's some more stories) and we are outside puffin' when this long, black, window tinted caddy limo pulls up. Now wouldn't you figure someone like Jerry Garcia or Carlos Santana or Grace Slick or somebody like that had arrived ... we did. But no... who piles out of the limo but a half-dozen Oakland Hells Angels! Now I'd long before gotten over my youthful terror of the Hells Angels, having been in a band early on that contributed to a Hells Angels movie sound track... had met some of 'em and found them downright friendly and courteous (I'm sure as long as you don't antagonize them)... anyway I go up to one who I thought I remembered as Terry the Tramp and asked him if he remembered me and he answers: No, fuckhead (or something like that). OK... but I'm curious and stoned so I ask: How is it that you Hells Angels guys are riding around in a limo like rock stars? And he answers: Cause when we go out to get fucked up we are driven...no scooters... dig? I'm thinking of course, makes sense... then I push it one more time and ask: How do you guys afford a limo? ... At first he gave me a look like I was the stupidest thing he'd ever laid eyes on ... then he grinned, and without saying another word he reaches into a pocket and pulls out a little packet of white powder, hands it to me and stomps off (these guys don't walk... they stomp)... so... about three days later the Crank starts to wear off... "

I told Chuck that one of my favorite Chuck Wagon songs was "Red Hot Women And Ice Cold Beer," and that I was thinking of putting it on the CD I was planning on releasing at the same time as this book, and he said that that would be fine with him, but it wasn't a Chuck Wagon song, and he'd prefer me to put on one of theirs. I was surprised it wasn't one of theirs, as I'd certainly never heard it anywhere else, and they'd clearly *made* it theirs.

Chuck wrote: "Red Hot Women wasn't ours... we lifted it from The New Riders' 'Who Are Those Guys' album... it was written by Cy Coben (he's still dead) Delmore Music Co. A.S.C.A.P... Cy Coben (an elderly, diminutive, old school Jewish gentleman from New York City- by way of Atherton) pulls up in a Cadillac 'El Do', he calls it, to the load-in door at the rear of The Keystone Palo Alto, where we are loading-in for a Fat Fry show- we were opening for Asleep At The Wheel... he's perturbed and demands to know who is in charge of "this wagon and wheel combo"... it was our Marx Brothers-like routine, in those days when asked that question, to point at one another and say him... him... him... so he gets more perturbed and explains (loudly) that he is the author of "Red Hot Women And Ice Cold Beer" and he had heard it on the radio and wanted an explanation as to why his publishing co. had not received a "notice of use advisory" and he demanded an accounting of record sales.... well I was 'the business guy' and didn't have a clue what a 'notice of use' was or much else of anything in those days… anyway... we asked him to calm down, nobody's trying to screw ya, we're busy right now, gotta show to do, stick around, see the show and we'll hash this out after... which he did. Old Cy, I guess by asking around (who, I'm not sure), figured out that we're a small-time outfit, on a puny label (our own), with sales in the 'ones of thousands' and he probably wouldn't get any royalties past what it would cost to fill his 'Eldo's' gas tank... so before leaving (after our show) he pulled me aside and said he very much enjoyed the show, he'd had fun and met a lot of nice 'kids' and said 'don't worry about the monies due... it's on the house and good luck to you boys'... that was one of the more gracious things I've experienced in the music biz'.

After Cy Cobern left, Chuck was approached by a guy who said that he'd liked them, and asked if they had any management, and if they'd be interested in some "professional guidance." Chuck was handling things in those days, and figured the guy for another stoner who just wanted to party with the band, that this was bullshit, so he "thanked him for his interest" and sent him on his way.

A little while later, one of the guys from Asleep At The Wheel asked him what he'd been talking about back there with Joe___. "Who's Joe ___?" Chuck asked. The guy said that Joe___ was a high-powered artist's manager, with all kinds of connections and successful acts signed to major labels.

"So a couple of weeks later, when I figure out how to get hold of him (that's how good at management I was)... I call and he says he's a bit busy…" and Chuck never heard from him again.

Chuck Wagon & The Wheels became an integral member of the KFAT family. In addition to playing more Fat Frys than anyone else, they also recorded several KFAT promos, which the station used for years. They were very much identified with KFAT, and happily so, until the end of Part Two of this book, when things went all to hell, and Chuck Wagon & The Wheels sided with the DJ's instead of the management, playing a gig to benefit the DJ's, pissing off the management so that their songs were eliminated from the KFAT playlist by an angry, embittered Leo Kesselman. But that's later.

One more note: there are two photos of Chuck & Co. in the next Photo Section: one on the roof, and one in the Production Room. The careful observer will notice that in both photos there is a pipe in Chuck's hands, but a different pipe for each photo, even though it is clear that they were taken the same day. It's easy to see in the Production Room photo, but in the other one, look carefully at the extreme lower left corner of the photo. It's a bong, and Chuck's face is obscured by smoke in the shot. I think the smoke might have come from the bong, but that's just a guess.

Chapter 98: The True Blue Chew News

Everyone at KFAT was a wit. It went with the territory. Some thought they were wittier than others. This is a chapter about me.

After a year, my sponsor, the Sportsman's Shop, declined to renew my contract, and the sales staff sold my show to Sierra Designs, who made high-end outdoor wear. With the new contract came a 17% bump in pay, from $75 to $88 dollars a week. Imagine my joy.

I'd always thought I was funny. Maybe not professional comedian funny, but I didn't know because I'd never tried, and people had always told me that I could. Now I had a radio show, and I could, and I would, and I did.

I thought <u>my Chew promos</u> were clever and entertaining, and I took a great deal of pride in them. I saw making the promos as the most fun part of the job. As I listen to the promos now, I hear the wit and enjoy the humor, but they all have the same problem. I was working at two jobs a hundred miles apart, and one of those jobs was a) close to home, b) enjoyable, and c) paying almost all my bills. The other job was a) a hundred miles away from home, b) took up a *lot* of my time, c) made me work through the night *at least* once a week, so I was missing a lot of sleep and d) paid me only $88.00 a week.

Between the two jobs, the commute and my relationship with Susan and my life, I was missing a lot of sleep, and I can hear the exhaustion in those promos when I hear them now.

Because of my over-lapping schedule, I needed to do one less show a week, but I was under contract for five shows. I spoke with Laura Ellen and the sales rep, and an accommodation was reached. One show a week would be a novelty show, and one would be a repeat. The novelty shows could be anything entertaining, like a fifteen minute episode of a classic radio show from the 30s or 40s, like "Flash Gordon," or "Burns & Allen" or some other unusual fifteen-minute offering that only needed an Intro, an Outro, and an ad tacked on at the end. From now on, the Thursday show would be called the "Potpourri Chew." On Tuesday's, the repeat Chew would be even easier, as I had all the old tapes and needed only to change the ad at the end, and together these changes in my responsibilities to KFAT meant two less interviews, two less editing sessions, maybe one less trip to Gilroy each week, and some sleep.

Jeremy gave me an old Revox reel-to-reel deck, so at last I could do some work at home, and that was a relief. But I thought I was as funny as some of the guys out there, and I had a show on radio. I wrote a comedic newscast, along the same lines as "Weekend Update," which been a huge success on NBC's "Saturday Night Live."

A friend sent me a cassette of the same teletype sound effect that played in the background on "Weekend Update."

I wrote a bunch of phony stories as a newscast, recorded it in the Production Room, adding the teletype background, and introduced it as the "True Blue Chew News." It was okay, but on recent listening, I believe I have found the flaw besides the exhaustion in all of those phony news segments. With no spare time whatsoever when I was awake, I had no time to write, rewrite, get feedback, reflect on it, and rewrite again the material for the show. In writing this book, I was surprised to learn how many rewrites are necessary to get something... *just right.* A punchline just barely mis-written will bomb, while with a change as small as a different word or inflection, it might kill. Ask anyone.

I found that not only had I no time to write and reflect, much less rewrite these shows, there were times when the writing was happening in the studio minutes before I recorded it, which was usually right before it aired. Sometimes it was really close. I remember the times driving to Gilroy late at night with a clipboard propped against the steering wheel, pen in hand and writing a comedy bit while I drove. It had to get done- my show was on at 8:30, I had nothing else, and the show must go on.

It was fun for me and I hope they were well received, for all the work I did. I had a conversation about this once with Don Novello, also known as Father Guido Sarducci. Knowing that Don was very connected to "Saturday Night Live," I told him about my phony news show, and about my frustration with never having time to prepare a show adequately. Don told me that at SNL they had 15 or so writers. The show was 90 minutes long, and after commercials, actual air time was under 70 minutes. The show always featured two songs, so make that about an hour of comedy.

That means that, with fifteen writers on the show, all a writer there had to come up with was an average of four minutes of comedy a week. That was all they had to do that week, and they had all week to do it. They also had other pro's to work with, and all of NBC's staff and resources to provide costumes, props, sound effects and video clips, and they had makeup and lighting and special effects, and engineers to work with, and directors to direct it, and experienced, professional actors to perform it, and even then, they all worked only a few weeks before their next break. And the gig paid well. And if a writer didn't come up with something that week, others always stepped up with *their* material to fill up the hour. And they all had others to work with, to bounce ideas off of and act out sketches with. And if a piece wasn't ready, you worked on it and tried again the next week.

My resources were more limited.

In any event, I put out about ten of these comedic gems, and I will try to get the author to put some of the material on the website that comes with this book.

One story on "The True Blue Chew News" caused something of a stir. I had done several of these by then, and everyone should have known that it was humor. It was KFAT, for God's sake!

Does anyone remember Paraquat? It was a defoliant used by the Feds to spray on marijuana crops, causing the harvested crops to have a painful harshness when smoked, and who knew what side effects there were to exposure to it?

I had written a story for The True Blue Chew News about a story "appearing in the wire services" that it had been discovered that the Paraquat scare had been a hoax. According to the story I made up, it seemed that soon after taking office, President Jimmy Carter had wanted to make good on a campaign promise that he would reduce the illegal drug trade. In reviewing his options with advisors on how to accomplish this, one of the suggestions was to circulate a false story about a chemical being used to eradicate the marijuana crop, convincing smokers of the illicit weed that they were in endangering their health by smoking it.

President Carter immediately realized that they could plant the story, and if the dis-information worked, they would have reduced growth, production, trafficking and consumption of the weed without having spent a cent of federal money. Carter's advisor's liked the idea, it was immediately implemented, and within weeks it was estimated that over 60% of the marijuana trade had been eliminated.

My story went on that now that it was out that Paraquat never existed, that marijuana smokers who'd thought that their weed smoked harshly because of Paraquat were in fact only getting bad weed from their dealers, and should take it back to them, demanding a refund or an exchange for better weed.

All of this was spoken in my best newsman voice, with the teletype clacking behind me, but I also wanted to make sure that no one could possibly take it seriously. To ensure this, I sandwiched the story between two stories that had to let even the most casual of listeners know that this was a comedic bit.

The story that came before it was that that day marked an important anniversary, as it was exactly four hundred years to the day since the discovery of the alphabet. The story continued by recounting that the alphabet had been discovered in an attic in a castle in Switzerland, and had spread across Europe and elsewhere rapidly, and then enumerating several of the things whose development had followed in the wake of the discovery, such as contracts, bills and Hallmark cards. It couldn't have been real, and it probably wasn't even funny, right?

Another, a more ridiculous story, followed the Paraquat piece. When I was finished producing the segment, I put it in the "CHEW-show" slot in the studio and left to go to work, a hundred miles away. Later that day, all hell broke loose.

Some listener had heard the story and was alarmed, mistaking it for real. They called another radio station to ask if they knew anything about it. They didn't, but by the time the story got to the third or fourth radio station, the fact that the story had originated on KFAT was out of the story, and the next station or news agency that was called heard only that there had been a report of a massive deception by the Federal government regarding Paraquat. Much panic and confusion ensued, until the chain of circumstance was finally traced back to KFAT, and calls were made to the appropriate authorities there. Jeremy apologized, explaining that it was a humor show, and said that he was surprised that anyone took it seriously.

But by that time, every radio station in northern California with listeners under 50 was running announcements that their listeners might have heard a story about a hoax regarding Paraquat, that Paraquat was not harmful, but that story originated as a parody on another radio station, and it was not true. They repeated that Paraquat was a dangerous chemical, and should not be inhaled, and that it had happened on *another* radio station.

Jeremy immediately offered to add KFAT to the list of stations warning its listeners, and the conversation ended. By this time, Jeremy had called me to find out exactly what I'd said, so Jeremy could react to it. When I called him back later, to see if I was fired, Jeremy told me that it was all settled. Jeremy thought it was great.

The True Blue Chew News had only one other problem, and that was internal. KFAT's only, but not really, rival on the dial was KSAN, the powerful rocker up in the City. They had an actual news guy, and that was "Scoop" Nisker. He had the hip sensibility necessary for hip radio, he was a veteran newsman, and he was serious about it. He had done the news on KSAN since its heyday in the mid- 60s, and he was a respected figure in underground radio. Scoop used to end his news shows with the tag line: "If you don't like the news, go out and make some of your own." It was a well-known tag line in the Bay Area.

I was ending my "news show" with the tag line: "If you don't like the news, go out and make up some of your own." And that pissed off Travus T. Hipp. We've already talked about Travus; you'll understand that he was a genuine, if blustery, self-important but bona fide news man, and he took his news VERY seriously. When Travus heard "The True Blue Chew News," he didn't much care for my casual mistreatment of the news. And now he was insulted for his brother news man, Scoop Nisker, who he thought I'd insulted by using his tag line that way.

Travus complained to Jeremy about it, yelling up a storm of indignation until Jeremy called me at home and told me about the problem. I defended my position, defended my right to do a phony news show, and offered to speak to Scoop personally about the deviation on his long-time tag line. Jeremy liked the compromise, and Travus said he'd accept Scoop's response if I did, too. I called Scoop, who thought the fracas was unfortunate, but he had no problem with the use of his line. He said he thought he was a little flattered by it, but he wasn't sure…

In any event, Scoop was cool with it and asked me to give Travus his regards. Travus was graciously placated, and the matter was settled.

To this day, I'm not sure how well those shows were received. I got no feedback on those shows, either from the staff or the listeners, but those shows were hard work, man!

Chapter 99: Collections

By now, you will have heard of KFAT's perpetual inability to pay its bills. Somehow, some way, they would have to be paid, or the electricity would be cut off, or the phones shut off again, or the staff would walk out, or who knew what might happen? So, if this was a chapter about a bill collector, you would assume that it's gonna be some story of someone threatening Jeremy or KFAT over an unpaid bill, but you would be wrong. This is a chapter about Jeremy sending someone to collect the money that was owed to him, and too long overdue. Jeremy had an in-house collection agency, and it was called Raymond.

Raymond was the black ghetto kid that moved with Jeremy and Laura Ellen from St. Louis, and he was now the only black kid in his Los Gatos high school. He had grown, with proper and constant nourishment, into an over-sized, hulking young man.

To collect the money, Jeremy encouraged Raymond's most fearsome countenance, and forbade Raymond from bathing for several days, and told him to wear the grungiest, most disreputable- looking clothes he could scrounge from the Salvation Army or, on at least one occasion, a dumpster. Raymond's orders were to go to the client's office, hand the bill to the receptionist, and ask to see the person who signs the checks. Jeremy told Raymond to inform the receptionist of his intention to sit there until the bill was paid. If the bill was not paid immediately, he should sit tight and wait patiently. If told that the check-signer was not available, he was to offer again to wait, and in fact, to insist on it. Raymond was told not to threaten anyone, but looking threatening was desirable.

Later, when Lori Nelson came aboard, her son was big and athletic, and he would go with Raymond on these collection missions, and Jeremy gave them a healthy share of whatever they recovered.

Here is what Lori remembers of the KFAT Collections Department:

LN: "They had a non-commercial station in St. Louis, and one night Raymond came in and stole a turntable, and Jeremy slept there, and saw him do it and beat the snot out of him. And when he discovered he was a kid, he felt *so bad*, because Jeremy loved people, and he tried to help everybody he could. So he started helping this kid, and then eventually they adopted him.

"Raymond worked with my son Paul, and I hired and I let him have a percentage of everything he collected. They would sit at Mervyn's for five and six hours at a time to collect what they owed the radio station."

GK: I remember hearing that Jeremy told Raymond not to bathe before making a collection call.

LN: "Exactly. And my son was with him. My son was a big guy, and they'd work together. Mervyn's was one of the worst. And we would say 'No Money, No Air' and we'd shut their commercials off. And so they'd pay a little bit. Well, finally we had so many receivables, that we told Raymond and Paul to collect for a percentage. So Raymond and Paul made a lot of money. And we collected [from] a lot of deadbeats. And if we couldn't collect money, we'd go in and take merchandise... took it physically."

GK: How many times do you think that Raymond and Paul went out on such calls?

LN: "They worked probably every summer. Raymond was still in school when he worked there, and so he was there during the summer months, and they'd go every day of the week and call on a receivable.

"Dozens of such calls, and if they couldn't get money they got merchandise. And the salesmen were so interested in making the sales, but didn't follow up on collecting the money. So I had a computer report on who they were and who the contact was, and I'd give these lists to Raymond and my son, and they would go calling on these guys, and they'd bring the checks back, and we'd record it and give them their percentage of it."

They were big, dirty, scary, and motivated. They succeeded admirably. This was how you collected overdue bills in a Jeremy Lansman Operation.

Chapter 100: Quotidian

Quotidian is such a good word. It sounds high-falutin' as hell, but it only means "daily." You could use it to impress someone, but they'll probably think you're a pretentious jerk, which is why I'd never use it.

Daily life at KFAT had a sort of routine, as you would expect. Jocks came in, did their various jobs, hung out and left. Hanging out usually meant in a lounge or some neutral corner, but every once in a while, Tiny liked to hang out with the ladies of the office. Of course, they had their jobs to do, but this was KFAT, and there was always time for some idle chat. And when there wasn't, Tiny could still hang out in the office and watch life go by, one story below her on Monterey Street. Once, she was sitting in the window and saw the bank across the street being robbed.

Another time, it was hot and the hoped-for evening breeze hadn't shown up yet. She remembers, "it was really hot, and it was night, and I was sitting in the window looking out, and I hear this big crash, and there was this little appliance store that was underneath us to the side, and a Mexican grabbed a TV and ran across the street and ran down the alley. He broke the window and grabbed the TV and hauled ass...."

She felt sorry for the guy; she knew how he felt because everyone at the station was broke, and everyone there knew something about desperation.

Sister Tiny remembers: "Trade-outs! We were so poor. We lived by the trade-outs."

ST: "One time, it was so stressful 'cause I was so broke, and I had this hoppity horse for the kids to play on. It was one of those big balls with a horse head, and they could bounce on it without making any noise. And this Mexican chick steals the hoppity horse! It was at the bottom of the stairway, and I walked by and I noticed that it was gone. So I ran down the stairs and I looked both ways... and I had just seen it down there... and I looked and I didn't see it, so I ran down. I was just... it was *the end* for me... and I went down into the bowling alley, and there was my hoppity horse sitting under this chick's legs. She was having a hamburger. And I said, 'Is that my hoppity horse? You just pulled it out of a doorway, didn't you?' Because it had Finnian's name on it!

"And she kind of shuffles and she goes, 'No, it's mine.' And she had some shopping bags, and it was like, 'yeah, you dragged your hoppity horse with you to go shopping, did you?' I couldn't do anything about it and I was just (teeth grinding) nnnrrrggghh!

"So I walked out, and I got just a little ways down the street and I thought, 'Fuck this!' and I went back to go get it from her, and she didn't have it. She had hurried outside and stuck it behind a doorway. I knew for sure she was busted. No- I didn't find it."

Sister Tiny's rage had been building from a simmer to a boil; all those years of struggle to keep her kids alive and fed and clothed, and now two years at this goddam job for no money! She was sick of it.

"I said, 'Where is it, you bitch!' and she stands up and '*WHAT?*' and we're, like, scrapping in the middle, her and her friend, and we were going for it. She had my shirt torn off and we were like *fightin'* in the café of the bowling alley."

GK: Did you ever get it back?

ST: "Yeah! 'Cause right after we got into the fight, they broke us up and they were going to call the cops and everything, I went out the door and I looked up and down the street 'cause I knew she stashed it somewhere, and I found it in one of the doorways. I grabbed it and went back to the station."

Sister Tiny: 1

Skanky Thieving Bitch: 0

ST: "I got a phone call once after playing 'Johnny, Johnny, Johnny, Why'd You Beat Up That Queer?' which I never played again, from somebody in San Francisco, and he was screaming 'You hetero bastard,' and I was scared, and I locked the door for the first time. I was scared!"

"And someone else called me and said, really calmly, that they were gonna come up, and it wasn't for that song, it was a different time, I don't know what I did- I did nothing wrong, it was just some real creepy guy who said, 'Tiny?' and I said, 'Yeah', and he said 'I'm gonna come up there and bash your head in with a baseball bat.' It was really creepy."

Sully: "And remember… anyone could have pushed in."

Terrell Lynn was living in his truck in the back parking lot, and asleep the time he heard something like a scraping sound. Then he saw a crowbar come through the latch of his camper shell, and the door of the shell popped open. Knowing that he was in the middle of a robbery, Terrell Lynn grabbed a baseball bat that he had for just such an occasion, and screamed at the guy to "Get the fuck away from there!"

Outside on the back deck, smoking something, Sherman saw Terrell Lynn, completely naked, running down the street, waving a bat over his head and screaming at some unseen enemy. Sherman hadn't seen anyone else, and didn't know if his friend had been attacked, or had gotten some bad drugs, or had just lost his mind. Being of a casual disposition, Sherman elected to not get involved, thinking he'd wait to hear the explanation, if he remembered to ask. He also remembers Tiny chasing a guy down the alley with an axe after a guy tried to break into *her* car.

Terrell Lynn remembers, "Christmas day I had a brand-new battery, and I watched my truck the whole fucking shift, and it was gone. I wasn't even pissed. I thought, 'if you were *that* good… ' "

Readers will remember that some Mexican guys liked to make their cars bounce. Front end, back end, up and down, you've seen it. It takes batteries to do that- several per car, and so there was always a need for batteries in Gilroy, and stealing them from the cars parked behind KFAT was one way to get them.

ST to Sully: "And how many batteries did you lose out of your Mustang? Three or four?"

Sully: "Yeah, three or four."

Still, Terrell Lynn says, "I always parked in back," because parking in front of the station on Monterey Street had its own perils.

GK: Did you guys get a lot of parking tickets?

ST: "Oh, boy, did we! My uncle, who lived in Morgan Hill, he finally paid for like, 150 of them. From parking in front. 'Cause if we parked in back our tires weren't on. So we parked in front, and I would go to sleep, and by the time I woke up in the morning, and moved my car to the back…

Don't even *think* about how many tickets Buffalo Bob got, or how much he still owes on them.

Lori remembers the day the cops made a sweep and rounded up everyone on Monterey Street who was involved in an illegal gambling operation. Lori didn't know that the Green Hut was Gambling Central in Gilroy, but she remembered, "one day I was in the second story window (Office 2) and there was a Paddy Wagon sitting on the main street of Gilroy, and they were taking people out of the businesses in handcuffs, and putting them in the Paddy Wagon. There was probably about 40 FBI people- taking store owners out of their shops, padlocking their front doors, and putting them in a Paddy wagon.

"Well," Lori said, "it turns out that there's bookies going on down there, and they've been watching them for a long time, and they arrested them and threw them in jail. There were so many shops on that street where the owner was arrested and their front door padlocked. Apparently, the whole town was involved in it, and they made a lot of money off of it. And there was about 5 or 6 Paddy wagons, and they had them full of people."

Sadly, that also included the Green Hut. Poor Steimie. Poor Buffalo Bob.

TLT: "But that's where a lot of the town political dealings were done. And they came in one morning and they arrested *everybody*. And the town looked like Sunday morning at six a.m. And Sherman and I were walking around and going, 'what day is this?' cause there was no one on the streets, the stores were closed. They came in and got everybody."

LN: "… the DJs started realizing that they were going to come up here, and they're gonna find marijuana seeds everywhere. And so the DJs were in the main part of the control room, sweeping seeds away from the control board with the albums, and there were marijuana seeds everywhere, and they were cleaning up their act like crazy."

Sherman: "I got to the station 10:30 one night- I had to be on at 12, and Frisco had me read some copy and we got it in one take. Frisco said 'let's go to Augie's,' and we did, and we were having a beer, and a fight broke out among the Mexicans, and they started looking at us white boys. And we were drunk, having had a couple shots of whiskey and some beers. Frisco bought a 6-pack of Mexican sodas –Peñafiel- and I bought a 6-pack of beer, and we crawled out of there."

Sherman still remembers the time the gun store, six doors away, got broken into. "Someone stole the guns, and the cops thought it was us." He was taken to the police station, questioned and let go, but still…

Sully remembers the day someone brought a case of beer to Buffalo Bob, his shift was over with half of the case left, and he took it all with him out the door. Sully thought that taking the beer with him and not leaving any behind was rude, and mentioned it over the air as Bob was leaving. Of course, the Gilroy cops were listening to KFAT, and they thought it was rude, too. And knowing Bob as they did, it was an invitation to illegality. They stopped Bob on his way home and found an open beer. Being friends of the station, they took the beer and let Bob try to find his way home. But he was pissed at Sully, and she would be on her guard until he forgot about the betrayal. And the cops kept the beer, too.

I asked the jocks if they were aware how unusual it was to have so many female jocks in the schedule, and especially one in charge.

CA: "Absolutely."

Sully: "Yeah, there weren't that many women on the air."

CA: "There was nobody on the air during the day."

TLT: "I never thought about it."

GK: Ladies, did you think about it?

Sully: "Yeah, because I had come from college radio, and the only person like that was- if you were a woman on the air, you got compared to Dusty Streets or Bonnie Simmons (big-time KSAN jock), because they were, like, the only women on the air. And I think that's one thing about the production- women were not assumed to be in production at all."

Sully had been seeking out Frisco, asking him for lessons in production, and he gladly complied.

Frisco liked Sully a lot, and they were both bright, funny people who became good friends.

He taught her all the basics, and then he taught her a lot of tricks of production.

Tiny wanted nothing to do with Production, but she remembers, "She came up to the station, I remember. She wanted to get a look at me. Bonnie."

Bob spent many of his days at a cabin on nearby Coyote Lake, owned by Gary Steinmetz, either drinking or drying out. There was no phone there, so no one could reach him, which was fine with Bob, as he liked to do both his drinking and his sobering up in peace.

Another part of daily life at KFAT was dealing with the children. Both Laura Ellen and Tiny had kids at the station, but most of the jocks were strangers to kids. Cuzin Al had a couple, and Buffalo Bob had some grown kids somewhere, they'd heard, and one was in jail…

Elsie was cute and adorable, but not getting a lot of attention. Kids will find a way to get attention, say the experts, and Elsie used to enjoy pissing off Buffalo Bob.

ST:" She would walk right up to Bob, spilling her coke on the…he was just like *nnngghh.*

Sully: "She would walk up to somebody and look at them and then spill stuff."

ST: "Let me say something about kids. When we all first started at the radio station, and my kids were there, and Elsie was there, I remember OZ and Sully and Bob being very upset and talking (mumble, mumble) about how the kids had to go. Because there were raisins on the carpet."

Sully: "Oh, yeah. Elsie put raisins all over the place…"

Sister Tiny will never forget, and neither will her son, Finnian, the time that someone left a child-sized turd in the vegetable drawer of the refrigerator at the station. Laura Ellen was furious, and blamed Finny, and Tiny, hating that she still needed to live at the station and couldn't afford this sort of behavior, was furious with Finny, too, although he swore he didn't do it. They all found out later that it was Elsie, but by then Finnian had paid the price. Living at KFAT was fucking up her kids, and everyone knew it, especially Tiny.

TLT: "I remember Elsbeth standing on the stool and looking at me, and picking up the tone arm in the middle of the song, and it was in the middle of one of my first day shifts or something, or I was doing fill-ins, I don't know- and she stood up on the stool and looked at me, and she just dropped the tone arm."

Part-timer Jim Conroy remembers coming up the stairs for his first shift as a KFAT fill-in. Waiting at the top of the stairs, Elsie watched him come up, then, when he reached the top and said hello, she kicked him in the balls. While he was writhing, Lara Ellen came out to apologize, but…

Sully: "She should have been in nursery school or something. She was bored to tears. For a long time she was just there by herself, and she got into all kinds of mischief and drove everybody crazy."

If Jeremy wasn't around to take the kids in the back and make them scream until their throats hurt, well sometimes those kids were just on their own. Everyone remembers the cardboard carton that got flattened and used as a sled by the kids. A few listeners I've heard from recall hearing the excited children's screams in the background as the carton carried the kids, shrieking in joy and fear, down the stairs to the bottom, with at least one dog barking after them. Then they'd carry their sled back up the stairs and fly down again. They were children, and they could do this for hours. How endearing.

Chapter 101: I Love Susan and Mick

Susan and I were getting married on July 18th, 1978, and we invited the staff to come to Sausalito for our wedding. We both loved the Rolling Stones. In fact, after we kissed for the first time as husband and wife, we walked back down the aisle to the opening chords of "Gimme Shelter." For the alarmists out there, just before the song went into its violent opening vocal, the tape, which I had recorded and edited in the Production Room, cut to "I Believe When I Fall In Love," by Stevie Wonder.

Knowing how much we both liked the Rolling Stones, a friend had let us know in advance that, for our gift, she had gotten us tickets to see the Stones at Red Rocks Stadium in Colorado the day after our wedding. I was going to take my show off the air for two weeks. I spoke to Laura Ellen and my sponsor about it, and it was okay to do that. The two weeks would be added to the end of my contract. I'd announce the break on my shows the week before, so there'd be no promos for a while, either.

We were excited about getting married, of course, but we were also dealing with our parents, and a lot of bullshit, and we were also excited about getting through the damn day, to the airport, and on to Denver. We couldn't wait to see the Stones. And it was even better than that! The Stones were playing at the Oakland Coliseum on the 26th, and we'd be home in time to see them there, too. Out-fucking-*standing!*

Then I had an idea. I might be a pretty average guy, but I was an above average Stones fan, and I was on the radio. And it would be really cool if I could meet Mick Jagger.

Before I left for my break, I cut a special promo and scheduled it to run, starting on July 21st, five days before the Stones show in Oakland. There had recently been a scandal in the recording industry, as reps from some of the labels were caught giving money and cocaine to DJs and execs of large city radio stations in exchange for airplay for the artists they were pushing. It was a big story. Bad for the radio industry, bad for the recording industry, *very* bad for the jocks and execs implicated. It had happened recently and gotten a lot of press.

I began this promo as I always did, with my usual opening riff of "Venezuela, There You Go" by Leo Kottke. Usually, when the opening eight bars of the song changed into the body of the song, I potted down the song and introduced myself with, "Hi, this is Gilbert, and on the next Chewin' the Fat…" Not this time. This time I opened with the normal musical bed, and the usual, "Hi, this is Gilbert," but instead of following that with what was coming up on the next show, I said, "Ya know, I usually do a promo for my show, Chewin' the Fat, about now. But I'm not gonna do that this time. What's in it for me? I still get paid the same pitiful amount whether you listen or not. So this time, I'm doing a promo for myself.

"Record companies: where's my cocaine? Don't you know that's how it's done? Don't you know that's standard procedure? Where's my cocaine?

"Concert promoters: where's the tickets? I like country and rock, and classical and jazz and blues and… send me the tickets, please.

"Publishers? I like books. All kinds of books. I like mysteries and science, and fiction and biographies and history and art, and just good literature, so send me some books, all right?

"Let me think. Is there anything else? Oh, yeah! Hey Mick Jagger- I know you like country music, so maybe you're listening to KFAT. Give me a call, will ya? Thank you all very much."

The music volume rose, then faded, and the DJ played the first song of the next set.

The wedding was a beautiful day in a beautiful location in the Sausalito hills, overlooking the San Francisco Bay, and most of the staff had gotten new flannel shirts and jeans, and Jeremy was sporting new work boots for the occasion. I thank them all for attending, and for their best wishes and gifts and I still have the ice cream maker that Cuzin Al gave us. Susan and I had hired local KFAT favorites Back In The Saddle to perform. I believe a good time was had by all.

In any event, Susan and I went off to Colorado to see the Stones, then did some camping, and when we got back to Marin County, there was a message waiting for me from Laura Ellen. Mick had called, and he wanted to meet me.

Which sounds incredible and how impossibly cool and all, but it really wasn't all that. Mick, or somebody with him, had either heard the spot or heard about the spot. I mean, it's not like no one knew about the station.

In any event, knowing what I know about the Stones, it is certain that no one got invited to meet Mick without Mick's either having thought of it or approved it. I doubt that Mick Jagger got on the phone, but probably one of his reps called the station, spoke with Laura Ellen, and she and I were invited backstage after the show to meet Mick. No, not Susan, sorry. Just me and Laura Ellen. Go backstage and ask for Art.

After the show, I arranged to meet Susan later at our car, I found Laura Ellen at the backstage entrance, and we went in. Following directions, we were led to a patio made up as a cafe-like setting; people were milling about, chatting, having a glass of wine or a soft drink. Laura Ellen and I found Art, and he said he'd get back to us, to please wait. The band had just come offstage, and they needed some time before coming out to meet people. We hung out and looked at people, and after about forty-five minutes, Art came up to us with Mick Jagger, introduced us to Mick, then left. We chatted for maybe two or three minutes. I had nothing to say to Mick. I'd given it some thought, and I had something to say as a fallback in case I got speechless in front of Mr. Jagger, but I needed something better than what I had. But all I had was that fall-back line: "I know you like country music, Mick, any chance of you guys recording any more of it?" It was lame, I know. But I said it.

He said he'd always enjoyed the music and listened to it, and he never knew where any of his influences would lead him. Playing more country was always an option. Or something like that. I don't remember what Laura Ellen said to him, but it was almost certainly not as lame as what I'd said.

Then Art came back and said that he needed Mick "over there," we asked Mick if he'd pose for a photo with us, he said sure, and I handed my little Kodak Instamatic camera to Art and we smiled as he took the shot of me, Laura Ellen and Mick. We shook hands, and he was gone. Laura Ellen and I stood around for a few minutes, realized that our part of the show was over, and we weren't needed there. Besides, I had a new wife waiting for me.

Shitty photograph? Yeah, and I'm sorry. It was one of those cheap-shit plastic Kodak 110's. But I'll tell you this: There were about 65,000 people at that show that day, and some of them had cameras of such power and sophistication as to give them a close-up of Mick's lips from an eighth of a mile away, but I had a Kodak 110, the cheapest camera ever made, and the only way to get a close-up with Mick with one of those was to be there, up close.

Chapter 102: Spooks

I liked to interview authors for several reasons. I like to read, I like to be prepared for an interview, and authors kept coming to San Francisco, looking for media outlets like television, radio, newspapers and magazines to be interviewed, to publicize their books.

KFAT obviously didn't have an office in San Francisco where I could conduct interviews, and the chances that someone on a promotional tour would have the time or the inclination to drive down to Gilroy to do the interview were slim. Which was great for me, anyway, because I wasn't in Gilroy. I lived a hundred miles away in San Rafael, and taught most days in the two high schools of San Rafael, about a half hour north of the City. What worked best for me, as well as the authors, was scheduling the interview for 5:00 p.m. at their hotel.

This meant it would be the last interview of the day for the authors, and if it happened at their hotel, they would invariably find that the publisher had provided wine and shrimp cocktail for them in their room, and they quickly relaxed, which suited my informal style of interviewing. This way, with non-standard questions, I got people to open up and talk in a conversational tone, which was more interesting to listen to than the academic tones many of the authors used for interviews. Some of them spoke like they were giving a lecture, and that wouldn't fly at KFAT.

Also, I usually read their books, so I was well prepared, and after weeks on the road answering the same questions (supplied by the publisher), the authors were motivated and engaged with my own, well-researched questions.

One of the press kits sent to me was about a book called Spooks, by Jim Hougan. Hougan had been the Washington editor for *Harper's Magazine*, and had written extensively about Watergate and about the United States intelligence community. I loved that stuff, and Spooks sounded great, so I told the publisher's rep that I wanted Hougan for my show. It was a big book, but I'd read it and had prepared a lot of questions. I was more excited about this interview than for most because I always needed more material, and most interviews gave me one show, but this could give me maybe several shows.

I loved spy stuff, and knew my listeners would like it, too. Hougan was exposing secrets that were astounding. Like that "they" can spy on you by listening to you through your telephone- even when it's hung up and not being used. Same with your television. That was news back then, and Hougan was exposing it.

What was also new and astonishing was what they could do with lasers. Voices are waves of air that move through space in specific configurations formed by your vocal chords, just as a microphone is a membrane that responds to those sound waves by moving in a very specific way, and these movements in the membrane are converted into electrical waves that are sent to an amplifier where they are made larger, and then to your speakers, which are membranes that vibrate in the reverse process of taking an electric signal and creating movement in the speaker cone, which in turn moves the cone in a specific way to create a wave of air in a specific configuration that is picked up by your ear, moving the membrane in your inner ear in such a way as to convert the movement of the membrane in your ear back into electrical signals which are sent to your brain and interpreted as sound. Thus, vibrations from a conversation would make a window move just... slightly, and the CIA had a way of aiming a laser beam at a window and converting the infinitesimally small vibrations in the glass into electronic signals that could be amplified and converted to sound, and could be monitored from across the street at a listening post.

Oh, sure- you know about that laser shit *now*. But back in 1978, that was pretty radical stuff, and the CIA was reportedly none too happy about the book. Or Jim Hougan.

There were lots of similarly startling revelations in his book, and I knew these would be great shows, and I had a lot of questions for this guy. This was a gold mine for me, and I was primed and ready.

Then, on the morning of the interview, I got a phone call from someone who said he was with Mr. Hougan's publisher, William Morrow, and he regretted to inform me that Mr. Hougan had a family emergency, and been called away unexpectedly. Mr. Hougan was sorry, but he had to cancel the interview.

Bummer in Gilbertville. Several shows was a lot of shows for me, and this was a big disappointment. In fact, I was so sorry to miss this opportunity that I went back to the promotional material sent out by the publisher, and found out where Hougan's next, and last, stop would be. It was Los Angeles, which was great.

I called the San Francisco publicist I usually worked with on interviews, explained what had happened, and she was surprised that they hadn't asked her to call me. It was her business to handle those things, and she wanted to call New York and ask about it, but I couldn't remember the name of the representative who'd called. I told her I didn't care about that and asked her to find out where Hougan was staying in Los Angeles, I'd like to fly down to see him. She was confused; she thought the call was strange, and even stranger that someone there would even have my phone number. She was reluctant to give out that information, but we'd worked together for two years and she knew I was no stalker, so she gave it to me.

I called Hougan at his hotel in Los Angeles that night, and he agreed to meet me at LAX the next day. Hougan was flying out at three o'clock, going home to Washington, and he would get back to his hotel after his last interview on morning television, check out by eleven, have lunch and meet me at noon at the airport. That gave us two hours, which I knew I'd need. I'd told Hougan I had a lot of questions, and he was okay with that.

Hougan's airline gave us a conference room at the airport. Well-researched and prepared, I had Jim Hougan for two hours and I used up almost all of it. He was a real good guy about it, and I got five good shows out of the interview. I was happy with what we'd done, and it was a pleasant surprise for the author, but the most memorable moment of the afternoon took place after the interview was over. Hougan was stretching and relaxing after talking for almost two hours, and I was packing up my recording gear and notes, and we were chatting amiably. I said that I hoped everything was okay with Hougan's family, and he looked at me quizzically and asked me what I meant.

I explained that the fellow who had called me yesterday from his publisher to cancel our appointment said that Hougan had had a family emergency, and I hoped Mr. Hougan's family was all right.

Hougan looked at me and said he didn't understand.

"What don't you understand?" I asked.

"But *you* canceled on *me!*" Hougan said.

"What?"

"Yeah," he said, "I got a call yesterday from some guy who said he was with KFAT, and he regretted to inform me that you'd had an emergency that had just come up that morning, and that *you* couldn't make the interview."

I assured him that there was no one at KFAT who would even know who I was interviewing, or when, and that nobody there knew or interfered with my schedule.

Hougan said that on the way to the airport, he'd told the San Francisco representative for the publisher about the call, and she'd thought it odd at the time that I hadn't called her directly, but that she'd dismissed it as something minor and then thought no more about it. She hadn't told me about that in our conversation.

And he also told me that this was the only time on the tour that an interview had been sabotaged. And now let us wonder together who it was that didn't want the CIA's secrets discussed on KFAT. Who could it have been? In fact, this episode has left me with three questions:

1. If the CIA orchestrated this ruse, did they follow the situation to see if the interview ever happened?
2. Do I have a file now at Langley?
3. What was Dick Cheney doing that day?

Chapter 103: The Garlic Capital of the World

Again, dispute blocks our path to a perfect understanding.

Gilroy is known far and wide today as "the Garlic Capital of the World." And it sort of is. All the garlic that used to be grown and processed in and around Gilroy has now given ground to the exigencies of a more ruthless and complex international marketplace. If this book was written by an economist, we might now consider the ebb and flow of capital, the calculation of various fiduciary influences and the effects of capital on their infusion into market forces and the consequent cost in infrastructure of salaries, benefits, taxes, levies, depreciations, deductions and added expenses to production, packaging, maintenance and shipping, as well as algorithms relevant to the vagaries of the changing climate and how that might affect market performance. Or we might be discussing the relative merits and disadvantages in the implementation of counter cyclical compensatory fiscal policies as espoused by the current administration, if it were one of those guys writing this. But it's just me, so I say: things have changed, and these days you are probably eating Chinese garlic. Deal with it.

Back in the 70s, garlic was grown in the San Joaquin Valley and processed and packed in Gilroy. Gilroy Foods was a huge, non-stop processing plant, employing a big chunk of the local populace, making it easily the largest local employer. The jobs they offered generally did not require a college degree or any degree. Growing and packing was the dominant industry in Gilroy, and it proceeded thusly for decades, happily, efficiently, and virtually unknown to the rest of the world.

Did anyone ever think about garlic before the 1980s? Did anyone know where it came from or went to be packed? Was anyone ever taught that in school? I could still tell you what Indonesia's principal crops were at the time of my time in high school because I had to write a report on it, but was anyone aware of Gilroy's importance? No, I say!

Decades ago, they'd put up a small sign on Highway 101 that said you were passing Gilroy, "the Garlic Capital of the World." The sign was placed on a short stretch of Highway 101 near a bunch of fruit and vegetable stands. By the mid-70s, the sign, about four feet by six feet, had been there for years and had faded, and had so long ago become part of the background that no one remembers it. The sign was all but invisible, but not just because it was so faded.

As best as I can remember, the sign was placed just off the shoulder of the highway, in the middle of a brief-but-dangerous section of the road with several severe curves. People unwittingly took those curves at the highway speeds they had been confidently driving at, and many an accident occurred there. So many people had died in that short corridor that it had earned the name "Blood Alley." Yep, "Blood Alley" was maybe a couple of miles long on Highway 101, and it went right past Gilroy, and that's where that old sign was.

As the road was somewhat treacherous there, the average driver would have gone into high alert by the time he or she passed Gilroy's little message to the world. Most drivers were busy clutching the wheel with both hands and slowing down and paying attention to the road so they wouldn't die, so it might be forgiven if most drivers chose to concentrate on the road while driving through Blood Alley, leaving little time for luxuries like reading small, faded signs.

Lori remembers: "…every day when I'd come home, I'd see a car on fire in Blood Alley."

But in fairness, the sign *was* there.

Blood Alley is long gone, and there is a new path 101 now goes past Gilroy, and it's all modern now, and most of the fruit and vegetable stands on the side of the road in Blood Alley are gone, but there is a big outlet mall on the side of the new section of 101, and you could shop there for bargains in a wide variety. The old sign is gone, but we weep not, for it was unseen even in its day. Unseen and unheeded, it's message ignored daily by tens of thousands of travelers. But the boast about "The Garlic Capital of the World" lives on because at least one person remembered it.

Back in the early days, Karl Hess provided "the Morning Nudes" to Fatheads. Karl had been a radio announcer at his local college, and KFAT was his first and last professional gig, and he wouldn't stay there long. He was a clever, clever fellow, and KFAT allowed him to indulge his humor and his individuality. To this day, early KFAT staffers remember him fondly, praising his wit and his station ID's.

An ID is what the FCC requires at the top and bottom of every hour, and it must be explicit. As Jeremy reminded his new staff in their first days, "It's K- F- A- T, Gilroy. Not K-Fat, not *in* Gilroy. Just K-F-A-T, Gilroy."

Of course, other than at the top and bottom of the hour, any sort of ID would work. All stations play ID's more frequently than only at those two times. Stations depend on selling ads, and the rates they charge depend not on how many people *might be* listening, but on how many people remember the call letters of the station when they are asked by pollsters. And that was why there was a "Ratings Book," which would tell subscribers which stations had how many people listening at which times, and if possible, as much as possible of the demographics, like age, gender and income. The more people listening, the higher the ad rates, so stations needed people to remember the call letters of the station they were listening to, and they played ID's as frequently as they dared. Stations were chronically afraid that when interviewed regarding what radio stations they listened to, people would forget the call letters of their favorite station. Sounds paranoid, I know, but it happened all the time.

And that is why stations announce their call letters all the time. KFAT, not so much.

But Laura Ellen and Jeremy knew about this part of the game, and ID's were produced, put into the appropriate slots in the 7-Up cases, and used. ID's were put on the logs, and left it for the jock chose which one to play. If the jocks liked an ID, they played it more often, and everyone remembers Karl's ID's as among their favorites.

He liked doing these because he could be creative, whereas with the news he couldn't. This was playtime. He could relax, write them, rewrite them and experiment with what worked; it wasn't like being on the air live. There, what he said was out and gone, but in the studio he could re-record, edit and improve. One morning after the Nudes, he wanted to do an ID, but what could he say? He had to say it was in Gilroy, but what could he say about Gilroy that would at least *sound* like he was proud of this shit-hole.

Being from the area, Karl remembered the billboard on 101, so he knew that some time in the past someone had said that Gilroy was the garlic capital of the world, he hadn't heard anyone dispute the assertion, and now he knew what he wanted to say.

But it had to sound… *impressive*. He needed *Sturm und Drang*. So Karl went to the Production Room's collection of sound effects and production records and found a record by the U.S. Navy Chorale with the song, "America The Beautiful," and near the end of the song he found a brief, soft section, building up to a huge choral swell at the end, and rinsed in echo, in a booming, self-important *basso profundo* voice he recorded: "This is K F A T, Gilroy, the *Garlic… Capital… of the World!"* as the music swelled and the massed chorale sang in stirring, ringing rising stentorian tones:

"From.….Sea.….To.….SHIII.…...NINNGGG.….….*SEAAAA...*"

Karl's KFAT ID would turn out to be the most enduring legacy of his time at KFAT. It was funny, it was fun, the jocks played it a lot and got some mileage out of it for a couple of years, and then representatives from the city of Gilroy called.

Laura Ellen was in the office when Lori told her that a Mrs. Christopher was on the phone. Everyone knew that Mrs. Christopher was the most important woman in Gilroy, and she instinctively braced herself for an onslaught from another outraged party. Someone was always offended by something one of the jocks had said or played. By a show of hands, readers: is anyone out there surprised at another offended caller? Raise your hands out there... Anyone?

But this call was friendly. Mrs. Christopher wanted some help. Let me give you some background. If you buy garlic, and it is grown in this country, then it might well say on the label "Christopher Ranch Garlic." Costco sells it, I know, and so do many of the major supermarket chains in California. The Christophers were far and away the biggest garlic growers among the several families that grew it in the area. Christopher Ranch was huge, and Mrs. Christopher was without doubt the First Lady of Gilroy. She was the most active socially, and she was Gilroy's most prominent citizen. Even Laura Ellen knew who she was.

But why did Mrs. Christopher know who Laura Ellen was? Why was she calling? How bad was it? Laura Ellen's experience had shown her that no good would come from this call, but she braced herself, took a deep breath and picked up the phone. Hello, this is Laura Ellen. How could she help Mrs. Christopher?

Mrs. Christopher had been a part of a recent series of discussions among her garlic-growing ladyfriends in town regarding the promotion of Gilroy garlic, and it had been determined that they should have a picnic- no, a party- no, a Festival! It would be a celebration of their uncelebrated but essential crop. Several of the ladies had met and wanted to put on this event, and Mrs. Christopher had volunteered the use of a couple of her fields, and the ladies would all cook up some of their family recipes that featured garlic, and they'd rent some picnic tables, the kind that have benches built on, and they'd invite people to come and eat and have a good time.

They saw it as a big picnic- with strangers. It would be informal, but their plans seemed somehow to lack a certain... something... that would make it... fun. Ah! They should have some music- like a band. With a fiddle...

Music? Who in the group knows any musicians? Anyone? Who knows what kind of music to hire? Who knew how to find the musicians? Or how to rent and set up the stage and get all the electricity and the speakers and microphones and stuff? And who will run it? Anyone?

What about that radio station in town? They might know some musicians. Let's call them!

Laura Ellen had long since given up on having a better relationship with the town. Remember the big breakfast that KFAT threw for the merchants of Gilroy when they first got there? Only Gary Steinmetz of the Green Hut showed up for that. But that was back in 1975, and this was 1979. Things had changed.

KFAT had been rejected by Gilroy, and the station had now outgrown it, anyway. But who wouldn't want to be wanted by your home town? This would be good for Gilroy and good for KFAT: Laura Ellen was thoroughly enthused by the project and asked Lori to work on it. KFAT would promote the event on the air, and she'd find the perfect band that would be Fat but wouldn't offend anyone. They'd need a stage and a sound system, and maybe some permits, and… she'd give the details to Gary Faller to make it happen, and she'll talk to the police about the permits. Laura Ellen and the cops knew each other by now, and with the Christophers behind it, they'd help all they could. She knew lots of bands and she'd choose one. Maybe country-rockers. With a fiddle. This was perfect, and Laura Ellen would not let them down. Laura Ellen hired Terrell Lynn's band, the Radar Rangers.

She handed the phone to Lori, who worked thereafter with the ladies, and soon KFAT was promoting the day as the "Gilroy Garlic Festival," inviting its listeners to come down that Sunday, eat some garlic, listen to music and meet the KFAT jocks and other Fatheads. Admission was free.

As a liquor license would be needed to sell beer, a beer distributor had to be brought in. Terrell Lynn used to drive for a beer outfit in Santa Cruz, so the call was made, and the distributor happily promised to bring the beer. Knowing the KFAT crowd, he would bring a lot of beer.

It was to be an open event with no tickets necessary, so there were no advance sales to give any clue as to whether or not anyone would show up.

Signs from the highway off-ramp gave directions, the Christophers had a field that was laying fallow that year, so they flattened it with a steamroller and posted a ranch-hand at the field's entrance to direct cars to the designated parking area. Next to it was another flattened field, and at the far end of that field they set up the stage. It would be a daytime affair, so no lighting was needed, and by ten the gear was set up on stage, the band had already sound-checked, and they were waiting for someone to tell them to start playing. On the edges of that field were a half-dozen tents, with a few picnic tables and bench seats in front. The cooking was going on in the tents, and as the first people began to show up, the ladies began serving the food, and all of it at reasonable prices. Everyone had been clear about their goal for the day: this was to be a party, not a rip-off. The ladies were hoping that three or four hundred people would show up, and they'd have plenty of food for them all. The ladies were ready. KFAT was ready. No one was ready.

But what a party it was! People showed up early and in great numbers. The band was playing, and more people showed up. The designated parking field was soon full, then the field next to it was also over-run. More cars came, and then the next field over filled up with cars, and they kept coming. The parking guy had long since abandoned any hope of order, and frantically ran about trying to keep as many cars from ruining as many working fields as would be ruined if he abandoned his post.

The place was packed, and KFAT T-shirts were everywhere, and more cars were lined up, waiting to get in, waiting for someone to figure out where to put all those cars. And damn near all those cars had a KFAT sticker.

Frantic calls were made, family was sent out and more supplies were delivered to the tents, and soon more food supplies were sent for again. All afternoon food was constantly bought, brought, cooked, sold and served. Controlled communal chaos abounded inside the tents and outside the tents, and on the field in front of the band and in the improvised parking lots that used to be garlic fields. But it was a happy chaos everywhere. And as for chaos, well, the people at KFAT knew something about that already.

The Gilroy ladies and their friends and family were overwhelmed, and it looked to the planners that it was about to go out of control. But it wouldn't. The Fatheads were mellow, happy to be there, happy to see what the jocks looked like, meet some of them, and meet other Fatheads. It was a great party, the people were digging the band and wishing there was more food, but not sweating it at all. Thank God one guy had known what to expect, and there was plenty of beer. The booth selling KFAT T-shirts sold out every shirt in the first fifteen minutes.

Afterwards, it was estimated that between eight and twelve thousand people showed up that day. In the end, a field or two might have been put temporarily out of order, but it was only dirt and some crops, and the dirt was okay and the crops would grow back. The garlic growers of Gilroy were amazed at the turnout, happy with their new friends at KFAT, and the decision was made that very day that this would have to go on next year and every year. Next year they'll find a park or someplace to put it, *with a lot of parking* and electricity and a clean-up crew, and whatever it took, but they were *definitely* doing this again next year! And KFAT would definitely be part of it!

And they did, and KFAT was. To this day, the Gilroy Garlic Festival is still a well-known highlight of the summer festival season in California, attended annually by over a hundred thousand people from the Bay Area and all over the world, and a permanent site has been installed. Parking accommodations for so many has been procured nearby, with shuttle buses transporting the crowds from the parking lots to the festival.

KFAT is long gone, but The Gilroy Garlic Festival endures, with music, garlic-saturated food and good cheer. As it should.

Chapter 104: Under His Arms

Wanda was at home and Awest was browsing in the stores in Santa Cruz, looking for oddities. He and Wanda collected oddities, and they enjoyed the browsing, the textures, the colors, anything interesting. Their money was tight, and they wouldn't buy anything unless it was outstanding. Awest had designed several things for KFAT, including, of course, the logo, the bumper stickers, the stationery, some ads, and a few cartoons, but Jeremy hadn't paid all that much for them. The sales people had business cards, but they used the drawing of the fat cowboy that had been taken from the first poster, done before Awest found KFAT.

The T–shirts were selling like hotcakes and were paying the bills and allowing him and Wanda to live, to browse, and occasionally to buy. This was a favorite store of theirs, full of incense and exotic and unusual things, and he was happy to be there, on no schedule and heeding no care.

Suddenly, Awest felt a presence next to him, on either side, and he looked up as he felt something slip into each armpit. On either side of him he saw two mountainous men whose hands had gone into his armpits, and the mountain on his left leaned over and whispered, "Be cool." Then they lifted him just barely off the ground and carried him out of the store, to a van parked at the curb just outside the store. The windowless side door of the van slid open, and Awest was gently placed inside, the two men followed him in, and the door closed.

Inside, a woman introduced herself as Sharon Barger, and Awest blanched, knowing that this was the wife of Hells Angel founder and leader Sonny Barger, who was in prison for murder. She said she knew who he was, that she and the Angels admired his work. He thanked her, he thinks he remembers.

She told him that KFAT was the official station of the Hells Angels, and they love the station, and they see a lot of KFAT T-shirts around, and she was glad that the sales were going so well. Then she told him that she appreciated all of his work on the shirts, and the efforts he had made in selling them, but that they were taking over the KFAT T-shirt sales now.

She wanted the screens, and told Awest not to worry, she knew he wouldn't have them on him, that they'll come by for them soon. Then the door slid open again, and he was dismissed.

The next morning, Awest had Wanda, the screens, Hezekiah the parrot and everything they could stuff into their VW van, he drove to Los Angeles and he never came back.

Chapter 105: Nothing Left To Burn

Frisco was getting worse again.

Terrell Lynn: "when he was living in the Production Room, he'd put his sleeping bag in the middle of the floor so you could walk around him and get to the turntables and other equipment." Sometimes Frisco passed out in there with no sleeping bag or blanket.

He was becoming irrational, and the people at KFAT were becoming alarmed. He was blitzed more than ever, but at least in the past you could always tell what Frisco was riding, whether it was speed or booze or downers or...

He'd stopped going home to his mother's trailer a long time ago, and was living in a back room at KFAT; he seldom went out, except to score. Speed, booze, pills, something to eat, something to drink to wash the food down, and for later.

When Tiny finally moved out, he moved into Room B, a back room with no through door, no way out the back. The door to the room was at the front of the room, and Frisco laid a mattress in the far right corner, in case some errant shaft of light might enter, it would find him last. The light bulb in the overhead fixture was burned out- or he sold it. He found an old lamp someone was throwing away, and he brought it in there, then cadged a bulb from someone. The next time someone saw him, he'd put aluminum foil on the windows in there, he said to protect him from the rays sent to harm him. Rarely going out now, he was holed up in there when he wasn't in the Production Room. Yes, he was still producing spots for the station, and they were almost always on time, but they no longer had the sparkle, the wink, the upraised single digit that trademarked Frisco's early work. That was a shame and all, but that wasn't what most alarmed the staff.

Rocket Man remembers the night when someone brought a gram of cocaine to the station to hang out for a few hours and party with the jock. The visitor dumped the whole gram onto a plate and said, "Who wants to go first?" Frisco jumped up and pushed his way to the plate, saying "I do!" and put his hand out for the straw that the man had brought with him. Frisco grabbed the straw, leaned over the plate and snorted the entire gram in one sniff.

There were three other people in the room when he did that, and they were all stunned. No one had ever seen that before, and they were speechless. Rocket broke the silence, saying, "Well… next time *I'm* going first."

Now you couldn't tell one day's high from the last. Or the next. No one could tell anymore what he was riding; it was all blurring into one constant babble. And then Frisco put tin foil on the windows of the Production Room, where he worked at night, so it wasn't the sun. It was the rays.

Then he painted pentangles in the room where he stayed, and said he needed them to protect him. He sometimes spoke as if he weren't there. But if he wasn't there, where was he?

Everyone knew that Frisco had devils, but now he was talking to them. Then they started talking back to him, then they became arguments, and then raving diatribes. The devils pushed and pulled at Frisco, and Frisco pushed back, when he was sober enough to push back.

Concern became alarm, which became fear. What should we do? Was he sane? Was he dangerous? What should we do? Everyone at KFAT loved Frisco Bo, but everyone knew that the life at KFAT was killing him, or at least letting him kill himself.

Sad that it had come to this, Jeremy arranged for one last dry-out at a cheap recovery place, and one day Sherman and Rocket went into his room while he was passed out, and packed his clothes. He really had little else. No photos, no souvenirs, no trinkets or mementos. He had lived in that room for months, and hadn't put anything up on a wall. He had no bedding or decorations, no stereo, no records. He didn't own a rug. He had a blanket, a pillow and a towel. And all those whiskey bottles, all of them full of Frisco's urine, all with their caps screwed tightly on. With a last grasp at civility, if Frisco couldn't make it to the toilet, at least he had avoided the ultimate debasement of pissing on himself.

Frisco needed help, and Jeremy and Laura Ellen arranged for it and paid for it, but it had to be the last time.

After Sherman and Rocket Man had packed up his few clothes and put them in a paper sack, they woke him up and, taking advantage of his stupor, they gently led him, luckily still fully clothed from the days before, to the car that would take him to dry out, and away from KFAT for the last time.

Listeners knew something was wrong at KFAT that day, but none of them knew what it was. And the jocks weren't taking calls.

Chapter 106: Win Ben Stein's Funny

As an interviewer, I loved authors, and I interviewed a lot of them. I knew all the publisher's reps in San Francisco, and they all liked me because I almost always said "yes" to their requests to interview their authors, and I was always prompt, polite and prepared.

Movie stars always wanted—needed—to talk about their latest movie. They'd cover the plot for a minute, and then go on to discuss the details of why they were attracted to the script or the cast or the director, and why the film was so good. It was all the same bullshit; they did it in every city on their tour, and everyone who played the game knew it.

Musicians were usually inarticulate about the creative process, and what could I ask them except how much cocaine they'd consumed while recording the current album?

Ah, but authors… Sure, they'd talk about their book, but writers were so much more verbal, which was good for radio, and they were so much more interesting. That was why they had been published- because the subject was interesting, and they were the experts on it.

Oh, you could talk to the author of a novel, but you'd be right back where you'd be with an actor. They would talk about the creative process, but it bored me. And if it bored *me*…

But the author of a non-fiction book with an interesting subject would make for a good interview. Even better- some books would have made better magazine articles, but they were there to push books, so they had to summarize. I had twelve minutes of interview in a fifteen-minute show, and this format was perfect. I wish I'd thought of it.

My show had now been sold by The Dragon Lady to Miller's Carpet Care in Campbell, about ten minutes' drive west of San Jose, and they were very accommodating folks to work with. They obviously knew the kind of insanity that they were buying, and they made it a point to allow me as much freedom in the ads that I wrote for them as they could take.

Of course, if they had a special sale or promotion happening, they'd want me to talk about that, and I would. But if it was business-as-usual, then I could say what I wanted about them, and so far they'd been okay with it.

Please recall that when my sponsor had been The Sportsman's Shop on Pacific Avenue in Santa Cruz, I'd had some fun when I completed my show by talking about a miniaturized Russian submarine that had been "caught" recently, "…directly beneath the Sportsman's Shop at 1225 Pacific Avenue, on the mall in Santa Cruz. That's… the Sportsman's Shop, 1225 Pacific Avenue, on the mall in Santa Cruz."

I interviewed Bill Murray when he was so new at Saturday Night Live that just that day he'd signed with a manager for the first time. I got a couple of Bill Murray KFAT IDs that day. He was a delightful guy, very outgoing, playful, and full of mischief, which I was able to exploit by explaining to him, after the interview, that I had a lot of freedom in the ads that I wrote for Miller's Carpet Care, and I asked him if we could make something up that would be fun. He was thoroughly into it and he told me to keep the tape rolling.

On the air, near the end of the show where I should have gone into a commercial, Bill kept talking about the time that the cast of SNL was awarded the Peabody Award, television's most prestigious award, and NBC had flown the cast out to Los Angeles to accept the award.

Bill said that he hadn't gone directly to L.A. with the rest of the cast. He'd join them in L.A. for the ceremony, but first he wanted to visit friends in the Bay Area. He said that his friends were the Hells Angels, and he went on to say that they had all gotten deep into the party, and that he hated to leave the fun, so he rented a bus to take everyone down to L.A. with him, partying all the way.

After Saturday Night Live got the award, Bill partied with the cast in Los Angeles, and then partied all the way back up to the Bay Area with the Angels on the rented bus. When it was time to return the bus, it looked like a war zone of spilled beer, crushed french fries, empty bottles, pizza boxes, burger wrappers, napkins, a towel, beer cans, vomit, and some disgusting mixture of fluids and smells that no one wanted to think about, and Bill knew that the bus company would clean the bus and send the bill to NBC. As this was a new gig with a major network, and he had just joined the cast and didn't want to look irresponsible to NBC by sticking them with a big bill for cleaning the bus, he was in something of a bind.

"So what did you do?" I asked obligingly, at this point slowly potting up the music, and he said, "What could I do? I called the crack cleaning team at Miller's Carpet Care in Campbell, and they came out and cleaned the bus so thoroughly that there was never a question about getting the deposit back to NBC."

At which point, with the music up at the end of the interview, into the studio mic I said,

"That's Miller's Carpet Care, in Campbell," then I thanked the listeners for listening, and ended the show. That was the ad, the show was over, and Gordy had a song cued up.

I presumed Miller's Carpet Care was happy with the ad, although I never heard any feedback about it, either from the sponsor or from station management, such as it was.

A few weeks after that I got a promotional package from a publisher who wanted me to interview the author of a new book, a fictional story about life in the film community in Los Angeles. The book was called "*Dreemz*," and it was by Ben Stein.

Mr. Stein was later to gain notoriety as the hapless homeroom teacher in "Ferris Bueller's Day Off," and his deadpan, nasal voice was famous for its lack of color or texture as he called out, "Bueller…? Bueller…? Bueller…?"

In the late 90s, his sardonic wit and acerbic, formidable intelligence was cleverly showcased on Comedy Central's quiz show "Win Ben Stein's Money," which also featured the then-unknown Jimmy Kimmel as his sidekick and moderator. But this was before all that, and Ben was known then mainly as being the suspected source of the leaking of the story about Joan Rivers' husband Edgar committing suicide. It was a scandal for a few days…

But back to the promo package for *Dreemz*. Promotional packages contain several elements that are included to make it attractive to a booker to book this guest on their show. A photo, of course, so newspapers and magazines can include that in their review. It also contained an introductory letter about who was being offered, and would include several instances of why you (I) should book this personage for their (my) show.

The remaining elements in the promotional package were the book itself, a biography of the author, and a set of suggested questions for the interviewer. These questions were a clever item to include, as most interviewers had no staff to read or research the book. Who had the time to read all the books that looked interesting and might make a good interview? I'm not going to say it, but it's possible that Johnny Carson didn't read any of the books whose authors he interviewed. Nor did Jay, Dave or the rest.

Carson would have a staff for such research, but the others, and especially the working stiffs like myself, had no staff, and it was up to us to do the research and come up with the questions for the interview. Publishers knew that people didn't have the time to read the books, and without reading the book, how would an interviewer know what to ask? No one wanted to look stupid, and no one wanted to invite an author onto his show and simply open the mic and hope the guest would carry the show. So the publishers included a set of questions, usually eight or ten, that the interviewer could ask the guest, and the guest could respond to, and the interviewer would look intelligent, well-prepared and professional. Without those questions supplied to them, I doubt many authors would get on the air. It was classic win/win, and everyone did it.

I was one of the rare interviewers who read most of the books whose authors were coming onto my show. This enabled me to ask questions that were unusual, and showed the author that I had read the book, which the author always liked. Most authors were sent out on two- to six-week tours, and of course at almost every stop these people were often asked the same set of questions at each radio and television station, newspaper and magazine where they were interviewed. The authors all knew the game, and they put up with it. I'm sure it got boring, but that was the job, and that was the business of getting the word out about their book. It was how it was done.

The publicists liked me because the authors liked me. I would pin them down and ask questions like, "On page 243, you said that…" and the prepared-to-be-bored author would be surprised, and they responded enthusiastically. And as the local publicists would always ask them afterwards about each interview, how it went, and what they thought of the interviewer, the authors always gave me high marks out of gratitude for my effort. Not after that dick Tony Sanchez ratted me out to the publicist for giving him coke, but mostly I worked well with the publicists in San Francisco. It worked for them, it worked for me.

But this is about Ben Stein.

Because KFAT had no office in San Francisco, I had no place for the authors to come to me where I could interview them. So I interviewed them at their hotels.

Because I was teaching during the day to make a living, I would set up my interviews for five o'clock so I could do both, and this proved to be good luck for me.

Because I couldn't see them until five o'clock, I was always their last interview of the day, so the interviewee had just gotten to go back to his or her hotel, they would usually sip some wine, relax and munch on shrimp cocktail; some would take off their shoes before we talked. The increased level of wine and relaxation on their part meant that I could get stuff from them that they weren't usually asked for, and their thoughts were fresh, and that came over the air. My interviews were more spontaneous than most, and most of the authors really liked me. As did Ben Stein. Which was good, as I wanted more from him.

The biography that came with his promotional package said that he had been a speech writer for Richard Nixon, and I saw an opportunity. I was *always* in need of material for shows, and this meant that if he would co-operate, I could get two shows from him, one about life in the film industry and one about being a Nixon speech writer. If he'd agree.

We got along well, and over his wine and shrimp cocktail, he agreed to the second interview. My first question for him was how he'd become a speech writer.

He answered, the interview progressed smoothly. I had been prepared for the second interview, I think he enjoyed it, we concluded amicably, and at the end, I had enough for a second show. Then I told him that my sponsor allowed me a certain amount of leeway with the spots that I wrote for them, and told him that I always enjoyed making something up for them. I told him about what Bill Murray had come up with, and he suggested that maybe he could come up with something "at least as interesting." I agreed, and we worked out the outline of a story. I released the pause button and introduced his proposed flight of fancy with our agreed-upon opening, "Is there anything that happened with the President during your years with him that isn't generally known to the public?"

He pretended to hesitate, going back and forth about whether he should reveal this thing or not, did some hemming and hawing, then a little dissembling and some more back-and-forth, a couple of maybes and maybe nots, then, after concluding that it had been so many years, that the public had a right to know, and he went into the story, and when it was over, I knew I had a classic.

We ran the interview about life in Los Angeles, and two weeks later I scheduled the show about speech writing. But it was a stand-out show, and I wanted to play it up big to get the most fun out of it.

Every week, in addition to preparing for the interviews and conducting them, I also had to edit them and get them ready for broadcast, and part of the gig was to write, record, produce and log my promos. I loved doing the promos, and had the most fun with them, and this was no exception. The promo that was scheduled to run for that show said that on the upcoming Chewin' the Fat, this Wednesday, I had an interview with a speech writer for Richard Nixon, and that "during the show, my guest will reveal something that happened during the Nixon administration that was *so* amazing, and that no one had ever spoken of it before, and it will be revealed for the first time during this upcoming Chewin' the Fat. It's a Chewin The Fat exclusive! It's an amazing, unknown story about Richard Nixon, so be sure to listen to Chewin' the Fat this Wednesday morning at 8:30."

I knew this would get the attention of all the legitimate news bureaus at television stations, newspapers and magazines, and they'd all be listening in, ready to report whatever startling story was revealed on Wednesday morning at 8:30.

The record, the cassette and the mic were all going into a reel-to-reel deck. There is a song by Leo Kottke that I used as the bed for my promos as well as for my shows. The song had a very distinctive acoustic guitar intro that went "Doop DOO-Doo, Doop DOO-Doo, Doop DOO Doo, dooo-dooo-doo." The music played under my vocals in the promos, and also started my show. After the intro, I would drop the volume of the song about 75% and talk over it for my intro. I had my guest's answer to my first question cued up on the cassette deck, and after I asked the first question into the mic in the Production Room, I would release the pause button on the cassette deck, and the answer would be recorded onto the reel-to-reel deck. I lowered the music during first answer for about ten seconds until it was off, and the interview would continue without the music until the answer to the last question, when I'd slowly raise the volume on the music again, then push the pause button on the cassette deck, lower the volume of the music and open the mic in the Production Room, and do my outro and commercial, then fade the music out as the DJ started the first song of the new set.

That was how every show and every promo started and ended. I liked it, it worked, and it was very identifiable. All of my promos had to be identified by date, so the right promo for the upcoming show could be scheduled and broadcast. I alerted Laura Ellen that this Wednesday's show was special, and she scheduled extra airings of the promo. That promo played all that weekend, and on Monday and Tuesday.

As I said, normally my show would start off with Kottke's guitar going, "Doop DOO-Doo, Doop DOO-Doo, Doop DOO Doo Dooo-dooo-doo," then the main body of the song would start and I would turn down the volume and talk over it. But that's not how this show started. I had written Gordy a note to explain that this show was going to start with a vocal, not music, and not to be alarmed. I needed to let Gordy know about this, as he would often put the tape on the deck, cue it up to make sure there was something on it, wait for 8:30, hit the Play button, and go out for donuts. This show wouldn't start off with music, and I wanted Gordy to know about it.

As soon as Gordy hit the Play button on the reel-to-reel deck in the studio, in a serious tone, I said, "Hi, this is Gilbert. Due to the controversial and newsworthy nature of today's show, Chewin' the Fat will be brought to you today, commercial-free, by Miller's Carpet Care in Campbell."

Then the music started. Doop DOO-Doo, Doop-DOO Doo, Doop DOO Doo Dooo-doo, then the music faded as I said, "Hi, this is Gilbert, and you're listening to Chewin' the Fat…"

This was the normal intro, and from there I proceeded as I normally would, introducing the subject, and then my guest. Then I said, "My first question for Mr. Stein was, 'How did you become a speech writer?' and I slowly faded the music as Ben said, "Well, I was writing for a magazine named…" and the show was on the air. Everything went normally. I knew that no one in the news business could have taken the chance on missing a good story; maybe, as promised, a great story. Oh, they were listening all right. I knew I had the news departments from ABC, NBC, CBS, Reuters, AP, UPI, *The San Francisco Chronicle, The San Francisco Examiner, The San Jose Mercury-News* and everyone else in the news game. They were all listening, waiting.

Ben Stein spoke in a dry, academic tone, and his enunciation is very precise. Erudite, if I may, and there was nothing whatsoever in his voice or demeanor to indicate that a prank was afoot. Nothing was indicated other than that I was interviewing a serious, scholarly man who was giving direct answers to direct questions. It was a dry, professional interview.

Without announcing that this was the last question, I asked, "Is there anything that might have happened during your years with the President that isn't generally known by the public?"

At this point, as you know, Ben started hemming and hawing. He said, "Well, there is one thing....No, no I couldn't tell you that. But it's been so long that.... no I couldn't.... well, it's been a long time, and there's no harm in.... no, I couldn't- and well, I guess I could..."and he finally decided that he could share this with us.

He said that there had been a cult of car-worshipping hippies in a commune in San Clemente, California, and that during the Arab oil embargo of the early 70s, when people were waiting in long lines to buy gas for their cars, they had blamed Richard Nixon for the shortage of fuel and the subsequent high cost of petroleum products.

Stein went on, saying that San Clemente was Richard Nixon's home, and was known as the Western White House when the President was in residence there. As a presidential residence, Stein said, it had undergone all the varied and extensive security checks and protective measures that the Secret Service could devise, but no one had ever thought of, much less prepared for, an assault from underground.

Stein went on that it was an engineering marvel that these hippies had somehow managed to dig a tunnel from their commune to exactly beneath the President's office in the Western White House. Somehow, they managed to figure out exactly where to burrow upward, and Nixon was alone in his office when they dug through the floor and surprised him, quickly taking him hostage before he knew what was happening.

They were furious with the President and wanted to drown him in used motor oil, which they would pump in from their commune. They had the President tied to a chair and gagged, and they began pumping oil into the room.

Ben said, "It was at this point that King Timahoe..." and I wanted to be a part of this story, so I interjected, "You mean Nixon's Irish Setter?"

"Yes, right," Stein agreed, continuing, "Well, King Timahoe knew that something was wrong, and he began furiously barking and scratching at the door, which got Ron Nessen's attention, and he called out to the President, and tried to open the door."

When the President didn't respond, and the door wouldn't budge, Nessen called in the Secret Service and they broke down the door, finding Nixon bound and gagged, and up to his chest in motor oil. They released him, and it was estimated that if they hadn't gotten there for another ten or fifteen minutes, it would have been too late.

"That's amazing," I said.

"But that's not all!" said Ben. "Remember, this was during the oil embargo, and the nation was in crisis for lack of fuel, and the President had scheduled a secret meeting for the next day with the Arab oil sheiks responsible for the embargo. Nixon was hoping to end the embargo, and the meeting was scheduled for ten o'clock the next morning in his office right there. But the office was an oily mess.

"Now, we didn't learn until much later about any secret taping devices the President had installed in his offices, and we had no idea why he insisted on having the meeting there in that office, but he did, and we were left with a real dilemma. The President insisted on using this room, but the floor and walls were covered in dirty motor oil up to about three feet high. And the meeting was scheduled for less than twenty-four hours away!

"My God, what did you do?" I asked, and started the music softly and built it up very gradually as Ben Stein brought it home.

"What could we do? We called up Miller's Carpet Care in Campbell, and sent Air Force One up there, and they sent their top carpet-cleaning crew down there. They worked all through the night, and just as the Arab oil sheiks were pulling up in the front of the building, the Miller's crew was being shuffled out the back. But the room had been cleaned spotlessly, the meeting took place, the oil embargo was lifted, and the crisis was over."

"My God, that's amazing," I said, the music getting louder, "why, you could almost say that the Miller's Carpet cleaning crew were unsung national heroes."

Music a little louder as Ben said, concluding, "Yes, you certainly could."

Then I hit the pause button on the cassette deck, waited my customary 3 or 4 seconds, turned down the music, opened the mic and said, "Well, that's it for today with Ben Stein. Chewin' the Fat was brought to you today, commercial-free, by Miller's Carpet Care in Campbell." I closed my mic and faded the music out as Gordy started another set.

I was listening for the amused, accepting chuckles from my media colleagues, but I didn't hear anything. In fact, I never heard a word about this, from anyone.

Chapter 107: More Than Once

More than once?

A few of the staff were fired a time or two for some infraction or dereliction, and a few quit in frustration a time or two. But all came back a time or two or three. Terrell Lynn barely remembers one of the times he was fired, recalling only that he was hired back by Jeremy, who "never told me anything about the music. I got hired back because he said, 'I miss your laugh on the air.'"

The book is already too long to justify the myriad retellings of who got fired for what, and how many times, but the undisputed kings of being fired at KFAT were Buffalo Bob and Unkle Sherman. Neither man remembers how many times they were fired, or for what. The need to simplify suggests that one was fired for offenses incurred while drunk, the other for insubordinations sometimes caused by drunkenness, sometimes not.

Buffalo Bob and Sherman had something of an unofficial competition for who would get fired the most. Sherman was a strong contender, but Bob was always ahead.

As the only semblance at management at KFAT, it fell to Laura Ellen to do the disciplining. Discipline? What do you do after you've asked, told, suggested, pleaded, warned, threatened, bribed, cajoled, implored, begged and insisted? You fired, and while that fate befell many of the jocks, it only regularly befell Bob and Sherman.

Leo: "I like Unkle Sherman. I fired him four times. He just doesn't follow orders."

Laura Ellen remembers trying to reason with Bob: "Well, he was totally unreasonable, and he would talk… he's unbelievable! But when he was sober, he was a great DJ. What he ought to be angry about was when he was drunk once, and I made a tape of him. He never knew what he sounded like. And the tape didn't lie, and I remember I gave it to him, and the reaction was pretty horrible."

GK: He listened to it? He didn't blow you off?

LEH: "He didn't listen to it here, but I guess Tiny told me that he just cracked the cassette open, threw it out the window, he was so upset."

GK: But was he upset at you or at himself?

LEH: "He probably blamed me, found a way to blame me. He won't take responsibility for how he was. And I was telling him that his friends were bringing him beer 'so they can listen to you and make fun of you.'"

Bob Cassidy was a problem for everyone, obviously including himself. He had lived at KFAT, worked there for three years, and he was increasingly a danger to himself and others. Ozone still remembers the headlock in the hall and the time Bob had him by the throat against the stacks. He played what he wanted and didn't put away his records because he was either too drunk to remember, or too drunk to care.

Laura Ellen knew that he was like the crazy uncle in the family- the one you avoid until you can't. She couldn't avoid the problems he was causing for the rest of the staff, and she certainly couldn't avoid the frustration she felt by his presence. He wouldn't let her forget his presence, he wouldn't follow her instructions, and he told everyone how he defied her.

Yes, Bob had been fired several times from KFAT, but he'd always been hired back, and at some point Bob became convinced that he was a valuable asset to KFAT, and that he would be a valuable asset to almost any other radio station. He began to believe that he could go to another market and succeed there. Bob would show them *his* brand of radio. His shit would shine in Los Angeles, he thought. He could get out of this cracker shit-hole, make a name for himself and some real money.

Bob was tired of the bullshit at KFAT, and he began fixating on moving to another market, another station. He kept drinking and thinking about Los Angeles. Then he began saying that he could go there, he *should* go there. Then he began saying that he *would* go there. Laura Ellen told him that she thought it was a good idea, and she encouraged him to look into it. Bob was indignant- and determined. He was *gonna look into it!*

He told her that again the next day, and Laura Ellen said, "Tell ya what, Bob, here's some motivation to look into it. You're fired." Now he'd look into it, for sure.

Laura Ellen had been looking into something, too. His name was Mark Beltaire.

Chapter 108: The Trouble With Mavis

Buffalo Bob was too much for Laura Ellen, and this time she made sure he wasn't coming back. Bob thought his shit was hot, and he took it to Los Angeles. He was going to let LA get a crack at the Bison Boogie. Tiny was seriously depressed by the loss of her friend and occasional lover, and she was getting real tired of living at the station. She had just met Sahn Berti, whose father was a bit player in films, and who owned a large property outside of Los Angeles that he rented out for films. Tiny loved Sahn and missed him when he went back to Los Angeles, and also saw in him a way to provide her kids with a steady father figure, a good, safe place to live, and regular meals. Tiny was thinking of giving in to Sahn, and moving to L.A. As much of a pain in the ass as Bob was, Tiny missed him terribly, and it wasn't as much fun without him around. He was a pain, but he was family. KFAT wasn't as much fun for Tiny as it used to be. It had been a great ride, but it was getting old.

Laura Ellen needed a person to replace Bob, and she worried over the increasingly likely loss of Tiny. And there was another loss to make up for: Frisco was drying out again, and this time he wasn't coming back. Everyone was getting edgy; the KFAT staff wasn't the family it had once been, and now they needed at least one new jock, and it would help if he knew production and could do Afternoon Drive. As Bob got more abusive and hostile and talked about leaving, she knew she'd need someone to replace him sooner or later. She was hoping for sooner, and Laura Ellen knew a guy. She'd been thinking about him a lot in the past few weeks. And then sooner became now.

Mark Beltaire was a jock from Detroit, but he had been a rock DJ in St. Louis, and he had always admired the format at KDNA. And he knew production. What he didn't know was the music, but Laura Ellen knew that no one else in America knew the music, either, and few enough would be willing to come to Gilroy or work for KFAT's idea of wages, so she had few choices. The call was made.

He was hesitant, but Laura Ellen told him he could live in a house she owned on the beach in Santa Cruz. She told him he'd be ten feet from the sand and fifty yards from the ocean. Detroit was dirty and cold, and he said he could start in a week. Knowing a new guy was coming to do Afternoon

Drive and production, the staff stretched their shifts to cover whatever holes were left open in the schedule by Bob's departure, and they warily awaited his arrival.

Laura Ellen knew that anyone coming in as a newcomer would be lost in the format, and automatically outside as a staffer. The staff had been there for three years, and had gone through so much together, and no one new was going to be exactly welcome, but she hoped they'd give him a chance. Fat Chance. But whose fault was it?

Mark Beltaire arrived in San Francisco and Laura Ellen met him at the airport, and they made their way to Santa Cruz, to unpack, then it was over the hill to Gilroy, and I will let the reader imagine his reaction to the smells, the anarchy, the crappy gear and the cold shoulder that he found there. Welcome, amigo, and get used to it.

Beltaire was an experienced radio man, but nothing in his experience could have prepared him for KFAT. He was in his early thirties, had gone slightly dough-y, with thin brown hair already receding and showing no signs of stopping. He came with his own set of insecurities, not the least of which involved his place on the staff. He saw immediately how close-knit they were, and he tried to fit in. But he didn't know the music, he hated the equipment, and he noticed that the town smelled funny. And the drive over the hill was beautiful and all, but when you have to do it twice a day, every day… But Beltaire loved the music he heard, and wanted to make this work. He would be Brother Mavis. He would do production and Afternoon Drive.

"And that was a thorn in my side from the beginning," Sister Tiny said, "because I missed Buffalo Bob." The Bison Boogie would be hard to replace.

Someone in management at KFAT realized that Beltaire was going to have a tough time fitting into this oddly functional family, and wanted to give him a good start with an unhappy group. On his arrival, he announced to the staff that he had gotten them a raise. The jocks would now be making $3.75 an hour. Now they went from completely broke to broke. Yippee! Thank you, sir. May I have another? No.

Jeremy and Laura Ellen were almost the same odd couple, but people were noticing little cracks in their relationship that hadn't been there before. It had always been difficult to get a handle on those two, and things now just seemed… off. Sully asked Tiny, "Did you notice how Jeremy…" and Terrell Lynn said to Sherman, "Jeez, Laura Ellen just said…"

Neither of them was exactly the touchy-feely sort, but now they seemed to be even more distant, and Laura Ellen already had enough on her mind, and really needed Mark Beltaire to work out. The staff was still unhappy even with the raise, things kept breaking, the sales staff might be robbing them, and Gilroy Foods was processing cauliflower.

Yucchhh. Laura Ellen just didn't need any more problems.

Chapter 109: The Size Of An Egg

The problem was that Jeremy wanted a $100,000 computer.

For one of my shows, I interviewed a scientist from NASA, who mentioned a microphone they had developed for space that he believed to be the smallest microphone in the world. It was the size of a pinky fingernail, and almost as thin- and Jeremy heard my show and wanted one. I made a call, arrangements were made, and Jeremy waited excitedly for about a week until the package arrived. He was delighted, which I hadn't seen before; he was almost child-like, and I was happy to do it. But Jeremy was like that- electronics were his life. They were his work and his fun. He communicated through electronics, and now there were these new computers that could be bought, and he wanted one. No one had ever seen one.

This would be nothing new in today's world, but in 1979, this was radical. Bob Simmons remembered Jeremy's computer because he had never seen one before, either. He'd heard of them, but didn't know that people could buy them. Jeremy was always at the cutting edge of what was either available or soon would be; he'd had a hand in some of it, himself. But any computer was cutting-edge back then, and Jeremy knew they could be programmed to do much of the work that was being done on the Telex machine now. He had to have one.

New as they were, the company had to send out a representative from Chicago to show Jeremy how to use it, and Teresa would also get instruction, as she would do the office work on it. The company sent Rita, who was used to the corporate world and instantly loved the fun and the anarchy at KFAT, and she wanted to hang out as long as she could. She was also attracted to Terrell Lynn, and wanted to fuck him, just as Terrell Lynn was having similar thoughts about Rita.

Sherman had his eye on Teresa, but she was nervous about this wild man, and she wouldn't go out with him without another woman along on the date. Enter Rita, and Terrell Lynn convinced Teresa to go out with her and him and Sherman, so he could get with Rita. Sometimes we are asked to make sacrifices like that.

Teresa had recently been divorced, and so had Sherman. Teresa was gorgeous, but she was friends with the hated Geneva, and Sherman was… Sherman.

They were both interested, but wary. Dinner for four included a bottle of wine, then another bottle of wine, then it was back to the station where someone had stashed a bottle of wine, but Sherman knew where it was. The third bottle and all four people went back to the house shared by Sherman and Terrell Lynn, and it was consumed before both couples went to both bedrooms. Sherman and Teresa each had other relationships brewing, but by morning they were a couple, and they are still together as of this writing. Teresa laughs now when she remembers, "I didn't want to hook up with Sherman. I was drunk."

Thus, Terrell Lynn also got his pipes cleaned, as did Rita, who no longer wanted to go back to Chicago, so you would think that everyone was happy, but that would not include Brother Mavis.

Remember that Teresa had another relationship in the wings? She had gone out with Mark Beltaire twice, and he had taken a very possessive attitude about her. He was insanely jealous when he found out that she had gone out with, much less *slept with,* that drunken lunatic Sherman Caughman. In fact, it wasn't going so well in general at KFAT for Mavis- not with the format, not with the staff, and this new disappointment made an already unhappy Brother Mavis *really* unhappy.

As the overnight guy, Sherman kept overnight hours, and was still sleeping, but Teresa, needing to keep "office hours," was at the station when Brother Mavis showed up the next morning for work. She knew that she had better tell him about what had happened the night before, because for sure either Tiny or Sully would, or he'd hear it on the air from Gordy, and she asked if she could speak with him in the empty Production Room. What transpired next indicates that Bro did not take the news well.

On the air, Terrell Lynn remembers watching what was going on through the window, and he saw Mavis grabbing Teresa by the arm, not allowing her to leave. She was struggling to get away from him, and he wasn't letting go. Terrell Lynn raced out of the studio and down the hall into the Production Room and got between them, screaming "Let her go! Let her go!" and Mavis ran out of the Production Room, down the stairs, and out the front door of the station. His blood was up and he was on his way to Sherman's house.

Because Sherman was doing the overnights, he'd covered his windows to keep the daylight out. In the darkened room, he wasn't sure who was yelling at him, jolting him out of a deep, drunken, sex-soaked sleep. It was Mark Beltaire, and he was raving, "She's my woman! I was with her! She's with me!"

Pissed off and angry that now he'd never get back to sleep, Sherman demanded to know how the guy had gotten into his house, then remembered that the door was never locked. Still, this was bad shit, and this dirtbag asshole was in his bedroom yelling at him for seeing the woman he had just started seeing and wanted to keep seeing. Beltaire stood at the foot of the bed, half yelling, half crying, saying that Teresa should be with him. Sherman sat up and yelled, "Well, *fuck you!* If she was your woman she wouldn't be going out with *me!* Get the fuck out of my house before I get out of this bed and whip your ass!"

Knowing Sherman would do it, Mark stormed out, but there was no denying the logic of what Sherman had said, and he skulked back to the station. Knowing that Teresa would soon hear about his visit to Sherman, he was miserable, but he just couldn't let it go. For a week, he glowered and mumbled at Sherman and pointedly ignored Teresa. Now he was truly isolated at the station. Laura Ellen didn't need the trouble and neither did Teresa, so Sherman said he'd take care of it. He knew some guys.

The guys he knew were Mexican laborers that came by the station to hang out once in a while. They liked Sherman and liked that he was serious about a Mexican girl, and they were sympathetic when Sherman told them what an asshole Mavis was being. Over beers, sympathies were expressed, suggestions were made, and options were considered, until a plot was hatched and money changed hands. It was August 3rd, 1979, in case you're interested.

The next night, as Mark worked alone in the Production Room, bent over the editing block, he must have heard something, and looked up just as a 4-inch steel bolt came flying through an open window, flew right past him, barely missing his head, and crashed into the carpeted wall of the room, making a loud, angry, clanking sound when it fell to onto the console. What? What was that? It shook him up.

TLT: "Missed Mavis by about an inch. And Sherman paid this kid to do it, and he told me about it, to watch out."

Sully: "I remember that, because I remember that I was horrified- not that he'd be violent like that, but also corrupt a young person to do it. I wasn't surprised, but I was horrified."

It shook up Brother Mavis, too, but not as much as <u>the earthquake</u> the next morning.

That happened at 10:05, right after Terrell Lynn went on the air. It was a 5.7 quake, centered in Coyote Lake, about six and a half miles away. Tiny and Laura Ellen were in the Production Room, and they remember the tables bouncing around in there. Terrell Lynn was on the air, playing "Galvanized Tub" by Little Jimmy Dickens, and Mavis had just walked in the door. In fairness, it's possible that life in a geologically stable area like Detroit had not adequately prepared Bro for an earthquake, but in the event, just as he walked into the studio, the station started shaking and moving under his feet, and Brother Mavis stood stock still, paralyzed. *What... the... FUCK?*

The needle was skipping over the record, the tables were bouncing, the walls were shaking, and the records in the library were tumbling off the shelves.

Everything around him suddenly made no sense, and Mavis ran back out of the studio and turned a quick left and headed for the nearest exit, the back door. He got to the door and jerked it open, unfortunately without accurately calculating the distance between his face and the door, and he jerked the door into his forehead, *SMACK!* dealing himself a sudden, severe blow. It shook him up.

After the blow, he stopped and stood there for a moment, dazed, looking startled, then he seemed to snap out of his stupor, and, coming to his senses, with a jerk that showed he was back in real time, he reached for the door again- and jerked it into his forehead a second time!... *SMACK!*

With all the noise and the clatter and confusion caused by the earthquake, Terrell Lynn heard the first *SMACK!* and turned to see what was happening at the back door, where he watched, amazed, as Mavis stood there looking stupefied, and then shaking his head to clear it, he slammed the door into his forehead a second time, and Terrell Lynn heard the second *SMACK!* as the door hit Mavis' face, and then, even with the sudden chaos of sights and sounds caused by the earthquake, he watched, fascinated, as Mavis stopped, looked dazed again for a moment, seemed confused as he turned this way and that, then turned to the door again, righted himself, took a breath, reached for the door and slammed it into his forehead a *third* time, *SMACK!* Thereupon, Mavis stood still for a couple of breaths, then turned around and slowly wandered back down the hallway in the other direction, with a confused look about him.

Terrell Lynn: "...and I was stunned, I was mesmerized. I'd never seen anybody do that."

SMACK! SMACK! SMACK! Terrell Lynn stared at Mavis, then turned his attention back to the stacks, as the records were jumping off the shelves, onto the floor. Back in the studio, the song was still trying to play as the earthquake hit, with the needle now jumping around the record. Little Jimmy Dickens was trying to say "I'm out of business" when Terrell Lynn shut down the turntable, opened the mic and said in a shaky voice, "Well I'm sorry, but *we're* out of business. We just had one hell of a shaker, boy... we just had a *rocker...!*"

Then it was over, and people went out the back door, knowing that after-shocks were likely, and the flimsy upstairs studios of KFAT was not where they wanted to be. Out the back door, the air was still and it was eerily quiet. No birds were singing, there was no traffic noise. The overhead telephone and electric lines were still swinging wildly between the poles, and they waited a few minutes on the ground out back for everything to settle down. When it seemed to be over, a simple survey of the damage back inside showed that little had changed other than that some records had been thrown from the shelves, and all the cassettes were thrown from the 7-Up cases.

The electricity was on, the phones were ringing, and listeners were calling.

Terrell Lynn cued up the song that had been playing when the earthquake struck. He felt it was a nice way to say "we're all carrying on." There was a lot of excited chatter as they were picking up the records and talking about the event, when the cops showed up.

The first place the cops had gone to, as soon as it stopped shaking, was the Continental Phone Company, to see how bad the damage was. The phone lines at the police station were all down, and they needed all their communication options working in the aftermath of the quake. "Sorry," they were told at Continental, there were "no working lines coming into or going out of Gilroy," and no one at Continental Telephone knew how bad the damage was, or how far away. No one knew where the quake had been centered, no one knew how big it had been, no one knew anything. They were told there was no communication between Gilroy and anywhere else, and the cops were discussing how fucked they were when an engineer at Continental told them to go two doors down to the radio station.

He told them that that guy, Lansman, had some pretty advanced ideas about how to send a telephone signal, and that he wasn't tied into their system. The engineer said "that guy was doing something different with his phones," something he didn't understand, but to go over there and ask. They'd work on getting the service back, but right now there were no working lines into or out of Gilroy. The cops went two doors down, up the stairs, and found Laura Ellen talking on the phone. All the lights on her phone were lit and she was talking to listeners. She saw the cops and got off the phone, and they explained that all the other phones in town were down, and KFAT had the only working phones in Gilroy. Could they use them?

Of course they could.

The phones worked just fine, the cops spent about an hour there, the crisis was handled, and they were seriously grateful. "No problem..." said Laura Ellen "...always glad to help."

Sitting in the office where someone had brought him and put him in a chair, Brother Mavis sat quietly with a bump on his forehead the size of an egg, and he didn't know how it got there.

It was a Jeremy Lansman Operation.

Chapter 110: Alligator Tears

For Sister Tiny, Brother Mavis "was a thorn in my side from the beginning, because I missed Buffalo Bob. They hired him to be the Program Director, and also he was the Production Director. He put on Joan Baez in the afternoon drive time, and let it run, the longest song, and all these long, slow songs. I was living in the station then, and he would go in the Production Room and do production."

Cuzin Al: "While it was tracking?"

Sully: "That's right, that's right!"

Sister Tiny: "Yeah, and finally I said to him—I couldn't take it no more—and I said, 'You know, really, it's Drive Time, and maybe Joan Baez is kind of slow... don't you think?'

"And he just looked at me, and then he went to talk to Jeremy, (baby voice:) 'Tiny blah blah blah, Tiny said this...' You know what he said to Jeremy? 'It's either Tiny or me.' and Jeremy fired him. That's when he got fired. I got rid of Brother Mavis!"

TLT: "He would come up to me and say, 'Terrell, no one's gonna tell you this, but today they're going to fire you.' The next day he'd come up and say, 'I think today's the day they're going to fire you.' And after two weeks, he comes up, and he's got alligator tears running down his cheeks, and he goes, 'They fired me'!"

LEH: "He never got along with the listeners. He never made that... he never got the trust of the listeners."

Chapter 111: Gordy Good-bye

I haven't talked about this very much. I've mentioned it, I've alluded to it, but I haven't pushed it very hard. Gordy was sick a lot.

He said his stomach was killing him, and eventually it would, but in the meantime, it hurt a lot. The staff would cringe when he talked about it on the air, but a few ladies would call to talk to him about it, and he liked the attention. He continued to talk about the discomforts within his bowels during his show, but then he talked about it less. Everyone at KFAT was glad that he was talking about it less, not knowing that it was getting worse.

He had talked about his troubles for over three years, but he kept the worst of it away from his friends, the listeners. But he couldn't keep it away from his friends at KFAT. For years everyone knew that you had better go to the bathroom before Gordy went, because only the reckless and the insensate could go in there after Gordy.

Sister Tiny: "Gordy used to go on and on and on about his operation and his… about the soap opera that he watched, and his operation, and about his kidney stones… And he had all those women riveted to their radios. But it turned me off. I didn't like it."

Then there were days that Gordy was too sick to come in, so Terrell Lynn volunteered to take over his shift, in addition to his own. Gordy was family.

He kept the worst of it away from the listeners, but it was getting worse, and he couldn't keep it away from Laura Ellen. And then there came a time when Gordy was too sick to go on the air.

Gordy needed medical attention, and he needed to move back to San Jose to get it. The State would have to pay for the reconstruction of Gordy's stomach, and he needed to live at home with his parents to do that. The guys at the station packed him up and loaded Sister Tiny's pickup, the Red Dog, and they drove him up to San Jose and unloaded all his stuff in his parents' basement.

Terrell Lynn took over Morning Drive, a couple of shifts were moved around, and a woman named Susan Uhouse began working some of the overnights.

The station was bombarded with calls about Gordy for a few days, and then Terrell Lynn stopped putting them on the air, and went on with the music. It was all about the Flow.

Chapter 112: The Center Does Not Hold

Jeremy left town. One day he was just... gone. In the office, Teresa overheard that he was in Colorado. Laura Ellen was moping despondently in the office the day after Jeremy left, and told Teresa it wasn't just Jeremy and the station that were on her mind, she was pregnant by the guy handling the T-shirt sales after Awest had fled to L.A. Everyone could see she was in bad shape.

That was when Sister Tiny told Laura Ellen that she was going to marry Sahn and move to Los Angeles.

Half of the key staff was gone, the rest were in a constant snit, sniping at management and the sales staff. The sales staff was at war with the air staff, Laura Ellen had no friends anymore among the jocks, and the only friend she had at the station was Lori Nelson. The jocks were talking about the station's problems on the air, and listeners were calling in to support them. Terrell Lynn was still doing the Fat Grams, which were now full of listener's support for the jocks, and Laura Ellen was thinking about taking the Fat Grams off the schedule. Jeremy was gone, Elsie needed attention and an explanation, and everyone needed something from Laura Ellen, who was abandoned, isolated, depressed, overwhelmed and exhausted. She had a five star clusterfuck on her hands, and everyone was looking to her for a solution. She had to do something about it, somehow. She had to fix it all. What should she do first? Laura Ellen was trying to breathe.

Chapter 113: Just Like That!

No one knew what went on in Laura Ellen and Jeremy's bedroom, and no one wanted to speculate. They weren't what you'd call demonstrative with each other, and sometimes one or the other would sleep elsewhere. Laura Ellen was touchy-feely with Elsie, but not with Jeremy. Jeremy was pleasant, or he was distracted or he was gruff or abrupt or happy or unhappy, but he was mostly a cipher. Who knew what went on between them?

Later, after it was over, there were rumors about Jeremy having a breakdown.

LEH: "Well, I suppose Jeremy was always pretty close to having a breakdown, anyway. He was never far from it, I don't think. Jeremy was always operating on the edge, and he may have had a breakdown, but I was too close to it to notice. I didn't notice that he was... I mean, I know that he was under a lot of pressure. There came a time at KFAT that we were both under a lot of pressure to make the thing work, and we were just... the way it was, it just wasn't gonna work..."

GK: Did he just say- That's it, I'm gone?

LEH: "Well, it was one of those breakups where we had grown further and further apart, and I remember saying, 'you know, Jeremy, this just isn't working any more,' and he said, 'you know, you're right.' And that was it- he just packed up and left. It had been coming for a long time. We both knew it, we weren't admitting it to ourselves. No drama. It was pretty quick. We both knew that..."

Sister Tiny: "I think when the radio station got real stressed out, because I remember Laura Ellen talking to me, saying, 'Oh, man, it's all screwed up, and it's really hard to take...' I think they loved each other. But I don't know what happened or why they broke up."

Cuzin Al said: "I don't either. Nobody even talked about it."

ST: "What happened was there was a huge rip-off. When the rip-off happened, it was Lori, when the money got stolen... Teresa and Rebecca Fenker had proof that they had stolen the money, McLain and Lori. They were in league together, the station was making *tons* of money at this point, and *all the paychecks bounced.*"

ST: "Laura Ellen said to me, or somebody, it might have been Jeremy, that there was no money and a bunch of money got stolen, or just was gone, they don't know what happened to the money. All this money was missing because Jeremy knew nothing about business and trusted totally his business manager.

And his office manager. And at *that* point, the relationship started falling apart. And Laura Ellen made noises like, 'It's getting tough…'

ST: "I think that there was just so much stress that it was over. Laura Ellen had enough, because I remember her crying and her saying, 'I just can't take it.' Jeremy was gone all the time, he was always on the tower, the radio station was taking its toll, and they—Lori and Bob McLain and Geneva—they all knew what the fuck they were doing… "

By this time, Lori and Cuzin Al were a couple, and Al swears that if Lori got any money, he doesn't know where it went. "If Lori got a lot of money, I don't know what the fuck she did with it, 'cause she was the worst money-handler I've ever seen."

ST: "You know, depending on your lifestyle, it could be gone right away."

CA: "Yeah, but Lori didn't do drugs, so she didn't spend it on drugs, she did not even own a car, she had no credit."

ST: "I'm not sure that Lori was in on it. But I'm sure that happened with Geneva, because Jeremy told me that."

Sully: "Terrell Lynn had a friend who was an accountant, and who used to go to companies specifically to find out what was wrong, and she volunteered to do it for free, and we said at one of those meetings, 'Terrell Lynn has a friend who is an accountant, she does this for a living, she has volunteered to come in and look at all the books, and find out where the money is going.' And the next thing we knew we were out of there."

The rumors soon generated accusations, and all the jocks were pissed off at both of the office staffs, and both office staffs knew that the station would run smoothly and profitably if it wasn't for the jocks. Everyone at KFAT was confused, pissed off and stressed out; no one knew what was going on, but everyone in this frustrated, failing family knew KFAT was unraveling. In any event, one day Jeremy was gone. Just gone. No one at KFAT knew exactly when he had left, and at first no one even knew where he was. He was gone- just like that.

Rumors spread among the jocks like wildfire, dominating every conversation, and they wanted to talk to Jeremy, but were told that he was working on a mountain and couldn't be reached. But Teresa was the spy in the ointment, and she told the jocks that she knew Jeremy was in Colorado, up on a mountain, building a transmitter for a friend, but McLain and Laura Ellen were talking to him every day. More lies! More furor!

The hopes and dreams the jocks had harbored since before the transmitter move had not gone away. Instead, their hopes had turned to frustration, the frustration turned to anger and distrust, and by now had become a roiling, bubbling stew, ready to boil over. Leo had been involved and then retired to his pleasures in his mansion in the hills above Los Gatos, but he knew that Laura Ellen needed him again, so he wasn't surprised when she called.

Chapter 114: Jeremy Gone

Laura Ellen was an able administrator for a staff of dysfunctional freaks, but she was no businesswoman, and Bob McLain was a steamroller, used to getting his way or rolling over anyone who stood in his way. The station always needed money, and the disconnect between the air staff and the sales office had everyone taking sides.

What made KFAT Fat? It was the music and the jocks. What would make KFAT profitable? It was Bob McLain and his staff. None of the jocks knew who had the ultimate power, but with Jeremy out of the picture and the two staffs at war, Laura Ellen's diplomatic skills were punctured by all the pain in her heart. Laura Ellen was overwhelmed and Bob McLain seemed to be running the show from San Jose. Laura Ellen was talking a lot more with Leo, who she trusted implicitly. Leo was a ruthless man, and no fan of radio, but these were her friends they were talking about, no matter how they may feel toward her. The jocks were family, but they all distrusted her now, and she was having a hard time even talking to them. No one on the air staff knew what Leo was saying to Laura Ellen, but they all knew he was no friend of theirs. They knew that Lori Nelson was her only confidant at the station, and as the two women grew closer, Laura Ellen seemed to give up more power to McLain.

Lori was from the business world, and Laura Ellen was lost, and increasingly dependant on her. Laura Ellen had to bow to the logic of the business side, and she knew changes had to be made, but what would they be, and who would have to change- or go?

Inside the whirlwind of chaos, McLain was shouting that the jocks had to be replaced. He'd offered the compromise of letting them stay if they could stop fucking with the ads, read them live like professionals, and take the dirty songs out of the library, but they'd refused to do that. Laura Ellen knew that for those people to stay, they had to change.

Change? What were the odds? For most of the jocks, this was their first radio gig and they had learned on the job, this job. Their brand of outlaw radio was all they knew. What they knew made KFAT unique and wonderful, and they would fight to keep it that way. Word went out that a meeting would take place.

The jocks were having trouble paying rent, and they remembered Bob

McLain buying a boat in Hawaii. They were making $3.75 an hour and the Sales Assholes were getting rich on their backs. Something had to change, all right. Laura Ellen started the meeting and quickly handed it off to McLain.

The meeting was acrimonious, as the jocks demanded to know when they'd get a real paycheck- "you know, a paycheck with a little something left over after rent and food?" The barbeque at McLain's almost a year ago hadn't helped anything, and since then the Ozoner was gone, Buffalo Bob was gone, Gordy was gone, and Sister Tiny had just left. The old days were gone, and what replaced them wasn't much fun, and it still wasn't paying the bills.

Everyone who spoke at the meeting had a point to make, but no one's point changed anything, and the meeting ended in a stalemate. Both sides were pissed off and nothing was accomplished. Bob McLain took note.

Chapter 115: The Big Axe

After the meeting, Bob McLain was on the warpath. He had this job, and he wanted to make money at it. It was a good signal with good coverage in a good market, the station had good name recognition, but those lawless, drugged-out freaks at the station were keeping him from making it work for him. The station always did well for its local sponsors, but the real money was with the nationals, so fuck the locals. The money was with the nationals who wouldn't touch this station the way it was now, so *fuck* those assholes at the station. *He'd* show them.

Now that that meddling hippie, Jeremy, was gone, McLain knew he had the upper hand. He wanted to tighten up the sound at KFAT, shorten the playlist, and discipline the staff. Of course, *that* staff would have to go. The playlist would still be good, you know, but… limited. He knew that station loyalty was a myth, and a certain percentage of hardcore listeners would disappear, but if he could give the rest some of the good artists that the station had already been playing, they'd stay tuned…

And he needed a guy, *his guy*, to run the station. He knew the guy.

Doug Droese had taught broadcast classes at San Jose State, and had been the Station Manager for San Jose rocker KOME when McLain had been there, and that was a tightly run, successful station. He and Droese had crossed paths in San Jose for years, and McLain knew Droese could manage KFAT, once the shit-heads were gone. McLain told him that the station was out of control, and he wanted him to wield a big axe there. He needed someone who could put together a list of artists and a "clock" to play them by. He also wanted him to get the station a staff that knew what a clock was, and what an ad schedule was, and would keep to the plan. Droese said he could do that.

Laura Ellen was increasingly disconnected in ways she never knew existed. She agreed to McLain's insistence that a new managerial force be hired. She was beat up and she needed help, and this was it, she guessed. She remembers, "Here was our theory: well, everybody was complaining that they weren't making enough money, and we said, 'if you want to make more money, you have to become more of a mainstream radio station. To become more of a mainstream station, you probably have to have a mainstream PD.' So we brought in a mainstream PD and they had a revolution."

GK: Who did he take over from?

LEH: "He took over from me. I said 'okay, I'll bow down.' If everyone wants to make more money, I'll step aside, 'cause I still don't want a commercial sounding station, but if you want more money, the only way we could think of making more money was to bring in somebody who knows how to make it more mainstream."

Word about a coming change leaked to the staff through Teresa, who was keeping her ears open in the office. She remained the sole source of information from the distant shores of the two offices, one in San Jose and one a few feet away from the control room, but separated by an impenetrable wall, whose bricks were betrayal, and whose mortar was distrust.

Everyone on the air staff knew that something was coming. Something had to happen, and they all knew it was coming soon. On Wednesday, the 31st of October, 1979, the air staff got the word: a new Program Director had been hired. His name was Doug Droese, and he would be in to start his job tomorrow. What about Laura Ellen? How bad was this thing? Laura Ellen would still be Music Director. The jocks knew it was bad, but maybe it was still salvageable if Laura Ellen was still there.

Then Laura Ellen came into the studio and said that there was going to be a staff meeting for everybody on Thursday morning. *Everybody!* If you weren't on the air at the time, you were to be at that meeting. Then she posted a notice about the staff meeting. Terrell Lynn volunteered to call me to tell me to be down in Gilroy for the meeting the next day. It didn't matter when your shift was, we heard- be there!

Droese came down to Gilroy the day before the meeting to look around and make an impact. He took some records out of the stacks, and took others from the hands of the DJs as they took them off the turntables. The jocks were frightened: they wanted to play the cool songs, but they didn't want to be responsible for getting them taken out of the station. Finally, it got dark and Droese left. The slaughter was over for the night.

Sully: "Laura Ellen was there, and she was not in good shape, you know... she was pretty upset and stuff, and then we kept trying to get in touch with Jeremy, and we couldn't get in touch with him. *And everybody was so tense*, because we were so freaked out because we were just waiting for the other shoe to fall. And we kept calling, we kept talking to McLain and Lori and saying "What's [happening], and we've gotta get in touch with Jeremy..."

TLT: "That's what they *said*- they couldn't get in touch with him, and Teresa said they talk to him every day. They just *told* you guys they couldn't reach him."

Sully: "'Cause we were trying to get ahold of him to say 'this is what's happening, and we're scared this whole place is going to fall apart.'"

That Wednesday, the day before the big meeting, Teresa alerted them that

Jeremy was on the phone with the station right then from Colorado, and they came into the office, now enemy territory, saying they knew he was on the line, and demanded to talk to him, and when Sully was given the phone, she told him what was going on at KFAT.

Sully: "But we were all really, really tense, and we finally talked to him on the phone... and we said 'You gotta do something.'"

TLT: "He said that you and I could do whatever we wanted. If we could bring the ratings up by January or February, when the next Arbitron came out... And that's when they fired us."

Cuzin Al: "That's right! That's right! I remember that!"

TLT: "We talked to him on the phone, and told him we had a deal with him, and he said 'You guys go ahead and do it. You can take it, and if you can change it around and make..' and it *was* making money."

ST: "But he never told Laura Ellen that! He never told Bob McLain that, and he never told Lori that."

TLT: "And Laura Ellen told me, she goes, 'Well, you can't prove it.' I was like, ohhh... I've heard about media jobs, oh, fuck, here it is."

On Thursday, November 1st, Doug Droese came in and called the meeting to order, telling them that there were going to be some changes at KFAT. He was there to tell them about the changes, and he was there to make the changes, and to make sure that it all worked the way he was hired to make it work.

He gave everyone <u>three xeroxed letters</u> on KFAT stationery, each letter dealing with an issue. One letter gave the new schedule of shifts. Weekdays it would be Sully, 6-10 a.m.; Terrell Lynn from 10-3; Laura Ellen from 3-7; Sherman from 7-12; Susan Uhouse from 12-6. Weekends would mix staff, and bring in Rocket Man, Susan, and some fellow named J. Cotter, who no one remembers. Also, Cuzin Al would now do a six-hour shift, from 6 until midnight on Sundays.

My shows would continue as scheduled..

The second letter dealt with the frequency and exact placement of the phrase, "the Fat One" each time a DJ back- announced his or her set. Also suggested were the frequency with which they had to mention their names, and when to take ID breaks.

The third letter was the hardest-hitting, and it broke the back of KFAT. It was titled "Re: Music Policy." It had three segments, the first of which listed the twelve artists that they were allowed to play, in which order, and only after the station ID at the top of the hour. They were:

The Eagles	Greatful Dead [sic]
Ry Cooder	John Stewart/J.D. Souther
Bonnie Raitt	Stephen Stills/Buffalo Springfield
Jimmy Buffett	Waylon Jennings
Atlanta Rhythm Section	Van Morrison/Taj Mahal
Pointer Sisters	Kris Kristofferson

The second category listed artists to be played "only on a rotating basis—at :20 after break." They were:

Bob Dylan	Jackson Browne
Marshall Tucker band	The Band
Linda Ronstadt	Leon Russell/Merle Haggard
Willie Nelson	Mickey Newbury/Tom Waits
Crystal Gayle	Charlie Daniels
Neil Young	Allman Brothers

The third category said simply, "Play Oldie Rocker every other (even hours) :40 break."

Examples:	Everly Brothers
	Rick Nelson
	Fats Domino
	Creedence
	Elvis Presley
	Freddie King
	—ETC—

The letter concluded with: "Remember you are trying to entertain 1st and inform, educate & expand listener's horizons 2nd."

The letter had no please, no thanks, and no signature.

This was shit! This was *SHIT!*

Everyone was shocked. Shocked and dismayed. It was as bad as it could possibly be. Sully noticed that there were only three artists in the first two categories that were even *country!*

Droese also told the jocks that they were no longer allowed to drink beer on the job- and anything stronger was also strictly forbidden. No drinking? And the rest? This was an outrage! This would not stand!

This was *SHIT!*

This was *SHIT!*

They didn't know what to do. They were hired to do a job, and that job wasn't sticking to this list. The music on the list was okay and all, but it wasn't what made KFAT Fat. If you've read this far, you don't need a list of the artists who were now banished from KFAT. The jocks were morose.

On Friday, Droese was back for another meeting. This time he came with a man named Joe, who the staff—through Teresa—suspected McLain had recommended as a replacement for Terrell Lynn. A confrontation, by which I meant a meeting, was convened, and few words were wasted before acrimony met inevitability.

TLT: "They were going to fire me, so finally Doug Droese, he couldn't face it, there was this whole room of people, and I said, "OK, man, let's go in the other room, you and I'll talk this out."

"He thought I meant 'let's you and me go outside and fight,' so he said, 'Well, look, I want to keep you on weekends, maybe for a little bit, but I don't think the daytime…' so I said, 'You know, you are so full of shit that you don't know what you're doing, you don't know what you're talking about, you don't know anything about this station. Fuck You.' "

"He gets on the phone [to the office in San Jose] and he says "Someone better get down here and reaffirm my position at this radio station."

GK: "So what happened?"

TLT: "Nothing. So I didn't get fired, nothing." Joe had said nothing at the meeting, and he followed Droese out the door, on their way to see McLain in San Jose. That was on Friday, and nothing much happened over the weekend other than the spreading of new rumors. On Monday afternoon, Terrell Lynn was informed that Droese wanted to see him the next morning, Tuesday, immediately after his shift. Everyone knew that it was to fire him.

Weeks before, Terrell Lynn had been asked to host the Fat Fry up in Palo Alto that Monday night, and as usual, Sherman would be at the controls in the studio to hand it off to Palo Alto, and then take it back when the show was over. Sherman would be on until midnight, when newcomer Susan Uhouse would take over.

Terrell Lynn got the message about the meeting with Droese that next Monday morning, he did his shift, then left for Palo Alto.

At the Fat Fry that night, Terrell Lynn told the audience and the listeners that the station was "coming apart."

Photo Section Two

Sherman At A Marijuana Festival

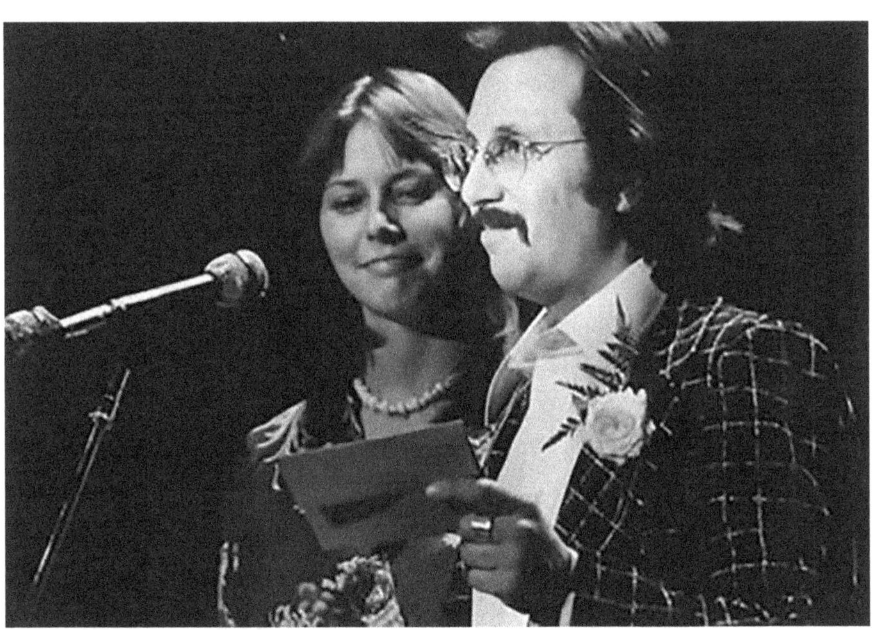

Gilbert As A Presenter at The 1st Bammies Awards
(This is a Jim Marshall Photo)

Sully, Amber and TLT On Back Deck With Chuck Wagon & Co.
(What's That Smoke From On The Left)

TLT and Chuck Maultsby At Work In The Production Room
(More Smoke)

Rocket Man And Frisco At Work

Teresa Cortez

Gilbert, Mick Jagger And Laura Ellen

Gilbert With Dr. Tim Leary
(Please Note The Distortion In The Photo – Discuss

94½ FM KFAT
I FOUND IT
AND IT'S HARD TO FIND TOO!

I Found It!

"Captured Russian Submarine"
(Gilbert Admits The Story May Be Factually Challenged)

Travus T. Hipp

Cuzin Al Taking a Joint While Lori Nelson Looks On

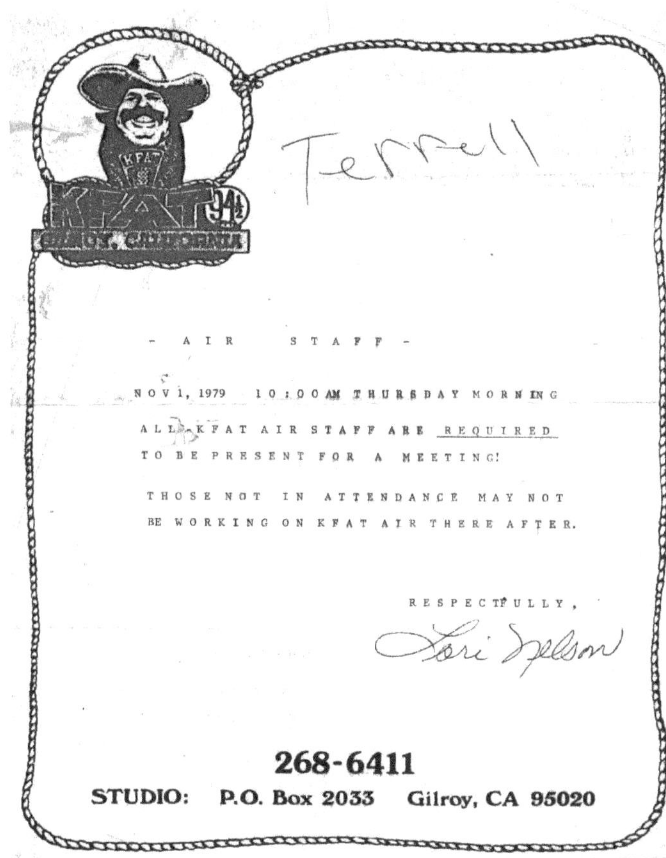

KFAT 94½

Terrell

- AIR STAFF -

NOV 1, 1979 10:00AM THURSDAY MORNING

ALL KFAT AIR STAFF ARE REQUIRED
TO BE PRESENT FOR A MEETING!

THOSE NOT IN ATTENDANCE MAY NOT
BE WORKING ON KFAT AIR THERE AFTER.

RESPECTFULLY,

Lori Nelson

268-6411

STUDIO: P.O. Box 2033 Gilroy, CA 95020

The Notice For The Dreaded Staff meeting That Caused Chaos

RECEIVED FROM Teresa Cortez for self and Uncle Sherm — Nov. 21 1979 — $ 100.00

REMITTANCE APPLIED AS INDICATED BELOW			PREVIOUS BALANCE OF UNPAID PRINCIPAL	CASH	X		ACCOUNT NUMBER	ML 206754 ML 206756
Amount of Paym't			$ 300.00	M. O.			BANK NUMBER	
Days Charges			CR. 100.00	CHECK			DUE DATE	11-13-79
Credit to Principal			$ 200.00	PMT.			PAYMENT NUMBER	2
			BALANCE UNPAID PRINCIPAL	CHANGE				
TOTAL AMT. PAID			THANK YOU	BY				

DAVE RICHARDS BAIL BONDS
827 Front Street, Santa Cruz, CA 95060
BY Dave Richards

No. 1410—LAW PTG. CO.

Bail Deposit – Someone Knew Trouble Was Coming

Format change denied

Four disc jockeys fired

By WALT GLINES
Dispatch staff

Have you been hearing different things on KFAT radio lately?

Well, there are three reasons to choose from.

· The county station with a rock music flare is missing 25 percent of its albums and police are investigating what has been labeled a theft.

· Four of its disc jockeys, including three each with four years experience, were handed walking papers last week.

· Depending on who's talking, the station has somewhat changed its format to be more rock oriented.

Also, the fact that it's rating month must be considered. Advertising time and revenue are mostly determined by the number of listeners. Stations are using all sorts of ploys to attract attention and listeners at this time.

KFAT CONTROVERSY — Ex-KFAT disc jockeys, from left, Uncle Sherman, Terrell Lynn Thomas and Cathy Reddy, known as Sully to listeners, relax Friday afternoon in a Monterey Street bar after an unsuccessful attempt to retrieve their personal belongings from station offices. They were fired last Tuesday, accounting for the glum looks.

in KFAT shake-up

Latest report to police is that 1,200 albums, valued in excess of $7,200, have been stolen from upstairs offices at 7459 Monterey St., Gilroy. The records were at the station at 9 p.m. last Monday but had disappeared by 4:30 a.m. the following morning.

Lori Nelson, general manager for the past four months, flatly denied there has been a change in format at the station that has put Gilroy on the map almost as much as has garlic.

Three of the four fired disc jockeys — Uncle Sherman, Terrell Lynn Thomas and Cathy Reddy, better known as Sully to FAT listeners — claim they were told not only what and when to play, but also what to say while on the air. The fourth DJ, Susan Chonae, was unavailable for comment.

Teresa Cortez, office manager, said she and MaryAnn Jelly, a receptionist, were also fired last week.

The disc jockeys say the firings were completely unexpected and that they received only a week's severance pay. The DJs have hired a Santa Cruz lawyer in an attempt to recover personal belongings they say are inside KFAT offices.

An attempt Friday afternoon to pick up the items was unsuccessful even with a Gilroy policeman on hand. Nelson refused to release any property, stating that was an option given to her by local police. Police, in turn, said Nelson was advised to release any purely personal items and that the station could hold onto property whose ownership was uncertain.

Also missing are the Federal Communications Commission licenses of the disc jockeys.

One of the Articles Written After The Great Record Heist
(Photo Taken At The Green Hut)

BY ENDORSEMENT THIS CHECK WHEN PAID IS ACCEPTED IN FULL PAYMENT OF THE FOLLOWING ACCOUNT

WHEATSTONE BRIDGE ENGINEERING CO.
DBA K F A T 94.5 FM RADIO
P.O. BOX 2033 268-6411
GILROY, CALIF. 95020

NOV 6 19 79

PAY TO THE ORDER OF THREE HUNDRED SEVEN DOLLARS AND 12/100 ———————— DOLLARS $ 307.

SHERMAN CAUGHMAN

Not Sufficient Funds NOV 13 1979

SAN JOSE, CALIFORNIA 95128

Insufficient Funds – Again!

At Sherman's After The Record Heist

Kathy Sully Roddy

Rebecca Fenker and Terrell Lynn Thomas

Sherman and Teresa

Bumper Sticker to Promote Concert Raising Money for Legal Fees

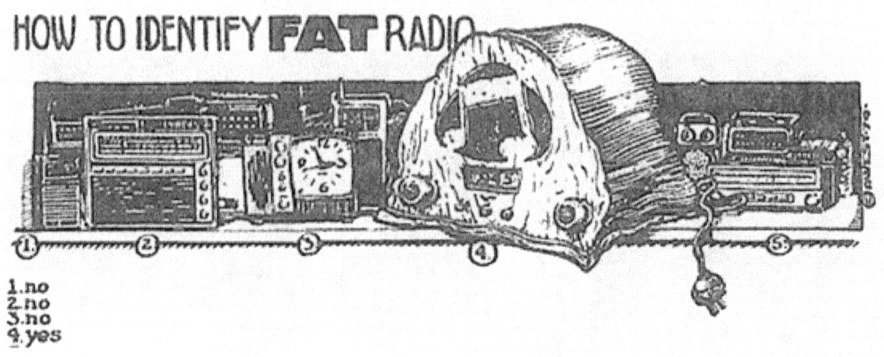
HOW TO IDENTIFY FAT RADIO
1.
2.
3.
4.
5.
1.no
2.no
3.no
4.yes

CAN YOU FIND THE FAMOUS DISC JOCKEY FROM KFAT RADIO (FM 94½)?
A.
B.
C.
D.
E.
F.
G.
LAST TIME WILLIE TALKED TO ME, HE SAID HE WASN'T GONNA BE BACK IN THE BAY TILL–
BULLTURD HE'S IN L.A. RIGHT NOW WORKIN' ON DOLLY'S LATEEST...HIS MANAGER CALLED ME–
WELL HIS BASS PLAYER TOLD ME THAT AUSTIN WAS A NICE PLACE TO VISIT...BUT QUITE FRANKLY, I'M SURE YOU'D AGREE–
THEN I TOLD DAVID ALLEN COE WHO THE HELL HIS REAL MARKET IS, AND DO YOU–
...Could one of you guys front me a brew?
WELL IF HE WASN'T SO DERN COCKY...NOW TAKE SOMEBODY LIKE THE MARSHAL TUCKERS–OR THE CHARLIE DANIELS BAND FOR EXAMP–
©AWEST-79-
ANSWER:(F)

AND NOW ITS TIME FOR A FAT FRY!
©AWEST-79-

Itz time four de fat Fries!
© AWEST-79-

CAN YOU GUESS...?
HOW MANY THINGS ARE WRONG IN THIS PICTURE OF A FAMOUS RADIO STUDIO (ANSWERS BELOW)
PORK-O
KFAT
94
2.2.2
ANSWERS: 1. KFAT IS NOT AN A.M. STATION...

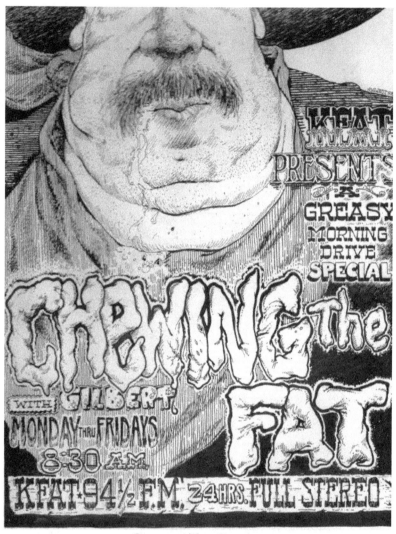

Chewin' The Fat Ad
(I Don't Think This Ever Ran)

Ticket for KFAT's First Live Show

Chapter 116: Wall Street Looks At Gilroy

Jeremy was gone, Laura Ellen was heartbroken and exhausted, Mark Beltaire had left havoc in his wake, Leo was on a rampage, Bob McLain was screaming about being in charge, Doug Droese was the Big Axe everyone feared, "Joe" was hiding somewhere in the wings, and everyone on the air staff was freaking out. The lines that had been drawn were now borders, and crossing them was an uncertain mission into hostile territory.

And then someone called and told a jock about an ad in that day's *Wall Street Journal*. KFAT was for sale.

Chapter 117: The Great Record Heist

Cuzin Al remembers that the breakup of Laura Ellen and Jeremy was the key to it all falling apart. Not that it had far to fall.

Laura Ellen and Jeremy were no longer a couple, Jeremy having moved to a mountaintop in Colorado where he was lost in Techie Heaven, building a television tower for a friend.

Doug Droese had been brought in to clean up KFAT, and everyone at the station was freaking out. All their friends who hung out regularly were asking, "What's going on?" and "What're ya gonna do?" Droese was pulling records out of the stacks, the cool records, leaving the harmless, inoffensive stuff, and taking the Fat stuff and disappearing it. The air staff was more pissed off than ever. They were trying to reach Jeremy again, but being told that he was "out of reach," and they knew that was bullshit, so now Lori and Laura Ellen were keeping their conversations away from Teresa.

Sully remembers: "You'd play something and Doug Droese would walk in like a ghost and pick it up and take it away after you'd finished playing it."

Everyone knew that all the old KDNA records that hadn't made it into the KFAT library were stored in the barn behind Laura Ellen's house. In a barn, with no insulation. They'd heard that the records were stored horizontally, which was bad for them, and heard that they were now moldy, warped, useless plastic discs, no longer playable. The jocks were horrified, afraid that that might also become the fate of the KFAT library. No one wanted to see these precious records, this fragile national treasure, stacked up with those others, moldy, ruined and abandoned…

Meanwhile, record by record, Doug Droese was taking KFAT away.

Everyone knew that a firing was coming, and while no one knew how many were going to go, all felt that the first to go would be Terrell Lynn. After the Ozone Ranger, he had always been the most vocal and intractable, and more than anyone else, he seemed to be speaking for the group. To McLain, he was troublemaking Ground Zero, and he had undoubtedly pointed that out to Droese.

Sully and Terrell Lynn were a couple, and Sully remembers Terrell Lynn being so upset that he couldn't sleep at night anymore, and it had been getting worse, week by week. One night, after tossing and turning sleeplessly into the early hours, he got up, dressed and went out, walking the dark, deserted streets of Gilroy, where he met Sherman, also too upset to sleep, and also out walking to clear his mind of all his worries about the turmoil in his heart and at the station.

They knew that there was no job security whatsoever in the radio business, and that Droese was going to fire them all. New management often meant a new format and a new staff, and they knew that Droese had been hired to make the changes McLain wanted made, and McLain had that power. The KFAT they knew was badly broken, and the jocks knew of no way to put it back together. And everyone knew that Mark Taylor, formerly the Ozone Ranger, was still in Gilroy, driving a forklift at Gilroy Foods, still unable to find a radio gig once he told people where he'd been. And when you apply for a job in radio, *everyone* will want to know where you've been.

After all the years of fun and struggling and partying and being a family, the station had become an intolerably tense and unhappy place to be. Going to work wasn't fun anymore, but those fun days weren't so far in the past that anyone had forgotten them. They all felt betrayed by the new regime, and they had been there first. They had created what was Fat and now they were being shuffled out.

Terrell Lynn believed that Droese had been hired "to give us such a hard time that we would leave."

"But they didn't know what we'd put up with," agreed Sister Tiny, "they had no idea how much we had put up with already!"

Then, on Monday, a few days after Droese showed up, Terrell Lynn was given a message: Droese wanted to see him tomorrow after his shift. That was it, they all knew. That would be the end of Terrell Lynn.

Hosting the Fat Fry that night, his comment that things at the station were "coming apart" had caused a firestorm of unhappy callers. To Fatheads, the music was already noticeably thinner, and the listeners calling now were all frantic about the station. The staff knew they had the support of the listeners, but how much would that mean when push came to shove? At the station, it was getting more tense each day.

Sully was inconsolable and admittedly "over-emotional" listening to the Fat Fry at home. She heard Terrell Lynn announce that the station was coming apart and, crying, she called Droese at home, where he was having a dinner party. "And I have to give him credit," she remembered, "because most people would have just hung up on me, because I was going (sobbing, breathless): 'What's going on here? I don't want to get fired... what's going on... you *can't fire Terrell Lynn...*'

"And I was on the phone with him for, like, 45 minutes or an hour, and getting progressively more hysterical. 'What are you doing to our station [mumble, mumble]…?' "

She kept him on the phone, crying, begging him not to fire Terrell Lynn, and pleading with him "not to do anything which would make the station go up in smoke.

"And he admitted to me then that he had been hired specifically to get rid of certain people, and then we knew that Terrell Lynn was getting fired."

Sully: "I hung up the phone and said, 'that's it, Terrell Lynn's being fired.' And we also saw it as a domino thing, that everybody would go, and that the records would not be protected, and they'd end up moldering in somebody's garage or barn."

From the Fat Fry in Palo Alto, Terrell Lynn called Tiny in Los Angeles and told her that the station was imploding and that she'd better get up there in a hurry. Sherman was using her truck, the Red Dog, so she had to borrow a car. But while she was driving up to Gilroy, the things that had been bubbling, began to boil; everyone knew about Terrell Lynn's meeting with Droese the next day, and everyone knew what it meant.

The staff and the callers were increasingly frantic at the changes the station was going through. But for the jocks, this was their home, and they had eaten and breathed its music, and the energy they had survived on had made KFAT what it was. They knew the music, all the unique stuff that made KFAT Fat, and it was disappearing from the stacks daily. If all the Fat stuff was gone from the stacks, why then… *anybody* could come in and DJ. The staff knew that radio was a volatile medium and that stations changed formats all the time. They also knew that whole staffs could be, and were, replaced. They certainly didn't want to lose their jobs, and they all thought about Mark Taylor, who was un-hirable, exactly as that prick Bob McLain had predicted. But there was an even larger consideration.

What would happen to the music?

KFAT's library was unique in all the world, and no matter what scab or flunky would be brought in after their departure, no one would ever know the music like they knew it. But that wasn't even the main problem. With Doug Droese taking all the Fattest records out of the library, what would happen to those records?

Everyone at KFAT knew about the records in Laura Ellen's barn. All of those KDNA records. Those records had been in that barn since Laura Ellen and Jeremy had brought them here from St. Louis, and the staff knew that in those four and a half years, they'd been ruined. If that happened here, Fat Music might be lost forever.

No one would ever hear "Hawaiian Cowboy" again. "Pussy Pussy Pussy,"

"Lovely Cupcakes," all the train songs and the Texas Swing and the Zydeco and cowboy yodeling and all the songs that KFAT played and no one else. It was unraveling, and the staff felt helpless to stop the oncoming tragedy.

They'd been a family, but the family had shattered, and no one knew what would happen, or what would become of them. They wanted desperately to speak with Jeremy again, but they were told repeatedly that he was out of reach on that mountain in Colorado. Teresa knew that the office was still speaking to him almost every day, but much of what she used to overhear in the office was now being kept from her.

The air staff had both offices, in Gilroy and San Jose, aligned against them, and they felt like they could no longer wander into the two rooms that comprised the station office, and they were confined to the studio, the Production Room and the back rooms, where no one was living now, but one of which still served as a lounge. No one knew what was going on, and they were frightened. "It's like we're mushrooms," Sherman said, "They keep us in the dark and shovel shit on us."

The DJs played what was left that was Fat, but the listeners noticed the change. More were calling in and complaining, but with their knowledge of the library, the jocks did the best they could, playing whatever Fat music was left. Then Droese would come in and take it. The staff was watching KFAT disappear, one record at a time.

That Monday night, while Terrell Lynn was hosting the show up in Palo Alto, Sherman, Teresa, and a friend named Rebecca Fenker were at the station, frantic. They'd been drinking beer and listening as Terrell Lynn announced the station's trouble, and then Sully, who'd heard it at home, came to the station, where more beers were consumed to assuage the despair. Then there was panic, which begat more beer, and then there was plotting.

And Cosi Fabian, the news woman, was there and said, "Oh, you've gotta get the records out of here if this ever happens." When Sully wasn't sure, Cosi asked, "What's the matter, Sully, are your lace curtains showing?" and the shit hit the fan.

Cosi was already a problem for Sully, who knew about the night that Cosi, high on heroin and Courvoisier, pounded on Terrell Lynn's door for an hour to fuck him, and now Sully was with Terrell. Cosi and Rocket used to get high on heroin and fuck each other, and Sully had dallied briefly with the Rocket. But this remark set her off.

This was the spark of an English woman insulting an Irish woman for being Irish; Sully was ready to burst into flames, and Cosi scooted out of the way. It was already late at night, the group had been drinking heavily, and it was decided that action must be taken.

Sully: "So, we freak out. We'd been under a lot of tension, we'd been drinking beer, we think the records are going to be destroyed, we panic and we say… and I remember saying… 'We've gotta get those records out of here!' Because I was stupid. Because I was STOO-PID."

And that was when Sully and Sherman and Teresa started pulling the records out of the stacks. Which was when a local kid named Steven, who had just been hired as a janitor, showed up and they asked him to help, while Rebecca went out to get Tiny's truck from Sherman and Teresa's house. When Rebecca returned, she backed in at the back of the station, the truck bed near the stairs, and it took about an hour to load all the records into the bed of the beat-up red pickup truck that had a KFAT sticker on the left rear bumper. They had no boxes at the station, and as this was a spur-of-moment operation, no boxes were available. Okay, no boxes. It didn't matter. What mattered was getting the records the hell out of there- now, while they were still allowed inside the station. They might not be allowed there after tomorrow.

Estimates say that about half of the KFAT library was what fit into that pickup that night. When loaded, the drunken crew of radio freedom fighters quickly put a blue plastic tarp over the records, Rebecca got behind the wheel, Teresa rode shotgun, and they drove out of Gilroy. It was late at night, they were drunk, and they headed west, over the hill.

On the way from Gilroy to Santa Cruz, once you're over the hill, there is only one brief stretch of highway. Most of the way, it's a two-lane country road, and quite beautiful if I may say. But there was that brief stretch of highway—just a short run on the highway—and that was where the Red Dog met the Highway Patrol.

It seems that the tarp, which had been hastily attached, hadn't quite been *securely* attached, and it had been flapping in the wind off the back of the truck. On the dark, curvy country roads, that hadn't been a problem, but once they got on that brief stretch of highway, the tarp began flapping wildly, and Rebecca pulled over to fix it, with Teresa screaming, "Don't stop on the highway! Don't stop on the highway! We're stealing something here! Don't stop!" But stop she did, and she got out of the truck, so Teresa followed to hastily re-secure the tarp. And that was when a courteous Highway Patrolman pulled over to see if he could assist them.

"No, thanks, officer, we're just gonna tie this down." Just the same, ladies… and the officer went to help them tie down the tarp that had left most of the back half of the pickup bed exposed. Which would be when the officer saw all the weird, old records, and they were all stamped "STOLEN FROM KFAT." That, by the way, would be the pickup bed above the rear bumper whose left side had a KFAT sticker on it.

"What are all these records?" he asked.

"Well, we're just moving some of these records into storage. We're all right, officer, thanks."

The officer helped tie down the tarp, anyway. With the tarp secure, the officer drove off, and the ladies drove into Santa Cruz, where calls were made to certain people. For some reason, no one wanted a truck-load of stolen records in their house or apartment past midnight that night. Desperate, Teresa called her brother in San Jose, almost an hour away. After hearing Teresa's frightened plea, he reluctantly agreed to take the contraband, "but only for a couple of days." Better than nothing. The records were taken to a garage in San Jose.

Late that night, Terrell Lynn had finished emceeing the Fat Fry and gotten pretty drunk, having much to remember and much to forget. A friend drove him home from Palo Alto, and he staggered into his house around three a.m., drunk, emotionally spent, spaced out and in need of sleep, as he had one more shift in seven hours. As he crawled into bed, Sully woke up enough to mumble, "we got the records," before falling back into sleep, and Terrell Lynn wondered what she meant by that for a moment, and then he fell asleep with the alarm set for 9:30. Sully was still asleep when Terrell Lynn woke up and shuffled off to his fate at the station. When he got to work, he understood Sully's last words to him.

Laura Ellen came in and found Terrell Lynn at work, spinning records. She looked in, and that was when she saw the library's mostly empty shelves. She was astonished, and gasped out a couple of questions: "What happened here? Where are all the records?"

Terrell Lynn knew that Laura Ellen would be coming in, and still had her eyesight, and he had prepared a response. "I don't know. I found it like this, this morning. I think Doug Droese took them."

Laura Ellen went in to the office and told Lori what she'd seen, and they both came back to stare at what was left of the library, with Laura Ellen asking, "Why didn't you phone me? Why didn't you phone someone?"

TLT: "I told her, 'I thought this was normal.' I said that 'Droese was taking records and I thought…' I had to maintain that façade."

Laura Ellen called Doug Droese, to ask him about the missing records. Right after Droese said he knew nothing about it, Laura Ellen called the police and told them that thousands of old records had been stolen. I mean, if Doug Droese hadn't taken them, and Terrell Lynn said that *he* hadn't taken them…

She told them that there were a couple thousand of mostly old, obscure country records, and they were all stamped, "**STOLEN FROM KFAT**."

Many believe that Laura Ellen was soon sorry that she had called the police, but the call had been made, and the police began an investigation into the disappearance of thousands of old records. Now, having watched enough episodes of all those cop shows to know what was coming next, I am assuming that the next step in the investigation into the disappearance of thousands of old records stamped "**STOLEN FROM KFAT**" was when the detective in charge of the investigation into the disappearance of thousands of old records stamped "**STOLEN FROM KFAT**" told the other cops at Roll Call the next morning about the disappearance of thousands of old records stamped "**STOLEN FROM KFAT**" the night previous, and that would be when the courteous, helpful cop who had pulled off the highway the night previous to help two females in a beat-up red pickup truck with a KFAT sticker on the left rear bumper, and whose truck bed was loaded with thousands of old records stamped "**STOLEN FROM KFAT**," remembered the two females in the red pickup truck with the KFAT sticker on the left rear bumper, whose truck bed was loaded with thousands of old records, stamped "**STOLEN FROM KFAT**."

That would be when the case would have focused on the two females in the red pickup truck, whose truck bed was loaded with thousands of old records stamped "**STOLEN FROM KFAT**." Oh, yeah- and the truck was beat-up and had a KFAT sticker on the left rear bumper.

About this time, gathered again back in Gilroy, several hours and several degrees of sobriety later, the rebel forces began to reconsider their actions, and panicked. They had now committed a crime, and they still had possession of the goods. They didn't want to be caught with the goods, so more calls needed to be made. Frantic calls were made for advice on where to put the records, and a good friend of the station, Bill Denton, got one of his hooker friends to rent a storage unit in Aptos, a few miles south of Santa Cruz. For some reason, Bill knew about storage units that featured late-night access.

As soon as possible, in a borrowed, nondescript van, the records were driven from San Jose to Aptos and unloaded at the storage place, and everyone went home.

It was at about this time that Laura Ellen was involved in her first interview with the Gilroy police, who told her about the two women in the beat-up red pickup truck with the KFAT sticker on the left rear bumper and the thousands of records stamped "**STOLEN FROM KFAT**," and Laura Ellen told the cops about the tensions at the station, and about her suspicions about who had taken the records. After all, just because Terrell Lynn had looked her in the eye and told her that he had no idea what had happened to the records, she now knew that he had been lying. I love Terrell Lynn dearly to this day, but I would not like to be asked to swear in court that he was a good liar, so let's assume that Laura Ellen knew enough. Let's assume that she also knew who had a beat-up red pickup truck with a KFAT sticker on the left rear bumper. And she told the police what she knew, and what she suspected. In other words, Laura Ellen was naming names. She must have named Kathy Roddy, Terrell Lynn Thomas, Sherman Caughman, Teresa Cortez, and possibly Michelle Sahn, aka Sister Tiny, who might be here and might be in Los Angeles.

It was about this time that the police were able to narrow the investigation down to Suspects Nos. 1-4, and possibly 5.

It was about this time that Sister Tiny got back into town. Welcome back, Tiny.

At about the same time, still during Terrell Lynn's shift, the KFAT staff reassembled at the station, and went into the Meeting Room, the one with the most chairs and a sofa. They were all tense and frightened and angry and frustrated.

Lori called Leo for legal advice, and then came into the room where they were gathered and told them all to leave. She said that this was a serious matter and that the police were now directing the investigation. She also told them that the FBI had called, because after all, this was a federally licensed and regulated business, and any crimes against it are federal crimes, so the FBI wants to know what's going on, and you all had better leave.

Later that day, they were informed that they had been fired, the reason given that they had held an illegal meeting.

By this time, the police had obtained photographs of suspects Nos. 1-5, and were showing them to the helpful highway patrolman. In his initial description, he had described a dark-skinned woman with brown eyes, probably Latina. The photos presented to the officer were all in black and white, and the officer picked out Kathy Roddy, aka Sully, saying, "that's the one who was driving the truck." So they'd want to pick up Sully. But he also picked out Teresa, and if Teresa was involved, they also wanted to pick up Sherman. Now, remember that Sherman had a somewhat nefarious past, and had dealt with the police, and against the police, and had taken great pains to avoid the police in his not-too-distant past, and he wanted to be interviewed by the police about as much as he wanted leprosy. Excuse me, I forgot again: Hansen's Disease.

Soon the police knew where all the suspects lived, and Kathy got a call at home from Teresa, telling her that the police had them, and this was her one phone call, that the police were coming for them, too, and that was at just about the time that the door opened and the police came in and arrested Kathy Roddy, aka Sully.

Sherman and Teresa were already in custody, and Rebecca was next. How'd they know about Rebecca?

Terrell Lynn had been keeping out of sight, and just out of range, and it was that night when he finally heard about the arrests, and he raced to the police station to get them out. From the counter near the entrance to the police station, he could hear everyone talking in the back, and he told the officer who he was and that he wanted to get his friends out. The officer called in another cop, who happened to be a Fathead, who said, "We know you're involved, too… and if you don't want to be back there with your friends, I'd leave right now, pal."

A couple of days before all of this went down, Sherman, Teresa, Sully and Rebecca saw something bad coming and knew they'd better get some help. They called Buffalo Bob, who had found no success in Los Angeles, moved back to Santa Cruz, and found another job suitable to his talents: he was now a bounty hunter. He worked with a bail bondsman in Santa Cruz, a Fathead, and he gave them the number, and said he'd call him now and tell him to expect their call. They called and went to see him, and all agreed that he did not need to know why, but they might be needing him soon, and they put down a deposit for the bail they thought they'd need.

They'd also need lawyers, they knew, but they felt they'd find them when they needed them. Sully spoke to Bill Denton, who promised to cover her legal fees. Then it was back to Gilroy to see how this was going to play out.

Tuesday night, Terrell Lynn called the bail bondsman, told him what had happened, and he came over the hill and got them out of jail. Very sober and badly shaken, and meeting back at Sherman and Teresa's house, they knew they had to fix this mess. They had to get the records back to end this nightmare. Here are some pictures taken at that meeting.

About this time, Sully was back at home, waiting for Terrell Lynn, and Sister Tiny was at Rebecca's place, in touch with Sherman and Teresa by phone. Now that she had been brought up to date, Tiny knew she had ample reason to panic, herself. She hadn't even been in town when the Great Record Heist went down, but the heist had been pulled *using her truck.* After conferring with the people over at Sherman and Teresa's house, they came up with a plan, but were afraid to call Sully, fearing the phones might be tapped. At her place, Sully also had a plan, and no idea that Sherman, Teresa and Tiny had come up with one, too.

Sully: "After the panic subsided, we realized that we had to get the records back to the station. And so we wondered, now- how are we going to do this? Because we kept trying to call the radio station, but the phones were off the hook because so many people were calling, and so we kept trying to call and we couldn't get through. So I thought, 'Okay, we'll call one of those anonymous tip lines,' but I didn't want to do it myself. So I had my sister do it, who unfortunately, her voice sounds *exactly* like mine. It sounds so much like my voice that sometimes when I get a message from her, I think it's me."

Meanwhile, Tiny, Teresa and Sherman went to Tiny's best friend, Mary Angeli's house. Tiny gave Mary a phone number, and told her to "call the number and tell them this." "This" was the storage locker where the records were. Totally freaked out, neither woman wanted to make the call. Mary's brother was there, and volunteered to make the call.

Tiny made Mary's brother memorize the one sentence. Saying she wanted to help, Mary wrote out a copy of it, word for word, for him to read from.

The call was made, and the message was delivered. Tiny left and Mary called Laura Ellen and told her what had just happened, reading to her the word-for-word message that had been delivered, and by whom.

I will leave the reader to imagine Laura Ellen's dismay. This was a woman who had a very deep heart, a heart already badly damaged, and it was now broken. The breakup with Jeremy hadn't healed... and all the years of stress... and now this. It was the staff against Laura Ellen, and she had no allies other than Lori and Leo. And now this.

Laura Ellen was bereft. Her life was breaking down and spinning out of control.

As it was for the former KFAT staff. All were morose, even Sherman, who ironically, had not been fired, as he had an alibi, and no one had implicated him. But the core of the KFAT staff was gone. Sherman alone was left at the station, other than Cuzin Al and me, and maybe Rocket Man, too, Sherman didn't know. No one knew anything in those first few days.

The police went to the storage locker, opened it, and found the records. Unfortunately, even though the highway patrolman who had helped Teresa and Rebecca had told the cops that the records were loose, they hadn't thought to bring any boxes with them. So they sat around the unit, waiting for one of their guys to come back with a lot of boxes. Which was unfortunate for one of KFAT's favorite friends, folk legend Ramblin' Jack Elliott.

Ramblin' Jack was more than a Fathead, he was a friend of the station. Everyone there knew and loved and respected Ramblin' Jack, and at this point in the story I shouldn't have to mention that KFAT played more of his music than anyone else, so I won't. For whatever reason, Jack had a storage unit in Aptos, and not knowing anything about what had happened at KFAT the night before, Jack chose to visit his unit at just *that* brief slot in time when the police were sitting around, waiting for boxes, telling police stories and KFAT stories, when a suspicious looking guy with long hair, a cowboy hat, jeans, cowboy boots and a KFAT T-shirt, was humming to himself and coming their way, in a direction heading for *that* storage locker.

Ramblin' Jack was questioned, protesting that he had no idea what they were talking about. And yes, he *did* have a storage locker right there, no, not that one- *that* one.

He really had no idea of what had happened and had no information about any of it, and was soon released, and he says he has no hard feelings about the whole thing.

The staff had been arrested; they were out, but they were still in trouble. Maybe big trouble.

Tiny wasn't in trouble with the police, but she wanted to talk to her best fucking friend, Mary Angeli...

Tiny: "Mary was more interested in being affiliated with Laura Ellen and

KFAT than she was in my friendship, which had been for years." Mary "went psychotic" after that, and refused to speak to Tiny. Forever thereafter, in person or when Tiny called her, she refused to speak to Tiny ever again.

The listeners were frantic. "What's going on?" They called constantly, so KFAT kept the phones off the hook.

There were no longer any DJs at KFAT. It was over. Who knew who they'd hire, or what records would be in the stacks? Even if they got the records back and kept them in the stacks, who knew them? Nobody.

Perhaps hardest of all for the jocks was something they hadn't even considered, but was just starting to sink in, and it hurt. For almost four years they had held this job. None of them had ever come close to having a job for four years. And what a job it had been! They had tried something new and kicked its ass. They had been celebrities for a long time, several years. They'd been celebrities in Santa Cruz from the moment they'd started there, but less than a year later, the signal had changed, and they were known up and down the coast, in San Francisco and points north, south and east. They were the KFAT DJs.

That had been their identity, each of them, and they were proud of it. For some of them, this had been the first time in their lives that they'd been proud of anything. They were KFAT DJs. That was who they were. And now they weren't.

They'd never have jobs like these again, because there'd never be another KFAT. No one would ever be this loose again. The Seventies would be over in a month and the Sixties were long, long gone. They were the last gasp of the Sixties, and in America, and in American radio, they were anachronisms, leftover rebels after the revolution had been lost. Their time had passed.

Where else could they work? They knew they were considered outlaws and trouble-makers by everyone in straight radio- and outside of KFAT, that's all there was- straight radio. They were going to be, like Mark Taylor, unemployable. But they didn't want those straight jobs at those straight stations anyway, where they were told which horrible record, maybe one of a couple dozen horrible records, they would be allowed to play, in the order that they were told to play those horrible records? No, they couldn't, and they wouldn't. The ride was over.

But the ride wasn't over for Laura Ellen, it was just this part of the ride that was over for her. She still had a radio station. She had a skeleton staff, mostly weekend people and fill-ins. Her main staff, that which had been her family and the staple of the station's success, were gone.

Now she needed a new air staff so she, and KFAT, could go on.

A wealth of documents from this period may be accessed here.

Chapter 118: Meanwhile. . .

Among those of the air staff who thought that Bob McLain was the Devil—and that would be all of them—even they had to admit that they were grateful that under McLain's direction, at least he had hired a janitor to come in and clean the station. That was new- it was a nice thing, but of course no one was going to capitulate to McLain and "tighten up" as they were urged, so the nice thing went noticed and appreciated, but neither acknowledged nor reciprocated. It was too little, and probably too late.

About 9:30 that Monday night, as the Fat Fry was on the air, local teenager Steven Eldridge, recently hired as a janitor, showed up and found Sherman at the station, as he'd expected, and "a few other people, including Kathy Roddy, who he knew as Sully, and someone named Rebecca," according to the statement he later made to the police. He was told that the station was "changing the style of music that they were playing, and they were going to store the records for safekeeping, and asked [him] to help them load up some records." He knew Sherman and Sully worked there, he assumed that they were in charge, and he agreed to help. He helped Sherman and another, unidentified man carry records out the back, down the stairway, making dozens of trips and putting them into a red pickup truck that he had seen in Sherman's possession before. The red truck, he told the police, had a KFAT sticker on the left rear bumper. Sully and another woman helped to stack the records in the truck.

He later swore that after "about two hours," and after "maybe a thousand" records were loaded into the pickup, he became suspicious that the records were being taken illegally, and after Sherman asked him "how much it would cost to keep him quiet," he told Sherman that he wanted no money from him or anyone, and he "wanted nothing to do with this whole thing." Then, at 11:30, wanting to get away from the whole thing, he left the station.

When he came back at 12:30, the red truck was gone, Sherman was still there, having just handed the controls to Susan Uhouse, who would be on until 6:00. He did his cleaning and left the station by 3:00, saw no one else, and didn't know where the records were.

The "unidentified man" that the janitor described soon called Lori Nelson at the station and identified himself. His name was Robert, and he had shown up after work to give Sherman a ride home after his shift. It seems that the truck Sherman was borrowing from Tiny was in use elsewhere that night, so Sherman had called him at work and asked him to stop by on his way home. While there, Sherman gave him a box with several reels of tape from a Marijuana Festival that Sherman had attended, and Sherman didn't want found with his stuff at the station.

Robert told Lori what had happened at the station the night before, and that he had the tapes and wanted to give them to someone at the station, and not get into trouble. He had a wife and an infant, and he wanted no part of anything illegal. Lori told him to come by and return the tapes, and he said he would. Then Lori called the police and told them about this conversation, and they told her to tell Robert to come by the police station and make a statement, which he did, after getting an assurance from the interviewing officer that the officer would do everything in his power to minimize Robert's involvement in the crime. Robert told it all.

At 6:30 that morning, Sully called everyone she could, to come to the station for a meeting of staff and supporters, who would gather to protest the firing of Terrell Lynn. The meeting was set for nine o'clock. Lori recalls that when she got to work that morning, as she later told police, she "found about 25 people gathered there," and she told them all to leave, but they were resistant. She remembers that when she came inside the studio, and noticed that the library had lost about half of its records, she went to get Laura Ellen. When Laura Ellen came in and saw the bare library shelves, she asked Terrell Lynn what had happened to the records. He told her that he didn't know, that he thought Doug Droese had taken them. That was what he'd been doing, wasn't it?

Lori then told the crowd to disperse, that she was calling the police and the FBI, as the management of a radio station was a federally regulated business, and any interference in its operation was a federal crime. She also said that everyone in the crowd who worked at KFAT was now fired, that this was "an illegal meeting," and that they had better vacate the premises- now! At this, the crowd dispersed.

The jocks were told to come back in an hour for their final paychecks. Knowing how angry both KFAT offices were at them, and knowing that they had committed a crime, they knew they were going to need money for some lawyers and maybe more bail money, so they all came back for their checks, which Lori gave them at the bottom of the stairs, not allowing them into the station.

Being old KFAT hands, they knew that a check in the hand was worth two in the bank, so they rushed off to the nearest branch to cash them, where they were told that they couldn't cash the checks, that the account had insufficient funds. Then, it was off to the Labor Board.

They all had personal items at the station, and Sherman wanted to get to his- soon. Among the staff, only Sherman had had experience with this sort of stand-off, and he knew what to do. When his step-mother had thrown him out of his home back in Arkansas all those years before, she had put all of his clothes and all of his possessions on the porch, and locked the door. But she had forgotten to put out his trumpet, which he wanted. She wouldn't let him in, so Sherman went to see his friend's brother, who was a cop. He was sent from there to see a judge, who said, "Damn, boy- go back and get your trumpet!" and sent him to the cops for a "civil standby" where a cop accompanies the complainant to a location, and supervises the removal of expressly permitted items. Sherman went to the Gilroy police that afternoon and requested a civil standby for himself at the station.

By now, Steve the janitor had contacted Lori and told her what he had done, and with who. Lori thought that Sister Tiny was involved in all this, but Steve told her that there was no one there named "Tiny, Sister Tiny, or Michelle," and there was no one there who looked like her. Just those he'd described, and a woman named Rebecca. Lori called the cops with the information. And the cop, the phone propped under his chin, looked up from his desk and said, "why, here's Sherman, now."

Of course, the cops all knew Sherman, and they knew all about Sherman. Sherman had come in after two p.m., by which time the cops already knew the details of the theft. An officer took Sherman into the back, sat him down and said, "Look, man- give the records back. I want to hear Bob Wills." Sherman said he didn't know where the records were, the cop sighed and led Sherman to the front desk, where another officer was dispatched to accompany Sherman to the station, but they were refused entry by Lori, who told the officer that Sherman would be welcome to come back after they'd had a chance to go through Sherman's effects and do an inventory to make sure that no station property would leave with him. The officer thought that was reasonable, Sherman agreed, and they left.

They came back a few hours later, and reporters from *the San Jose Mercury News* and *the Gilroy Dispatch* were there to interview them about the crisis. After being admitted, briefly, to the Station, Sherman, Terrell Lynn and Sully retired to the Green Hut for refreshment, and the reporters followed them there to interview them. A photo was taken of the three at the bar, to accompany the article.

After the interviews, and knowing that the clock was ticking on the garage in San Jose, a call was made to Bill Denton, who knew a guy. A storage business was called in Aptos, about ten minutes south of Santa Cruz, and one of Bill's hooker friends was dispatched to rent a unit there.

Terrell Lynn drove to San Jose to pick up the records from Teresa's relieved brother, and he moved them to the storage unit in Aptos. By the time he got back to Gilroy, as predicted, Sherman and Teresa were arrested, as were Sully and Rebecca. Terrell Lynn clearly had to have been a part of this, but as he had been at the Fat Fry in Palo Alto when the caper had gone down, he was off the hook for the moment. But he still made himself scarce until it played out. He called the bail bondsman, and went into hiding.

Chapter 119: And Right About Then...

Every record album has two sides, and so does our story.

Lori Nelson has her own memories of the Great Record Heist, and there are some interesting variations. Her opinion of the heist was that it was undertaken purely as an act of anger at what the DJ's thought was the imminent loss of their jobs. She was right about what was planned for their dismissal, but she believes that it all started when the jocks found out that the station was going to be sold. But it is beyond speculation that more than half of the records were gone and the phones at KFAT were ringing constantly. Everyone wanted to know what was going on, and all the callers supported the jocks.

After coming in and finding the empty shelves, Lori and Laura Ellen had immediately obtained legal advice from Leo, and were advised that they had fired those people with justification. Lori was afraid that if they got back into the station they might steal the turntables, too. Done and done. They were gone and would stay gone. The police were informed and an investigation was under way.

Cuzin Al and I were not involved in any aspect of the heist, and were still employed in our normal shifts. Sherman insisted that he was not actively involved in the planning or execution of the heist, so while he was clearly of a sympathetic nature to the discharged staff, he had had no involvement with the heist that they knew of, and Laura Ellen and Lori were advised by Leo not to fire him without just cause. Everyone in the office felt that sometime soon, Sherman was sure to self-destruct and say something inflammatory on the air, thus laying the groundwork for his final dismissal.

And now, the news of the goings-on at KFAT had become the news. *The San Jose Mercury-News* put it on the front page. *The Gilroy Dispatch* featured the crisis prominently. The phones were going non-stop crazy with people asking what was going on. Then they got some calls that said the records were being given away in the streets of San Francisco. Every lead had to be followed, and the police were on it. The FBI called from their San Francisco office and wanted to know what was going on down there. Lori was in constant contact with the Feds, and, of course, the FBI already had files on Jeremy and

Lorenzo and KDNA and KFAT, and had a few other items of interest that they'd like to speak to some of them about.

Then the FCC got in touch. It happened that these goings-on had occurred during a "ratings period," and the FCC wanted to know if this was an actual meltdown or just another inappropriate joke or ratings stunt from those out-of-control hippies in Gilroy.

Lori assured them that this was neither joke nor stunt. And the phones were still going crazy. Laura Ellen worked a sixteen-hour-shift to keep KFAT on the air- and without the missing records. It was a heroic task, if not an impossible one. With Lori filling in as best she could, and spelling Laura Ellen whenever she could, both women were living at the station. Laura Ellen, as always, had Elsie with her, and started working at least twelve hours a day on the air. Calls were frantically made to anyone who could do the job. And the police still had to be dealt with. Lori remembers that "we were in siege at the station, and trying to get DJs that weren't involved with the theft."

Then Lori remembers getting a call: "I'm tired of hearing non-KFAT music, and I'll tell you where the records are." The caller said they were in a storage unit in Aptos. Her first call after that was to the police. Then, Lori and the cops called the FBI, who coached her on what to do.

After the briefing by the FBI, Lori drove out to Aptos, where she found the storage unit and approached the manager and said, "I have reason to believe that there are stolen record albums stored in a unit under your management." He said he knew the unit she was referring to, and he asked, "How much is it worth?"

Lori asked, "What did you have in mind?"

"Twelve thousand to start."

"Well, I'll have to talk to the owner. I'm just the General Manager."

She left the premises and went to the first phone booth she found and called the FBI and told them what had happened. They told her: "If you say that we told you this, we will deny it, but go to that storage unit, take the serial number off the lock, go to a locksmith, tell them that you lost the key and would they make you a new key to get into the lock, then when you get the key, go back and open the storage unit, look in and see if the records are there. And if they are, call us back."

She did, she did, she did, she did, she did, she did, they were, and she did.

Lori recalls that immediately, "umpteen FBI agents swarmed the place like a drug bust." They surrounded the door to the unit and read it its rights. That's right- the Feds stood in front of the door to the storage unit and read the door its rights, the whole "You have the right to remain silent, you have the right to..." Then, having fulfilled their constitutionally mandated responsibilities to the door, they broke the lock and went in and confiscated all the records and took them to the police station in Gilroy, with that whole waiting-for-boxes-and-questioning-Ramblin'-Jack thing in the middle.

On November 1st Doug Droese showed up at KFAT to start his job. By November 7th the insurrection had erupted, been quelled, and KFAT had lost most of its core staff. All in a week's work, eh, Doug?

Also, in case you're confused, some people remember it was the cops at the storage unit, others remember it was the FBI. It was a long time ago…

But for a fact the cops now had the records, which of course meant that KFAT didn't have them. There were too few records to play, and all the really Fat stuff, the stuff the audience tuned in for, was in a police evidence locker. Laura Ellen told the police that she did not wish to press any charges in the matter, but they said it had gone beyond that now. She was told that, regardless of KFAT's unwillingness to press charges, the state of California wanted to press charges against the record thieves. Charges, they declared, would be pressed.

Now, *that* was not going to expedite getting the records back. She told the police that not having the records was affecting their revenue, and they told her to see a judge.

The judge told Lori that she could have them back as soon as she made out an inventory of every album in the evidence locker, listing the artist's name and the name of the album. She drafted a grab-bag crew of anyone within range who was willing to lend a hand, and they worked at the Gilroy Police station for two days, and made the inventory.

Meanwhile, back at KFAT, the phones were ringing constantly, day and night, and no one was answering them any more. All the lines were lit up, all the time.

Chapter 120: Ex Post Heisto

Susan Uhouse's role in the record heist was unclear, but she'd been there that night, so she had to go, too. I still had my show, and Cuzin Al would still do Sunday nights. No one knew how to reach the Rocket, so they thought he was still on board, but they wouldn't know until he showed up- or didn't. They needed every other shift covered, and it was up to Lori and Laura Ellen.

The calls never stopped. A radio staff stealing the station's records had never happened before, so the he press was calling, too, and they were sending reporters to the station because no one was answering the phones.

KFAT was still a business. No, make that: now more than ever, KFAT would be run as a business, and it would need listeners, so someone had to talk to these hostile listeners. Someone suggested that the management should go on the air to answer questions from callers, and get their side of the story out.

The former staff was listening to KFAT all the time, too, and they were listening as Doug Droese went on the air, hoping to keep some listeners, and Terrell Lynn recalls, "he said he was hired to give us such a bad time that we would leave."

GK: He said that on the air?

Sully: "Yeah. I had a tape of it. After we all got fired, they had him do a call-in talk show and we taped it, and…"

TLT: "And so we'd call in, and our friends would call in, and people we didn't know would call in, but then a caller called in and said, "As I understand it, the DJs themselves sometimes bought the records and put them in the library, and they bought records themselves just to enhance the station.' And Laura Ellen went, 'uhhh…' and after, like, 20 seconds of dead air, 'No, that's not true.' I mean, it was this pregnant pause. It was just such invariably a lie."

Legal matters for the former DJs would cost money, and a benefit was quickly organized to support the "Legal Defense Fund," and flyers and bumper stickers were printed up by sympathetic listeners, imploring, "FREE THE FAT FOUR." A show to benefit the Fat Four was arranged at the Catalyst in Santa Cruz, and acts were sought to play at it.

Doug Droese brought in some of his students and former students from San Jose State to work until something more permanent could be put in place. They all hoped they'd get to keep the jobs, especially as KFAT was very much in the news at the moment, and the amount of people listening was enormous. But they didn't know the music that the listeners tuned in to hear, and the callers persistently berated them.

GK: How long did they last?

LEH: "They didn't last very long. Some of the staff came back, like Sherman came back begging for forgiveness; some, like Terrell Lynn and Sully never came back."

Laura Ellen remembered that the replacements, "didn't know anything. It was just awful. They didn't know anything. They didn't know who the artists were, they didn't know what they were doing. They'd do whatever you told them, but… the problem with the format… and it's always been… is that you really have to have a knowledge about the music in order to be able to do it."

GK: So they played Willie and the Dead and stuff they knew?

LEH: "Yeah, but they didn't mix it right. They didn't put it together in any way, shape, or form because they just didn't know… And that's the days when the staff was so small that we were all doing six hour shifts. You know, there were four DJs. So I would do six a.m. to noon… I don't remember who else was there at this point.

GK: Did you spend a lot of time training these people?

LEH: "No. I didn't have any time to train them. Had to just put them on the air. I worked with them when they were there, but it just obviously wasn't going to work."

Fatheads were a loyal, knowledgeable group, and it was clear to Laura Ellen that these kids couldn't be DJ's for long. They just plain weren't going to work out. And neither would Doug Droese, and with an agonizing last grasp at control, Laura Ellen sent him packing.

The response to the ad in the *Wall Street Journal* was tepid, and it was not renewed. There had been some interest from some interesting people at first, but nothing much came of it. Little Jimmy Dickens- "May The Bird Of Paradise Fly Up Your Nose" came by. Bob Hope's wife contacted Jeremy, and wanted to buy the station for John Denver. She was the agent representing him. But he didn't have enough money to buy it. Emmylou Harris wanted to buy it, but didn't have the money, either. There were no serious offers out there. It was still Laura Ellen's bastard baby.

Chapter 121: The Protest On The Wall

All those years ago, in a moment of anguish, Jeremy took a can of spray paint and wrote WHO STOLE MY SCREWDRIVERS? on the wall outside the studio, and no one ever removed it.

The message became part of the KFAT landscape, but no one ever took it for granted. It was a symbol for the staff, a banner showing that insanity prevailed here, that passion prevailed here, that the normal rules did not apply here.

Terrell Lynn remembers: "The last time I saw that wall, the lettering was still there, but there was all the letters [that] had come in where people had peeled their KFAT decals off their car and stuck on pieces of paper and mailed it to the station."

GK: Why'd they do that?

TLT: "Because we all were fired. And this was their protest. So there was just like sheets of torn decals stuck on the wall."

Chapter 122: The Island Who Said Yes

Now, for a change of pace, let us consider the strange case of Hewlett Crist, a Texas musician and record producer. Born in San Antonio and raised in Laredo, he found his way to the Bay Area playing in Doug Sahm's band. Sahm was a Texas musician at the time of the British Invasion, and whose record company was more interested in English acts than cowboy bands. Sahm accommodated them by calling his band The Sir Douglas Quintet, and they had a hit with "She's About A Mover." Sahm toured behind it, and went back to his roots as soon as he could. He has since played with other Americana stalwarts such as Freddy Fender, Augie Meyer, whose Farfisa organ sound was so influential, and Flaco Jimenez as the Texas Tornados. Sahm played in a few groups, using several musicians, and one of them was Hewlett Crist.

Hewlett is one of those curious types: bright, intellectually curious, musically gifted, country-oriented and likes his independence. He wound up following Doug Sahm out west, played in a bunch of bands and recordings, and stayed there long enough to raise his five sons, married to the same charming woman he met years ago when he first got to San Francisco. For a while, he had a successful school in the Bay Area for training musicians in recording techniques and music law. Needing to learn more about entertainment law for his school, he took a job as a paralegal in an entertainment law firm. He learned a lot about it, but he always kept his hands in the production and arranging end, until he abandoned the business in 1982, after finishing work on a Grammy Award- winning album by Zydeco star and Fat fave, Queen Ida.

In 1979, Hewlett was living and working in San Jose. His studio, Red Dirt Productions, recorded music demos and vocals for commercials. Of course he knew about KFAT, and when he wanted to advertise his studio, knowing about all the musicians who listened to the station, he theorized that KFAT was a pretty good place to advertise. So he did.

He'd been there a couple of times to work with someone on his ads, and he remembers the first time he wanted to cut a spot by himself. No one there knew who he was or anything about him, so he showed them he knew his way around a board, and they let him come in and work on his spots. They were always vocal spots, with no musical bed, so the process of recording his ads was simple, and he was in the station and out again, real quick-like. He liked the station, he listened to it, but he never thought much about it.

He had it on in the background and all, and he knew that it was an extraordinarily good station, but he had a life, and KFAT mostly mattered to him if his spots brought in customers. He had a wife and a bunch of kids, and everyone had to be fed and etc. He was at the station one afternoon to record a spot when Laura Ellen came in and asked him if he wanted a DJ slot on Sundays.

HC: "Anyway, Laura Ellen knew all that shit about me, and she says, 'Oh, you'd be so good... and off the wall...' But I knew she was bullshitting me, because the only reason she was interested in me was because of my Texas accent. I knew that was her real area of interest, plus I could fill some air. And Sundays wasn't, you know, prime time."

It was Sunday mornings, from ten until two, after the Gospel Hour. Uh oh... problem.

Rick Nagle used to hitch-hike in from Gustine, an hour away by car, and he'd bring a sleeping bag with him because he'd usually spend the night there. Hewlett remembers when he got there and Nagle was already there, having spent the night on a couch in a back room.

HC: "Rick, if need be, hell, he'd just stay there all night on the air. He'd just get out-of-it, and eat peanuts and shit, and stay on the air for hours and hours, if he had to."

But this isn't about Rick, it's about Hewlett, who was unusual at KFAT for one specific trait: he always came in an hour or more early and pulled his records in advance. Nobody else at KFAT ever did that. Cuzin Al brought a bunch of his bluegrass records with him every Sunday, and he knew a lot of what he wanted to play, but Hewlett had more initiative.

Hewlett had been on the road and studied the inside of the industry. He knew a lot about music and music law, but he had also studied the art of the stage show. The songs on any album are placed in an order that works best for the album, and that order is usually studied and discussed vigorously among artists, managers, producers and record company executives, who are all very much invested in the art of sequencing an album.

Hewlett told me about a guy in Las Vegas at the height of the interest in lounge acts in the late 60s, "and no matter who represented you, you were not going to get booked into a lounge in Las Vegas unless you brought your material to this guy to evaluate. And he was so good that it was widely known that what he told you to do, worked. And if you got the gig in Vegas, $1,500 of your first check went to him." Hewlett remembered using what he learned from this guy with Queen Ida on two shows in Austin, who was there for a taping of "Austin City Limits."

HC: "Ida flew me in to Austin, and she had a gig there a couple of nights before the taping, and we had some rehearsals, and we came in and tightened up her set. And the first night we played the club—a big place—she did what she usually did in the way she chose her material. And the next night we used the newly designed set list, and everybody was BLOWN AWAY by the audience response to those seemingly very subtle changes."

And Sundays, with his carefully chosen selections, he planned out his four-hour shift into four one-hour sets. He came in early and pulled the records for all four sets, and he put them in four piles. He was doing all this "for cigarette money," and he was giving up his Sunday mornings and afternoons, and he hadn't come looking for this job, Laura Ellen had asked *him* to take the gig, so he played what he wanted.

HC: "I would take an hour or 45 minutes to pull records, 'cause you had two turntables, and I'd go and carefully pull my records, and sometimes I'd take more than an hour, and I would pull my records very carefully, and I would sort of design what I'd play by how it felt. And it was kind of like designing a set for a concert... So if I decided I was going to have something with Texas in it, you know, I'd have to look high and low to still fit in to that flow that I wanted to design, and I would treat an hour show like it was a set on stage, in the way that I chose the material."

Oh, Laura Ellen had been right, of course: he was good. But he played what he wanted, and she wasn't so sure how Fat it was. Most weeks, he'd start his show with an hour of songs about Texas. He paid no attention to the callers when they all asked him to do something else. He played what he wanted.

HC: "I particularly would come in and, for the first hour, I had to work Texas in, or something about Texas. Boy, I'm telling you that would piss people off, and the phones would light up and stay lit the whole hour. I figured you didn't have to kiss people's ass on the air- what you want is participation, so that worked really well.

"So then what I'd do is I would... some Sundays I would do that, and other Sundays I would get everything that I could think of that would have the word 'Watson.' Like Gene Watson, the Watson Brothers, Watson this, Watson that, Watsonville, California. There was a big VW dealer in Watsonville, and I did a commercial for them, and what I would do, I'd mis-pronounce Watson, and always say Waaatson (like in "cat"). And I'd say Gene Waaatson, and the Waaatson Brothers. Normally, the "a" is pronounced like in "what." He did it prominently before and after each set and during the commercials, and it infuriated the listeners, who called repeatedly to correct him of the odious mispronunciation. Fun with words. And there was more fun to be had with words.

HC: "Yeah, and you know there were some pretty sharp people who would call you up on the air and bullshit you. Man, there were some slickies out there. And there was this phone, and if the buttons come on it's line 1, 2, 3 or 4, and you'd hit the button when you're off air, and say, "this is Fat." And there was all kinds of things- chicks would call up and offer to come over to the station and tighten you up, and there was all kind of shit. When you picked up the phone you never knew what it was.

"And they had the most wonderful promo bank of any station I ever known of. People would come in and do a station ID. "Hello this is Merle Haggard, and you're listening to…"They had… wonderful… and they had this comedy section…. Remember the Japanese Roll Call? My daddy loved the shit out of that one.

"So I had to do a voiceover on a commercial for Emmylou Harris, she was going to be at Spartan Stadium at Stanford, and I played one of her songs on the air, and I said, 'Boy, Emmylou Harris, she is fine. You know, I would drink her bathwater.'

"And pretty soon Boom! Boom! All the lines light up. One would say, Hey dude- that's pretty cool, and the next one would say, well, that was the most sexist, horrible… and here this voice comes on the phone, and she says 'is this the person who said they'd drink my bathwater?'

"And I says, 'Is this the person whose water I said I would drink?'

"And the voice says, 'Yes, it is. At first, I thought that was very childish and very crude. And then after a couple of minutes, it hit me from a different direction, and I had to giggle.' Then she said she just had to call."

Crist still smiles when he remembers that call. Every jock has their individual memories, but every single KFAT jock I spoke to for this book rhapsodized about the KFAT library.

HC: "I mean, it was UNBELIEVABLE, their library. And I'd come in early, because I used to go bass fishing on Pinto Lake near there, and I'd go over there- I'd get off my deal at Fat at two, and drive on over to Watsonville, and go hit that Mexican joint and get some burritos and stuff, and go fishing about five o'clock. I had a little boat out there and I was a bass fishin' freak, that was my therapy for that time.

"And I'd pull all my own records like everybody else, but every once in a while, some of this new market-slick stuff would come through, and I said, where does this shit come from? And there was a Jeannie Fricke, country artist, and this was so bad, they had this cassette in their little cassette library, besides all the station ID's, and they had this cassette drop-in with [Monty Python's] 'Oh No! Not this record...! followed by the sound of a needle being scraped across a record, and I played that and then I opened the window, and I opened the mic and I said "Here it goes!" and I let it sail out the back [into the] parking lot. It was terrible.

"And some old boy found it and brought it in. He heard that, and he came down and was looking around. And he came up the steps and he said, 'You really did it! Man, that's cool. Here.' It was Fat."

That was wild for a fan to come down and park out back and look for the record, just to see if the jock really did it, but that was what happened, and as much as that amused Hewlett, the fundamental problem was that the record he had sworn never to see again was now back and right there again, in his hands. Thanks a lot, pal.

Hewlett had some fun on Sundays, but he was an island at KFAT. He came in and left, had almost no contact with any of the staff, and had no knowledge of what was going on at the station when he wasn't there. He'd come in around nine a.m. on Sunday mornings, went on at ten, and left to go fishing at two. Rick Nagle, who followed him, was often there when he got there, and followed him on the air when he left. He followed a succession of people who did the Gospel Hour, and he never developed any relationships with anyone else on the staff. He never went to shows with the staff, and even when he emcee'd a show, he did it by himself. That the station was in turmoil was unknown to him. He did not know about the trouble brewing between the jocks and McLain; he didn't know any of the old jocks or any of the new jocks. He hardly knew Laura Ellen. He had seen her only on the day that she'd hired him, about a year ago. He came in, played what he wanted to play, and left. But over that year, what he played while he was there became a problem, and he knew he was doing it, but he didn't care. What were they going to do, fire him? He'd only even *seen* Laura Ellen once.

He had a mail slot at the station, and every once in a while there'd be a check there or a note from Laura Ellen. The notes invariably said 'You can not play Utah Phillips on Sunday. Especially after the Gospel Hour.' What she meant, of course, was "Moose Turd Pie," because one line in the song was, "Got his shit together, so to speak." That was okay at night, but after the Gospel Show? Laura Ellen thought not.

HC: "And of course, every Sunday I played 'Moose Turd Pie.' Every Sunday, faithfully."

GK: "How often would you get notes about not playing it?"

HC: "Probably 2 out of every 3 weeks."

GK: "And yet you played it every week."

HC: "Sure. It was Fat. Damn right I played it every week."

GK: "You weren't afraid of getting fired?"

HC: "No, of course not."

GK: "I just think it's interesting that all the other DJs got requests for Moose Turd Pie, and they said no. You're the guy that said yes."

HC: "I guess so. I played it every Sunday. I don't know about the other guys, I know that I played it every Sunday, faithfully."

Calls came in asking for it, and he said "yes." I guess he wasn't afraid of Laura Ellen, or whoever was running the madhouse. And he'd only seen her that one time, right? No.

HC: "I saw her one time afterwards. She had to come down personally to tell me that management was just really upset 'cause I continued playing 'Moose Turd Pie.' The problem was it was right after the Gospel Hour. I said, 'Okay, yeah, fine..' but I'd play it anyway."

Hewlett Crist lasted a year at KFAT, and he didn't know about the trouble that was already there, or the big changes that were coming.

"The second time I saw her, she came to me, not when I was on the air, but when I was there one day doing commercials, 'cause I still did my own commercials for my own damn thing. She said that the station was having some problems, and this and that and everything, and they would love to keep me on the air and everything, but they were at a point right now where they just couldn't keep paying. And I understand that there were some people that just stayed on the air regardless... And it wasn't for the money, it never was for the money. It was a very unusual art form, KFAT, and it had its own special audience.

"And I said, 'Well, Laura, I like doing the show, and it's not about the money,' I don't even remember what it was, it was chickenshit, I didn't care. I was making nice money off of production projects. 'But I tell you what- if you're having problems and things, I'm probably needing to loosen up my Sundays,' I says. 'Now, if you have any other times during the week that you need somebody to come in, I'd be glad to.' I said, 'it's not about the money, but I got to open up my Sundays for a while.'

So if you were wondering who was on the air on weekends, because I haven't been talking about it much: why, Hewlett Crist was on the air on Sundays, from ten a.m. until two p.m. He was on for the year of 1979.

Now I'm going to tell you one more thing about Hewlett, and it's more about the Production Room. Hewlett had recorded and worked in dozens of recording studios, on all kinds of equipment, and I wanted to know what Hewlett remembered of the Production Room. He was using the equipment when he cut his spots, right? And he knew his way around recording studios, right? Right, and his insights would be valuable, right? So I asked him about the Production Room.

HC: "I never saw it."

GK: "Didn't you cut your commercials there?"

HC: "No, they put a microphone out in the hall."

GK: "Excuse me?"

HC: "They put a microphone out in the hall."

GK: "There was no music track on your commercials?"

HC: "No."

GK: "And they put a microphone out in the hall?"

HC: "Well, not the hall. It was like a little area back out of the way, that they- wait- did they have an isolation room in the Production studio? Where you could go in and shut the door? They put me in some little… room with a microphone, and I just… "

GK: "That was the Production Room."

HC: (laughter) "Then I've been in it. So that was it." (laughter, both men.)

GK: "There probably isn't a more eloquent commentary on the sophistication of the Production Room than that you didn't know until 27 years later that you'd been in it."

Chapter 123: Chew Out

I wasn't having much fun at KFAT any more. My friends were gone, the job took most of my time, and I was still getting peanuts for pay.

I also had a new career in the real world, and I was succeeding in that. A publicist for the Calaveras County Fair and Jumping Frog Jubilee had contacted me to interview the World Frog Jump record holder, who lived in Silicon Valley, KFAT's prime listening area. The interview would publicize the event, and I had a good time with the frog jumping champion. The publicist liked the way I conducted the interview, and hired me to be the on-air interviewer for a filmed documentary he was producing for another client. A few weeks after that second interview, he called me and offered me the six-week job of doing the publicity for the Calaveras County Fair to be held that May of 1979. I told him that I had never done anything like that before, but the man had confidence I could do it, he had a contact list and a PR outline for the event that he would provide, and he would be available by phone to consult when I needed him.

I took the job, did well, and the man offered me a full-time job at his PR agency in San Francisco, which I accepted. A job that paid a living wage? What about KFAT? I would do both.

Then, my father died in October of 1979, I'd inherited enough money to buy a house, and my wife and I started looking for a place in San Francisco, away from Marin for the first time for me in 20 years, and for Susan, ever.

By January of 1980, I wanted out of KFAT. The new people were showing up, and they seemed like a good lot, but they weren't my friends; it wasn't the same, and I was always tired. With the benefit of hindsight, I understand why I did it.

I could have told Laura Ellen what was coming, but I didn't, and I regret it to this day. I have mentioned my regrets about this to Laura Ellen twice, and both times she has graciously nodded her acceptance. I'm sorry about it, but things were strange back then.

During the shake-up at the station, my contract with my sponsor expired in mid-January. The sales staff wanted to know what my position was before they sold a new contract for my show, and the sponsor wanted to continue paying for my show, and they agreed to keep paying for the show until a new contract was signed. On the last show of January, I began the show as I always did, and ended with a 60-second spot for the sponsor, as always. Then, I came back on with no musical bed and announced that I had great regrets about this, but I was no longer able to keep Chewin' the Fat on the air. I said I loved the station, believed that it would, and should, continue, but without me. The station, I said, was like a mistress to me, a great love of my life, but my life had other, more pressing responsibilities elsewhere, and that this was the last Chewin' the Fat. I thanked the station, the staff and the listeners, and as I neared the end of my farewell, I began potting up the volume of the slow, mournful intro of "You Gotta Move," by the Rolling Stones.

I'd said what I wanted to say, so I said good-bye to the listeners, and raised the level of the song to the normal broadcast level, and turned off the microphone, as Mick Jagger sang:

You may be high,
You may be low,
You may be rich, child,
You may be po',
But when the Lord gets ready,
You got to move

The song would play through to the end, and whoever would be DJing would play it out and segue into the next song. Nothing would interrupt the Flow. Then I put the tape in its box, wrote "CHEW- Friday" on the side of the box and slipped it into the CHEW-Show slot in the 7-Up case. I went back into the Production Room, packed all my gear into two cartons, one big and one small, and took them downstairs to my car. It was almost dawn. I was tired, I was sad, and my bed was a hundred miles away.

Chapter 124: Leo and Lori

It was a sticky, bile-filled mess at KFAT, and Teresa was the spy in the ointment. She knew what she heard, and it was different from what Laura Ellen was telling the jocks.

Despite the crisis at the station- the programming changes McLain was insisting on and the jocks were resisting, the money that seemed to be unaccountably disappearing before it reached her, and all the rest of it, Laura Ellen used to have Jeremy there to take on some of the responsibilities. But there were a lot of things that Jeremy couldn't do, like talk to people or make decisions, and those responsibilities still needed to be addressed, and Leo Kesselman was their friend, lawyer and advisor, and once before he'd stepped in and made some changes that saved the station from bankruptcy.

Jeremy was gone, and Laura Ellen needed to listen to someone other than Bob McLain or Lori. Laura Ellen was out of her depth, and most of her friends were now enemies, who she couldn't talk to about this stuff, anyway, and she needed a smart, trustworthy voice with a lot of business experience. Leo had saved the station over a year ago, and he'd enjoyed the challenge. But once he had made those changes, and once the crisis was over, the challenge was gone and he went back to his other interests. Now the station was in crisis again, and once more Leo dove back into the vortex.

I left KFAT at the end of January, 1980, and also left a bad taste in Laura Ellen's mouth. I'm sure I wasn't crucial to the KFAT sound *per se*, but I was a part of the family, and I'd been there since the beginning, four years ago. I lived in Marin and all, but I was a part of the family, and I'd been a dick about leaving with no advance notice.

It would probably be hubris to assume that my leaving had much impact on Laura Ellen, but it didn't help, and Leo knew it. Yeah, they could always get someone else to do a Public Affairs show, but I was one of the original voices of KFAT, and my leaving was another leak in a sinking boat. And who would do the work I was doing for what the job paid?

February passed and KFAT battled on. Laura Ellen had gone through the 12- and 16- hour days, and then the college kids and the bullshit with Doug Droese. Leo was involved and he was a friend, but Leo was an arrogant bastard, and although his arrogance was currently in what he thought of as her best interests, dealing with an arrogant bastard was never Laura Ellen's idea of fun. Friend or foe, Leo was a fucking handful.

Leo was the station's consiglieri, and as such, he asked Jeremy and Lorenzo and Laura Ellen how they felt about selling KFAT. Leo thought about buying Lorenzo's share of the station, leaving him and Jeremy as the owners. But Jeremy was gone and wanted out, and he really wasn't really into those talks seriously. Lorenzo clearly wanted out, too. Laura Ellen was only a minority owner, and she was pretty dispirited…

By March, Leo's renewed ad in *The Wall Street Journal* hadn't gotten much response, so Leo started asking around. One evening, a friend told him he knew a guy he should call, so Leo took the number. In the morning, Leo called Harvey Levin.

Depressed, defeated, overworked and lonely, Laura Ellen got the news from Leo that the station had an interested buyer. "Let's move on this," he said.

Laura Ellen took a breath, then quietly said, "Yeah, okay."

Part Three

Heart Transplant

"I gave them the concept, and in April of '81, I was back at the station."

—Larry Yurdin

Chapter 125: Act 3, Scene 1: Harvey Took The Call

Harvey Levin had worked in radio sales for several years and several stations in the Bay Area, and been very successful. He was still a young, vibrant man, and he was on his way, having already bought his first station, KUIC, up in Vacaville, a growing bedroom community for Sacramento.

He was tall and gangly, in his mid-thirties. He had a full head of black hair, was a good looking guy. Very intense, very intelligent, very poised, with a smile that made him look like he had a secret. Handsome, funny, well-educated, he had been in the radio business for a long time and knew everyone there was to know in the business, except, apparently, for Leo Kesselman, and when they met, it was hate at first sight. Harvey knew radio, and thought that KFAT was the best station in the world, and he couldn't believe his luck that he had a chance to buy it. Despite his personal antipathy for Leo, Harvey wasn't going to let a suckwad like that ruin it, and he pursued the purchase. He came to the station, so he knew what he was getting, but he still wanted it.

He knew that another power boost was out of the question, but the signal was more powerful than the one at KUIC, and while that station couldn't get into San Francisco, this one could- and San Jose, too, which was growing rapidly with the prosperity that came with the high-tech industry. He could be a player in a major market, and it would be with *this* station. And it would be *his* station.

When he came to Gilroy that first time, he told Leo that he had no intention of changing anything much, except for most of it. The temporary staff would have to go, of course, and he wanted to rent space in the old Gilroy Hotel on the next block, and build a new studio. He'd install new equipment, new carpets, new everything. And he'd paint the walls and buy curtains. And build a new Production Room with proper gear, and proper shelving for the records, and a couple of offices. With air conditioning. But the sound, the music- that would stay the same. He'd just get some better people in there.

Harvey was an easy man to like, but Leo was not, and their antipathy grew with each meeting and conversation. In fairness, it's easy to dislike Leo Kesselman: he's arrogant, smug, intractable and demanding. Marty Manning remembered his first impression of Leo as "cynical, amoral, sarcastic." And Marty *liked* Leo. But it takes two to tango, and I asked Leo about his relationship with Harvey, and how it started. Here in his words, as he wrote them, is Leo's response.

LK: "First of all, the negotiations were contentious with a lot of 'fuck you' back and forth... Harvey was used to getting his way and this negotiation was not easy for him... secondly, as soon as the deal was signed, he tried to start running the station… this was both illegal and stupid.... he hung around and watched and got progressively more pissed off at the way the station was being run, by me. At the end, I threw him out of the station... 86'd [him]…told him to stay away until he owned it... almost came to blows... lots of threats from Harvey about how the station was being run, too many ads, etc."

It was a protracted negotiation, and Leo wasn't giving anything away. Laura Ellen wanted out of this station, she told him again. She was mentally exhausted, betrayed, abandoned by her friends-and-family, and left behind by Jeremy to deal with the station and raise Elsbeth. Whatever it took, she told Leo, get me out of this. Leo knew that this was for his friends, but that no one wanted an extended battle. Leo remembered that his friends had bought this little pissant station four years ago for $150,000, and all along Jeremy wanted to keep an open mind about selling it after the required three years had passed. Leo was a competitor, a negotiator, and he hated Harvey. As a point of pride, and as he was representing his friends, who were also his clients, he stretched Harvey as far as he could take him, then took him for 3.1 million dollars. The negotiations were giving Harvey headaches, so he swallowed and signed the papers. It was a lot, but he knew he could get a really good deal on the money to finance it, and the signal itself would do nothing but increase in value. It would take a while to go through the FCC transfer process, but everything seemed to be in order, and no one foresaw anything that would stop the sale. The embittered negotiations were a series of agonies, almost impossible for Harvey to endure, but they were over, and now, at last, the headaches would stop.

From the start of the talks and the transfer of ownership, the bitter battles between Harvey and Leo would leave a scar. Leo wasn't only "cynical, amoral, sarcastic," he was also vindictive, and until the transfer, he was still in charge at KFAT. Banned from the station by court order, Harvey listened, and, between headaches, fumed helplessly. The sale had gone through, but until the legal transfer, Leo was still fucking him.

In July of 1980, the sale had been progressing nicely. In August, Harvey found out about Leo's personal use of the station's income—where he'd run ads in trade for cash or goods and keep it all—and he freaked out. He demanded that Leo stop, but Leo shrugged, grinned, and said nothing. And the trade-out ads continued to run. Harvey just wanted to get his hands on the station, but he became so stressed out by that prick that now the headaches were getting worse. Leo was running ads in return for merchandise for himself! With no cash coming into the station! Harvey couldn't prove it until he got hold of the books, but who else could it be? A lot of trade-outs were running, and *someone* had to authorize them, so *someone* at the station had to be doing quite well for himself. The first someone had to be Leo, and the second one was Leo. Shit! What if the prick signs and takes payment for a sales contract that would run for six months- long after the transfer! He notified Leo in a letter where he again demanded that Leo no longer make those deals, and Leo proceeded to demonstrate his resistance to Harvey's suggestion by making even more trade-out deals.

Leo remembers: "After my argument with him, I went and bartered some more. I bartered $80,000 in merchandise in one month. With everybody. Washing machines, dryers, appliances, lumber, plants, nursery goods. Went to me and others, determined by me.

"I had to ban him from the station. He had already bought it, but it hadn't closed.

He was talking to individuals, which you shouldn't do, and he was yelling at me to do this and do that. I told him to stop bothering me until he took over, which would be November 1st, 1980.

"He continued to bother me, threatening me he'd sue, telling me not to run any barter accounts. We were doing this wrong, that wrong. We were at the verge of fisticuffs when I threw him out. I told him I'd barter if I wanted to.

"Since I have always been poor at taking orders, I immediately started to place that kind of ad on the station, all to run before he took over. From August to October, over $100,000 in ads. One salesperson was bartering ads without the station's or anyone else's permission. He was caught by me and stopped… Harvey made him sales manager (I believe) and over 3 months, got the station for over $200,000... this is a rumor... just what I heard."

Leo had stopped that one guy from making trade-out deals for himself, but that guy wasn't acting alone. Leo remembers "one incident where a salesman was writing barter contracts and taking the merchandise and selling it for cash, which I was told had to do with a drug habit. I was going to fire him, but the sale of the station was about to close, so I let him stay. He'd be Harvey's problem."

So Harvey had some problems. He had rebellion in KFAT's management, anarchy in its sales office, and upheaval in its air staff. And Harvey still had to find someone to tame the chaos when the ownership transferred and he'd be allowed back inside the station. He'd been in radio for fifteen years and seen many stations change ownership, but he'd never seen anything like this!

As Jon Stewart once said, this was a catastrophuck.

He needed help, needed to talk to someone about this out-of-control wreck of a hybrid. He had a lunatic asylum on his hands, and who could he call to talk about it? He had no answer to this, and then he got a call from the guy who had run the asylum before: Larry Yurdin.

Larry had his ears to the ground constantly, always alert for rumblings in the radio world. He still made his living in that world, and to be a player, he had to know what was what, where and when. Of course he'd been following the KFAT story off and on since his departure from Gilroy in December, 1975.

Harvey took the call and told Larry what was going on at the station. Larry remembered Leo from the meeting in Los Gatos when he and Jeremy were first discussing the plan for the station. Leo was the prick who'd argued against going with Larry's format, so he, too, had little love for Leo. So Harvey told Larry the horror stories.

Chapter 126: The Levin's Meet The Staff

Larry heard that Leo was fucking Harvey by whoring spots and taking the goods for himself. Everything he did just pissed off Harvey so much that he even told Larry about the headaches. When Leo filed a court order banning Harvey from the station, the headaches started getting worse. The transfer would be official in less than a month now, and a call from Leo totally surprised him.

Leo was inviting him to a lunch at Mountain Mike's in Los Gatos to meet the staff. He'd rented a private room at the restaurant, so they could take as much time as they wanted to get to know each other and talk about the station. Taken aback, but pleased, Harvey said yes, and a time and a day was set. "Oh yeah," said Leo, "and bring anybody you like."

Harvey was incredibly proud that he was about to be the new owner of KFAT, and he'd wanted to bring his wife to see the station, but he'd been banned from the station by court order. This was almost as good, and certainly a nice gesture by Leo, so he invited his wife to come with him to Los Gatos to meet the air staff.

Leo remembers: "Everybody was on their best behavior. There was an open bar, but nobody was drinking… didn't want to offend the new owner and his wife. I had no such inhibitions, so I brought a couple of hookers to meet the new owner… I was paying for the party, and me and the hookers were drinking, and pretty soon they were under the table… under everybody's table."

GK: Under the table? Doing…?

LK: "Giving blowjobs to the staff. Then the staff started drinking, and then they got pretty drunk, and Harvey's wife said 'I'll divorce you if you don't fire every one of these people!' She was shocked and disgusted. She stomped out, he stayed a while and then he left, and then everybody got really drunk. Harvey never partook. That was my farewell to the staff."

Chapter 127: Larry Took The Call

The bitter negotiations finally concluded and the FCC began the process of transferring ownership. There'd been no problems with the investigations into Harvey's financial stability, into any criminal past, connections with other media, and any past history that might render someone unfit to own a radio station. The ownership transfer went through. Good for Harvey, good for Lorenzo, Jeremy and Laura Ellen, all of whom were looking forward to being out of the KFAT business.

Walking away with 3.1 million dollars would be fun for the three owners. For Harvey, that money would come with a serious obligation, but not for a while. He'd gotten a sweetheart deal from some investors who knew him and liked him, and the note wouldn't get heavy for two years, and he was confident he could make it turn a profit in that time. He could do it with this station. This was the best station in the world, and he'd keep it Fat, but trim some around the edges. His background and success came from radio ad sales, and he knew why advertiser's blood pressure rose whenever the name KFAT came up, and he thought he could fix that. It was a serious commitment to make payments on the note every 30 days, but he had two years until it got heavy.

The rest of Harvey's media empire consisted of quiet KUIC up in Vacaville, an area that was growing yearly. KFAT was a volatile station, still settling down after several explosive months, with a new staff, and devoted listeners who were testy and suspicious after all the smoke had cleared and the changes made. He knew that Fatheads were an excitable bunch, prepared to scream if he changed their station too much. Yet somehow, the station would have to generate the money to pay for itself. He had two years.

While the papers were still at the FCC, Harvey had been busier than ever, running KUIC and making plans for KFAT. Busy as he was, when Larry Yurdin called, he got right on the line.

KFAT was Larry's baby from the old days, and you don't just wander off and stop caring about your baby. He'd tried giving birth to this format at Houston and Austin, and been thrown out of both towns, and almost made it work here. KFAT might have been born in a barn, but it had struggled and thrived, and now had a maturity and a voice of its own, and Harvey knew immediately that Larry was proud of it, even though he'd left it over four years ago. It had been a rough ride then, and it had been a rough ride since.

Larry had been bruised by radio for too long, and was still just a step ahead of broke, but couldn't leave the medium for a real job. He had by now become a radio headhunter in Los Angeles, going into business for himself, but the business was struggling. Over the years, he'd found so many people and even whole staffs for so many radio stations, that surely he could find a way to make a living at it. Yeah, doing what he'd always done- but collecting a paycheck and going home for the weekend would suit Larry just fine. He'd put the word out in the industry about his new business, the Radio Talent Bank. No more working with crazy people; now he'd be working with business people.

He had some bad memories from his days in Gilroy, and some good. He loved the music and he loved hanging with the jocks. He loved the sound of anarchy, but he respected the station that paid for itself, and that's what still makes him shudder at the memory of KFAT and Jeremy. "I couldn't work with him, because Jeremy's attitude was one of contempt for commercial radio, and that would make him shoot himself in the foot as many times as he could.

"I had bad feelings about having left it. At first, I didn't like what they were doing with it. Over time, this mellowed into a more cohesive sound. Self-indulgence was minimized by evolution of time- listeners requests, what works, collective sound, etc."

Hearing KFAT only when he drove between LA and San Francisco, he'd heard the jocks "arrive at the right direction circuitously. Gradually. [They] became more and more listenable, began to mirror the increasingly large audience, responded to listeners and ratings… and then I realized what a loyal audience they had developed."

Larry was hearing about KFAT from others in the industry. Interestingly, it was sometime in this period, around 1981-82, that he first heard KFAT's format described as "Americana." He noted it, but paid little attention to it then, as he wasn't concerned about what it was called, only whether it was working or not. He was proud of what he heard, but he also heard the rumblings on the radio circuit that KFAT's "management and air staff… were on different planets."

He'd heard that "Leo was whoring spots, [and the] staff [were] wondering why there were so many spots and they still weren't getting paid." [I] heard horror stories about Leo and staff rebellion. Leo was flooding the air with cheap commercials- 'Dollar A Holler' spots."

Larry remembers his first meeting with Lorenzo and Jeremy and Laura Ellen in Los Gatos. He remembered that Leo was there, and tried to talk his friends out of Larry's idea of a format. He'd heard that Leo was now the Station Manager. And now a new guy—a real businessman, not a nut—had bought it, and Larry was nothing if not full of advice for the new guy.

In the fall of 1980, Larry remembers: "I was working at the Radio Talent Bank in LA, placing people at stations. I had followed Harvey and KFAT in the trades, and I contacted Harvey. I was interested in helping him staff the station. There was an R & R ("Radio & Records"- a major trade publication) convention in LA. I had a hospitality suite. I met with Harvey there."

Larry had followed Harvey's career in the Bay Area, and admired what he'd read and heard about him. He knew about his success at KNEW, turning it into a successful country station in San Francisco, when everyone said that a country station would never work there. He'd made so much money he could afford to buy KUIC. Larry admired that, too.

"Harvey told me the seedy stories about Leo doing horror stories to KFAT. He wanted to expand, broaden the country format that he'd had success with."

Larry said the first thing he told Harvey was to ask Laura Ellen to stay on, train some jocks, do a show on weekends, make it casual. She'd give credibility to the new staff, and a sense of stability and approval to the listeners.

At the convention, Larry also told Harvey he'd "made a mistake in using country radio as his model for KFAT. KFAT was never a country station." In fact, he'd just had that very conversation an hour ago with Marty Manning, from the first staff, who was now working in Texas.

LY: "I talked with Harvey prior to the convention, mentioned Marty Manning, and talked Marty into coming to LA from Texas to meet at the convention. I asked Marty to be PD, he said absolutely not. He'd been around too many situations, and saw himself as a production and personality guy. He didn't want to have to manage people." Would Marty Manning want to come back to KFAT? Marty wasn't sure, but Larry could be persuasive.

Larry remembers that "David Chaney had done radio in Albuquerque, and had come to Jeremy's house before KFAT went on the air, while I was working out of Jeremy's garage." Larry put Chaney on weekends at the early KFAT, "and I thought he was a nice kid." After Marty Manning started the mass exodus from KFAT back in 1975, Chaney also left in the next ride out of Gilroy, getting in the car with Christa Taylor and Bobby Eakin, and dropped off in New Mexico. Now, four years later, he was working in Los Angeles at KLOS, a big-league Rocker. "Harvey was looking for a mainstream broadcast guy who would understand KFAT and would facilitate some kind of marriage." Larry and Chaney were both in LA, and they spoke occasionally.

LY: "After Marty Manning, I suggested David Chaney, but I had some reservations about his maturity. Different physical quarters is one thing- David could handle the same location with the sales staff, but not this situation. He wasn't mature enough. I said 'how about if we split the role?' and Marty was willing to be Operations Manager, which to me meant sharing."

With Marty on board as morning man, and thinking Chaney's experience in LA would help him run the show, he asked Chaney to come back to do Afternoon Drive and be the Program Director.

"I expected Marty to ride herd over Chaney. That wasn't his understanding, or he wasn't able to. David Chaney tried to turn it into an LA rock station." Larry's new business wasn't cutting it, but Larry still wanted in on the new KFAT.

"My Radio Talent Bank wasn't going to be economically viable, so I went off to Grand Rapids, Michigan, to do an AM talk station, staying as a consultant to Harvey."

Needing a staff, and listening to Larry, Harvey approved the two hires. Now, Marty Manning was coming back, and so was David Chaney. With his ear always to the ground, Larry knew Christa Taylor was just back from Hawaii and looking for work. A call, and that was done. KFAT had some good people coming back, and there were others he'd heard about. Amy Bianco was already part-timing there, and she'd be really good with some direction. It began to look like it might work.

Harvey loved KFAT. Larry remembers: "It was his favorite station, and he wanted to keep it what made it special, but make it more commercially viable." Larry was happy, which was a rare thing. It looked like it would work. Now maybe Harvey's headaches would stop.

"Harvey asked me to continue to keep an eye on the station. I was getting paid for placements, but nothing after that. I offered to consult to Harvey for free. I did, to a degree, from Michigan."

Larry watched from afar, and whenever Harvey needed him, Larry took the call.

Chapter 128: A Year To The Day

On November 1st, 1979, there was a staff meeting. The notice for the meeting was posted the day before, and everyone had to attend. The notice was signed by Lori Nelson, who for the moment had a lot of power at KFAT, as she was the only one down there with any business sense that Laura Ellen could trust. The staff was to meet the new Program Director, who turned out to be Doug Droese. The notice said that "THOSE NOT IN ATTENDANCE MAY NOT BE WORKING AT KFAT THEREAFTER." Droese was gonna run KFAT his way.

One year later to the day, the station had been sold, the money put into escrow,the transaction reviewed, approved and certified by the FCC, when the money was released and on November 1st, 1980, KFAT's ownership transferred to Harvey Levin. Harvey was finally in charge.

Leo remembered: "I was the very first person fired. By Harvey. I found Harvey to buy the station. My relationship with him was somewhere between loathing and hate. He thought I wasn't running the station right in the last months before he took over. He continued to bother me, threatening me he'd sue, telling me not to run any barter accounts. We were doing this wrong, that wrong."

That was then, this was now, and a new staff was coming into focus. Some were in sight already, some just out of sight. Some were old KFAT staff, some were new; there was a new sales staff coming together and new management, but no one heard the clock ticking.

Chapter 129: Marty To Dale To Gilroy

We shouldn't gloss over Marty Manning too lightly. Marty is one of the few players in this story with a serious career in radio, both before and especially after KFAT. As of this writing, he is still very successful in his market in Phoenix, and has had a long, much- respected career in radio. We'll let him ponder what such a respected professional broadcaster such as himself was doing at KFAT- twice!

At the convention in LA, Marty remembers, "Harvey liked me and hired me on the spot. We hammered out a deal." Already convinced by Larry that Marty was the guy, Harvey had heard his tapes and happily hired him. Marty knew about Harvey, too. He knew that Harvey had succeeded in San Francisco with a country station when everyone said it couldn't succeed. He knew about KUIC, where he remembered that Harvey "took a slack property and made it work. Now he had a property that he had some passion about. He loved this station, wanted to make it work. He needed someone who knew how to work."

But despite Larry's pleas, Marty wouldn't be PD. He knew that KFAT was still an unconventional station where the normal rules didn't apply, and he remembered that Larry was crazy. He didn't want to have to discipline anyone or watch over what they played, or when. He wanted the gig, and he loved the music—the playlist was the best he'd ever heard—and he wanted to do Morning Drive. Larry suggested sharing the managerial duties, and Marty felt he could handle that if the owner wasn't crazy, and he'd just settled that matter by meeting Harvey.

But there were problems. The first was that at that convention, Harvey wanted Marty to call himself Sam Antonio, and he flat-out said no. If they wanted him to have a new name, he'd think of one, and on the flight home, he thought of Dale Evans. There were bigger problems than names to consider, and number one was his wife and how she'd deal with KFAT and Gilroy; that would take some selling. He wasn't sure of what he could tell her about KFAT; he'd been away from it for years, they were stable where they were, and she would be resistant to a move.

MM: "I didn't really follow KFAT's progress, and I came back and it was different. Jeremy and Laura Ellen were gone. I was always very fond of Laura

Ellen, and I had to get out of living at their house because I was developing a crush on her." He'd spent lots of time talking to Laura Ellen, with Jeremy ignoring both of them. Now they were gone.

When Marty got home and told his wife about the new job in California, she cried. And when he told her he was going to be Dale Evans, she cried some more: she "didn't want to be known as Mrs. Evans."

Diana Manning "was a serious Texan. Leaving Texas for first time, she looked back at the Texas border and cried again. She'd been out of Texas just once in her life." He remembers the trip west, and the "wild ride on Pacheco Pass," the curvy, scary downhill ride after you get off the highway, about 45 minutes away from Gilroy. "The windshield smeared… driving a truck full of stuff… sort of raining… endless line of lights… punching roof of truck… dogs freaking out…"

Once in Gilroy, Marty left his wife at a Motel 6, in a room smelling of mold and new paint, and had dinner with people from the station. When he got back to the motel, she was crying again. But there was hope, and he told her about it.

Harvey had hired an engineer, John Higdon, to build the new studio down the street. Harvey had taken the office space above the Old Gilroy Hotel on Monterey Street, just a block from the old station. Harvey wanted Bill Goldsmith to work with Marty on it: for Bill and Marty to design it, and John to build it. John had built over 30 stations by then, and he was a friend of Jeremy's, so Marty knew the new station would be built right. This would be both a challenge and an opportunity for Marty. Also, having just met him, Marty knew that Higdon was a little crazy, and that helped. They sat down one afternoon, "smoked a joint, and put down pieces of cardboard, and I said I want this, how do I do it?"

JH: "Get a cabinet maker…"

MM: "I found a cabinet maker in Morgan Hill who was great, and they made it out of birch plywood."

Soon, the "studio was ready, but the antennas were delayed." Higdon jerry-rigged a phone line "out the window to a pole, one block to the old studio, and [we] broadcast that way. It was completely illegal. We faked readings to the FCC."

So, of course the FCC, who were always interested in knowing what KFAT was doing, came by to look things over. The FCC guy was a typical low-ranking bureaucrat.

MM: "Plaid pants and a horrible cold. He was miserable. He looks around, saw the new rack of gear, and asked, 'where are your logs?'

"I'll be honest- here's the deal. They've been ordered, they're not here yet. When they're here, they'll go up. Frankly, it's just too much of a pain in the ass to do the readings. He said OK."

Confucius say: It bodes well.

Gilbert say: We'll see.

Chapter 130: From Mild To Wild

I guess he'd changed. Earlier he'd been too mild, and now he was Wild. He is going to be an important player in these pages, so let's get to know him.

Bill Goldsmith has the typical story- one you've read repeatedly and recently in these very pages. He was barely old enough to own a transistor radio, which he hid under the covers to listen to the far-away stations at night. He was born in Colorado, and at seven or eight he was listening to Chicago and Oklahoma City. "Transported" is the word he uses to describe those nights. He thought DJs were the coolest, man.

"If there was one job in the world that I could pick, that would be the very best one. 'Cause these guys sounded like they were having the time of their lives, and they probably made tons of money doing it. I found out more about *that* as I went along, but… "

His parents moved to Gilroy, where his father started a business that became one of the biggest seed companies in America. "And then when I was a teenager, I was living in Gilroy, and I worked out in the flower fields, in my dad's place, in the summer. And I had a little transistor radio that I carried around with me, and I listened to the DJs on KLIV, San Jose, and I'd be out there with my hoe all day long, pulling weeds out in the hot sun, and I was listening to this guy who was sitting in a nice air conditioned room somewhere, probably surrounded by pretty girls, sitting back, talking to folks on the phone, chatting on the microphone, and playing records. And I thought: 'That's the gig I want.' I was 14 or 15. When I was 16, I discovered drugs and all that kind of thing, and I decided that I wanted to be a hippie, and started listening to those hippie radio stations."

He convinced his parents to send him to the Columbia School of Broadcasting in San Francisco, but that was mostly through the mail, and it was preparing him for a career in radio that was stuck in the 1950s. The school had no idea of what was going on in the world, and he soon dropped out, but finished high school.

It was 1970 and radio was still exciting, and still calling to him through the airwaves, and he went to the only station around, KDON in Salinas, and "this crazy-looking old guy, who reeked of alcohol, was chasing this younger guy out the door and screaming at him at the top of his lungs. And that was the Program Director. And the drunken old man was the guy who ran the place for a friend of his, and so he hired me, right then. I had never been inside a radio station in my life, and I walked in and I was on the air at the radio station within ten minutes.

"It was a rock station, and it kept floating somewhere between playing hit singles, and playing all this hippie music. And I was leaning over on the hippie music direction, and Charlie Brown, he was, tops, five, ten years away from cirrhosis of the liver, he was also a hopeless pedophile. Young boys, he would bring them into the station and make them a DJ for the day so that they would sleep with him. I was 18 by this point, and too old for him."

Bill was PD there almost immediately, and he was out of there in eight or nine months. It wasn't his kind of music. For that, he listened to KLRB in Monterey. One day there was a guy on the air who just didn't seem to fit in, so he called to ask what was up, and the man who answered said that he had just blown out about half of his staff, and Bill should come right over, and then Bill was on KLRB.

One day he was hanging around in the Production Studio at KLRB, and someone fiddled around on the radio to see what else was out there, and they heard something out there, *really* out there. They listened to the song, curious, and then another, even more out-there song came on. Without knowing what it was yet, Bill had found KFAT. They kept listening and liking what they heard, and then a DJ with a Texas accent came on and said it was a new station, it was KFAT, and he hoped "y'all like it." It was August 4th, 1979. I know! I'm amazed, too! Both Sister Tiny and Bill Goldsmith had found KFAT at the same moment. Consider the forces at work here…

Bill called up and spoke to a guy named Larry, who invited Bill over to do a weekend shift. After the shift, Larry told him he was "too normal." Bill was shocked and offended. At KLRB, *he* was the weird one, "the one they felt they had to constantly reign in." He said, "well, okay- your station is kind of weird, anyway," and he left. He still listened, but that was the last time he set foot in the station for several years. But when he came back, everything there had changed, and so had he.

After KLRB, Bill went to Honolulu to work for a few years, then did some radio in Boston, and then back to Gilroy, living at his parent's house with his girlfriend and their four children. Yeah, things had really changed, but some things never change. Needing work, he was soon back at that first station he'd worked at, in Salinas, and hating it when John Lennon was shot.

KDON was still playing the hits, and had only one John Lennon song in rotation, and that was Lennon's still-current single, "Just Like Starting Over." Lennon's death was a shocking, meaningful loss to him and his listeners, and he couldn't play that song. He was stunned. He knew this was a major, culture-wide tragedy, and he wanted to do something special to honor the fallen hero, but there were no Lennon or even Beatles songs in the house, so Bill called over to KFAT and asked Laura Ellen if he could come over and tape some music. Laura Ellen said yes, and the next morning he played what he had put together until the Station Manager called him up and told him to stop. Bill argued, and the Manager reminded him that what he was doing was not the format, and he insisted that Bill stick to the format. But *this was special*, Bill insisted. "Well, I don't think we should be doing this!" he was told. "We should be playing the hits! We play the hits here!"

"So I finished up my show, walked up to him and said, 'Dan, I think I need to get my check. I don't think this is going to work.' And I quit my job and went down to KFAT and said 'Hey, I want to work here.' And this was right after Harvey had bought the station."

David Chaney was looking for people back then, and he hired Bill on the spot, and Laura Ellen was still there as Music Director. Bill Goldsmith would be Wild Bill. Chaney was still PD and doing Afternoon Drive, Laura Ellen was on from seven p.m. until midnight, and they gave Wild Bill the mid-days. A couple of weeks later, Laura Ellen was fired and Bill was given seven 'til midnight, and made MD. Soon Chaney was fired, and Larry was back at the station, full-time.

BG: "Larry and I talked and compared visions for the station, and it didn't sound like me being MD to Larry's PD was going to work, and Larry thought so, too," and that was when Larry went looking for help outside of the area. Bill was moved to weekends after Larry fired Unkle Sherman again, no one remembers why, this time. Then Larry went to Harvey and told him about Michael Hess, aka Dallas Dobro.

With Bill back to weekends again, he started looking to "get the hell out of radio." He liked the work, but he'd never made much money and he was living at his parents' house with a six-member family to support, and he knew he knew only one thing more than radio. His life-long interest in electronics had developed into an expertise, and should serve him well. He kept the weekend gig, but got a job at an electronics firm in nearby Morgan Hill, designing circuitry for the early home satellite equipment.

He still did mid-days on the weekends. Harvey wanted to move the studio down the block, and Bill's knowledge of electronics would be invaluable; Larry wanted Bill to stay, at least through the transition out of that rickety old studio. Harvey and Larry needed Bill to help with the new studio, and Bill remembers, "They were still above the optometrist's office. We were there for probably 6 months. Harvey had been down there once or twice, absolutely hated the place, didn't want to have anything to do with it. Didn't want it as part of his radio station. He thought it was just an ungodly pigsty. It hadn't been vacuumed for 5 or 6 years, literally. The walls hadn't been washed or painted or had the graffiti scrubbed off them." Bill remembers the wall that still asked WHO STOLE MY SCREWDRIVERS?

I asked Bill about what most excited the new staff.

BG: "About moving into the deluxe new studios down at the old Gilroy Hotel. We hauled what little equipment we were bringing down there, 'cause it was all going to be brand new equipment... and we hauled everything down there, and closed the place up and shut all the windows and locked the doors and went away, and didn't come back for two or three weeks, in the middle of summer. [When] we opened that door down there, the place had been shut up solid in the middle of summer for three weeks. The stench... it was like a physical presence when you opened the door. I had never smelled anything quite like it in my life. It smelled like it was a rotting corpse in there. It was bad. And we aired it out, did our salvage mission, and someone asked me to take pictures of the graffiti in the bathroom.

"So I went in there and spent like, fifteen minutes or so in the bathroom, making sure I got all the graffiti, 'cause it was everywhere. Every square inch of the bathroom was covered with stuff. Layers of it. It was some pretty historic shit. And so six weeks after that, I came down with Hepatitus A. Six weeks is the incubation period for Hep A. I know I got that from that bathroom. I know it for a fact. It had to be."

Chapter 131: Pre-Flight Amy

"I was living in LA and transferred to UC Santa Cruz, and in the process of deciding whether to transfer or not, I went to see UCSC, and my boyfriend and I at the time were very into music, and we were driving around the mountains of Santa Cruz, checking out the radio stations, and all of a sudden [we] heard the Grateful Dead on the radio, and thought, 'Oh! Cause we were Deadheads.'

"And this was late '77. Summer of '77, probably July. I think Sully was on the air, or Laura Ellen, it was probably off of "Workingman's Dead," maybe "Dire Wolf"– and we were 'Whoa, what's that?' And after that they played Jimmy Rogers and something after that, I don't remember, I just remember that it was totally interesting. And we were camping down in Big Sur and couldn't get it the whole way, so we wound up going up to the Watsonville area to camp, so we could listen to the radio station, which is a typical KFAT story."

Yeah, it was a typical KFAT story; you've heard them before. This one started in Los Angeles, where her parents were musicians, and music infused the rhythm of the house. And like those other stories, she listened to radio in bed, late at night when the distant signals came in. And like the others, she had to be sneaky about it.

Amy Bianco listened to Wolfman Jack, who broadcast from Baja* with a power that no station in the U.S. was allowed. It traveled across the U.S. border and up through the Plains, and it traveled north and south, and it traveled to Los Angeles, where little Amy Bianco was listening to what was then considered wild radio. I know Amy now, and I'll bet she was cute as hell.

She listened to stations from San Francisco and Denver, and then, in Los Angeles, they started running "Suspense Theater," which sealed her "falling in love with the mystery of radio." At eleven, her parents gave her a Radio Shack cassette recorder, and she made her own radio station, with music and interviews. Amy called it KRZY Radio.

AB: "Krazy Radio (laughter). I may have been about 11, and that's where it started. And then, fortunately, in the late 60s, early 70s, there was some good progressive radio in LA. KPPC, later KMET, that really was interesting. I remember they had the Persuasions doing their IDs. We had some interesting personalities on the radio- kind of reflective of what was going on in San

* Okay, it wasn't *exactly* Tijuana- He was in California, but the tower the signal went from was fifteen miles south of Tijuana, Mexico, where it was legal to broadcast at a much higher power than was legal in the U.S. The interesting thing for me is that I now live about a mile from that tower, and it's still broadcasting into the U.S.

Francisco, but of its own thing, because of the music scene in LA."

By the end of the 70s, she was serious about going to the University of California at Santa Cruz, and she looked into the radio department there. She transferred from UCLA, where the school radio station was only on cable, and was being phased out, but at UC Santa Cruz, there was an opening for anyone willing to take on whatever responsibilities needed taking on. She says that Santa Cruz "has always been its own little universe. We had (listener supported) KUSP, and then the fledging station at the University, which was KZSC, and KZSC was suffering badly from neglect. It had some good people there, but the U didn't care about it, and they were in the process of trying to cut it loose. You know, the usual sort of political stuff that goes on. This was in the days before the licenses were so valuable, and... "

A group decided to get involved, and "I went on the air from two 'til six p.m. My first radio show was 'Spaghetti Western' and I was fascinated by what was going on over at KFAT, because I listened to that the whole time. But that was 'commercial radio,' and I saw an opportunity to learn about radio while I was at the U, so while I was studying my English career, I volunteered at the radio station as much as I possibly could. I lived on campus and I was there all the time, and within six months, I was PD."

She went on the air when others failed to show, which was most prevalent around mid-term exams and finals. But the show had to go on, and if there was only Amy available to do it, then Amy did it. A lot. Doing all kinds of music. But she "learned a lot about music in that year and a half or so. And one of the people at the station was Elizabeth Gipps. She did a show called "Changes," and she had been doing it for years. And Elizabeth was old-school hippie radio. I mean, she..."

GK: Wait! Wasn't that Jeremy's mother?

AB: "Yes. She went by 'Elizabeth,' and she would go interview at the Mall in Santa Cruz, and interview people at the St. George Hotel, and bring in nose flute players, and (laughter) and I learned a lot about the counter culture through Elizabeth. She had this great way about her- just this old hippie thing. And one day she said to me, she had this great way of talking, kind of like Marjorie Main meets Janis Joplin, and she said 'Amy, Laura Ellen needs somebody over at KFAT to do fill-ins. Have you ever thought about that? You oughta send her a tape.'

"And I thought OK, and I was still on the air from two 'til six in the afternoons, and so I stuck a tape in one day, and I made it a little more commercial. The thing that was cool about college radio back then was you could go off for hours and do these weird fiddle and banjo things. And so I made that a little commercial.

"When I did that show, specifically, I had spent a year and a half learning.

I consciously studied. I didn't know who the hell Jimmy Rogers was when I moved there. I knew who Bill Monroe was, because my mother was from Tennessee, and I knew some commercial country, but I didn't know the roots of it. That year and a half- I love to study, that's one of my weird things- I just threw myself into it. And I was playing banjo at the time, and I was interested from that perspective, and I just dove into it, learning as much as I could about the music, and I learned from listening to KFAT.

"And they really did educate- and a lot of the DJs had this great... Laura Ellen, Sully, Terrell Lynn, Sister Tiny, they all had this way of educating you subtly. I mean it wasn't patronizing, it was funny... you could tell they had it in their hearts, and that was what turned me on. It made me know you could be comfortable, that you didn't have to have this formal approach."

GK: What do you remember about that tape you sent?

AB: "I still have it. I have the receipt. I sent it over on a Greyhound bus, and she got it the next day and called me and asked me to come down, and I made it over there. I had transportation problems, but I made it there, and she put me on the air that weekend. I did afternoons.

"Now, college radio... and being stuck on the air for eighteen hours at a time when people didn't show up for their shifts, as students are wont to do... that also made me... helped me become comfortable, because in the space of about two years, I did thousands of hours of radio, and I would go some days, when everybody would get 'sick,' like during finals, and I would go from doing a soul show to doing a classical show to doing an acid rock show... "

Amy had been listening and learning from KFAT, she was playing a blend of country, blues, western and a lot more, she was used to working extremely long shifts, having to do different shows with different themes, sometimes with no advance notice, for no pay, at great sacrifice, for not enough listeners. In other words, she was already Fat, and she fit easily into her weekend shifts, untouched by the chaos.

Laura Ellen was there, but the station was in turmoil. Rocket was still there from the last staff, and he knew a lot of the story, but he was around at unpredictable hours these days, and no one knew much, anyway, and so rumors spread fluidly and with alarming alacrity. The station was for sale, or it was sold or it was off the market. Big changes were coming, imminent, or no changes were expected. Or at least, not for a while. Or...

The main staff had mostly left, leaving only a few irregulars around. An interim cast from some college had come in, knew nothing about either the music or about good radio, and were now leaving, and KFAT needed a staff. It was late 1979, and Amy fit in just fine, thank you, doing weekends and fill-ins. On the air, she called herself... Amy.

I asked her if anyone gave her direction about her shifts when she first got there, and she laughed. Laura Ellen told her to "keep it fresh and to mix it up," and then put her on the air.

GK: Did Laura Ellen ever ask you to do anything different from what you were doing?

"(laughter) No, that never happened! She'd tell me to keep the cat off the table. There was a cat there named Maxine... And try to lock the door down on Monterey so no one bothers you at night."

She did weekends and fill-ins. She took some time off in late summer to go back east to attend a college media conference, "and I don't know if it was a regular shift until I got back from my summer trip, but I remember being there in 1980 when John Lennon died, that was around the time that I got a full-time shift, and we were all in the studio, and I think there was an earthquake that day, too- there was something weird going on."

Chapter 132: Everyone vs. Chaney

David Chaney was making everyone unhappy, including himself. He was frustrated and floundering, and everyone knew it. But Amy remembers the joy of working with a true heroine of hers, Laura Ellen. And then Laura Ellen was gone.

AB: "It happened while I was a part-timer, and I was pretty upset that Laura Ellen wasn't going to be there anymore. And I remember worrying terribly, and I may have distanced myself. I was getting a little stand-offish on the air, and Chain Man said to me, 'Listen- don't blow it. You're going to stay.' And I was worried about being fired, because everyone was worried."

Chaney had just fired Rocket Man. *Rocket Man!*

Sherman: "I was doing afternoons, Ranger Rick Nagle worked ten 'til two, I went on at two and Chaney showed up and he fired Ranger Rick as soon as he got off the air. I knew my time was near, and as soon as I got off the air he fired me. By this time Marty Manning was there, and I was still cleaning out my box and getting ready to go, and Marty remembered me from before, [from the first staff] and he called Chaney over and he said that he had made a mistake in firing Unkle Sherman, and that he needed to keep a couple of these guys. He said Sherman has been here a long time and he's solid, and etc. Chaney said 'Fuck it- he's outta here!' "

Marty Manning says: "David was a pain in the ass... didn't belong there... completely wrong. Young guy worked in LA radio and thought he knew everything... filled with commercial radio experience..."

Marty remembers Chaney was "the wrong guy."At a meeting, Chaney said, "let's do this," and Marty said, "how about if we do that?" and after the meeting, Chaney said to Marty: "Don't ever challenge my authority again!" Marty thought Chaney was "a small player in LA, he came in thinking he'd show these bumpkins how it's done."

Chaney was a train wreck, and then there was a rumor that Larry was coming back to straighten it out. Chaney was—and still is—convinced that Larry was going to be brought in to fire and replace him. Larry insisted then, and insists now, that while Harvey was imploring him to fire Chaney, it was Larry who defended him and urged Harvey to give Chaney a few weeks under his direction. He'd turn him around. He'd turn it *all* around.

But remember: Larry is crazy. He's as enthusiastic as a kid on Christmas morning, God bless him, but the crazy stays around all year long.

Chaney was too big of a problem at KFAT. "He was confusing," Amy remembers, not fondly, "and that was why I didn't have a lot of respect for him. He was really struggling… "

Not knowing the music "was part of the problem. His idea of country music was 'Country Honk' by the Rolling Stones. So I think he struggled with that, and I think he may have had some problems with being comfortable with the root of KFAT… why KFAT was so popular."

GK: How effective was Chaney as an administrator?

AB: "Not terribly. I don't think he was clear. He had a hard time giving clear instructions. In other words, there's constructive criticism, and there's just criticism. And there's times I would get… I'd remember getting memos that I didn't understand. They'd say, don't say this, say this instead, but why? What are you getting at? What is the point? I think he was kind of reactive. I try to make communication facile, because I don't like conflict.

"I think maybe there was a bit of a... I don't think he respected country music, and that is a problem when you're coming into a station that- granted, was not your typical station, but you gotta have respect for Loretta Lynn as well as Stevie Ray Vaughn. You gotta understand that these things all have their place."

GK: So his shift was more rock 'n' roll and less country?

AB: "Definitely. And then it gets a little muddled for me when Larry comes into it, because I remember them both being involved, and the memo process, and then we went through this horrible process when Harvey took over, of this consultant, focus group stuff…

GK: Focus groups?

AB: "Yeah, there were focus groups. Larry… was involved in that. Groups saying "what do you think of this song? Of that song? I think that's part of what Larry was doing… and the whole transition, when Harvey took over, who I always had good experiences with. Harvey liked me a lot, and it's not like he was hands-on, it wasn't like I had a lot of experience with him. He was always kind of complimentary. Maybe understanding that being critical probably wasn't going to help me, 'cause I was kind of sensitive, and I was pretty young.

"I also really enjoyed what I did, I knew I was lucky. I knew that it was going to be… that I was a lucky person to have this as my job. That to have the freedom that we had, even when things started getting more restrictive, because Larry came back and he and Bill Goldsmith started creating a format that…

GK: Bill Goldsmith worked with Larry on programming?

AB: "Oh, yeah. He was Music Director."

Chapter 133: All But Gone

Back in 1975, Larry met with Jeremy in a stairwell in Madison, Wisconsin, to discuss what KFAT could be. Then they met in Los Gatos, this time with Lorenzo and Leo. Leo hated the ideas being discussed by his friends and this lunatic. He argued against it, and Larry argued for it. Larry prevailed and rarely saw Leo after that.

LY: "Harvey's concept was closer to mine… that's why he hired me again when he got it."

At first, Harvey consulted with Larry in a philosophical sense about the station, about what could and what should be done with it. Harvey was in the *business* of radio, looking for practical, applicable programming and managerial advice, and Larry was a radio programmer and a pragmatist, and that's what Harvey needed right then. Before the transfer, Harvey started going down to Gilroy to talk about his ideas, maybe use some right now. But he and Leo fought with increasing bitterness until Leo got a court order that threw him out of the station until the transfer was complete. Harvey's calls to Larry were productive enough when the station was going through its change-of-licensing process, but the closer the transfer got, the frequency and the urgency of the calls increased. Harvey had lots of ideas, lots of questions, and only one real plan. He knew he wanted it to sound like the old station, but with more control, make it more ad friendly, whatever it would take to meet that payment in two years. He wasn't worried about the current payments every month- those he knew he could cover. Back in 1980, three million dollars was a lot of money. He wanted KFAT to continue as a great radio station, but it had to pay for itself, and he would do what it took to combine his two needs. The only thing Harvey knew for certain was that his first act as owner of the station would be to fire that fucking Leo, that fucking fuck!

He wanted to fire the asshole before the asshole quit, but at least he had the satisfaction of knowing that he'd pissed Leo off by hiring Larry.

By the time of the transfer, KFAT was in a manageable chaos. He had a pretty good air staff now that the college kids were gone, and there were people there with real talent. He liked that new girl, Amy, and told her so the few times he came by when she was there. Sherman was gone, and he liked that. Chaney had fired Rocket Man, but Larry hired him back, and he was pretty good and he didn't cause any problems, so that was good. I was gone, but Cuzin Al wanted to stay.

Marty Manning, now Dale Evans, was *really* good. Everyone knew a good Morning Drive DJ is key to a successful station, and Marty/Dale was good. Harvey was glad that he'd listened to Larry about that guy; him and the rest of the new people made it a pretty good staff. Now they had Paul Krassner doing a talk show. But Harvey knew that Chaney was the problem.

Larry remembered: "I made a big mistake. I served as a consultant for Harvey, and recommended David Chaney. But I also recommended that Marty Manning come back."

Worn out and broke again, Larry had gotten out of radio and gotten himself a real job. Imagine, if you will, Larry in a real job. Soon, radio called to him again.

LY: "I worked for a headhunter firm, and didn't know what I was doing, so I accepted a job for myself running an AM talk station in Grand Rapids, Michigan, but also consulted for Harvey."

With the station now Harvey's, the phone calls to Larry weren't enough: Larry had to come to Gilroy. Harvey was frantic and the headaches were getting worse. The guy Larry had recommended over the phone, Chaney, was a disaster, Larry had to come in and fire him and find someone else. Larry said he'd be there the next week.

Larry remembers the call: "Harvey wanted to fire Chaney on the spot; he didn't understand the whole thing. I put him off for a week until I could look at things, and saw that Chaney was the problem. I went back to Harvey and told him that his instincts were right. Harvey asked me if I was interested in programming the Station, and I said no, but he talked me into it.

"Chaney was convinced that I was there to take his job and get rid of him, and the paradox was that I saved his job. For a week. Some on-air were good, others were a problem. Chaney had fired Sherman, I thought that was a mistake and I immediately hired him back. Harvey didn't like Sherman. I thought he was the spirit of KFAT."

By then, Chaney was all but gone. Larry just needed a little time to look things over. Sherman had almost caught up with Buffalo Bob for Most Times Fired, but Bob was long gone, and then Larry called and Sherman was back! The record was now within his reach. He knew that Larry liked him, but he wasn't sure about Harvey. Harvey was sure: he didn't like Sherman. Sherman was Larry's guy, so Sherman was back. Then Larry called Rocket Man and hired him back, too.

Larry was back in charge, even over Chaney, and there were new rules: "I tried to turn it back to pure Fat, with the illusion of spontaneity, rather than the reality of chaos. At that point I did color coded cards. Not to restrict choices, but anything that anybody who is good enough to be hired to work here wants to play, belongs on the air, but we can't keep that balance all the time. I wanted to create a structure where if something wasn't normal for the station, we could discuss which box to add it to, to discuss where in the rotation it belonged. Different balance. Create a mosaic where the elements can interact. And it should prevail over the whole day equally, as opposed to block programming.

"If you are inspired, and do something that fits the spirit of the law and not the letter…

If you break the rules and it's not noticeable, I won't come down on you. Only if it's noticeable."

Chapter 134: The New Friend

For much of 1979, Russ Martineau had been out of work, and he "wanted to get into the radio business." He'd worked in college radio and at his college newspaper. Russ remembered, "I called the sales manager at KNEW, and told him I wanted to get into the radio business and he said, well, you don't know anything about the radio business, I had no experience. He said 'I don't have anything' but 'call this guy Harvey Levin out in Vacaville.' So I called Harvey and got an appointment with him, and I met him in Berkeley for lunch. I was late 20's..." Russ Martineau continues:

"So we had lunch. Lunch lasted less than an hour and a half... he asked me a few questions. At 3:30, we were in a carpet store in Suisun, California, selling radio time. And I just loved the guy, just from the second... and he couldn't get anybody to work for him out there in the middle of nowhere, so that's why he needed a warm body. We were literally selling radio time two hours after lunch!"

Harvey wanted Russ to work for him, telling him he'd make money and they'd have a ball, and by that evening, Russ had a job in Vacaville, but still lived in San Francisco. Harvey was married and living in the Oakland Hills, but growing increasingly distant from his wife. Harvey's evenings were frequently open, and he and Russ would work until after the traffic thinned out, drive south together and go out drinking, with Russ spending the night at Harvey's. At year's end, Harvey was increasingly preoccupied, getting more involved with buying KFAT, so Russ moved to Vacaville.

Once Harvey heard about the possibility of buying KFAT, he'd moved quickly, giving it all his time, and soon after that he owned it, and all this was happening while Russ was learning about the radio business for the first time, on his own. Harvey was focused on the negotiations and he was getting headaches from that asshole Leo, and found himself so consumed in KFAT business that he was sorry about it, but he had to let his pal Russ learn the radio business on his own. So up in Vacaville, Russ learned the business as best he could. With Harvey's direction by phone, he was doing all right. He was a salesman, remember? He was eager to succeed, and he knew he could do this. He could sell anything. With all of two month's experience on the job, Harvey promoted him. Meet Russ Martineau, KUIC Sales Manager.

Russ worked long hours, made a good deal of money, and remembers that "working with Harvey was great fun."

With the transfer about to happen, and with a station in chaos both on the air and in sales, Harvey needed someone down there that he could trust in charge of the sales side. Bob McLain was a hustler, out for himself, and Harvey knew all about McLain from his years in Bay Area radio, and he didn't trust him and he wouldn't have him. Harvey wanted his own guy in there, and his new go-to guy was Russ, so McLain was gone. Now, meet Russ Martineau, KFAT Sales Manager. Also good to know: Russ had always liked his booze, and he and Harvey shared that, but about now he found cocaine, and he was very enthusiastic about it.

So, the future was bright for Russ, except for three problems: Russ had almost no experience in the radio business, Russ really liked to party, especially with his new best friend, Harvey, and Russ was about to get some news that he wasn't going to handle very well.

Chapter 135: The Path To St. Harvey

The negotiations for KFAT had been heated and hateful, but they were over. Harvey owned it now; Leo had been fired, Larry was staffing the station, and that was going well, and he had a handle on what he wanted from a sales staff. With the improvements he'd be making at KFAT, he could finally attract a competent, aggressive and honest sales staff, and the station could be paying for itself in six months. The odious pressures of the sale were finally behind Harvey, this was a happy and exciting time for him, and the headaches should have stopped by now. He had fought the battles and won the prize. The station was stable and on track for success, but the headaches kept getting worse. A few weeks after the transfer, he made an appointment with his doctor, who then sent him to a specialist at Stanford for tests, and the next day he saw the specialist again for the test results.

Harvey Levin had brain cancer. It was inoperable. He had maybe six months to live.

Chapter 136: Russ Gets The News

Russ was a twenty-something, good-time, semi-serious career-oriented guy in the most normal of ways. With a little college radio in his background, he thought he'd like to get into radio sales, so he called KUIC. Immediately, Harvey and Russ became great friends, spending a lot of time together, both on station business and for fun, and Russ knew that he was hitching his star to Harvey. The day Harvey got that first call from Leo, he told Russ about it, and they were both excited, as one of the things they had in common was that KFAT was both men's favorite station. Harvey loved KFAT and he swore to keep it Fat. He thought he had two years. For the last time: fat chance.

As the KFAT sale was progressing, and taking more of his time, Harvey told Russ that the way their relationship was developing, this new station would be Harvey's, but the next one might be theirs, together. Russ knew that he wanted to be in business with Harvey for the rest of his career.

Russ was Sales Manager and co- General Manager of KUIC, and Harvey depended on him more up there in Vacaville as the KFAT sale was heating up and the battles ensued. Russ missed Harvey, seeing him now only on weekends at his beautiful home in the Oakland Hills. Harvey was married, but the marriage was in trouble, and Harvey spent as much time with Russ as he could. While the sale was going through the FCC processes, Harvey was spending all his time on the station. He was on the phone with Larry, hiring and talking to people, looking at building plans for the new studios, looking for a sales staff, and by the time he had some time, he went to see his doctor about his headaches, who put him in Stanford for more tests.

Russ remembered: "I'm sitting in my office in Vacaville, and I knew that Harvey was going into the hospital for some tests, and he called me and he said 'I got it.' I said 'Got what?' 'I got the Big C.' Harvey said that he had inoperable brain cancer, and maybe six months to live."

Russ again: "And I remember that night I went out. There was a big sign on the freeway in Fairfield, and it said "Happy Birthday" and I was drunk and I climbed on the sign and pulled off the word Happy, and went to his room in the hospital the next day and taped the word Happy up on his wall… climbed up on this sign in the middle of the night… I was devastated, just as he was.

I mean, he was my mentor. By that point I thought… at that point you could own seven AMs and seven FMs, and that was the law, and I'm thinking I'm going to be in business with this man for the rest of my life…"

Harvey was devastated, but still very much alive and in the moment, and by the end of that visit to the hospital, Russ was the Sales Manager *and* the General Manager of KFAT. Personally, I like Russ, but he was now about sixty feet out of his depth, and his jerk gene was about to kick in.

Chapter 137: Big Enough

Larry Yurdin: "I came to Gilroy, sat down with Harvey the first day he was diagnosed with cancer. Harvey said that I shouldn't pay attention to the cancer, that he was going to fight it."

Holy shit! The guy had just bought the station, had been the new owner, what- a few weeks? And now he had brain cancer? And the station was in turmoil again? And he was asking Larry to be Station Manager again? Larry said no.

Harvey put Larry up at a motel for a week, and gave him a car while he looked around. Harvey wanted him to find out what was wrong and recommend what he should do.

There were lots of problems at the station, and one of them was Harvey himself. Larry remembers that Harvey was a tall, handsome guy, with a great way about him, but "he could be acerbic, abrupt. I said to someone, 'Being one of the few Jews in a very cowboy environment, Harvey was doing a lot to advance the cause of anti-Semitism' because personally, he conducted his manner in the stereotypical east coast Jew manner… Always obsessed with petty money decisions. His people skills weren't great.

"He was abrupt and issued edicts. Part of my job was to be a buffer. The isolation from the studio was a good thing. In my twice-a-week meetings with him, I was able to be the ambassador from Gilroy and discuss what needed to be done, and to interpret things.

"Harvey wanted a very tight format. I came up with the color coding which saved the freedom of choice for the jocks, but on paper provided Harvey with what he thought was a tight format. I wasn't deceiving either side, but the spin allowed me to create the hybrid I thought the station should be. I packaged it for the staff, and Harvey could see the charts. The staff didn't know how much was me, how much was Harvey. The smart move was in not letting either side know what was from whom."

Marty Manning, a.k.a. Dale Evans, was a smart, experienced radio guy by now, and Larry listened when Marty told him that "Chaney was blowing it." Looking into every corner of the station's operations, he found what he already knew, and everyone else knew: David Chaney was tearing the station apart, and to put the station on a productive path, he had to go.

LY: "I had seen Marty's and Chaney's roles differently. Chaney had to go. Chaney was more immature than even I had thought, and he was taking everything personally."

Chaney felt betrayed by all the chaos around him. He felt he was in an untenable position. When he got there, he recalls, the station had been through chaos and he was made responsible for getting the station immediately on a stable, commercial track, but "the reality was, when I came back as PD, there were people who were hostile to the regime, Harvey's regime, because he was trying to tie the station up, concentrate the essence of it and make it more marketable and meet this big nut that he had to meet." Chaney felt the hostility was already there when he got there, and it was Harvey's fault, not his.

He wanted to do a good job for Harvey, but the odds were stacked against him. In any event, Chaney had no friends at KFAT, and now Larry was there, and Chaney was convinced that Larry had come back to fire and replace him, so he was even more pissed off. He'd been betrayed. He had only been there a few weeks.

But Larry insists that when got there, he told Harvey to hold off on firing Chaney- that he needed a week to scope out what was what, and he'd make his recommendations in a week. He'd put it all in a written report.

When he gave Harvey his report, it said "they needed to bring in at least one larger-than-life strong KFAT personality. I said I wanted to find that person. With a little direction, Harold was good and should stay. Amy was very good and with a little direction could be a star. Marty Manning was his usual excellent self. He was a little bored, and could use a partner to liven it up- someone he could play off of."

Larry wrote that "Marty was a gem, but not being used effectively."

Bill Goldsmith had come to audition when KFAT first went on the air, and Larry told him he was too straight for the job, but now he was back, and he was good. He wasn't as straight as he'd been, and the station wasn't as bent. He was good on the air, he was reliable, a good techie, a husband and dad, so they made him Music Director and he called himself Wild Bill.

Larry says: "Sherman was the real spirit of KFAT and it had been a mistake to fire him. Though he was a little out of control and Harvey hated him because he thought he was a loose cannon. I said that's part of what you're buying with KFAT, but you've got to create a context in which you can live comfortably with your loose cannons. Guys like Unkle Sherman were 'Image Enhancers.' This was spin with an element of truth."

GK: What about Cuzin Al?

LY: "He [Harvey] didn't care- it was Sunday and not a large commercial day. I was pissed at Al- he was constantly making mistakes, he had mobs at the station, he had two turntables running at the same time, and was flakey and unprofessional. He was trying to wave the banner for sloppy radio.

"Cuzin Al was an institution, but he wasn't a radio person, and he did his thing and I loved his show, but he was pissing me off. I talked to him about it constantly."

But Al Knoth wasn't the problem, Chaney was. It was easy enough to fire Chaney, and replacing him was a priority as difficult as it was important, but Larry knew a guy. As always, Larry had had his ear to the airwaves and the talk in the trade. On the air up in Idaho was a guy named Michael Hess, who was doing a similar format for a cable radio station, and he was really good. He called himself Dallas Dobro. He was loose, he was tight, he was funny, he was smart, and he played the right music. He was perfect.

LY: "I needed a strong personality. Harvey wanted Marty off mornings, and to find a strong personality. I went to a Gavin conference and started talking, and someone mentioned a friend who had a Fat- type thing on cable in Idaho. I'd heard of him. Would he send a tape? Would he come to California?"

Larry contacted Hess and had him send an audition tape. Larry knew that this was the "larger than life personality" he needed; the tape came and he heard exactly what he wanted to hear, and he gave the tape to Harvey. "Harvey thought he was weird- not a morning guy. Not what he had in mind at all." Harvey hated it, and wouldn't hire the guy.

LY: "Harvey respected me because I could talk his language. He hated Dallas' tape, and I believed it was just what we needed, and I wouldn't back down."

Larry said, "Look- this is KFAT. He's funny, really an original, runs a tight ship, knows his music and puts it together in a really engaging way, a perfect sensibility for KFAT. If not morning, okay, put him on in the afternoons and let Dallas Dobro be Dallas Dobro."

But the Morning Drive guy was the anchor for any station, and Marty was great, but not happy. Larry told Harvey that a "better solution than finding a high-powered morning man was to team Marty up with someone who comes in as news director, but is part of a team. Give Marty someone to play off of. I've got just the guy." Larry made a mental note to call Michael Turner, if he could find him. The last he'd heard, Turner was seeking asylum deep in the mountains of North Carolina.

Again, Harvey asked Larry to stay, "and I said I hadn't thought about it. But in the past week, I've gotten my old passion back for what KFAT's all about. I said it's possible. We discussed money. Harvey said I had to fire David Chaney." Larry had asked for a week to look into things and get back to him. After the week, he reported to Harvey, and called Michael Hess, who said yes.

Uh, David, could you come in here for a minute? Larry wasn't the Big Axe that Doug Droese had been, but he was big enough. Chaney accused

Larry of stealing his job, and would not believe that Larry had been doing just the opposite. "I had come to try to save his sorry ass. I took over. Based on my report. This station works by being larger than life. Amy Bianco has to be something more. Wanted cartoon characters. Unkle Sherman gives you an image. Same for Dale Evans. Amy came up with Amy Airheart. Perfect."

Sherman remembers he was listening to the station at home when "the phone rang, and it was Larry, asking me to come back. Chaney had been fired."

Larry began "crashing somewhere with Travus." They had worked together 15 years ago in Los Angeles, where Larry's slovenly ways so pissed Travus off that he complained to the PD about it and wound up with his walking papers. Now they were room-mates. What goes around, eh?

Chapter 138: Staff Meetings, Dallas And Fate

Talking to Amy, I guessed that staff meetings were not a regular thing. "Yes, they were," she said, "and they were pretty horrible. When we moved to the Old Gilroy Hotel, they got a lot more official. And usually the meetings revolved around what we were doing wrong, or some change that we'd have to do. It was mostly records and what we're playing. Usually a joint was passed around, so I don't remember them. I remember there was one in Laura Ellen's big office around the time of the sale, but I don't remember much about it."

Marty was the Operations Manager, and he found that his title had about as much definition at KFAT as he wanted it to have, so he'd better take charge and do something. He held staff meetings, and he assumed more operational duties, and he went off of his full-time schedule, and that was when they brought in Dallas Dobro.

AB: "I was living in Laura Ellen's house in Morgan Hill, and Dallas rented her house, and she moved over to Watsonville," where she lived for the next 27 years. But… Dallas Dobro?

Michael Hess was living in Idaho, being Dallas Dobro on a channel that reached well over a hundred people. He was doing a loose, KFAT kind of thing, and he was doing it well. Of course he knew about KFAT, and when Larry called, he was pretty happy. So was his wife, Kathy.

They'd had their first baby in February, 1980, and looking around him, Michael Hess saw no reason for optimism. Idaho was beautiful, but paying the bills was attractive, too, and that wasn't working out so well up there. He was a clever man with a clever format, but of course the signal could never go as far in Idaho as it went down there in California. Imagine his glee when, in May, he got that call from Larry Yurdin. KFAT needed an Afternoon Drive guy and a Music Director. Then Larry heard his tape, and now he was in Gilroy.

Dallas Dobro was excited about the opportunity and all, but boy, he had no idea what he was stepping into. When he got to town he didn't know anything about the record heist or the chaos, or the publicity, or that the station had just changed ownership, management and air staff. But *you* know, and you know they were building a new studio a block away, and you know the fans were intensely curious about the new staff, and were listening, warily. Dallas didn't know any of that, but everyone else on the staff did.

What no one on the air staff knew yet, was what had just happened up at the Stanford University Medical Center, and which would have the greatest impact of any event in the entire volatile history of the station. With Russ mostly in the San Jose office and up in Vacaville at KUIC, only two people at the station knew about it yet, but one of them was the person to whom it meant the most, and the other was Larry.

Harvey didn't have much time to set his legacy into cruise control. He had to fix the air staff and the sales staff- fast!

Chapter 139: Ch- Ch- Ch- Changes

Chaney was gone and Larry was back in charge, and he brought in a system he wanted everyone to follow. A system you say? At KFAT? Well, Harvey wanted some structure, and Larry, if you'll remember, liked the *sound* of anarchy, but hated anarchy itself. Larry wanted structure, and he wanted stability. Harvey needed stability, so Larry took a deep breath and went to work doing what he liked best, with a format he had first envisioned, even though it had mutated for the better under Laura Ellen's direction. He still wanted the sound of anarchy, but he needed stability.

Bill Goldsmith was a stable sort of guy, not given to alcoholic or chemical excesses any more than normal, so Larry made him Music Director. Bill remembers: "Laura Ellen, who was running the station for years before that… before Harvey… had a very different style of executing control, or maintaining control over the radio station. She did that by who she hired and how she trained them, and she would basically get someone that she was comfortable with, try out them in a DJ spot, and basically leave them alone unless they went way, way overboard in some direction. Then she would talk to them and nudge them back.

"It was much more about hiring a whole crew of Program Director's that she would then turn loose and kind of loosely oversee. So she exercised very little direct control over them, and Larry wasn't comfortable with that at all. Larry wanted it to be *his* radio station, and [he] had a particular vision of what it ought to sound like… and he wanted it to sound like that all the time. And that's the way radio is done. That's the background that he had been working in for the last several years. And I spent a lot of time working in that background, and kind of understood it, understood that mindset.

"Well, it was a whole different set of people... The whole station, pre-Harvey, was considerably more off the deep end than it was after Harvey. Larry came in and really tried to sanitize the place. To pretty good effect, in the cases of some people. There were a lot of people that just followed the format that Larry set up to the letter."

And that, of course, was why Sherman had to be fired again. It frustrated the bejeezus out of Larry that Sherman just couldn't follow orders. That was why Leo had fired Sherman so many times, as had Laura Ellen. But Larry loved Sherman, too, so he hired him back again; Sherman was used to it. The competition with Buffalo Bob for who'd gotten fired more times was by now firmly and forever in Sherman's hands.

Bill continues: "And then there were people like me. I would stand there, because we were supposed to choose all the records off of little index cards, and that was so that everything would be properly rotated, and we would play a certain quota of familiar country artists, and a certain quota of rock artists, and all this stuff, and Larry had this regular radio system, and some people took it very seriously. I never paid any attention to it whatsoever. I would stand there at the wall and wait for inspiration, and pick out an album, and Larry would be standing right next to me in the room, and would, like, not even see this, not even see that I was doing my whole show the old way, and only paying token attention to his index cards. He'd bitch at me every once in a while."

GK: Was he an overbearing guy to work with?

BG: "In spots."

GK: Easy in spots?

BG: "Oh, incredibly, yeah. Larry was a real fun-loving guy, and enjoyed all the recreational chemicals as much as the next guy. My only real direct contact with him was the first couple of months after he came back, and then I kind of pulled out of that situation, I wasn't around there during the week. And when he was there on the weekends, he was usually just there to party. And the station was a whole different scene on the weekends, and I had a really good time down there."

GK: How much did the music change?

BG: "The station was much more eclectic, pre-Harvey. There was basically no structure to ensure the airplay of any particular artist or set of artists or genre of music.

"I thought that the station, before Harvey bought it, was very difficult to listen to a good deal of the time. So did a lot of people, because the ratings were actually miserably poor. So were the ratings after Larry came, for the most part. The station was never much of a ratings success. There were a couple of good books, good periods under Laura Ellen, where she had a staff that was really clicking, and the station really started to come together. But then things would happen, and people would go away, and the station would be left in a very chaotic, scattered state. And any time she was forced to have somebody on the air that was kind of new to the situation, who hadn't been there for three or four years, the station kind of tended to fall apart. Even if that person had a lot of experience elsewhere in radio, they couldn't do KFAT, and they would end up playing a lot of substandard music, and not put it together in an interesting way, and the station would sound like crap, and I think Laura

Ellen would probably agree with that.

"She was always great when she was on. She had several other people who really made it work. Sully made it work, Gordy made it work, Sherman, Terrell Lynn, Buffalo Bob. And during the time when those people made up her full time air staff, the station sounded pretty damn good. When things would happen, you know, arrests, the Fat Four thing, Sully Sherman and Teresa and Rebecca… That was a pretty big falling-out, when Laura Ellen had two of her full-time members of her air staff arrested by the police. So that was a disruptive point in the station's history. And after something like that, the station would sound crappy for a while."

GK: What about after Harvey bought it?

BG: "It was kind of hit and miss. In some cases sounded pretty good when some people were on the air, and pretty crappy a lot of the rest of the time."

GK: Who was good?

BG: "Frankly, during the bulk of the period when I was there, I thought the station sounded pretty shitty during the week, and really great on the weekends. Me and Sully, when we were on weekends, the station sounded really fuckin' awesome.

"Sully worked there for a long time on weekends during this period. Year and a half, at least, she did afternoons on the weekends. Probably after you (Gilbert) were out of the building. But she did both days, and Mary Tilson worked on the weekends, and I always really enjoyed what she did. And during the week, I don't know, Dallas would sometimes sound really great, but he was trying to be the PD and was really stressed out a lot of the time, and on occasion he didn't put a lot of attention into his show. Amy sounded good. And I love him as a guy, but Marty never really clicked for me. I never cared for the way he did the music. He was funny, but he didn't… he didn't care much about the music. He tended to follow the format pretty religiously. And the guy they had in the evenings was pretty boring, Weird Harold.

"So that was the air staff that was there for the longest, consistent time when Harvey had the station. Marty in the morning, Amy mid-days, Dallas afternoons and Weird Harold in the evening shift. And it just didn't sound very exciting to me."

GK: But I still don't have a sense of how different it was from Pre-Harvey to post-Harvey.

BG: "Well, internally it was run much more like a traditional radio station. With some degree of formatics, and a definite mindset of 'we're going to do something different than what the old KFAT was.' With the goal in mind of having good ratings.

"And we played a lot more familiar artists than the old KFAT had. Familiar country artists, primarily country artists. Willie Nelson, Waylon Jennings,

George Jones, Emmylou Harris, people like that. The old KFAT played all those people, but there was, like, a certain amount of each hour that would be those core artists, and we played some more familiar rock music- Creedence Clearwater Revival, the Rolling Stones, the Beatles. Nothing disco- era Rolling Stones.

"KFAT's trademark was the zany local spot, where you try to do commercials that fit the spirit of the radio station, whenever the client would let us get away with it. And I had some fun doing that. I didn't do much, pre-Harvey, I wasn't around much. Under Harvey, there was a more aggressive sales staff, there were more national commercials, and more kind of bio-engineering, local San Jose commercials. The station had always had a sales office in San Jose, but they were never serious about selling it."

GK: But the sales staff was re-energized under Harvey?

BG: "Yeah. But when Harvey came in, and he had paid a whole bunch of money for the station, and with Jeremy, anything he made was great because he had paid virtually nothing for it. And Jeremy spent another seven or eight dollars on equipment, and they were set, 'cause everything else he built himself. It was scary-looking stuff, man... I have a lot of respect for Jeremy, a lot of respect for Jeremy as an engineer. I never, *ever* wanted to fix anything he had built or ever worked on."

Bill Goldsmith was—and still is—a brilliant electronics engineer, as well as an excellent DJ, and he had worked on radio-related gear his whole life, but he couldn't figure out any of Jeremy's inventions. He was awestruck at how well it all worked, but he couldn't understand Jeremy's logic, how he got a signal to go from here to there, using that.

BG: "Harvey had an engineer come in and look at the studio and the transmitter, and he said the same thing: 'There is just no way that anybody is going to be able to work on this junk. You're going to have to replace it all. Everything has to go. Every piece of equipment in the studio needs to be replaced, everything at the transmitter needs to be replaced. We can salvage absolutely nothing from this. You have to buy everything new.' "

GK: Harvey didn't know this before he bought the station?

BG: "He knew it. He'd seen it. I doubt that he knew that he couldn't salvage *anything*. He did pay several million dollars... He didn't have that in his pocket, you know- he borrowed it. And so he had to make a three million dollar note every month. That really put the pressure on for sales. And we didn't have as good a production staff as they had had with Frisco and some of the people who'd done production. They were genius level guys when they were in the zone."

p.s. Bill now does *excellent* radio online at www.radioparadise.com.

Chapter 140: Christa's Back

In July 1981, Christa Taylor came back, so let's look back at Christa. She'd first heard about the station from her friend, semi-legendary underground radio engineer Don Mussell, who was, like Jeremy, another barefoot electronics wonder. Jeremy and Lorenzo had just bought the station and it had no staff yet. Don brought Christa to Los Gatos, where she told Jeremy she wanted in, was immediately given a job, calling as many record labels as she could find and ask for record service, and to continue doing it during her off- air hours.

This was August, 1975, a week before they went on the air. She recalls her time there with words like "primitive" and remembers the "long hours." She drew up the first logs at KFAT. She only stayed until December, when she left in the mass exodus of Texans and locals, but she laughs when she remembers the window that was open to let the air in and fondly recalling the trains whistling as they came through town. She not so fondly recalls the "heavy smell of garlic." It was Christa who bought the bowling pin lamp for the studio.

Marty Manning had been the first to leave, then a few of the others, and when Chris Feder showed up for the first time the next day, it was Christa who told him that there was no way he was going to be the new Program Director, and Chris left, looking for Jeremy. The next day he came back, and Christa was gone, along with everyone else. She left with Bobby Eakin and David Chaney. Eaken went back to Texas, Chaney headed for Albuquerque, Christa for Louisiana, where she stayed for a while, then went to Hawaii in 1977 and stayed for four years before coming back to California, working at KDON in Salinas.

By now, with the new owner, KFAT had a staff and some stability. Marty was back as Dale Evans, Dallas and Amy were there, and Chaney did afternoon drive. Then Chaney was gone. Larry knew that Christa had been learning a lot about the roots of country music at KDON, and he knew she'd do well back at KFAT.

Christa: "Larry Yurdin called and asked me to come back because he was firing Chaney. I did 3-7 p.m., between Amy Airheart & Dallas Dobro." Now Christa was doing Afternoon Drive.

Christa: [It was] "the hardest shift I've ever done, because they were so talented and so funny, and I was the straight girl. [I was] hired to program the straighter stuff. [It was] hard to feed an audience used to what Dallas and Amy played. [It was] hard to mesh humor and straight country music. Great place, but not easy." And this was Afternoon Drive on KFAT in a highly charged, emotionally raw time for both listeners and staff.

She remembers that back in '75, Larry was a new phenomenon to her who had painted a much prettier picture of KFAT than what was true, and by now she knew that Larry was just plain fucking crazy and hard to work for, and here she was working for him again. And now he was stressed out again. Oh, my. But this was KFAT…

KFAT was still special, so to deal with it all, Christa "tried to keep Larry out of sight."

She knew what it was when she'd left, and she knew what it had become. She knew only a fraction of the library, so she listened all the time and noted notable songs and segues. And the bowling pin lamp she'd bought all that time ago was still there, so she settled in to do good radio. She had a lot to live up to.

Chapter 141: Felton, Harold and Seaweed

It takes a bunch o' people to make up a staff, and we have some players you don't know about, so let's get to 'em. Let's start with Doug Abernathy, aka Felton Pruitt: "I got into radio because when I was a young teenager, fourteen, fifteen, listening to the wonderful broadcasts in Baltimore & DC, where I was raised in Annapolis, Maryland. Back in the early 70s there was some great radio being done there, progressive radio was being born, just like it was in San Francisco at KSAN… and I thought this must be a great job: these guys are sitting around drinking, smoking pot, playing great music.

"I didn't get into radio until I moved to California. August of 1977. I moved to Turlock, because the Grateful Dead sang about it in two songs, so I thought it must be this wonderful, beautiful place in California where all the cool people lived. I was mistaken. I'll never forgive Bob Weir for that. But I was in college there and they had a radio station there, and they needed DJs. So that was the first thing I did… was sign up to be a DJ. Best gig in the world: Friday evenings, six 'til midnight. Party hours, I actually had listeners.

"I tried to mix in Jefferson Airplane and Hank Williams and Bill Monroe. The proto-Fat format. That was the thing, as I was sitting in 1977 trying to do this on a college station, thinking that I was changing music and inventing a whole new genre, and then somebody told me that I ought to tune in 94.5 FM…

"I was a rock 'n' roller in December '77, and I didn't start at KFAT until November of 1981. I was a student, I moved to San Francisco, at SF State, and doing radio there, and that was a really fun gig, and once again it was Friday nights, seven to midnight. You could do acid on the air, and do your show on acid. Before, you were just drinking beer and smoking pot and playing records. Now you could do acid, and you were in San Francisco on the radio! How much cooler could that be?

"I just went to KFAT and started hanging out, and I met Larry Yurdin and Dallas Dobro, November '81. By that time Harvey had bought the station, and Larry was the PD, and I was bringing in really good pot, so Larry and I became good friends immediately. Larry liked to smoke, and Dallas liked to smoke, and so they liked having me around, and so in about a week I was the assistant MD. Dallas was the MD, and I was the assistant MD.

GK: With an air shift?

FP: "Not yet. How I got the air shift, there was Chuck Wagon & the Wheels was playing, and I guess Weird Harold was the DJ from seven to midnight, and then Unkle Sherman was supposed to come in at midnight to six. Well, I was good friends with Harold, too, and I was in there with Harold, and Larry was there, and we were all just sitting round for like three or four hours, smoking pot and playing records, having a good time, and Sherman had gone to the show, and he'd gone with Terrell Lynn, so Sherman and Terrell Lynn are at the show in Santa Cruz, and midnight rolls around and there was no Sherman, and we didn't have cell phones… and so Harold has a date and had to leave. Harold was like that… Harold had a date whenever he wanted one, he was very good… so Larry said, 'we'll just cover it' meaning Doug will cover it until Sherman gets back.

"So Larry was on the air, and had not been on the air, live, in years, as a DJ. I mean, he was a legendary PD, but he wasn't a DJ. So I was selecting records and spinning them, doing the segues, and when it got to be commercial time, Larry wouldn't even back-announce, he would just play the commercials, and then I would play the next record. So basically I was just engineering and there was no DJ. And after about an hour and a half of this, it's now maybe one, one-thirty in the morning, and Sherman still hasn't shown up and we haven't got a call from him, and Larry wants to go home, and so Larry says, 'Well, you're now a DJ at KFAT.' I had just cued up a song, and we had three turntables at the new studio, so I had three songs up and ready, and he said 'When those are up, you're on the air. Think of a name.

"At KFAT, everybody was something. Dallas Dobro, Sister Tiny, you had to have a KFAT name, and Doug Abernathy wasn't a KFAT name. So he said that when these songs are up, think up a new name, so I rolled a joint and I went into the other office and I sat on a sofa and smoked the joint, and I started thinking up my name. What could my name be?

"I remembered Sammy Pruitt was a guitarist for Hank Williams, and I remembered that Felton Jarvis had been a producer on Willie Nelson's records in the 60s, and somehow I made a mistake- I meant to be either Sammy Pruitt or Felton Jarvis, but I guess somewhere between the sofa and the joints and getting back in after the third song, it came out Felton Pruitt, and so I went on the air as Felton Pruitt, and started playing another record. Then, just as Larry was leaving, the phone call came. It was Unkle Sherman, and Larry was (snarling) 'Where the hell are you? You're supposed to be here an hour and a half ago!'

"And Sherman said 'I can't come in, I'm not coming in, I'm getting a blowjob right now.' And Larry says "Where's Terrell Lynn?" and Sherman says, 'He's right next to me, and he's about to get a blowjob, too!' So Larry goes, 'OK, you're on. You've got Sherman's shift."

So I guess this was one of those times when Sherman was fired. No one was keeping track and no one remembers. It was just another night at KFAT.

FP: "So I became Felton Pruitt that night and I got the graveyard shift. That was a Tuesday, and he gave me Wednesday, Thursday and Friday, 'cause he was pissed off at Sherman. And then I started doing weekends, and I ended up doing the graveyards.

"I was making six dollars an hour as a fill-in DJ, and I did that until my last day at KFAT. I never got put on a salary or a full-time thing. I remember turning in a time card that had over sixty hours on it, frequently, because I was there all night long and all day long. I loved being there, and I would do any shift, and I would do everybody's fill-in shift, I would do graveyard, and then engineer Marty's shift from six 'til ten."

It was a new KFAT, but it was still KFAT, and Felton remembers one day when, "it's like four o'clock in the afternoon, on a Wednesday, it's cool, we can get stoned, so Amy and Dallas and I were sitting there, and Dallas was on the air, and we were sitting against the window with this joint, and Belinda opens the door, and she goes, "We have some guests I need to show the studio.

"And she just looks at us, and we looked at her and said, "Well, this wouldn't really be a very good time." And it was too late, and so in behind her walked the mayor of Gilroy and the head of the Catholic Diocese of San Jose. And Dallas looks up and sees the mayor and someone who looks like the Cardinal, and he immediately grabs the turntable, and pulls up the spinning part out of its socket, and goes, those damn belts are burning again!" and for the next 5 minutes Dallas and Amy and I are pretending like one of the turntables is on fire and we're trying to fix it..."

What else would be smoking and smelling like that? Discuss.

His real name is Steven Seagrave. He'd worked in several stations in the Bay Area, and now he was doing evenings at KFAT. He called himself Steven Seaweed. Perhaps you'll remember when Patty Hearst got kidnapped by the Symbionese Liberation Army. At the time, she was living with her boyfriend, Steven C. Weed. There ya go.

He was a journeyman jock and he did his work, and made no waves, he was the only one hired by the new people, and he stayed on at KWSS after KFAT went off the air. He'd started at KFAT three or four months earlier, and he was a real professional jock and he knew how to work the politics, and he kept his job, and stayed with them for another five or six years, until he went somewhere else.

Right now he's on KSAN, he's doing classic rock. Steven has remained one of the most stable people in radio that I've ever met.

Harold Day graduated from broadcast school in Fresno in 1974. They thought Harold had talent, so they called a station and got him a job, "which I didn't take for six months because I didn't want to work in Coalinga. So I finally went to work in Coalinga, a little day-timer, KOLI, I went from Coalinga to Tulare, there was an AM/FM combination KGEN, Cajun Country…"

GK: Had you determined by this point that you wanted to make your career in radio?

HD: "I think so, because I found that it beat working. And the second job was a free-form album station, and I started off with a six hour Sunday show, part time. And I played head-bangin' music- Aerosmith and Blue Oyster Cult, people like that, Ted Nugent, what have you. And I also substituted the AM country station, playing Johnny Paycheck and 'Take This Job And Shove It.'

He "stayed in Coalinga for six months, then a year in Tulare… I moved on to a better paying job. I went to Hanford, to another AM/FM combination outside of Tulare. It was an Adult Contemporary station on the FM, and the AM was country, so here again I did some part-time work on the country, and got to know the artists that later on I would find in abundance on KFAT."

We're not gonna stay with Harold for long, but there are four things you should know about him:

1. He was a good, if unexciting jock at KFAT, he was responsible about showing up on time, and I never heard him do a bad show.
2. I am still in frequent touch with Harold, and he is one of my closest friends.
3. When I knew him at KFAT, Harold was getting more pussy than any man I've ever known.
4. I'm not going to tell you the fourth thing.

Back outside Tulare, "I got laid more at that station than any other station. Easy listening, bored housewives, and all that.

GK: Married women?

HD: "Married women, yeah. This one lady had an answering service, and we got together, and got to know each other, and she passed me around to her friends."

Harold was pretty happy with the gig's perks, but the music was boring. Then a friend said, "Hey- I discovered this station! You've GOT to work there! It's called KFAT. You've GOT to work there!

HD: "Some friends moved from Tulare to Mountain View, and it was right about the same time as I was having a falling-out with a lady I was with... So I moved in with them into Mountain View, and I started bugging the folks there at KFAT, trying to get on. This was in 1979. I bugged KFAT, and bugged them and bugged them until one day I got a phone call, saying, we need someone to do production. And I told them I excel in production. The caller was Doug Droese.

"However...the albums had just disappeared from the station, and a number of people were cut loose, and that opened the door for me. My first year there I did off-air production, Monday- Friday. It was wonderful, the most fun gig I ever had at any radio station. It was creative pressures, not time pressure."

Harold made up weird commercials like "Meedley's Wild Moustache Rides" and also one about odd- and even days for taking a pee. "That and just the straight-up commercials that I tried to put a slant to so that you didn't know if they were tongue-in-cheek or regular, or what." Harold did some very out-there spots at KFAT.

GK: And what kind of feed-back did you get from management?

HD: "Less than never."

He started doing Saturdays, and filtered into fill-ins, and when Harvey got the station, he started doing a full-time thing. Harvey told him, "you WILL BE Weird Harold." David Chaney once chewed him out about a song, but that was the only direction he received. "Actually, I was told that I had more training than most of the people who had been in there. So I had to relax. Well, I got a little more stoned than usual, but my whole radio career was walking into the studio just... lit. I started off mid-afternoon under Harvey's reign of the station. I wasn't as well-read as what they wanted. I didn't have enough banter... verbal manure, as Dallas calls it... and so I was put on the overnights. Which lent itself to the fact that I dealt with KFAT's inventory of assorted merchandise- belt buckles, T-shirts, caps, all of that. So I had the concessions, and I got 40% of the revenue.

"I was there from November '79 until about August or September of '82. At one of the Fat Frys, Russ was 86'd for drinking too much, and I thought it was funny, and I laughed, and I was let go within a week. They said they were trimming the budget and I made too much money. It was the most fun of any station I've ever worked for."

When he got fired, Sherman told him to go to Dan Healey, the legendary engineer for the Grateful Dead, and he got a job with him and his loose-knit commune up in Garberville, California, where he still lives.

I asked Harold if he learned much at KFAT, and he said he "learned how to party." He wishes to thank everyone for all the pot, the coke, the blowjobs, the pussy and the rest of the good times.

Chapter 142: Live Chew

In the spring of 1979, the owner of a San Francisco Public Relations agency offered me a job in his agency, and I took it. I was working full time doing PR, and also doing my show on KFAT. My father died in October of 1979, so I got some money, and my wife and I bought a house in San Francisco. By January 1980, with most of my radio pals gone, I took my show off the air. A few months later, I quit the PR job and thought about writing a book, but couldn't think of anything to write, so I—or someone like me—subsisted on an underground income. When my wife joined a cult and left town in March, 1980, I was thrown into the usual despair, for which I assure the reader I expect no more than the usual sympathy. Besides, I feel better now, thanks.

In October, 1981, Larry Yurdin invited me to emcee a KFAT show at the Saddle Rack, where I met Russ Martineau. With my usual supply of… supplies… and my usual gracious generosity, I invited Russ to share in the bounty, and Russ and I became friends. After that, Russ would come up to the City every couple of weeks, and we'd go to clubs and shows, with Russ often spending the night at my house. Many such… supplies … were consumed, resulting in long hours, late nights, foggy mornings and shaky starts. Good times…

My friends from KFAT were all gone, I wasn't speaking to anyone else at KFAT, so I didn't know anything about how Russ was to work with. To me, he was a fun guy, and we had fun. Russ kept suggesting that I bring my show back to KFAT, and I continually and sincerely praised my time there and my love for the station, but I refused to do it again- until the time Russ didn't ask me if I would do it, he asked me what it would take for me to want to do it. Looking back on that question after all these years, it sounds a lot like a car salesman asking a reluctant customer what it would take to put them in this car. Happily, I'm OK with that.

At the time I was out of work and drifting, so I thought about it and came up with only two conditions: no interference about who I would book, and no censorship. Russ immediately agreed, and could I start in a week? I'd be on Sunday nights, live, from ten until midnight, with interviews and listener call-ins. The interviews could be live in the studio or on the phone. I'd have a jock to engineer the show, screen calls, play commercials, etc. Done.

My first guest was my friend, Andrew Ross, an editor at *the San Francisco Examiner*, and also an expert on cults. I had some interest in that. Also, it was another chance for "keep practicing" to catch on, so...

I was nervous as we got to the new studio an hour early to look over the new room I'd be working in, and get familiar with the equipment. The show had been promo'd for a week, and I was prepared for the interview, but not for the response from the callers- I was totally unprepared for that. For the entirety of my radio career, my shows had been taped, and I'd been live only during the isolated instances when I talked with a jock for a bit of banter. Conducting a live interview left no room for the editing that I'd come to count on. In the past, if I didn't like the way my question sounded, I only had to ask it again, and edit the offending portion out of the show. I took out whatever didn't serve the show, said by either myself or my guest. That was easy, this was real. This counted. This would take some getting used to. It's embarrassing to say this, but I had never actually factored in the whole "live" thing, and I was surprised to experience being there and talking at the same time that people were *listening*. And some people were engaged: some of the callers had relatives in cults, or friends who had been lost to a cult, and one caller had been in a cult and escaped. The callers were listening and commenting, adding to the conversation, changing its shape and direction, and it was pretty good radio. The people who called were all hurting, needing to share, needing to understand, seeking comfort. I don't remember what I was expecting, but I just wasn't prepared for that. It was a very moving experience for me.

I realized that KFAT had always been a game for me, all about having fun, that I hadn't been connected to my listeners, but this was real life that was calling in to talk. Yes, I know how ignorant and badly thought-out it sounds, but I was shocked that people were listening. I know it must seem incomprehensible to you that I would say this, and I hope it doesn't alter your perception of me negatively, but this was a real revelation to me, and it took me a few days to process it. After my divorce, my life had changed a lot, and I was partying a lot and I was a little emotionally unavailable just then as I was, uhh... *supplying* myself regularly in those days, and there is evidence that a steady ingestion of such supplies might render a person incapable of emotions. Studies have been done...

The studio was neat, orderly, and the all the new jocks were friendly, almost professional people, and I liked all of them. During my shows, a jock was scheduled to run the board and answer phones and take care of station business for those two hours. There was never a problem in finding a jock to work the shift, as the show was an easy two hours, it was a relief not having to play records, the shows were often interesting, and my... supplies... were always shared with my co-workers.

I remember going to commercials and once off-air, opening the mic into the control room and asking, "Harold? Conference?" Everyone who engineered my show knew what that meant.

I went back on the air in January, 1982, and in April they changed it to Sunday mornings, eight a.m. until ten. Russ told me that the ratings were good, the feedback said that people liked it, and more people would be listening on Sunday mornings than at night.

Now I was getting to sleep after dawn six nights a week, and Saturday nights I tried to get to sleep by nine p.m. As a schedule, that worked as well as you would expect, and many supplies were consumed to keep the host awake and Chewin' the Fat on the air. Keep practicing, indeed.

Chapter 143: A Ghost, A True Ghost From The Past

Amy says: "After I got to do my first overnight, Laura Ellen said, 'hey, she can do fill-ins,' so I did another week of overnights, 11:30 p.m. to 5:30 a.m. One day I was on the air playing who knows what, and Laura Ellen had always said to make sure to lock the door on Monterey, but I never checked it.

"At 11:30, Rocket Man says adios, and I was alone there. And now it was four a.m., and I hear the clomping up the stairs, clomp! clomp! and I thought, My God, what am I going to do? So I turned up the music really loud and I was sitting there, and then I turned around, and there in the entrance to the studio was this tall, sandy-haired, kind of handsome guy, standing in the doorway with a big brown paper bag in his hands, and he goes, 'Howdy!'

"And I go, 'Hi' and he says, 'I just got out, and I came to party with Miss Amy.'

"And I said, 'Out?' And he said, 'Soledad Penal Institution.' I said 'Wow!'

She got calls from there, too.

"Oh, yeah, lots of calls. So he sits down in a chair, he's just got out, and he's got a bag with beer, liquor, pot, coke, 'anything you want,' and he wants 'to party with Miss Amy,' and I was sitting there scared shitless, you know. I was twenty-two years old, and I said, "Well, I gotta work," and I put the headphones on and wondered, can I open the mic and say 'Help!'?

Remember that window to the Production Room? The curtain covering the window between the Production Room and the studio? It was rarely closed, but it was closed that night, and that made Amy even more alone and frightened.

"And all of a sudden, I heard a commotion in the Production Room, and this guy stumbles out. And it's Rocket Man, and he says he ran into Frisco on his way out, and they'd been partying in there, and I never knew they were there.

"And I said, 'Rocket- here's somebody you should meet.' And I foisted the guy off on Rocket."

As neither of my friends is available to ask, I am just going to guess that they had been in there, for those four and a half hours, discussing the joys and salutary benefits of sobriety. Yup, that's what I'm guessing.

Well, Frisco was back in town after drying out again. He had no employment and no prospects of employment, and he was living with his mother in her trailer again. Of course he came by the station to see the Rocket, or Sherman, or whoever was doing the overnights. That night it was Amy, but on his way in, he'd run into Rocket. Frisco ran into her a few more times, late at night.

AB: "My memory of Frisco is pretty vague. I remember him having shakes. I remember Frisco being my first… thinking 'what's wrong with that guy? Oh… well… he's got a drinking problem…'

"Well, it turned out that he had a heart problem, he had a drug problem, and he had all kinds of problems. But he was brilliant, absolutely brilliant. And a few times, and I can't remember what they were specifically… he would come out of the production room with this great thing and stick it in the deck and play it."

"Yeah, just some amazing little drop-in, and do great ads, too. What he could get out of that shitty little Production Room was amazing. I remember that he lived with his mom in her trailer…

"He was gone sometime, maybe in 1980, when I was still doing the part-time thing."

Chapter 144: Too Tall

First let me say that by the time he died, I was no longer afraid of him. I loved him dearly and grievously feel his loss. I wrote this chapter before he died in the fall of 2008, and I am going to keep the chapter as I wrote it, in the present tense. Thus:

To anyone who has met and conversed with him, Michael Turner is a remarkable presence. He is charming, intelligent and forceful, but he also has a sinister aura about him that I find hard to explain. Uhh... I might not really be over being afraid of him. He is the only person I know who actually exudes menace.

Back in an early chapter I apologized in advance for the dumping I was going to take on my friend, Larry Yurdin. Larry was, and is, a slob. He has a beautiful mind and a kindness of spirit that radiates, but he seems to have little regard for what we call "people skills." So I made a little harmless fun out of that, and we all moved on. Everyone knows he is charming and brilliant, and that's what you should remember about Larry Yurdin.

But, Michael Turner, now... I have been friends with Michael Turner for 25 years, and we have rarely been out of contact for long. I respect him; I guess I love the guy, I don't know. He's quite handsome in a cut-granite sort of way, but it's his eyes you notice, eyes that bore into you. It's hard to get in touch with the love when he frightens the shit out of me. Always has. He is a fascinating, startlingly intelligent, articulate, learned man. He is also a coiled spring, and at 6'3" and without a discernable ounce of fat on him, he seems to be in control of an inchoate, simmering anger so close to the surface that you will feel the heat when he invades your personal space. Which he will. To intimidate you. And it works. Always has.

But I think I love the guy. I would not let go of my relationship with Michael for any reason. I could tell you stories about him that I won't, for everyone's sake. He's kind of a small player for all the pages I'm going to use on him. Not small... small-*ish.*

He doesn't seem like the under-the-covers-late-at-night-with-a-transistor-radio kind of guys, so I asked him how and when he got into radio.

MT: "In the 60s, before Woodstock, I was a record pimp, which meant that my friend and I worked at a one-stop in Detroit. A one-stop is a place where, like, Capitol might have a distribution center, as would Mercury. But if they brought them together, one company would handle their distribution. So record promoters would work out of there. So it was kind of a front for stealing and dealing, and all my friends who were there were just having a ball and playing all the time, on record companies' dimes.

"So I took a job with Mercury and it was the best time I'd had in my life. I lived on the penthouse floor of the Jeffersonian, the nicest apartment building in Detroit. My next-door neighbor was Bobby Taylor of Bobby Taylor and the Vancouvers, who had a wonderful racially-set hit called "Does Your Mama Know About Me?" He and [Motown Records president] Berry Gordy hated each other, so Berry buried him at Motown.

"Bobby decided he was going to make a comeback, despite Berry Gordy. He said he 'had a buddy from Chicago' that he used to be in the Golden Gloves with, and this guy says he's 'got some kids who are really good. They're good singers and so my buddy is coming in with his brother and the kids, and they're going to stay with me.'

"I said, 'Bobby- you've got a one-room apartment, and you've got a wife and a kid, how many kids are there?' He said 'Five. Five boys from Cicero, or something like that.'

"So they rolled in and I thought, 'This is really a pipe dream. How is this going to get Bobby back on his feet?' He said, 'they're really good- wait'll you hear 'em.' By the way, 'could any of them stay at your place for a while?'

"We were right next door, and were attached at the balconies. I said they were probably too small, but there was no discouraging them. They moved in, and then they started staying in my place. Sleeping on the couch. I'd come home at night and there'd always be one or two in my bed- always the smallest, Michael, was in my bed. It was the Jackson 5."

Michael Turner always hung out in interesting company. He had been at the famous Millbrook estate with Timothy Leary and Richard Alpert, and had been part of that LSD community in its early days. Turner was a White Panther, and when John Lennon, Yoko Ono and Eric Clapton came to Ann Arbor to play a benefit for White Panther leader John Sinclair, it was Turner who emceed the event. He knew the dealers, the dames, the players and the playees.

MT: "We were at a legendary place called the Gar Wood Estate, a famous hi-speed boat designer, and a lot of races were held on the Detroit River. So Wood built his mansion on the river, and boats would come in on it, and there was a swimming pool and so on, and the place had fallen into disuse and there were squatters and hippies there.

"But they had a party there for the Rolling Stones after one of their shows, and they came in, and I was there and in the record business, and they were milling around and it was no big deal because there were always other English acts coming through town, and these guys were no more interesting to me than the Faces or Cream or anybody else, and I was used to partying with those guys.

"And people started attacking Mick... not attacking, but swarming. And I was sitting on a couch and he came over and sat on the couch and started a really intense conversation with me... I think probably just to get people out of his face, so they wouldn't bother him while he was having this serious personal conversation. We sat and jabbered for a half hour or more and that was it- no big deal.

"But in the meantime, a guy who owned a radio station in Detroit that was just getting ready to go alternative free-form said, 'Who is that guy who Jagger's talking to all this time?' And he happened to ask my best friend, who said, 'You don't know who that is? That's Michael Turner, who's been all over the world and done all kinds of things; he's got a great education, Jeez- he's the most fascinating guy in Detroit' and blah blah, and 'he's lived in India and China and Japan and everywhere in the Middle East...'

GK: Is that true?

MT: "Yeah. He said I was a fascinating guy- are you kidding? Of course Jagger would want to talk to him!"

"So he tells him this little thumbnail, and after Jagger goes on, and I'm just milling around- doing what you do at a party, talking with people, and this guy comes over, his name is John Small, he called himself Big John Small. He comes over and introduces himself, 'have you heard of me?' I said no, and he said, 'I run KEENER- WKNR- and we're getting ready to go free-form and I want to offer you a job.'

"Doing what?

"He said 'Disk Jockey,' and I almost hit him. I had never been so insulted in my life. I had regarded DJs as the lowest form of slime on the planet. And I was profoundly insulted that he thought that I would be willing to do that, and I told him so... I forget the exact words... and he said 'No, no, no! It's not like that at all!'

"He had my interest, and I got over wanting to slap him in the face for offering me a DJ job, and he asked, 'Do you know what Free Form is?'

"No, never heard of it." At that time it had only been heard in San Francisco. Didn't have it in Detroit. He said, 'I'm going to start a station like that, and you can play any kind of music you want. You don't have to play any music- you can just talk.'

"What would I say?"

"You can say anything you want."

"Really? Shit, fuck, piss, cunt?"

"Well, we don't want to lose the license, but anything else- you can say what you want."

MT: "I told him my political views and position. 'Perfectly fine. Do anything you want.'

"Okay, I'll give it a shot."

"Okay. Next Sunday. You come over to the radio station. I'll do a Sunday night show and you can watch me do it, and you'll see how it's done, and if you want to do it. And if you do, we'll find you a slot."

"So I went over and he was in his office, and he said Dan Carlisle is on the air… (But he didn't like me, and didn't want me to succeed because I had no experience). He said 'go sit with Dan and you can see how it's done.'

"So I sat next to Dan, who wouldn't talk to me because he hated the idea of a guy off the street who didn't love radio. He wouldn't tell or show, but I was watching. He finished and said, 'Okay, babe, what do you want to open with?'

" 'Gimme Shelter' had come out that week. He put it on, and no Big John. And I better get something ready, and I fumbled through a segue, and didn't know what to say, and I tried to be witty, funny, philosophical, poetic, sweat streaming down face and bombing, did really bad. I realized that it was best not to say much. Played music, tried stuff out- I didn't know the music. What I liked was jazz and classical. I didn't know rock.

"Small never showed, and the station went off at midnight. He called at midnight and said he and his wife were at home listening and he thought I was a fucking natural. I said, 'Listen, cocksucker, when I get my fingers around your throat you might not think that!'

"No- you're a natural! Terrific! Keep going. At midnight, you see the red switch on the deal? Play the star spangled banner, and flip us off.'

"I couldn't find the Star Spangled Banner, so I sang it. At the end, I said, 'Good night, motherfuckers, this is the last time you'll hear me on this motherfucking show.' And I clicked the red button and walked out, pissed off." But Big John liked it and wanted him on the air.

"Later, I thought this could be a good forum for my political ravings, so I started bringing in my boys and it got interesting and I liked it."

Turner was a success in Detroit, and people were talking about his shows from the start. Big John Small had been right, and Turner stayed there, doing his rants and playing music, when he heard about something called the First Alternative Media Conference in Vermont. It sounded more than interesting to Turner, it sounded essential. He sent back an immediate confirmation of his attendance to the conference promoter, some guy named Larry Yurdin.

GK: Did the station pay for your trip to Vermont?

MT: "Yeah. I demanded. I insisted. That was where I also met my best friend, Mario Madeas, who was the tax accountant for [mega-influential Atlantic Records owner] Ahmet Ertegun. He knew where the bodies were buried. And he said to Ahmet one day, 'Ahmet, we've got a new music going on, and it's not singles any more, it's album stuff, and you don't seem to be on to that. So I want to get out of this business I'm in, and get into running the promotion department for that kind of music.'

Mario wanted to go to the conference, but "Ahmet told him no, because he was an accountant, and Mario said, 'Yeah- I am an accountant and I know the figures,' and Ahmet said 'Okay, you're on.'

"Mario showed up with Dr. John the Night Tripper, and with [San Francisco radio newsman- see Chapter 98] Scoop Nisker. We three flew in together to Vermont. Mario brought the J. Geils band to play at the Vermont conference.

"It was very influential. Everybody who had anything to do with the new developments in radio were there. The Free Press- Art Kunkel was there. People who were doing alternative media rags- and every major city had one at that time. They were all there. And bands were there. Atlantic underwrote the convention. I raved about it. I told everybody."

GK: You were surprised to see so many people with a similar consciousness…

MT: "Yes! And that I was not just a lonely voice. And by this time I had become very influential at the radio station. So I was the leader of the pack, and when Big John made suggestions for bad promos, I told him to get the fuck away. After a while, they got tired of my shit. After a year, and I had taken over the station. I was too demanding. I had pure standards. I was into speaking for the people."

GK: So after WKNR, where to?

MT: "I took a vacation, I took my girlfriend to New York. I had been on the phone with Larry Yurdin and Yurdin had established a connection with WABC in New York. I was chatting with Yurdin when the afternoon guy called in sick, and they asked him who they should get. And Yurdin said they had the best guy in the country right there in his office. Would I do it?

"So I did the show, and the phones lit up. They offered me a job on the spot, and I said, 'I don't work in the afternoons. The people I want to talk to are not listening to the radio in the afternoons. They're sleeping. So I won't do an afternoon show.'

"Well, when do you want to do it?"

"This is Manhattan, I want to be on from ten 'til two."

"Well, we don't have a ten 'til two slot. How about nine 'til one?"

"That was cool, so I started doing the evening show. This was WPLJ. This was 1970. Shortly after, I got a letter from Pink Floyd thanking me for being the first person to play their record in the U.S. I had a dear friend who was a well-off dope dealer in Detroit, who went to London regularly, and would come back with records, stuff that nobody knew in the U.S. I would play it.

"That was New York and I loved it, and I assumed the same posture I had had in Detroit. I demanded standards from my associates. ABC was interested in the money, not the alternative culture. Their regular DJs weren't reaching their market, and they hired guys like me.

"One night I had Jane Fonda and a friend, who had just returned from Hanoi, whose message was that the U.S. soldiers turn their guns on their officers and leave the Vietnamese alone. I had them on the air, and I was in accord with their position. I knew Jane from Detroit, and the three of us got along well and we spent the night at my place, and the next day I got a call from a secretary at ABC, at my place, and I told her she wasn't supposed to call me there during the day. She said, 'Well Michael, the FBI is here.'

"So?"

"Well, it's about you and about the show you did last night with Jane Fonda and Genie Plamandon (wife of White Panther co-founder Lawrence 'Pun' Plamondon), and they're talking to Allen Shaw right now."

"So I went in early that day... and paranoia was running deep those days... and I was organizing marches then, and I had a marvelous office, and I went there and sat down and waited, and then I walked by Allen's office. I waved, he waved. He never mentioned it. He was heavily sedated every day, all day already, and I guess it never registered that the FBI had been there. No repercussions, nothing. The secretary listened to the proceedings and they demanded that Allen fire me and he never even mentioned it to me.

"John Lennon lived at the Dakota, and asked me to be in a movie with him. Dylan was on my show. At the time I was dating Nico a lot.

"I got fired because I was the spokesman for the alternative media guys. I was the leader of the pack at the station at ABC. The people at ABC were trying to make money- they were trying to reach out and speak to this segment of society, as they had an idea by reading *Time Magazine* that this group was rapidly growing and had money to spend, and they wanted some of it. The old ABC station was a classical station, and so they killed the classical component and made it a free-form rock station. But there were problems with that. NABET was fully entrenched at that time. National Association of Broadcast Engineers. I couldn't run my board according to union regulations. I'd have to tell them to put on a certain cut and how to segue- a cross fade or whatever.

"I was sitting one night with a guy who especially hated me, Dick Sisk, who was on the other side of the board during a long cut, and I told him to put this on and do a cross fade segue. He sighed. He did nothing. By the time the record had run out, 'Mike- what are you gonna do?' I said to turn on the mic.

"Friends, do you hear this piece of shit on the other side of the board who has just sabotaged our show? The show we're all in on? You're going to hear something you're all going to enjoy. I'm going to climb across this board and kick the shit out of this motherfucker. I'm gonna do it right now. And you're going to be able to listen.' And I did. People heard fists flying and faces slapped."

GK: Ever get any flak about that?

MT: "Yeah. But the technique they used to get even with me was, they had one black engineer, name of Carl, and they gave me Carl, the idea being that either I'd kick Carl's ass and drive him out... and they wouldn't have to deal with a black... or Carl would kick *my* ass and drive *me* out. And when I kicked Sisk's ass, NABET threatened to shut down every ABC station in the U.S., until Michael Turner is out of there.

"Well, Carl loved the music and he loved what I was doing, and his girlfriend was a major dope dealer. So guys were coming up– Duane Allman was a buddy of mine, and Delaney Bramlett would come up and cop smack from her. It was the hippest place in town to cop."

Chapter 145: Too Tall Redux

Summer, 1981, and Michael Turner remembers, "I was in North Carolina, living in a cabin in the Nantahala National Forest... where Eric Rudolph was recently captured, in Murphy... and Ann and I would go down to Charlotte from time to time, and when I got down there, there was a call from Larry Yurdin from some place called Gilroy. And I returned the call and he said, 'Hey, how'd you like to get back into radio?'

"I said, 'not a chance. I have no interest what-so-fuckin'-ever in radio on any level.' He said 'Yeah, but this is different. This is the last free-form radio station in the country, and you might really have a good time.'"

Turner told Yurdin again that he had no interest, that radio was "over for me in the early 70s," and Larry told him that there were really good people there, and that he'd enjoy it. By then, Turner was getting fed up with his mountain life, and looking out from inside the phone booth, he saw a town with paved streets, restaurants, book stores, a newspaper/magazine stand, a town with television reception, and then he thought about going back up into the mountains. He also knew that his girlfriend was about ready to return to civilization, and he told Larry he was thinking of coming back to California. He asked Larry if the new owner would pay for his move back, and Larry said, "Yeah, definitely."

MT: "The reason I didn't want to be in LA anymore was because everyone I knew was dealing. I'd go to someone's house and there was a big pile of coke. I'd had enough. All my friends were involved with music and it had been fun, but it wasn't fun anymore. I thought I'd try San Francisco."

Larry said that Turner could do anything he wanted in the news area, and that he should talk to Harvey, he'd like him. I know I promised no more of these, but I have to, so let's say it together: fat chance.

Harvey asked Turner to send a tape, and Turner said "absolutely not."

HL: "Well, that's OK, we don't need a tape. Tell you what: what will it cost for you to come back across the country and come to Gilroy?"

MT: "Where is Gilroy, exactly?"

"Well, it's just south of San Francisco."

"Hmmm. That sounds promising. At least two grand."

"Can't you make it for less?"

"No."

"Well, where shall I send it?

"Send it to Western Union in Charlotte."

Harvey agreed to send the money the next day. Staying over for the night, the next morning Turner found one thousand dollars waiting for him. He called Harvey to "point out the mistake."

HL: "Well, I think a thousand ought to do it, and that's all I'm going to send."

MT: "Well, Harvey, you just lost a thousand dollars, didn't you?"

There was a long, empty silence, and then Harvey said, "Okay, the other thousand will be there tomorrow. When will you be here?"

"I don't know- I'm driving back, a week or two, maybe."

Harvey said, "Make it as fast as possible."

"I got there and became news director, and I was terrible. Absolutely terrible. I was the worst, and they'd had the best- Travus was the best. He was Mr. KFAT in my opinion. I would have Travus on my show occasionally. I loved him, and I don't why he left.

"I was on with Marty in the morning. I'd do a little news update and kibbitz with Marty and take calls, and one time they sent a rocket to the moon or somewhere and I said 'I know you guys at Lawrence Livermore are listening, give me a call and let me know what's happening.' And the director of the project called. Dr. Klein called, and he explained it all on the air. I'd do ten minutes or whatever I wanted, whenever I wanted."

GK: Was Marty good to work with?

MT: "Not really- well, yes he was a real good guy to work with, and he knew how bad I was. I was trying to read news off the newspaper, I didn't know what I was doing and didn't care. It was minimum wage. Peanuts. No one was making money. I held Harvey in contempt. He was an asshole. He wanted to run KFAT like a stiff business- like Leonard Goldenson ran ABC. He would call and tell Marty he was playing the wrong things. I told Harvey to get the fuck off my back, 'cause you're not gonna get it. If you don't like what I do, then fire me.

"By this time, at this point, I was listening to another station and I heard about the promising 49er rookie, Keena Turner, just in from Purdue. I thought: 'The Mike & Keena Turner Show'. I called the 49ers and got hooked up with Keena, and he liked the idea. I launched his career. He and I became really close friends. Harvey said flat out that as soon as the 49ers are out of the race, Turner is gone. Me. The only reason I was still hanging on was because I was interested in the 49ers- I was going to practices, in the locker room, at games. And this was the magic year. I got Joe Montana when nobody could get to him. I was in the locker room after the victories. The Super Bowl that year was in Detroit, and I went there and sat with the (team owners) DeBartolo's, and Harvey could do nothing. He sat back and ground his teeth and could not fire me as long as it went on. I had Keena Turner, and that kept me from being fired. Harvey was going to fire me because he hated me and because I sucked. I didn't know how to do it, and couldn't say what I wanted to say. Harvey was listening. I was awful compared to Travus.

"Larry told me it was over when the team was done. It was a very popular segment. And Keena would come down to the station. Sweet guy. Sometimes us on tape, sometimes live on phone, sometimes live in studio. Dwight Clark comes in. The 49ers went all the way.

"Just to piss Harvey off, I set up a contingency plan to do the same kind of show with the Warriors' Bernard King. He was a lunatic- a cop had pulled up while he held a .45 to the head of a girl who was sucking him off. Another time… he was from Brooklyn… while playing for the Knicks, slumped over the wheel as the horn blasted and he had his crack pipe and a pistol in each hand, so they sent him to the Warriors, and he hooked up with a beautiful Ph.D who straightened him out. In the locker after a game with Bernard, and Joe Barry Carroll, 7'2" black man, whose nickname was Joe Barely Cares, and Bernard was BK… after the game- Bernard's locker is next to Joe Barry's- and he says, 'Hey, Mike- what do you think about Joe Barry's dick?' I looked over and it was at eye height, and I said, 'Not that impressive for a seven foot nigger.'

"Yeah, right- that's what I think, too! Man is hung like a fuckin' peanut!"

"So we got tight and started doing some good stuff, so Harvey couldn't fire me, because if there was no overlap… if there was a weekend in between, I was gone."

Chapter 146: Homage To John 11:35

Harvey died.

Harvey loved KFAT and wanted it to stay Fat, but without his passion, he knew how hard that would be. His will gave the station to his family back in Chicago, but with a twist. His family did not own KFAT, they owned the KFAT Foundation, which would run the station with an eye towards making it profitable, and any profits generated by the station would go to cancer research or in protecting the environment. This made KFAT the only non-profit commercial radio station in America.

Voilá! St. Harvey!

Chapter 147: Lost it

I know this guy's gonna be pissed off at me for writing this, and I'm sorry, but I can't help it. Russ Martineau was a dick.

I really like Russ, I always have. I may be the only person at KFAT who ever liked the poor son-of-a-bitch, and herein I will explain why I like him, and why he was such a dick.

Russ had no experience in radio sales, and none in management when he went to work for Harvey. Soon after he got there, Harvey became distracted by his pursuit of KFAT, and left Russ to glean what experience he could on his own. Then came the sale, the transfer of ownership and the awful diagnosis, and suddenly Russ was Sales Manager and General Manager of KFAT, a one-of-a-kind station which no one with experience in radio had ever been able to handle. Virtually every KFAT staffer who worked for Russ had bad things to say about him, but the one thing consistently said was that he was stupid, and I know that is not true. I spent many hours with Russ in the early 1980s, and then a whole night and day with him in 2007. While he may not be the brightest bulb in the marquee, neither am I and neither are you, so lighten up, out there, and Russ Martineau is clearly not a stupid man.

But in fairness, he did do some stupendously stupid things while at KFAT, and I believe that his reputation as stupid and mean-spirited came from the same stupid and mean-spirited place that we all have inside us. But the inhibitions that normally prevent us from entering this cesspool of self-destruction were overpowered in Russ's case by his frequent ingestion of cocaine and alcohol- and a surplus of each. And in that environment, stupid and mean-spirited thrive. But the staff cared not what Russ' problem was, or where that shit was coming from; Russ was friendless in Gilroy and engendered no pity whatsoever with the air staff. Russ was hated by all. And now for some reasons why.

Larry Yurdin says he will be forever scarred with embarrassment from the night he and Russ had front seats for a JJ Cale Fat Fry, which Larry had booked and was producing. Russ was very drunk and having so much fun singing along, waving his arms in the air and being a disturbance that he was taking the audience's attention away from the performance, so Larry tried to calm him down. Then Cale introduced a surprise guest—Neil Young!—and

Russ became even more enthused. Russ wasn't hearing any voice but his own as he decided to climb up onto the stage and sing with ol' Neil. Larry caught him by the back of his jeans just as he had one leg onstage and was preparing to swing up the other, and pulled him back into his seat. *Larry* did that?

Russ became incensed with Larry and fired him on the spot, which Larry ignored, and Russ stayed in his seat and sulked until he forgot about the incident. But he didn't make for the stage again, either. I assume that Larry later apologized to Mr. Young, who I assume was gracious about it. I know now that Larry knew that Russ was killing the station- and that his hatred for Russ burned white-hot.

Mary Tilson was an enthusiastic, happy, fresh-faced part-time weekend jock, and she was, and still is, quite good at it, now at KPFA in Berkeley. When Wild Bill Goldsmith says something complimentary about a jock, attention must be paid, and he thinks Mary was quite good. Mary had come from rural Michigan, where she did non-commercial listener-supported radio in a mountainous, sparsely populated, highly wooded area, where the signal carried a few miles, or more if you were up high enough. She favored acoustic stuff and some of the new country, the Fat stuff- within reason, of course. In want of adventure, she came to the west coast and, being of a left-leaning bent, she and her boyfriend gravitated to Berkeley, where she found work as a clerk in a law firm in San Francisco. Wanting to get back into radio, she started looking among the non-comms in the Bay Area, and a new friend told her that KFAT had recently lost its staff, she knew a lot of the music, and to call, they might be hiring. Mary Tilson worked weekends.

Too poor to afford a car, she took the long ride to Gilroy on the bus on Saturdays, carrying her sleeping bag and a change of clothes. There would be no shower, but who cared as she took the bus back up on Mondays in time to shower, change, and get to work. It was a brutal schedule, but this was KFAT, the absolute height of cool radio. But KFAT paid about as much as the bus cost, and she was poor- real poor.

Working only weekends and being out of the loop, Mary didn't know that KFAT's newly inspired sales force had some sort of promotion, and one of the jocks could earn a trip for two to Hawaii. It would be a dream: an all-expenses-paid vacation that no one on the air staff could afford. Mary was unaware of the promotion until she was told she won the trip, and she was thrilled. A year ago she was in the northern Michigan woods with "mostly Jack pine to hear me," and now she was in California, on KFAT, and she and her boyfriend were going to Hawaii. Yippeee!

She talked to her law firm, and they arranged a week in the near future for her to take off. When it was all set on her end, she called the sales office in San Jose to tell them which dates she wanted to travel, and was told that

Russ had already taken the trip.

Michael Turner had one of the most brilliant minds I've ever known, and also the most cynical, suspicious, intolerant, and acidly acerbic mind I've ever known. He detested Russ.

MT: "Russ came down from Vacaville and called a staff meeting. Said we need some good ideas for a promotion for the station. 'Too Tall- you've got a good imagination. What idea do you have for a promotion?' I said 'Russ, I've got it- just the thing to put KFAT on the map. It'll create a collectible that'll be sold unto perpetuity.'

"What is it?"

"It'll be a radio receiver that would have the KFAT logo on the side and tuned only to 94.5.

"He said, 'Yeah? What?' I said it would be a butt plug so people could shove it up their asses and listen to KFAT when they're driving their trucks or sitting in study hall. He took that as further insolence on my part, but it was actually a great idea. I thought it was a Fat idea. People who weren't into butt plugs would try it out. No one had any other ideas during that meeting. I detested Russ. I told him not to talk to me."

Larry came up with an idea and Russ fucked it up- in public and on the air. Larry had heard somewhere that some station had a contest where everyone who buys a ticket needs to come to a concert, and bring a suitcase packed for a weekend getaway that someone would win that night, and the getaway would commence that night- hence the suitcase.

Russ decided that he would be the emcee, not a jock, and it would be he who drew the tickets out of a bowl. If someone's name was called, they were eliminated. As the night wore on and more people were eliminated, Russ got drunker and drunker. A lot of people had left after they were eliminated, so there were only about twenty people hanging around when the last two tickets remained in the bowl- and they were mostly friends of the two remaining couples. Everyone there knew who the two couples were, and all eyes were on Russ and the two couples, as Russ pulled a ticket out of the bowl and excitedly screamed out the names of couple number two- they were going to a spa!- and a great roar of congratulatory joy arose as the happy couple came to the stage, and those around the other couple came to them and patted their backs and said how sorry, and too bad and all that, and the winning couple were on the stage, jumping and hugging and kissing and thanking Russ from the very centers of their hearts, and Russ was jumping up and down with them, and shouting and having too great a time to notice Larry pulling at his sleeve, trying to get his attention. After a solid minute of wild jubilation had passed on the stage did Larry manage to get Russ' ear and shout into it that he had announced the wrong winner. He was supposed to pull the ticket, throw it away and then pull the last one and read *that* name. Instead he'd announced as the winners the couple who had just been eliminated, and were now on the stage, jumping and crying for joy at having won.

Larry knew that a moral issue had just arisen, and as a moral man—but no fool—Larry had Russ announce the mistake, and a twenty-voiced chorus of general booing and hissing ensued, so Larry asked the rightful winners to come up onto the stage, and told the two thought-they-were winners to also stay. When all were assembled and the crowd sufficiently quieted, Larry told Russ to announce that both couples were hereby declared winners, and both couples would go on the weekend getaway.

Christa remembers the night a drunken Russ asked her to sleep with him. She refused, she was fired the next day, and she's still pissed at Russ for severing her ties to KFAT.

Then there was the time when Russ was so drunk that he was thrown out of a Fat Fry. The boss, thrown out of his own event? Harold Day thought that was funny, laughed at Russ being tossed out, and was fired that weekend.

Everyone had a story about Russ, and not one of them was nice. Yow. So let me say that I enjoyed Russ' company on many a night together. When we were both at KFAT, he never got out-of-control drunk or stoned with me, and he was always a gentleman with me.

And then I spent a full day with him in 2007, and he was a widower with a bright, vivacious, well-adjusted five-year-old son. He'd been clean and sober for almost twenty years, and I enjoyed my time with both of them. I told him some of the stories, including those above, and he remembers almost none of it. He knows he was stupid to get so carried away, but he remembers enormous pain at the deterioration and death of his closest friend and mentor. He knows he should have been sober and paid more attention to KFAT, but he lost control. He lost it, he regrets it, but he has moved on. He has had a modestly successful career in radio and publishing, lives in a beautiful house in the hills, and he was working on a project that looked both promising and potentially lucrative when I last saw him. I liked him in 1982, and I liked him in 2007. But he was a real jerk in 1982. Everyone hated Russ because he was a jerk and, single-handedly, he was killing KFAT. Ads were being sold and played, but the two-year, low-payment period was almost over and the era of mighty big payments was looming large and comin' up fast. Larry knew that Russ had lost it and that Harvey's family were now the owners and thinking of selling the station. KFAT needed something dramatic to save it, and it would have to be Larry who made it happen.

Once again, KFAT had chugged up the rusted, rickety frame of the roller coaster and was again falling forward, gathering speed, exciting the jocks and thrilling the listeners, but unknown to anyone, the ride was almost out of track.

Chapter 148: Stars Align

Larry wanted KFAT to survive. He remembers: "Harvey's will said that every effort should be made to set the station up as a self-supporting community-run commercial station, and that was what I was trying to do." He loved this format and knew that without that asshole Russ getting in the way, he could make it turn enough of a profit to keep the lawyers at the KFAT Foundation from selling it. He needed money and a loud noise to distract the Foundation from their threat to sell. Money? From where? A loud noise? How loud?

Immediately, he knew: he'd throw a concert. A benefit concert! Larry knew so many stars, and they all owed Larry markers, and this was worth cashing in every one them. Frantically, Larry made the calls, and everyone said "yes." Willie Nelson said yes, Neil Young said of course. Jerry Jeff was on board. Emmylou was coming along with Rodney Crowell. Local guys like Larry Hosford and Lacy J. Dalton were on board, as were Kate Wolf and Utah Phillips. Larry was waiting for a call back from Bonnie Raitt and Randy Newman.

This was starting to look big, so Larry called the Cow Palace and asked them to hold a date for him, and they said they would. Larry made more calls. Bob Dylan said yes, he'd come up and play for free to save KFAT. Most of Austin was coming, and Memphis would be sending several artists that weekend. Los Angeles checked in with Jackson Browne, who said he'd be there, and he'd call some friends. Michael Nesmith was gonna be there, you bet!

Word was spreading about the event, and Jimmy Buffett asked to be put on the bill. More stars heard about it and contacted Larry, who looked at his list of acceptances, and called the Cow Palace again to ask for a second day, and they gave it to him. Larry was feverish with calls to make, calls to return, lists to make. He needed a sound company and a lighting company. Who owed Larry, and who owed KFAT? So many people were calling and saying yes that Larry saw that two shows might earn enough to buy KFAT, and so the fevered pitch rose even more as more calls came in from more stars.

Larry had so many people on his lists of artists that this was going to be the show of the year in the Bay Area, and he called to let Russ know that the station was going to be safe, to stop the lawyers from signing anything, and Russ told him that it was too late, the papers had just been signed. KFAT was sold. It was over. To this day, Larry believes Russ and the lawyer took a bribe to sell KFAT, but no one knows. I never asked Russ, and anyway…

KFAT was, as The Eagles said, already gone. It was only a matter of time for the FCC to approve the sale, and they would, as the buyer was a big-time media corporation with lots of lawyers to dot the i's and cross the t's. The station would belong to the new owner around the first of the new year. KFAT would be KWSS, a mid-tempo middle-aged MOR station. MOR? Middle of the Road.

Larry called back all of his artists and thanked them, told them no thanks, and went to pack his bags while Russ called a staff meeting. Larry was leaving again. It was too much, and he was in too much pain to stick around and watch KFAT die. It was October, 1982.

Chapter 149: Slow Chooglin'

The station had gathered a stable air staff and sales professionals that had sales chooglin' along, but slow. The clock was still ticking on that two year free ride, and KFAT's ad sales needed to pick up. Frustrated by the mess and the way Russ was handling things, Larry was gone again and the Music Director was Dallas Dobro, and he was doing an excellent job. The clock format that Larry had left behind was tight, but the station still sounded good. Not *as good*, you know, but still better than anything else on the dial. Tightening the format cut out most of the spontaneity, but they still played a great mix, and it made the station sound more uniform. Gone were the individualistic quirks whereby listeners always knew who was on the air by the songs they were playing, but it was still an incredible library and they were able to use more of it than any other station would allow. But KFAT needed to make more money, and that had always been hard. Several salespeople have remembered that they'd often get a call the day after a sale, to cancel the sale. The client went home and told either a spouse or a neighbor, or sometimes their minister, about the sale and gotten an earful about the station's profanity and Godlessness. They called the next day and said they were sorry, that they had to cancel. It happened a lot, I was told.

Harvey's estate had been settled, and his family in Chicago owned it, but none of them had any interest in KFAT.

But the station still had to make money so it could pay its bills- and the monthly note that was about to get big. When Harvey's family met Russ at the funeral, he must have been the representative of the station who would least inspire confidence. But I don't know anything about that, and I believe that the family must have realized that KFAT was a strange station with a big debt, a volatile staff and quirky and vocal listeners, that it was 2,000 miles away, that they had no experience in radio and no interest in the problems that were certain to follow, so they voted to sell the damn thing.

In October of 1982, while Gilroy was still in Larry Yurdin's rear-view mirror, Russ called a staff meeting to tell us about the decision. There were about 20 or 25 of us in the room, and we all knew something was up. Russ had been acting… mysterious… for a few days, and all manner of rumors were flying. Russ acted casual while he waited for us all to gather and find places to sit, but we could see that it was an act. No one trusted Russ, but he was the boss, and everyone was respectful and subdued. After Russ welcomed everyone, he quickly said that he knew there was great suspense in the air, and now he could tell us that Harvey's family had decided to sell the station, and that there was almost no hope whatsoever of finding a buyer who would want to keep it as KFAT. When asked when he thought that would happen, Russ said he didn't know, that there were no serious buyers in the picture yet. At his best guess, Russ said he thought KFAT would probably go silent around the end of the year. There was a stunned silence in the room. People had to absorb this. No more KFAT…

Everything was to stay as it was now, and sales would be made to run until the first of the year, and any contracts currently in play will still run until the station changed hands. Russ would let us know when there was a development.

Radio stations with an established signal—and thanks to Jeremy's imaginary data, the station had an admirable reach—had been increasingly valuable, but with the rise to importance of Silicon Valley, such stations were now even more valuable, and with a quick sale and proper format change and serious promotion, this signal could turn a profit. But by this time, only the bigger players could afford to play in this market. Soon there was a player, and then it was confirmed: a big player, a giant corporation, was buying it. A corporation! *Definitely* not Fat!

KFAT was going silent, but first there were the good-byes to make, then the farewell party. It was a very rare occurrence when a radio station got to say good-bye. It was—and still is—a rare thing to have a day, an afternoon, a shift, or even an announcement, to say good-bye. People were usually fired after their shifts, and by the time they collected their stuff, the new staff was in the studio with a new format. It was going to be different with KFAT; well, it was always different with KFAT.

With two months to say good-bye, the air staff alerted our listeners about what had happened, and the calls started coming in. Fat Grams bewailed the news, callers incessantly wept and pleaded for hope. We stopped putting the callers on the air because everyone was saying the same thing, and we knew that everyone felt that way, so answering the phones was a constant "Yeah, I know… I'm sorry, too… No, I don't know of any plans to resurrect it anywhere… Yeah, there's always hope… Yeah, I know, thanks for calling."

Soon, more people started coming in than ever before. Visitors were doubling, then tripling, then we lost track. I remember weekend nights when it seemed like it had to be busloads. But everyone was respectful, and if they knew we were air staff, we were mobbed.

There was a lot of "Yeah, I know… I'm sorry, too. No, I don't know of any plans to resurrect it anywhere. Yeah, there's always hope… Yeah, thanks for coming." But the big night was going to be New Year's. It was widely rumored to be our final farewell, and we were going to go out in style- KFAT style. For what I thought would be my final "Chewin' The Fat," I made what I thought would be my final promo. I got all my friends from the second staff and called it "The KFAT Story." It was a two-hour retrospective of KFAT's history, and several of my friends from the second staff were there, and Larry, Jeremy and others called in. It is still played today on www.kfat.com.

People had been buying blank cassettes all over our broadcast area, and people were setting their recorders and letting them track. I know there must be thousands of recordings out there of the last few hours of KFAT. Well, we *said* it was the final few hours of KFAT, but in the last week, a few of us knew something the listeners didn't.

Those two months passed quickly, and everyone was in sad-but-brave mode. Everyone knew that they were a part of radio history, and a part of a lot of people's lives, and there was real satisfaction in that, but there was also a real, tangible sense of loss, a sense of bereavement. It was a time of great emotion.

Chapter 150: A Fat Farewell

New Year's Eve in 1982 was on a Saturday night, and while people had been coming to the studio to say goodbye for weeks, by that Friday night the visitors were showing up in droves. There were over a hundred people in the studios and offices, and most brought something to drink, smoke or snort. The mood all over the station was strange, because they were happy to be there, but sad about why they were there. It was festive, yet moribund, and no one was sure how they felt. If you worked at KFAT, you had booze, joints and assorted treats handed to you. Most of the time there was at least one joint being passed around.

The jocks, before taking their last shifts, had compiled their lists of songs to play, and they'd been playing them, thinking that for some, this may be the last time the song ever gets played on the radio. Good-bye and thanks to the Light Crust Doughboys, Frasier & DeBolt, Utah Phillips... It was a time of firsts and a time for lasts. It was the first time this staff had been as loose about a format as they were this weekend; KFAT was the last surviving outpost of the cultural revolution of the Sixties, and the end of a golden era in radio.

They played whatever the hell they wanted, and said what they wanted. No one was unprofessional, and wanton use of profanity to express their frustrations was not welcome- not by jocks, and not by the callers who kept the phone lines jammed. KFAT would go out with dignity. The jocks took calls, but said on the air not to call if all they had to say was how sorry they were. We were sorry, too. Everyone was sorry. And no, they didn't know of any plans to resurrect KFAT anywhere, and thanks for calling or coming. We knew their pain. We felt it ourselves.

The jocks stopped putting on the callers who wanted to say they were sorry, but they put on the Fat artists that called in. Old jocks called in, and they put them on the air. Marty Manning called from Phoenix, Larry called from wherever Larry was. But whatever time was spent talking on the air meant that much less music got played.

The jocks took calls off the air, and said thanks and good-bye. People were respectful and happy and sad, and no one was a stranger because they all had KFAT in common. We were all family during an unhappy time, brought together by this tragedy. At the end, Laura Ellen and all the staff already knew that KFAT had been a family to them, and the family had spread out and grown into a community, and now the community was coming home to say good-bye. People of all ages came, but mostly they were in their 20's and 30's. One lady was in her 60's and sort of Church People-ish, and she never took a sip or touched a joint while she was there, but she sure didn't seem to have a problem with anyone else having some. It seemed she was there by herself and she was having a good time, and then she was gone and the party didn't notice, but I did. Good-bye ma'am, and thanks for coming.

KFAT had certainly seen chaos in its day, but Saturday night, New Year's Eve, it was chaos in a good way. No cops, no reporters, no FBI or FCC. The TV News people had been there earlier in the day, and when the NBC station in San Francisco came down, they interviewed me. You can see what NBC said. Notice that they mention a surprise.

Dallas and Amy had drawn up a schedule to share the air with former jocks. There were about two hundred people in a space that shouldn't exceed 20, but it was a blast, and jocks former and current were coming by, doing sets, talking on the air, saying and playing KFAT off, hanging out with Fatheads. Musicians came by and some played us off. No one had heard Cuzin Al actually play that banjo before, although he had threatened to a bunch, and then Chuck Wagon came by with The Wheels and played, and Al played with them. Dallas and Amy had spent weeks preparing for the last two hours- so there'd be no regrets, nothing they forgot to say or songs to play, and no one un-thanked that deserved the thanks. These two were carrying the banner of KFAT until it was time to take it off the pole, fold it and put it away. With Larry gone and Russ incommunicado, Dallas was the Music Director, the Program Director, and the Station Manager. Amy was a wizard of organization, and she and Dallas were the focal point of what was still KFAT. They weren't going to let emotions get the better of them and make them forget what needed to be said or played.

Christa, still upset at being fired, was too emotional to get down to Gilroy from up in Nevada, but she grieved at home with friends. Tiny did a jock set, as did Sully and Terrell Lynn.

I recorded the "I'm as mad as hell, and I'm not going to take it anymore!" speech from the film, "Network," and gave the cassette to Dallas. As it turned out, it was the last time I ever used that recorder. I've still got it. Smithsonian? Anybody?

I put the last of KFAT at: KFAT's Farewell.

Here's a photo of me and a date, and some other people in the KFAT Farewell photos.

In brief, here's how it went down: the place was packed and the joint was jumpin'. It was New Year's Eve, and even though it was such a sad thing, ev everyone felt good, and was gladder than hell to be there. It was an important event and everyone felt like this was the only place to be that night- and they were there! By 11:30, all the music KFAT was going to play had been played, and Dallas and Amy started calling into the studio everyone in the house who was air staff. Harold called in from up in the Redwoods, where he wished he could "hop a log and come down there." Everyone made room for air staff, but after they came in, the rest of the space was jammed past anyone's ability to move. A mic was passed around and the now former DJs said what they wanted, and thanked who they wanted to thank, and if someone thought of someone else to thank, they shouted it out. Some got muttered "yeah, thanks" and some got applause. I was the last to speak before giving the mic back to Dallas, and I guess I was too emotionally involved to hear it then, but now when I listen to it, I am unashamedly proud that when someone handed me the mic and Dallas introduced me, I got a cheer. A good one, too! Thank you!

Sometime in the last half hour, Dallas played the clip from "Network," and that got a cheer. I thought about how it turned out in the movie, and then I thought about how it would turn out here. After the spot played, everyone was silent for several seconds, maybe thinking that, too. Then Dallas got it back on track and the farewells got wrapped up, which *you'd have heard* if you'd gone to KFAT's Farewell like I suggested! If not- Spoiler Alert!

Okay, so we're back in the cramped and crowded studio and it's almost midnight, and Dallas and Amy have timed it pretty well, and all the goodbyes have been said, and all the thank you's and the love you's have been thanked and loved, and all the music had been played, and as midnight approached, everyone jammed into the studio, and those that couldn't get in crammed into every nook and cranny in the rest of the station, listening on the speakers everywhere, and everyone knew that at midnight it was all going to be over- gone forever, and everyone knew that something very special, very precious would die that night and there would never be anything to replace it, and they were all there at that midnight moment, when this special thing, this voice, this spirit that was so much a part of so many lives, this piece of history, was ending, and they were all there to say goodbye and to be there to hear it and see it when it went silent, to be there for that, to share in that moment, in that place… and emotions were at a fevered pitch, at this piercing conflation of sadness and exuberance, and then… oh, wait! Did I forget to tell you that we weren't going off the air at midnight? That there had been a glitch in the approval of the transfer of ownership, and that there had been a delay in turning the station off, and that Russ said that we could play whatever we wanted until someone came in and pulled the plug? And no one knew what the delay was about or how serious it was, so no one knew how long they could still be KFAT?

Russ had called to say that no one would get paid, but if they wanted to stay and play music... No ads, just music. I'm sorry if I forgot to tell you that, and now I wish I had, but the air staff knew it, and we agreed to tell no one but our most intimate and trusted friends, and as for the public, it was kept as a surprise.

So at the stroke of midnight, Dallas, that deceitful scoundrel, with a record cued up on each turntable, said that this was KFAT, signing off, and he shut off the microphone. There was no static, there was no sound. Everyone in the room was going to spend the first post- KFAT moments of their lives listening to the empty hole where KFAT had once been. They had listened to the end of KFAT, and this was what it sounded like.

The studio was silent, no one said a word, until Dallas counted off the appropriate number of seconds, and he hit the Play button on the playback deck above the turntables, and bonging into the silence was a bell, being rung by hand somewhere in the distance, and someone shouting something, as the bell and the cry came closer, "Bring out yer dead! Bring out yer dead!" It was from the movie, "Monty Python and the Holy Grail" and a conversation ensued between a man who wanted to dump his old uncle on the cart with all the dead from the plague, but the uncle kept insisting he wasn't dead. "Almost," said John Cleese, "You will be soon."

"But I'm not dead..." and the bit went on for about thirty seconds, and then Dallas hit the switch for the turntable on his left and Jerry Lee Lewis was singing that he was "Rockin' My Life Away," and KFAT was back on the air, although no one knew for how long.

Dallas had called me early one morning near New Year's and woke me up. He informed me that KFAT would probably still be KFAT on Sunday, January 2nd, and did I want to come down and do a show? I thought my last show had already happened on December 26th, and I joked with Dallas as I struggled to wake up and make sense of what he was saying. I guess he forgot to tell me that he was recording our chat, and he cut it into a promo spot for my show. Apparently, I said yes, and you can hear my real final Chewin' the Fat promo here. On this show, Dallas and Amy were featured and we talked about their history and told some stories.

Then, after a week, someone came in and pulled the plug, told the jock they were done, and moved in the new records. It was a Top Forty station now. Ewwww...

No jock I ever talked to knew what the delay had been about, but I found out on my first interview for this book, and I told them.

It seems as if KFAT's old friend, Don Mussell (and I'm telling you this with his permission) remained a friend and fan of the station because he loved radio. He'd built several radio stations and he was still friends with Laura

Ellen and Jeremy, even though they were barely speaking to each other, and he really appreciated good radio.

It seems that when Don learned of the impending sale, he waited a discrete amount of time, and anonymously informed the FCC that the initials of the proposed and approved call letters of the new station, KWSS had a secret, sexual meaning that was inappropriate for any society that lives by standards of decency. Or something like that.

So the FCC halted the transfer, investigated it, and it took them about a week to decide that the complaint had no merit and let the transfer of ownership go through, which was when people came to the station and told whoever was on the air to either go home- or help moving in the furniture, the equipment and the boxes of records.

As the last jock left, he took the record off the left turntable and hit the switch for the turntable to his right, and played a song so there wouldn't be any dead air. That would have been… unprofessional. The first of the new jocks sat down, cued up a record on the empty turntable, then took a breath, and reached over to the tone arm that was playing the last song on KFAT, and scratched the tone arm back and forth, back and forth across the record, making a screeching, ear-scrunching sound, and he did that for about seven seconds, which is a lo-oo-ong time, and then, his statement made, he reached to his left and hit the play switch for the record he had cued up, and on came Queen's "Another One Bites The Dust."

Chapter 151: Distribution Blues

KWSS was on the air, and KFAT fans were bereft. They'd known it was coming, they'd tried to prepare by recording every last minute. Everyone I've spoken to who has the recordings of those last days on KFAT says that they never paused the tape to eliminate the jocks' talking: they wanted that, too. But now it was gone, 94.5 was KWSS, inaugurated by the insult of a needle being scraped across the last KFAT song before being replaced by "Another One Bites The Dust," and a crew of DJs with no discernible personality and no interest in Fat music. My God, they had *trained DJs* at their beloved Wide Spot On The Dial! And *these fucking DJs*- they were *for-shit*- and they played the most awful for-shit music, and only played the broadest of Fat artists, like the Eagles. Didn't even play Willie!

It wasn't until after several days had passed that Sherman or Terrell Lynn could listen to the new station. It was terrible, but they, too, had had time to prepare for it, and they grimaced at what they heard and knew that it was over, really over, and there they were, the two of them, still in Gilroy. Everyone else was gone. They knew they'd have to move soon, probably in separate directions, and as they listened, they waited for something, anything, Fat. They had a moment of optimism when the morning jock played a song by Emmylou Harris. They were glad to hear Emmylou, but then the jock back-announced the songs she'd played, identifying the song as being by "Emmylee."

"Fuck..." they both said, and almost without speaking they went down to the Old Gilroy Hotel and walked up the stairs to the new studio. The door was unlocked, and they came in acting as natural as possible, telling the DJ she was doing a great job, and they liked the sound and to keep it up. Once she saw they were friendly, she went back to her job while the two men picked Fat records out of the shelves, and put them in the boxes that were lying around from the new gear, and then emptied the now-ignored Hot Box into a carton with the rest of the records. They took as much as they could carry, and left the studio, and to this day, Terrell Lynn believes that the DJ did not know who they were. The boxes were bulky and awkward to carry, but they weren't going far.

Hans Teufel, the western-clothes-wearing chef at the Harvest Time restaurant in the Old Gilroy Hotel, downstairs from KFAT, who everyone called

"Tex," was a long-time friend, fan and benefactor of the station. He'd lamented with the staff at the loss of their jobs, and then again at the loss of the station. He'd had his radio tuned to KFAT exclusively for seven years, regularly sending over escargot, garlic mushrooms and other treats, and he considered himself a member of the Fat family, and he was. He hated what was happening, and he had promised them all that if there was ever *anything* he could do to help them, they had his unquestioned support.

"Unquestioned support?" Sherman and Terrell Lynn wondered? "*Anything?*" The two men brought the records to Tex, and he stored them in his freezer, where no one would look for them. Simple as that.

Well, you might wonder, if it was that simple, why not go back for more? And so they did. They came back to the station and, out of boxes, took armfuls of whatever Fat records were left. They extracted about half of the library, so we're talking about a lot of trips for these two men, and some big balls on Tex Toyful.

No one called the police this time. No one at KWSS cared as the KFAT records were removed from the station. Still wary of the police after the record heist, they left the records in the Harvest Time's freezer for a few days, and when they saw that there was no heat to dodge, they came back for the records and brought them to the home of another great friend of the station, John Sandidge, aka Sleepy John. The three men talked about it, then called all the jocks they could find, and invited them to come to Sleepy John's house two days' from then, at noon, for a distribution of the records.

This being the staff of KFAT, no one was surprised that folks were late, and it was already mid-afternoon when they convened. Sleepy John, always a true friend of KFAT, brought out a case of beer and set it on a table. Sherman, Terrell Lynn and Sleepy John spread the now-defrosted records on the lawn and explained the rules: fifteen minutes to look them over, then the selection process would start. There was to be no arguing about who gets what, or how many. There were seven people there, and each one would get a turn, and everyone would get one record per turn. Double albums were still one album.

They all announced the date they started at KFAT, and that was the selection order. The earliest hired went first, then… It started with Sherman, Terrell Lynn was next, and then Buffalo Bob, Sully, Tiny, Cuzin Al, and Mark DeFranco, aka Rocket Man. They took turns dividing what was left of the library, one jock at a time, one record at a time, and put them in piles behind them. As they got down to the last of them, a bit of laughter broke out, as people began making jokes- like how "many Hank Jr.'s I'll give you for that Chip Taylor." Then they thought of how painful it would be to listen to some of these, and the laughter faded into the sad finality of what they were doing.

KFAT came down to two things: the jocks and the library, and this solemn ceremony was the real, behind-the-scenes end of the story, the final, final good-bye to KFAT. These records and these people would never come together again, and they were making official the end of something remarkable that they had created and shared, and they were saying good-bye to each other.

When they finished, dusk was darkening the sky, and they stood around awkwardly, not knowing what to say or how to say it, until one and then another said goodbye, shook hands and hugged each other, and headed for the driveway carrying their records, the ones stamped "STOLEN FROM KFAT." It was really over. This was the very last function of the late, great KFAT. It was the last of a type of radio that would not be heard again.

The ride was over, and what they had gathered to do that day was the end of it. Life called them all in separate directions. Several have stayed in contact, a few of them in close contact, but it was all to be from different areas of the map, and different corners of their lives. Sherman and Terrell Lynn watched the cars disappear down Sleepy John's long, wooded driveway, then they looked at the lawn where ten minutes ago everyone had gathered. The legendary KFAT was smoke, dissipating into the air above Sleepy John's lawn.

After the last car was gone, Sherman and Terrell Lynn went back inside with Sleepy John and fired up another joint. None of them remembers what music they put on, but it was the first time they'd been together that it wasn't KFAT. No one felt much like talking, and then they left, too, Sherman to Santa Cruz, and Terrell Lynn to the Black Rock Desert in northern Nevada.

When they were gone, too, Sleepy John went out to the lawn and picked up the empties.

Photo Section 3

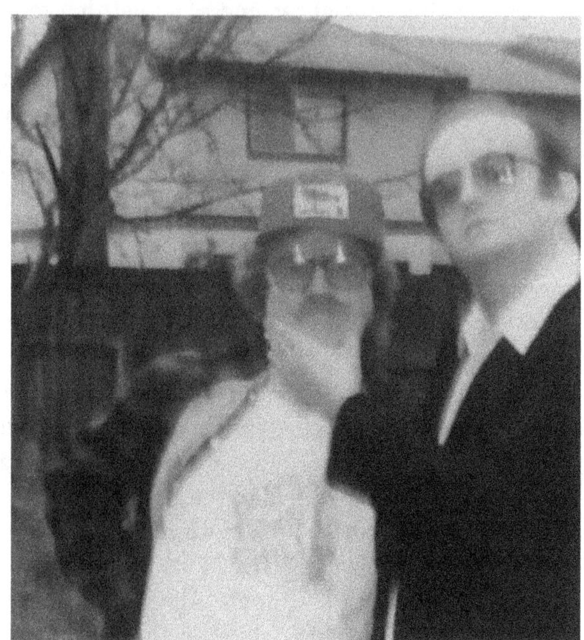

Larry Yurdin & Marty Manning
(Both Back for a Second Helping of Fat, 1981)

Marty Manning, Larry, and Michael Turner
in The Production Room

Michael "Dallas Dobro" Hess

Amy Airheart Bianco

Wolfman Jack with Harvey Levin
During the Negotiations for KFAT

Russ Martineau with Son

Christa Taylor

Mary Tilson

Gilbert Chewin' The Fat Live, 1982

Cuzin Al, Gilbert, Gordy, Robin, Elsie, Laura Ellen on back stairs after my almost-last Chew, "The KFAT Story"

Amy Airheart, Dallas Dobro, Bill Goldsmith, Harold Day, Cuzin Al, Unkle Sherman
(that's *NOT* TLT in the background)

Cuzin Al Plays in the Studio, TLT and Felton Pruitt look on

Sister Tiny at KFAT Farwell

Sully and Robin at the KFAT Farewell

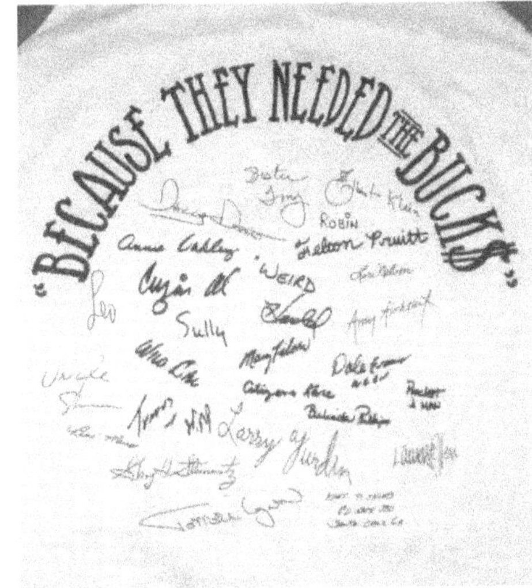

Someone Printed a T-Shirt for out Farewell and got Many of our Signatures

Raves

Emmylou Harris

Gilroy, Calif. Possibly the most fearless format in radio. You'd hear Ry Cooder, B.B. King, Tina Turner, Tammy Wynette and Hawaiian music – whatever the DJ wanted to play. Kind of like KBHR on *Northern Exposure*. I don't think KFAT is still operating, but I knew some of the DJs, and they'd send me tapes. I listen to those and pretend it's the radio.

Emmylou Harris Raves About KFAT

Stolen From KFAT

Poster for the Second KFAT Reunion

Sherman, Amy, Harold

Michael Turner and Amy at the Reunion

Laura Ellen at the Board at KPIG

Cuzin Al, Michelle Busk (Sister Tiny), Kathy Roddy, Terrell Lynn Thomas, Meet me for an Interview, July 2005

Cuzin Al and Unkle Sherman at the Hardly Strictly Bluegrass Festival

Gilbert & Cuzin Al, 2011

Gilbert, Sherman and Terrell Lynn, 2011

And Sherman is Still … Sherman

Thanks, Fatheads!

About the author

Gilbert Klein spent 25 years on Long Island, 25 years in San Francisco, and (so far) 20 years in Baja California. He has been a cub scout, teacher, carpenter, talk show host, nightclub owner, event producer, Most Eligible Bachelor, El Guapo Misterioso, and some things he won't discuss until he knows you better. Why? What have you heard?

Books by Gilbert Klein

Fat Chance (2012)
Football 101 (2012)
God Watches Over Drunks And Fools And I Don't Drink (fall, 2016)
The Music In Me (fall, 2016)

Main Frame Press
858 Third Avenue #320
Chula Vista, CA 91911
www.mainframepress.com

ined

9 780985 679002

CPSIA information can be obta
at www.ICGtesting.com
Printed in the USA
BVHW081447140820
586422BV00004B/340

CPSIA information can be obtained
at www.ICGtesting.com
Printed in the USA
BVHW081447140820
586422BV00004B/340

9 780985 679002